SCIENTIFIC ILLUSTRATION

The Art & Design Series

For beginners, students, and professionals in both fine and commercial arts, these books offer practical how-to introductions to a variety of areas in contemporary art and design.

Each illustrated volume is written by a working artist, a specialist in his or her field, and each concentrates on an individual area—from advertising layout or printmaking to interior design, painting, and cartooning, among others. Each contains information that artists will find useful in the studio, in the classroom, and in the marketplace.

Among the titles:

The Art of Painting Animals
A Beginning Artist's Guide to the Portrayal of Domestic Animals, Wildlife, and Birds
Fredric Sweney

Cityscape
The Art of Painting the Urban Environment
Fredric Sweney

Designing Greeting Cards and Paper Products
Ron Lister

Drawing with Pastels
Ron Lister

An Introduction to Design
Basic Ideas and Applications for Paintings or the Printed Page
Robin Landa

Graphic Design
A Problem-Solving Approach to Visual Communication
Elizabeth Resnick

Visual Discoveries
A Workbook for Artists and Designers
Robin Landa

The Art of Field Sketching
Clare Walker Leslie

Nature Drawing
A Tool for Learning
Clare Walker Leslie

The Art of Watercolor
Techniques and New Directions
Charles LeClair

Portrait Drawing
A Practical Guide for Today's Artists
Lois McArdle

Cartooning and Humorous Illustration
A Guide for Editors, Advertisers, and Artists
Roy Paul Nelson

Graphic Illustration
Tools & Techniques for Beginning Illustrators
Marta Thoma

In Print
A Concise Guide to Graphic Arts and Printing for Small Businesses and Nonprofit Organizations
Mindy N. Levine with Susan Frank

Notes for a Young Painter
Hiram Williams

Zbigniew T. Jastrzębski, an assistant professor at the School of the Art Institute of Chicago, has been a senior scientific illustrator at the Field Museum of Natural History in Chicago since 1969. He has widely published scientific and editorial illustrations, and many of his illustrations are included in a permanent collection at the Museum.

ZBIGNIEW T. JASTRZĘBSKI

SCIENTIFIC ILLUSTRATION

A GUIDE FOR THE BEGINNING ARTIST

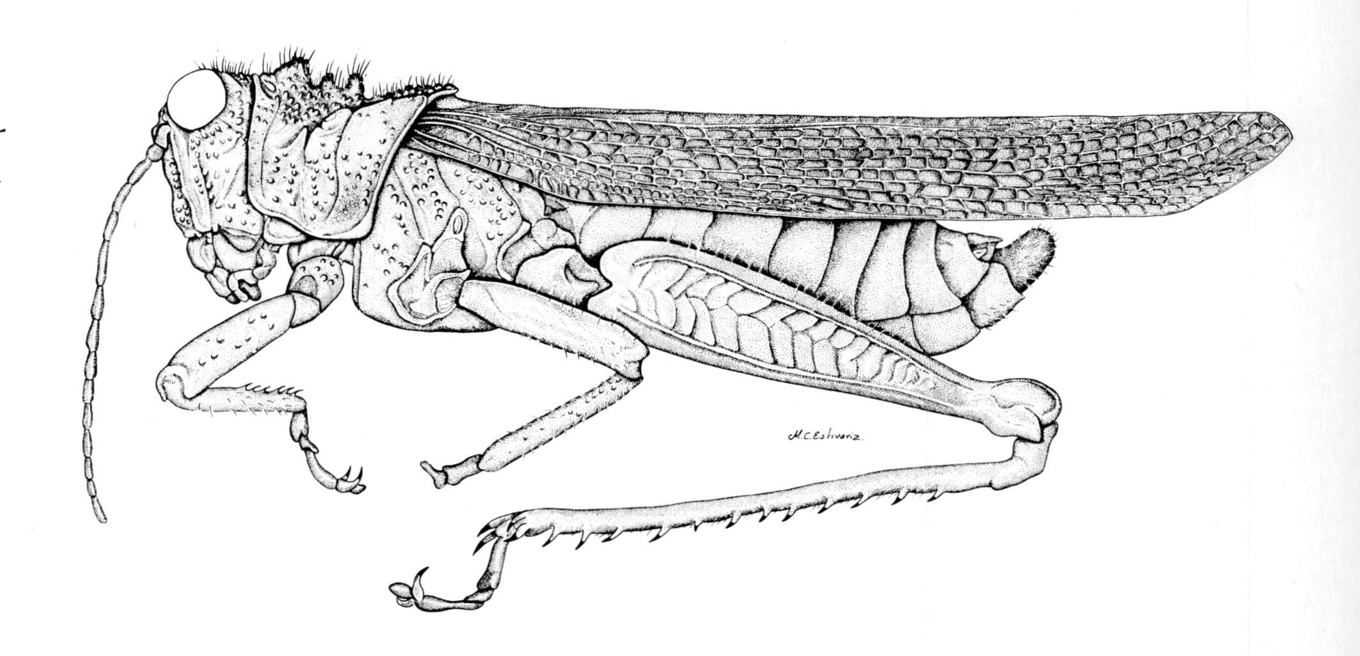

PRENTICE-HALL, INC., ENGLEWOOD CLIFFS, NEW JERSEY 07632

Library of Congress Cataloging in Publication Data

Jastrzębski, Zbigniew T.
Scientific illustration.

(The Art & design series)
"A Spectrum Book."
Bibliography: p.
Includes index.
1. Scientific illustration. 2. Art and science.
I. Title. II. Series.
Q222.J37 1985 502.2 84-22877
ISBN 0-13-795949-4
ISBN 0-13-795931-1 (pbk.)

A Spectrum Book. Printed in the United States of America.

10 9 8 7 6 5 4 3 2 1

Editorial/production supervision: Marlys Lehmann
Cover design: Hal Siegel
Cover illustration: Sarah Forbes Woodward
Book design: Alice R. Mauro
Title page illustration: Maria Cristina Estivariz
Manufacturing buyer: Frank Grieco

Prentice-Hall International (UK) Limited, *London*
Prentice-Hall of Australia Pty. Limited, *Sydney*
Prentice-Hall Canada Inc., *Toronto*
Prentice-Hall Hispanoamericana, S.A., *Mexico*
Prentice-Hall of India Private Limited, *New Delhi*
Prentice-Hall of Japan, Inc., *Tokyo*
Prentice-Hall of Southeast Asia Pte. Ltda., *Singapore*
Whitehall Books Limited, *Wellington, New Zealand*
Editora Prentice-Hall do Brasil Ltda., *Rio de Janeiro*

ISBN 0-13-795949-4

ISBN 0-13-795931-1 {PBK}

To the Lyceum of Fine Arts (Liceum Sztuk Pięknych), Kraków, Poland, with thanks for the best humanistic and art education I have received . . .

and to my mother

Contents

Introduction ▪ 1

CHAPTER ONE
WHAT IS SCIENTIFIC ILLUSTRATION? ▪ 5

CHAPTER TWO
AREAS OF SPECIALIZATION ▪ 17

CHAPTER THREE
TOOLS: THEIR DESCRIPTION AND USAGE ▪ 47

CHAPTER FOUR
PRODUCTION OF ILLUSTRATIONS ▪ 75

CHAPTER FIVE
PROJECTS FOR THE PORTFOLIO ▪ 165

CHAPTER SIX
OTHER ARTISTS' TECHNIQUES ▪ 255

Appendix A ▪ 289

Appendix B ▪ 294

Appendix C ▪ 297

Bibliography ▪ 304

Index ▪ 317

Preface

THE continual process of learning, preparing a portfolio, planning a career, and finding a profession in the demanding field of scientific illustration is in your hands. The goal of this book is to help you, step by step. Descriptions of how to draw and paint and what to do with the specimen are reinforced with explanations of possibilities that await. Drawing is a pleasure, especially when it is meaningfully interwoven with the growth of knowledge. There is always a need for more information, and knowing how to convey the information is the key to drawing and painting. Explanations of professional tools, of how to handle materials, and the procedures leading to the finished product, the illustration, are the basic subjects of this book. Preparation of you, the individual, as well as your skills, forms the solid base from which the understanding of know-how will emerge. Descriptions of drawing exercises involve you in the practice of observation. Combining both, the realistic drawing or painting will be easy to attain. Starting with an explanation of art in general and scientific illustration in particular and proceeding through the necessary steps of production will lead you to consciously plan a career.

Where Can a Beginner Study Scientific Illustration?

Depending on personal interests, the direction taken will lead to some area of specialization. Broad areas such as natural sciences or more narrow areas such as medical illustration or biology will dictate the appropriate college or university. Some institutions of higher education do offer complete programs, especially in medical illustration or so-called technical illustration. Call universities and ask for information. Numerous art schools offer some insights into scientific illustration, with a varying intensity of curriculum. Look into natural history museums or other research institutions and find out about adult education classes. Usually such courses are at the introductory level; nevertheless, you will get some idea of what is necessary for a scientific illustrator to know. In many instances it is possible to organize your own curriculum with the guidance of an advisor from your chosen university.

For a detailed list of institutions in the United States that offer courses, programs, and in some instances degrees, write to the Guild of Natural Science Illustrators, Inc., Ben Franklin Station, Post Office Box 652, Washington, DC 20044. The booklet titled *Scientific Illustration Courses and Books* lists state by state all presently available sources of instructions in biological illustration.

ZBIGNIEW T. JASTRZĘBSKI

Introduction

BACK in the days of the Renaissance, artists were routinely preoccupied with analytical observations of nature. The technical problems encountered during visual presentation of the observations, the two-dimensional design and composition, as well as the manipulation of viewers' minds were understood differently. How differently? Well, let us project ourselves back in time and understand the mentalities and realities of a Renaissance artist. We can look at their work and reach certain conclusions. We can study history, customs, music, and literature as well as laws, rules, and regulations governing the lives of the artists of the Golden Age. We do have some definite information preserved.

The most important fact is that the artists of those days did not regard themselves as fine artists in the same sense as many artists do presently. Artists of the Renaissance were craftsmen, and those whose craftsmanship was outstanding were considered to be fine artists—fine artists because they did their work finely, because they did their work well. This means that the knowledge of principles of drawing and painting were mastered by them. This means that they had the technical knowledge and had been able to satisfy all customers. What was their motivation? Why have artists struggled to master techniques of drawing and painting? The world has *not* changed; and the very same rules are applicable today. Problems involved in marketing, in getting the income desired from the production and sale of drawings provide the truthful motivation that pushes artists toward high levels of craftsmanship.

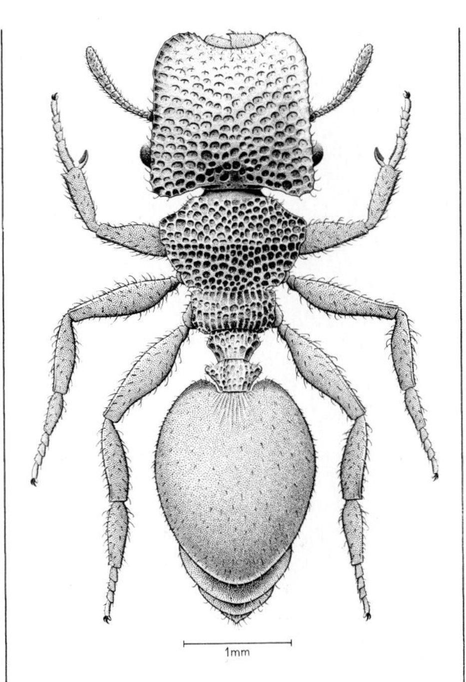

Figure I-1. *Zacryptocerus sp.*, reconstruction of an undescribed type species of genus *Zacryptocerus* from Lower Miocene of Dominican Amber. Pen and ink. Drawing by Armin Coray, Museum for Natural History, Basel, Switzerland. Published in C. Baroni Urbani and J. B. Saunders, *The Fauna of the Dominican Amber: The Present Status of Knowledge*, vol. 1, 9th Caribbean Geological Conference, Santo Domingo, Dominican Republic, 1982. Drawn in 1979.

Renaissance artists had realistic attitudes. Naturally, the sociological frame of the fifteenth and sixteenth centuries pressured artists into certain frames of mind, but the pressure has created very desirable products, and currently we tend to compare ourselves, in a reassuring way, to the intellectual products of the greatly admired Golden Age.

The meaning contained in the term *fine artist* as it is understood today is relatively new. It was projected upon public and artistic communities during the late nineteenth and early twentieth centuries, and it is currently applied to the area of arts which creates constant controversy reflected in everlasting changes of styles.

Many artists of today cannot exist as artists unless in their own minds they reach a plateau of fine arts by creating or discovering a new style. By doing this, they prove to themselves and to the community their own power of creation. Unfortunately, during that process, the craftsmanship necessary for production as well as for establishing communication between the artist and the viewer is completely lost. I hope you are aware of the extremities some styles or movements can reach. Why? Basic rules govern such situations. First and foremost is the great personal ego directing the individual and telling him that he must be different from all other artists. By being different he is possibly the greatest. The second is the market.

In our society something new, even if it is not necessarily produced well technically, has a better chance to be sold, thereby stimulating artists to strive for newer and more-difficult-to-understand creations. Most of the time communication between the artist and the viewer does not exist. To top this, the realization that our lives are short places us in competition with time, and we know that it is nice to be famous during one's own lifetime.

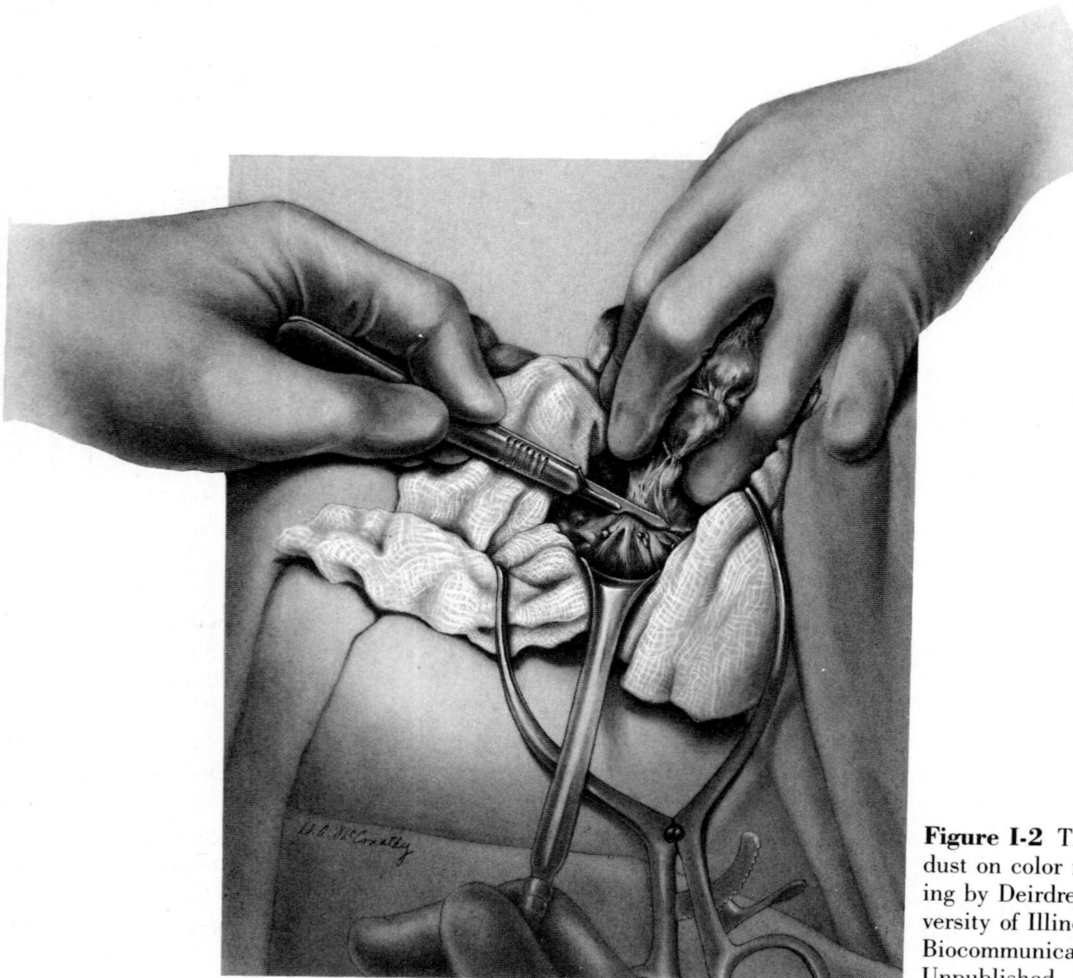

Figure I-2 Thyroidectomy. Carbon dust on color match paper. Drawing by Deirdre Alla McConathy, University of Illinois, Department of Biocommunication Arts, Chicago. Unpublished. Drawn in 1979.

We cannot easily say that some of the fine art is not really fine art. It represents the syndrome of the emperor's new clothes: It is very hard to point a finger at His Majesty in front of the whole entourage and announce the truth. The weight of criticism is heavy and hard to bear.

But criticism during the production of art and especially after it is finished is of great importance. It is a major tool stimulating the learning process. Through criticism it is possible to detect weaknesses and to improve the ability of observation. Through criticism it is possible to improve the technical skills needed for the presentation of observations in a coherent, informative manner. This is especially important in scientific illustration, as such illustration presents very precise and well-defined information. The information contained in scientific illustration is applicable only to the very particular research in the given area of science. The adaptability of an illustration for printing in scientific publications is of great concern to the artist, as scientists communicate with each other through such publications. At the same time it is very easy to recognize the art of a particular scientific illustrator by his technical ability in presenting the information.

If you look into *Scientific American* you will find that various illustrations retain the individuality of the artist. Personal style in scientific illustration is not of great importance to the viewer, the scientific community, but at the same time an artist cannot escape from his individuality. We have to remember that great masters were primarily skilled craftsmen who by the process of analytical observation of a given subject and a continual learning process enhanced their craft and were able to produce art that is regarded by us today as very fine.

Commonly, artistic talent is understood as something that exists within the artist and with which good artists are born. Such esoteric inner spirit supposedly allows the artist to be creative. Let's look at the definition of talent. *Webster's Ninth New Collegiate Dictionary* states that talent is "... a special often creative or artistic aptitude ... general intelligence or mental power."[1] Such definition should be taken as it is, without complicated explanations. As we can see, it is possible to have a special aptitude toward certain abilities but at the same time not always be creative. Creativity does coincide with intelligence. A person's intelligence and learned ability of control of the tools used in the area of art will make him or her a creative artist. Intelligent manipulation of the viewer's mind will accomplish more than uncontrolled inspiration. Many great people have been very creative but at the same time never drew a line during their lifetime. Many artists have labored but hardly can be judged as good or creative. It is quite obvious that talent is not necessarily associated with art. A logical, intelligent being is a talented person by the fact that he or she possesses an ability to think logically and therefore analytically. We all have such ability, although many of us are not aware of it. *Homo sapiens*—a being that knows and is wise—is a designation given to us by us. Now the only problem is to use our ability, properly combining it with technical know-how, and the outcome will be finely executed art. If a hand is exercised continually through practice of various techniques of drawing and painting, it will obey the directions issued from the mind. And if the mind is exercised in analytical observation of nature and in the processes used during drawing, the directions issued to the hand will become more rational, allowing for better control of the hand and the tools. If a rational approach is taken, the scientific illustrator will have a better chance of becoming master of situation, therefore producing exactly what is required and in the desired

[1]By permission. From *Webster's Ninth New Collegiate Dictionary* © 1983 by Merriam-Webster, Inc., publishers of the Merriam-Webster Dictionaries.

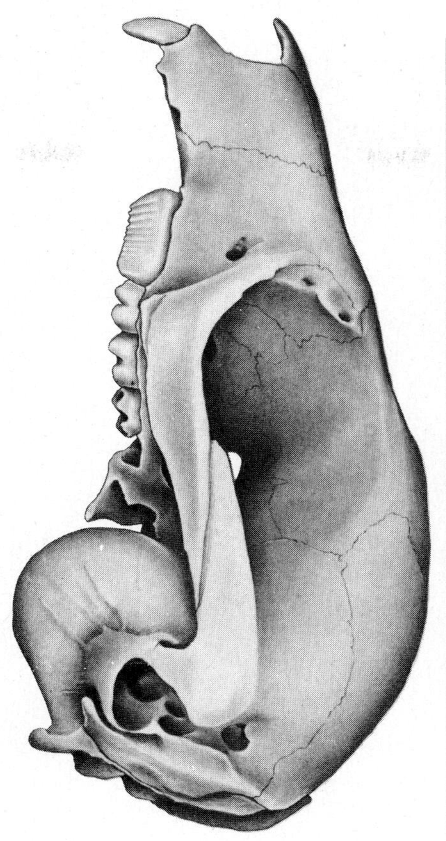

Figure I-3 *Bettongia lesueur* from Madura Cave, surface (probably recent) TMM 41106-20, skull, left lateral view. Pencil on Strathmore drawing paper. Drawing by Zbigniew T. Jastrzębski, Field Museum of Natural History, Chicago. Published in William D. Turnbull, "The Mammalian Fauna of Madura Cave, Western Australia, Part VI," *Fieldiana*, Geology, New Series No. 14 (1984), by Field Museum of Natural History, Chicago. Drawn in 1983.

technique. His work will be acceptable to science and will be of a very good quality, making it, in my opinion, work of fine art. Anatomical illustrations drawn by Leonardo da Vinci, Michelangelo Buonarroti, Raphael, and others so familiar to us are good examples of technical skill and analytical thinking. Many nineteenth and twentieth century scientific illustrations have to be greatly admired for their content and technique. The splendid illustrations found in *Anatomia Uteri Humani Gravioli,* written by Guliermo Hunter and published in 1774; or in *Vasorum Corporis Humani at Iconographia* by Paulo Mascagni, published in 1787; or in *Crania Ethnica les Cranes des Races Humaines* by two authors, Quatrefages and Hamy, illustrated by Bacquet and published in 1882 in Paris, are simply breathtaking and at the same time quite correct scientifically.

The closer we get to the twentieth century the less we know about scientific illustration and the less we realize the beauty of such art. Why? The art produced by the scientific illustrator is very specialized and is seen by only a handful of scientists; it is not exposed to the general public. In the present day of specialization, the art has been decentralized and subdivided into many areas, a few of which are editorial illustration, scientific illustration, medical illustration, sculpture, architecture, printing, advertising, and the area of "fine arts." The specializations are further subdivided to the point that it is possible to limit the area of work to one subject or one technique. At the same time, it is possible and very advisable to retain a healthy Renaissance approach. A working knowledge of various techniques and various applications of gained abilities into specialized areas will make the artist stronger and better. In order to achieve this, the basic principles of drawing and painting have to be practiced very patiently over the years, throughout the life of the artist. It is very acceptable to us to think that way about music. A good musician—for instance, a piano player or a guitarist—will practice many hours a day so his fingers will respond properly and he will be able to play what he wants and the way he wants. The very same method should be applied to drawing and painting. Our logic and

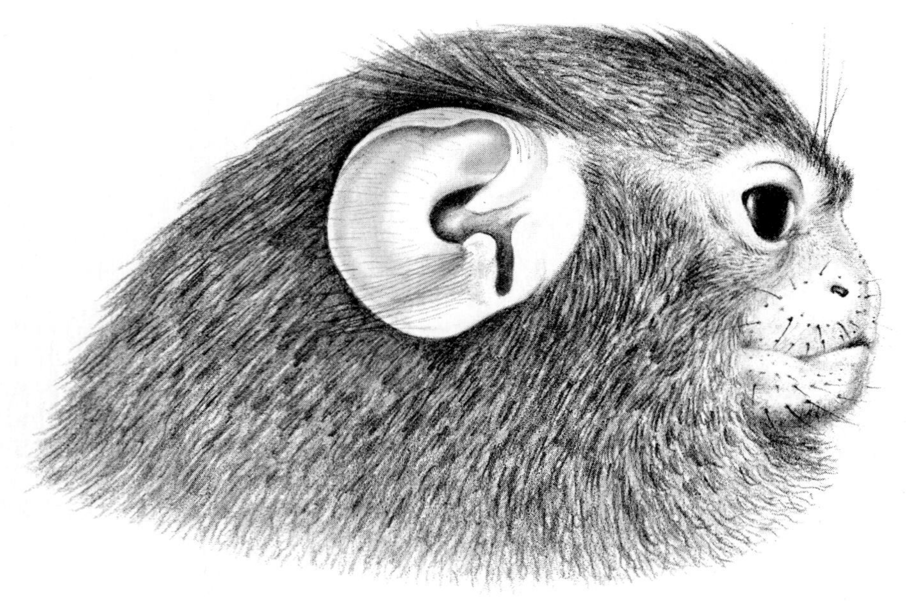

Figure I-4. *Cebuella pygmaea*. Pencil on scratch board. Drawing by John Pfiffner, Field Museum of Natural History, Chicago. Published in Phillip Hershkovitz, *Living New World Monkeys*, Vol. 1, by the University of Chicago Press, Chicago, 1977. Drawn before 1977. ©1977 by the University of Chicago. Copyright reserved.

ability to observe are only the beginning. Learning various skills takes patience. Through exercises and conscious self-cricitism it is possible to improve the quality of work.

There are currently more artists involved very actively in marketing their work than ever before, creating tough competition for all of us and thus prompting us to draw better. In order to stand up to the competition in the area of scientific illustration, the artist must know techniques of realistic drawing and painting and must be able to produce high-quality illustrations. Being precise is not enough. A scientific illustrator must execute his illustrations according to specifications given to him by the scientist; his work should have the same quality of technique and appeal as works of great masters have. As Darwin observed in *The Origin of Species*, in the chapter on the struggle for existence, "The land may be extremely cold or dry, yet there will be competition between some few species, or between the individuals of the same species, for the warmest or dampest spots."[2]

Let's go to work.

[2]From Charles Darwin, *The Origin of Species and the Descent of Man* (New York: Random House, Inc., The Modern Library, 1977) p.61.

CHAPTER ONE

What Is Scientific Illustration?

SCIENTIFIC illustration is an art in the service of science. It is a complex compound of information, craftsmanship, and cooperation between the artist and the scientist. It is one of many means of communication between various scientists. It is a visual explanation of scientific studies and findings. It is a mixture of centuries' proven techniques of observing, drawing, and painting combined with inquisitiveness into nature. It is the ability to produce an image of a particular subject appearing just like that very subject. It is a love of art combined with the desire for knowledge. It is a good eye sharpened by the process of observation. It is the constant practice of drawing and an ability to use any technique for the benefit of representing the truthfulness of nature. It is the great satisfaction of accomplishment, and it is hard work.

All scientific illustrations used as supportive material by scientists to illustrate their research depict designated information with exactness, present pertinent problems clearly and by doing this, untangle various parts of nature. What about exactness? We are all aware that as everything is relative, exactness is a relative statement. Naturally, all drawings must agree to scientists' specifications, and that is where the exactness will vary. It is important not to forget that natural science is an enormous field encompassing very diversified research. What is exact for the entomologist may be too exact for the anthropologist. It all depends on the field of specialization and the nature of individual research. Illustrations are drawn to help explain and clarify the research; therefore their exactness must comply with requirements. After all, this is the reason why such illustrative material is produced.

The researcher is a specialist in his field of study; the artist, the scientific illustrator, is an expert in the visual presentation of subjects necessary for the scientist's work. The scientific illustrator should be an expert in all drawing and painting techniques. The product, an illustration, is a result of mutual cooperation between researcher and artist. Ask yourself an important question: for whom is an illustration prepared? For yourself and your ego? The answer is obvious. Drawing is primarily produced for the scientist and his research, secondarily for the whole scientific community. It is definitely not prepared for the artist himself or the accidental viewer, although in truth, the care taken in preparation and the technique used for rendering will also benefit the illustrator. After all, a number of people will see the

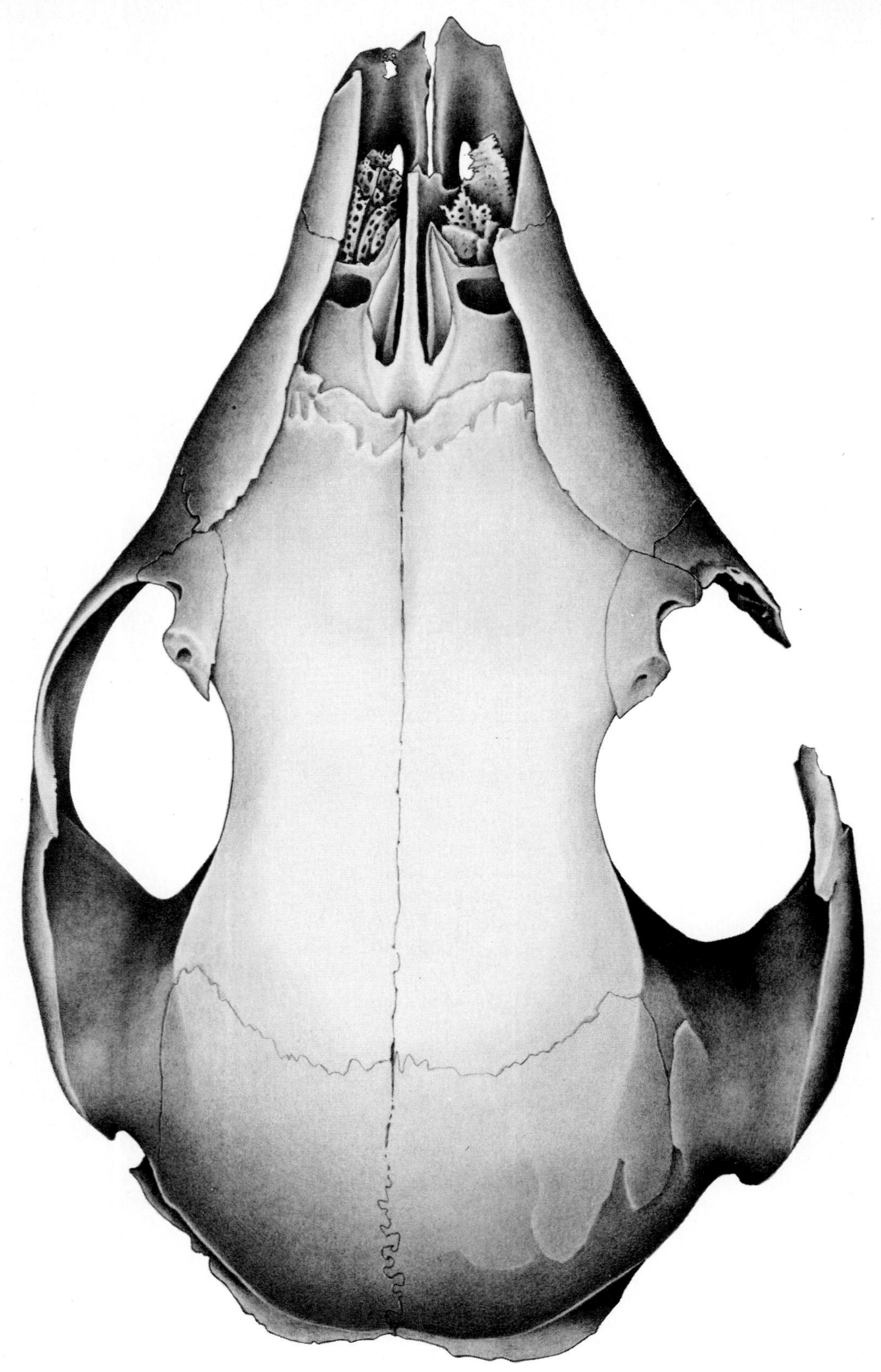

Figure 1-1. *Potorous platyops*, PM 4355, recent skull from surface of Webb's Cave, Western Australia, dorsal view. Pencil on one-ply Strathmore bristol paper. Drawing by Zbigniew T. Jastrzębski, Field Museum of Natural History, Chicago. Published in William D. Turnbull, "The Mammalian Fauna of Madura Cave, Western Australia, Part VI," *Fieldiana*, Geology, New Series 14 (1984) by Field Museum of Natural History, Chicago. Drawn in 1982.

final result and be able to judge the illustrator's skills.

This brings us to printing. All illustrations must be prepared for reproduction through printing. When a scientific paper is printed, the illustrations also will be printed. The scientific illustrator will be judged by others on the basis of the final product, the printed image. This is very important because the communication betwen scientists is established through the printing process, and both the scientist's as well as the illustrator's work will be judged by other scientists on its appearance after the printed matter comes off the presses. The illustration is good when it is clearly readable on the page of the publication. In order to achieve this result it should be drawn with this ultimate end in mind. Remember, the drawing the illustrator produces is not designated for gallery walls, although it may eventually make its way there.

Illustrations used in the sciences vary in their scope as well as in their final rendered presentation. The range is enormous, calling for a wide variety of techniques. One approach must be taken for simple line drawings such as diagrams and charts; a different approach is needed for extremely complicated halftone renderings. Good and visually interesting drawings are needed, but their technical refinement depends on the illustrator's know-how. Yes, visually stimulating drawings can be produced, but they depend on the subject to be drawn. Scientific illustrators do not improve on nature; they cannot diverge from the truth, although in many cases they will interpret the data given to them by the scientist. In such situations the interpretation is more of a clarification, accomplished with the help of sound observation and only under the direction of an expert in the appropriate field. The scientist is such an expert. He or she will explain and give all necessary information, and will patiently answer all questions. An artist's spontaneous "improvement" will not be truthful to nature and will become a part of an imaginary world.

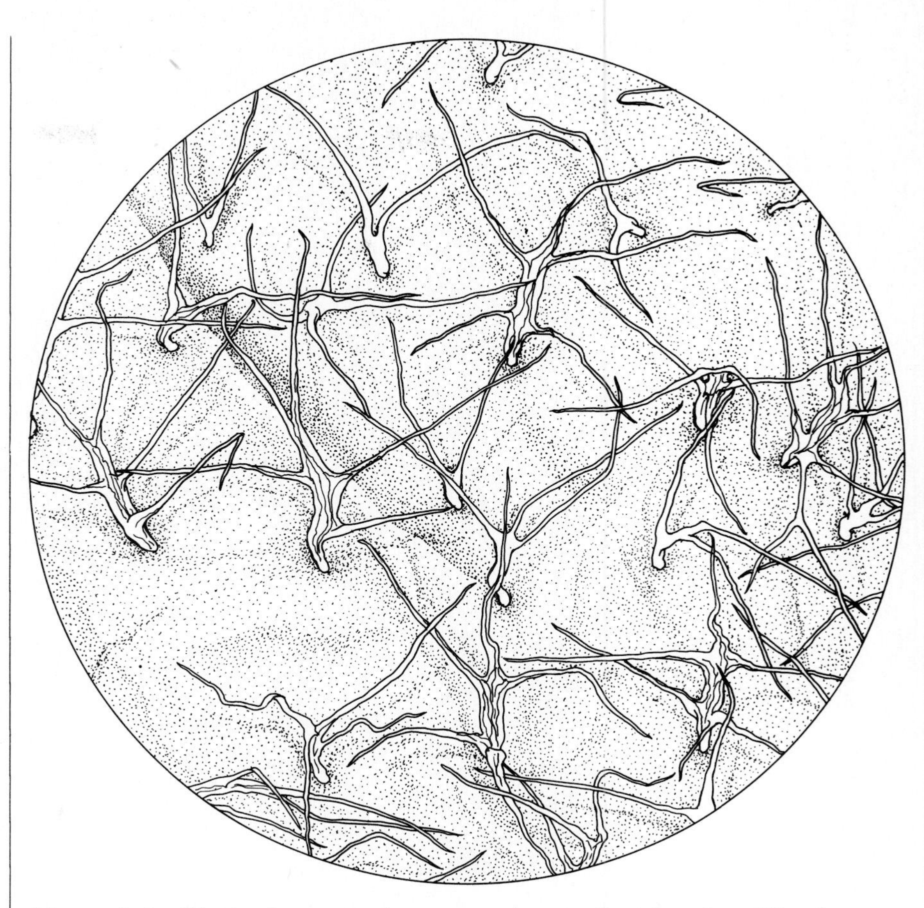

Figure 1-2. *Elaphoglossum auricomum*, scales on abaxial surface (12 ×). Pen and ink on mylar drafting film. Drawing by Zorica G. Dabich, Field Museum of Natural History, Chicago. Published in Robert G. Stoltze, "Ferns and Fern Allies of Guatemala, Part II, Polypodiacae," *Fieldiana*, New Series No. 6 (1981), by Field Museum of Natural History, Chicago. Drawn in 1980.

The real reason for drawing is to provide illustration for study. Scientific research must deal with facts, and as the illustration is a part of such a carefully monitored program, it therefore must do the same. This does not mean that all illustrations will be easily and properly understood by the layman. Each separate illustration, regardless of its visual beauty or apparent dryness, has a language of its own. Every illustration contains a definite message. Sometimes the information may be understood by a small group of specialists or, depending on circumstances, may be easily read by the broader public. The message is there and the informational context is extensive. Regardless of complexity or apparent simplicity of illustration, the "coded" information is more important than visual stimulus. Naturally, in order to present appropriate information in a coherent way, technical skill and patience are musts. The great diversity of subjects as well as the particularities of the given research add more challenge when combined with international codes for visual presentation. What is the international

Figure 1-3. *Fagopyrum cymosum*. Watercolor. Painted by Y. Ohta, Japan, World Life Research Institute, Colton, California. Unpublished. Painted in 1968.

code for visual presentation? It depends on the subject, the choice of technique used for drawing, the direction of illumination of the subject dictated by practical reasons, and agreement between scientists as to the uniformity of presentation. Naturally, as we humans cannot easily shake the imprint of past times and past habits, various codes of the past are still with us. The dotting technique known as stippling and the direction of light "from the upper left side" are firmly implanted. This does not mean that some codes of presentation of the subject, or the technique or illumination, are more beautiful than others. It simply means that codes are used by the world of science for the benefit of mutual understanding. It also means that the scientific illustrator will have to abide by them.

Standards do not develop quickly. It takes a few years as well as a few tries. As our illustrations are to be reproduced through printing, technology of the printing processes and the price of reproduction both influence the standards or the code of presentation used in drawing. Take stippling, for example. Stippling is making dots. It is a pen-and-ink technique based on the concept that the density of dots will produce a certain tone. The more dots, the darker the tone; the less, the lighter. Look at any printed halftone illustration. All continuous-tone printed illustrative material—such as a drawing in pencil, a wash, or a photograph—will be made up of dots. Such a reproduction is called a halftone, because for the most part it is not filled up with tone. Take a newspaper and look closely at a printed photograph: Part of the image is white and part is black. As expected, the spaces between the dots are white, while the dots are pure black. Only the changing density of the dot pattern produces the changes of tone. In order to change any drawing into an array of dots, a photograph has to be taken through a specially manufactured plate of glass. Such a plate of glass is called a "screen."

At present, our technology allows for production of extremely precise as well as various-dot-density screens, but in the late 1800s such a technique did not exist. At that time, the only way to obtain delicacy of tone was to draw an image in continuous tone directly on the printing plate. For that reason, limestone was used in the same way as it is now used in lithography. Sometimes etching or aquatint has sufficed. When, early in the twentieth century, photography had developed sufficiently and "screens" were introduced as an intermediate step in the process of transferring continuous-tone illustrations to the surface of the printing plate, an illustrator's direct involvement with printing plates was eliminated. But with time, that extra step in production combined with growing prices of printing small editions and prompted scientists and illustrators to cut corners. The new way of drawing became commonplace: imitating the "screen" effect through presentation of tone in already dotted form. New international standards were born when such drawing techniques were accepted by the majority. Similar processes have influenced positioning of the subject. When we look at the fish, we find the creature fascinating enough, tempting us to draw its representation in a pleasant- or interesting-appearing position. This is fine, but when the set of information about that fish is well defined, how can we convey a message in such a way that the viewer will not have any chance for personal interpretation and will receive only facts? During the transitionary period from 1800 to 1900, scientists discovered that in order to present particular information, the position of the specimen must be very well defined. A three-quarter view is interesting and will make a good visual impact on the viewer, but at the same time it does not allow for precise presentation and explanation of structural aspects of the subject. How long is that particular fish? Can all rays and spines in all fins be easily presented and recounted? How is it possible to count all of the scales after the drawing is printed if the subject, the fish, is not presented flatly on its side? Because of this, all fishes are drawn from the left lateral view. Such position of the specimen allows the illustrator to properly present the subject, depicting everything without the distortion produced by foreshortening. This is important for the scientist because such drawing illustrates his writing and his research. In scientific papers a particular specimen is described by words and numbers. An accompanying drawing must present that specific specimen with no discrepancies between the written text and the drawn image. As all specimens are numbered and carefully catalogued, the drawing can be compared with actual subject at any time. It is assumed that when such illustration is accepted by the scientist as a part of his publication, it is an accurate representation and therefore can be used by other scientists in place of the actual specimen. Naturally, the dawing is checked and rechecked during preparation. If it is necessary, it is corrected further after its finished stage. Would you, as a scientist, risk your reputation by accepting an illustration on the basis of good looks alone? Would you not be willing to recount and recheck all necessary details?

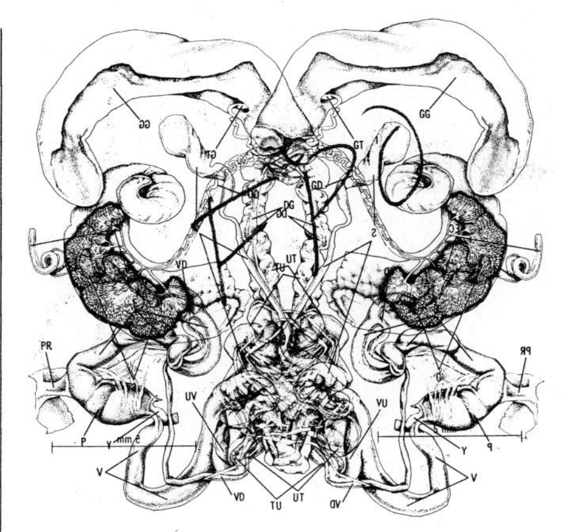

Figure 1-4. *Amphidromus incognatus Fulton;* palial and apical genitalia of snail. Pen and ink on plate-finish bristol board. Drawn by Linnea M. Lahlum, Field Museum of Natural History, Chicago. Published in Alan Solem, "First Record of *Amphidromus* from Australia, with Anatomical Notes on Several Species," *Records of the Australian Museum*, by Australian Museum, Sidney. Drawn in 1982.

Unfortunately, we humans do tend to make mistakes, sometimes overlooking the obvious. It is necessary for the sake of truthfulness to have the product, an illustration, coldly appraised by an expert. The position of every subject drawn for natural sciences is stated; it is the responsibility of the illustrator to comply with such requirements and render all drawings accordingly. Subjects received by the scientific illustrator to draw will vary in range and in scope. The areas of natural history, medicine, veterinary, and biology offer so much that no two specimens will be alike. Subjects will range from the anatomy of snails through the anatomy of human beings, from reconstruction of the earliest forms of life on earth through puzzle-solving problems found in the reconstruction of pottery for anthropology. Do you think you may escape from fashion design? Not at all. Who else will draw Indian clothing, deciphering the patterns used for cutting and sewing, than the scientific illustrator? Good knowledge of drafting will come in handy when it is necessary to depict large architectural structures, dwellings, and small houses, as well as fortifications. Subjects will come from outer space as well as from the depths of our oceans.

Scientific illustration is usually linked with the natural sciences. This is because more illustrations pertaining to natural sciences—botany, anthropology, zoology, and geology—have ventured into public hands. We are all very familiar with botanical illustrations but rarely see veterinary publications. Regardless of a scientific illustrator's specialization, all drawings require thinking. Sciences are divided into groups which in turn are subdivided into very precise areas of interests. This offers the illustrator the possibility to match up his personal interests with the pleasure of drawing. Let's take a branch of science with which we are all familiar—zoology. Zoology is a natural science, and it is subdivided into major areas such as insects, birds, mammals, amphibians and reptiles, invertebrates, and fishes. Each of these divisions is further subdivided into well-defined specializations, all presenting challenging situations to the illustrator.

Medical illustration is a separate, highly specialized section of scientific illustration focusing primarily on one complex subject—the human body. Complexities of that subject call for sound knowledge of human anatomy. It is impossible to draw a representation of operating procedures when a surgeon is describing the operation in progress unless the illustrator knows what the doctor is talking about. It is difficult to draw only from assumptions. The illustrator must rely on his knowledge of the subject and on understanding the terminology used during such a descriptive process. It is much easier to draw something that is visible and tangible. Knowledge of appropriate terminology will be of great help; it will be much easier to know what to draw when we know what is being asked from us and what to look for. Nevertheless, scientists are quite willing to issue instructions and explain everything in down-to-earth language, providing that the illustrator is interested in producing a proper drawing. Such a situation does place more demands on the scientist, as he will have to spend more time explaining the subject as well as describing the desired outcome. At the same time, an illustrator who does not know the terminology is placed under tremendous pressure and therefore must think fast, ask lots of pertinent questions, and take quick notes as well as draw general sketches of the subject right there in front of the scientist. Precision is a must—a couple of millimeters do make a difference. The granulations found on the surface of megaspores must be drawn with the same exactness as the cross section of the pottery sherd or the presentation of structural aspects of soft tissue.

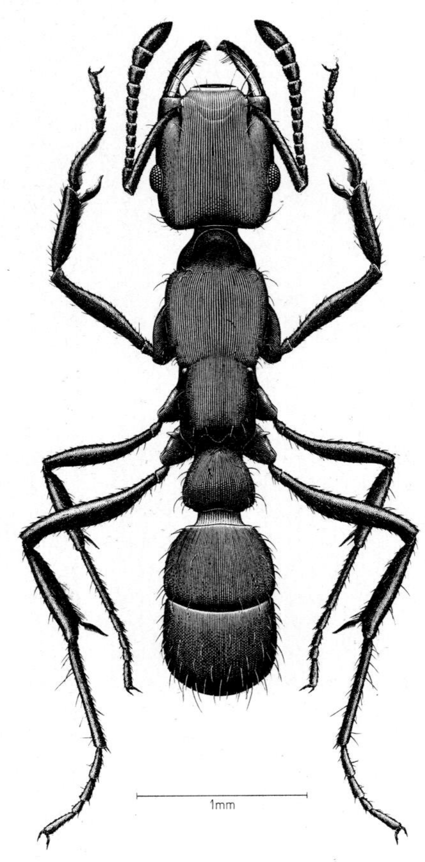

Figure 1-5. Reconstruction of *Gnamptogenys pristina* Baroni Urbani, Lower Miocene of Dominican Amber, dorsal view. Pen and ink. Drawing by Armin Coray, Museum of Natural History, Basel, Switzerland. Published in C. Baroni Urbani, "The Ant Genus *Gnamptogenys* in Dominican Amber," *Stuttgarter Beiträge zür Zaturdunde,* Series B, (*Geologie and Paläontologie*), Nr. 67, Stuttgart, 1980. Drawn in 1980.

Subjects must remain correct and easily readable, especially after reproduction of the illustration. Many times certain clarification of the subject is necessary. If discolorations of the surface of the specimen interfere with presentation of that surface, such should be consciously removed. If strands of tissue obstruct the important matter, they should not be drawn. As always, the illustrator should discuss such matters with the scientist. The whole project, be it a simple graph or a large quantity of three-dimensional illustrations, must be talked over with the scientist. Meetings must take place and precise communication established. Truthfully, all decisions are undertaken by the researcher, the only exception being an illustrator familiar with the problems and with some experience. Nevertheless, most clarifications are implemented on the sketch or sketches, although the scientific illustrator should be prepared for changes after the rendering is finished. In such a situation a lot depends on the changes in the research and changes in the scientist's mind, as well as mistakes made by the illustrator. A finished illustration is not untouchable, as technical problems of rendering may distort information and confuse the viewer. A process of analytical selection must take place in the illustrator's mind. Observation is a must. Thinking is a must. The illustrator must closely observe the subject, scrutinizing everything and asking himself a lot of questions. Remember, the final objective is an explanation presented in the form of the drawing. The illustrator should eliminate, add, or change values and contrasts very consciously when drawing. Drawing is a thinking process, especially when it is used as a visual explanation of scientific research.

When drawing, a super-realistic representation of a specimen does not necessarily mean that everything visible automatically should be included. Photography can achieve this better and faster than the human hand. A camera will not differentiate between accidental reflections of light, but a human mind will be able to sort wanted and unwanted features. Take a look at a specimen with a semitransparent outer layer of tissue. Some of the internal structures will be visible through the outer layer. Does this mean that when presenting visual information pertaining to the outside forms you really must include in the drawing whatever is visible? Just because the specimen is covered by preserving liquid, must you draw all the reflections caused by light? That is an obvious example of the selective process through which a scientific illustrator must progress during

Figure 1-6. *Chloraea densipapillosa* Schweinfurth. Crow-quill pen and India ink. Drawing by Josefina E. Lacour, Instituto de Botánica, Instituto Nacional de Tecnología Agropecuaria, Castelar, Argentina. Published in M.N. Correa, "*Chloraea* genero sudamericano de *Orquidacae*," in *Darwiniana*, Instituto Darwinion, Buenos Aires, 1969. Drawn before 1969.

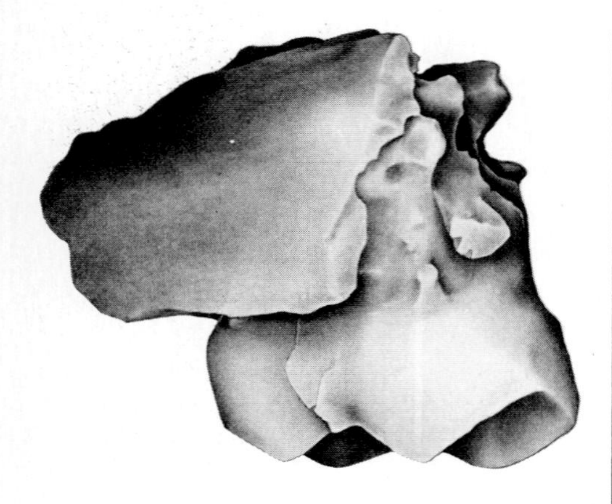

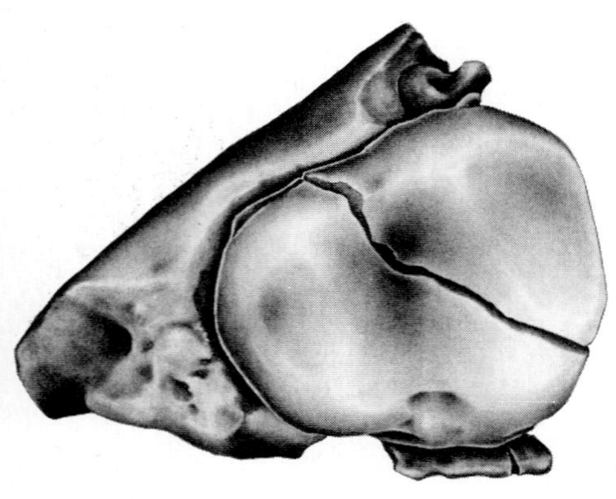

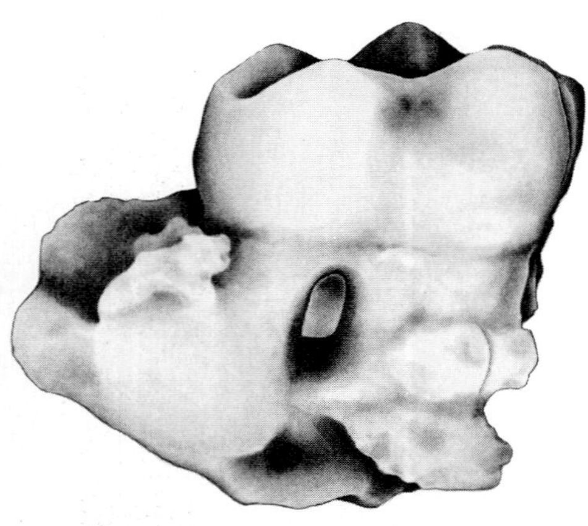

Figure 1-7. *Potorous platyops* from Madura Cave, PM 34419, right M^3 in maxillary fragment shown in labial (top), crown, and lingual views. Pencil on Strathmore drawing paper. Drawing by Zbigniew T. Jastrzębski, Field Museum of Natural History, Chicago. Published in William D. Turnbull, "The Mammalian Fauna of Madura Cave, Western Australia, Part VI" *Fieldiana*, Geology, New Series No. 14 (1984), by Field Museum of Natural History, Chicago. Drawn in 1983.

the preparation of the sketches and finished rendering. The same processes—observing, thinking, and making decisions based on received information as well as on explanatory needs—may yield a simple-looking outline illustration or a complex continuous-tone rendering. Sometimes simple, well-defined clarity contained in an outline may be harder to obtain than continuous-tone illustration. Most of the time the purpose of outline drawing is to present the essence of pertinent information. For example, when preparing explanatory drawings of clothing artificats, simplicity is a must. Such simplicity will be more beneficial than a detailed, superbly rendered presentation of the artifact. Why? Readability—a quick understanding of the subject.

As always, before the final pen-and-ink drawing is made, a series of sketches culminating in the final sketch is prepared. That final sketch is analyzed by the scientist and his decisions are implemented into the sketch before proceeding further. The difficulty lies in deciphering the unknown. The puzzle must be solved. How many pieces of leather have been used by the Athapaskan dressmaker? How has each separate piece been cut out? What are the actual dimensions of each piece? Clothing artifacts cannot be taken apart and then simply traced. Taking an artifact apart will destroy it completely. Not the slightest damage can be inflicted on the specimen. The illustrator cannot mark reference points using artificial means such as needles, pencils, or metal clips. Here the illustrator's skills of observation should help. Any natural speck of discoloration found on the surface of the artifact will aid in measuring and will serve as a reference point. Slowly but surely the illustrator will progress using just measuring tape, observation, and logic.

Small objects must be viewed through a microscope. The question is how small and through what kind of microscope. Generally speaking, a specimen measuring ten centimeters will be considered large, but one measuring five to seven millimeters will be termed small. As everything is relative, the five-millimeter one is of enormous size when compared to those measured in microns. Naturally, the smaller the specimen the more powerful the microscope that should be used for viewing. But, how, you may ask, is it possible to view an Eskimo fishing net through the microscope? The answer lies in the area of research. If the structure of fibers used to make the net is a field of study, then small sections of those fibers are viewed or even photographed through a scanning electron microscope in order to see everything clearly. On the other hand, when the general shape of the net is needed, or if a weaving pattern has to be deciphered and illustrated, then the microscope will not be of much use.

Most specimens which find their way to the illustrator's drawing table are relatively small, so enlargement is a must. At the same time the specimens that are too large to be placed under the lens of the microscope must be reduced. Because all drawings are precise representations of their subjects, the visual image is conceived by measuring alone or in conjunction with old-fashioned grid techniques. Such techniques are simple and very reliable, allowing the illustrator to manipulate the proportion existing between the subject and the drawing. When drawing small specimens, tracing devices are used in order to cut corners. Why? Time is money. You do know that illustrations are essential to the publication of scientific paper. The scientist is eager to have his statement printed and must submit his work, with all illustrations finished, by a certain deadline. If he does not make it, his paper will have to wait. You, the illustrator, can bungle up everybody's schedule. You may have not realized all of this, you may simply have assumed that things would work out somehow. But things do not happen by themselves; only you make them happen. For that reason as well as for precision of drawing, drawing tubes or, as such apparatus is sometimes called, *camera lucida* is widely used. Such an apparatus has beneficial points, allowing the illustrator to draw fast, but at the same time creates problems. The biggest problem is optical distortion. This becomes very visible when a relatively "large" specimen (one which is about 10 or more centimeters) is viewed and traced. When magnified 25 times, such a subject is not visible through the oculars as a whole object because of its size. Only very small sections can be scrutinized without moving the specimen under the lens. Unfortunately, distortion is not noticed immediately. It is noticed when the sketch is almost completed. Is technology imperfect? Well, the tracing apparatus is attached to the microscope, and the light must pass through a series of lens and prisms. Because a three-dimensional subject is viewed in sections, the viewing pattern is similar to the eye of a camera photographing a rotating planet. The image toward the edges of the lens will be slightly stretched. Only the direct centered view is truthful in its proportions. As the specimen has to be drawn in small sections, because you cannot see the whole subject at once, the multitude of small distortions will compound. A thinking illustrator will take a general measurement of the whole subject by hand, drawing such a sketch on

Figure 1-8. *Potorous platyops*, TMM 41106-619, from Madura Cave. Right P^4 shown in labial (left), crown, end lingual views. Pencil on one-ply Strathmore drawing paper. Drawn by Zbigniew T. Jastrzębski, Field Museum of Natural History, Chicago. Published in William D. Turnbull, "The Mammalian Fauna of Madura Cave, Western Australia, Part VI," *Fieldiana*, Geology, New Series No. 14, (1984), by Field Museum of Natural History, Chicago. Drawn in 1983.

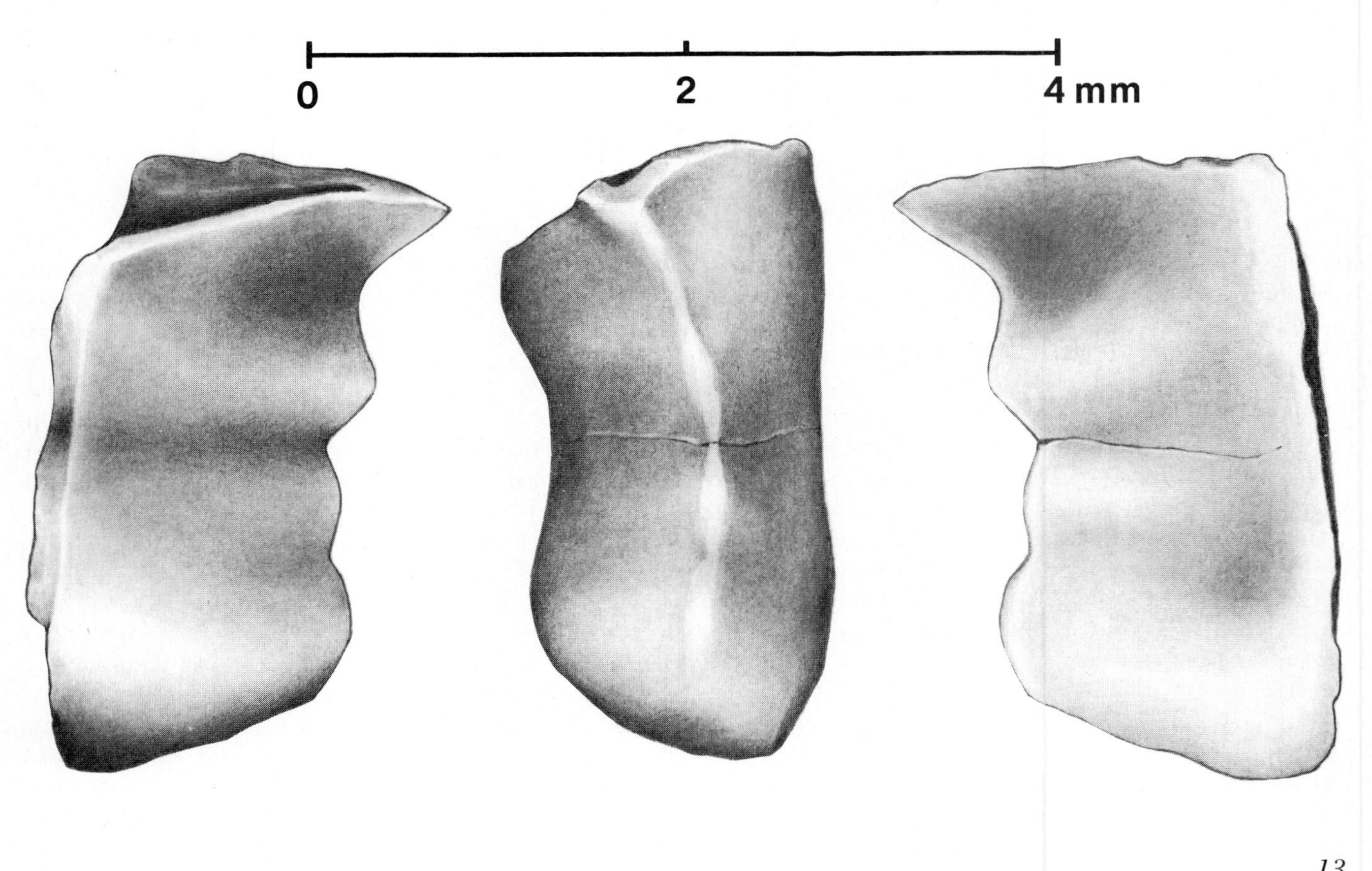

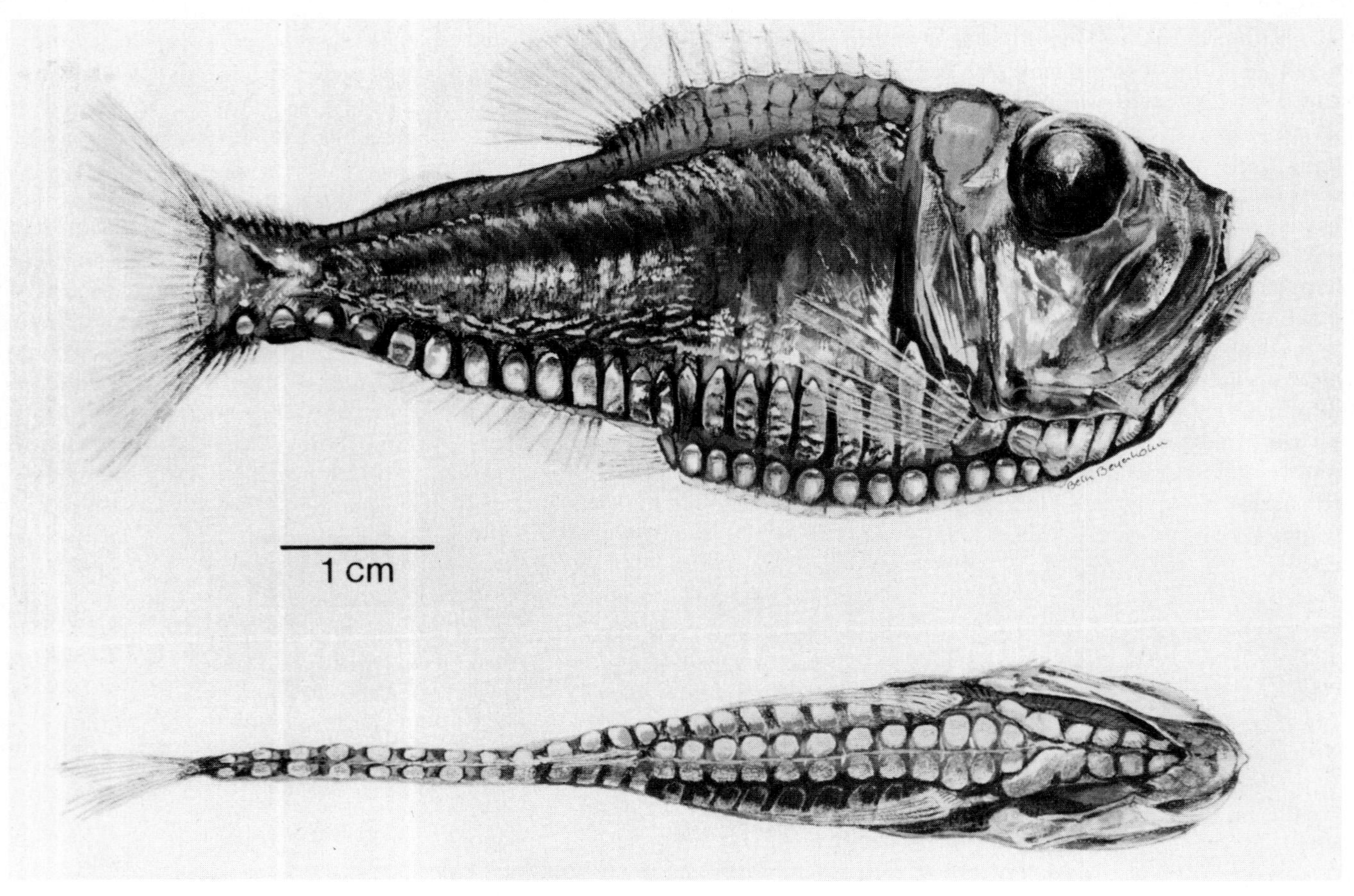

Figure 1-9. *Agryropelecus affinis* (hatchet fish). Tempera and crayons. Painting by Beth Beyerholm, Institute of Cell Biology and Anatomy, University of Copenhagen and Danish National Museum (Antiquities), Copenhagen, Denmark. Published in *Naturens Verden* by Rhodos, Copenhagen. Painted in 1982.

semitransparent tracing paper. By matching this sketch with the one produced when using camera lucida, the illustrator will be able to see by how much the camera lucida sketch is distorted.

Depending on the quantity of details, the scientist's say-so, and available viewing tools, the magnification or reduction of the specimen is established. Usually, when magnifying, the sketch will be much larger than the finished rendering. Small magnification does not show a great many details, but very large may produce too much information. Again, the illustrator will have to rely on the scientist's judgment and his definition of what detail consists of.

The final outcome of the original illustration is the reproduction, a printed image on the page. But from the illustrator's point of view it is more than just that; it is a message to everybody that you can draw and paint. The audience is limited, but it is worldwide. Unfortunately, you will have to wait for the printed image, and the waiting period can be long, from six months to ten or more years. Why such a wait? Because major participants in the endeavor have to finish their work; because there is a waiting line for publications. This may be a bit frustrating, especially if you are accustomed to the speed of production of editorial illustrations. You know that when, let's say, a book of fairy tales is about to be published, the project will have a specific date of publication. All involved will do their best to make sure that such a deadline is kept. However, in scientific illustration the end of the project, the end of the research, is most of the time unknown. The general subject of the research can be sectioned and a set of deadlines will be established, but as the whole point of conducting research is to enhance knowledge, there may be many stops, breaks, and reroutes during that time. Shortcuts will be omitted. In some cases the project may take a different course than planning has predicted. It may happen that some of your illustrations will not be used at all, and it may happen that you will have to procure more illustrative material than originally was planned. You will only be involved with drawing or painting, but every change in research will affect you directly. It will affect your work, it may affect your time and may speed or prolong publication of your illustrations. Always remember the purpose of your drawings: explanation—clear, visible, easy-to-see, easy-to-read explanation. Scientists do prefer a drawing that is technically well rendered. This is definite. But at the same time the researcher may choose a drawing of poorer technical quality if such an illustration is more explanatory. Many scientists themselves draw—some because they like to draw and others because they have to, if professional artists are not available.

In scientific illustration a style of drawing does not have any place. Techniques such as pencil, wash, scratch board, and washes are sometimes referred to as styles of drawing. Such terminology is misinformation. The quality in handling a technique used for drawing has nothing in common with style. It is possible to recognize a particular illustrator by the quality of rendering, be it good or bad, just as with handwriting. The technique of calligraphy has nothing to do with the context, although it is possible to recognize a person's handwriting.

Communication is at the base of all scientific illustrations. Take a fish, for example. It is presented in outline for clear readability. The technique, line with stippled tone, has been chosen by the scientist. The materials picked from those available are tracing tissue and technique pen. Now look at the "hidden" message. First, realize that all the scales are counted. Their shape, unreal as it may seem, comes from careful observation and tracing—measuring systems of transposing the information onto the paper. Missing scales are not drawn, but the missing scales have exposed lots of area around those in place. Usually when looking at the fish you will not be conscious of the size and shape of individual scales. Their illustrator has been forced to concentrate on outlines of every one. The drawing may appear unrealistic, but actually it is not. The transparency of the scales has been omitted for the sake of clarification of important matter, shape, size, and quantity of scales. Whatever was visible through translucent surfaces is not drawn. All fins are clearly defined. It is easy to see and differentiate spines and rays. The coloration, presented in dark and light values, represents the areas of real color which this fish had some time ago. The fish is drawn from its left side, because a custom or a code presentation of fishes demands this. The fins on the right side are not drawn, because they are in shape, size, and count the same as the left. A scale, in metric, is near the drawing. The scale represents a proportion. It is possible to measure the drawing with depicted distance and therefore to know the creature's real dimensions.

What is the difference between scientific illustration and drawing from nature? Sketches of animals, flowers, landscapes, and other drawings of material subjects cannot be included in scientific illustration. Regardless of the precision of execution, such works do not have information selected for the basic purpose of representing scientific data. Some of such drawings or paintings do come very close, but nevertheless important ingredients are missing. The position of the subject may be incorrect, visual interference will be forthcoming from the background, the imprecision in observation are all small parts of the whole problem. Such drawings and paintings are produced for the viewer or for the artist himself. They do not benefit anybody else except those who find a beauty in such works of art. Certainly such artworks cannot be used in functional scientific texts. Scientific illustration is based on nature, which is all around us. The border between drawings from nature and scientific illustration is thin but well defined. Artworks produced for the walls of a gallery or for the pages of editorial publications are not scientific illustrations, although the very same artist may produce both. Anything that is not prepared for active research is not a scientific illustration. The problem of separating and defining illustration produced for science is a difficult one and is reflected in the various job titles given to scientific illustrators. Re-

Figure 1-10. *Chusquea quila* Kunth. Crow-quill pen and India ink. Irene Coloiera, Instituto de Botánica, Instituto Nacional de Tecnología Agropecuaria, Castelar, Argentina. Published in M.N. Correa, *Flora Patagonica part 3 Gramineae* Vol. 8, by Colección Científica del I.N.T.A., Buenos Aires, Argentina, 1978. Drawn before 1978.

search institutions will refer to scientific illustrators as artists, illustrators, natural science illustrators, technical illustrators, draftsmen, staff artists, and lastly, scientific illustrators.

Educational illustrations found in textbooks or encyclopedias are even harder to separate from scientific illustration. In some cases such drawings are scientific illustrations. Most illustrative material in biology textbooks was originally prepared for research projects. Some of such illustrations are redrawn or modified explanations of a subject, but are still based on the original concept. Naturally, you may say, all illustrations found in books on anatomy will fall into our category, but this is not really so. For most of such publications this will hold true but unfortunately, so-called "anatomy for the artist" types of illustration have nothing in common with scientific illustration. Such illustrations do not offer proper readability and proper information. The approach may be similar, but the context of the drawing is not. The field of medical illustration is within the realm of illustrations for science. Such illustrations are highly specialized and focus primarily on one complex subject, the human body. The complexities of that one subject as well as constant medical developments call for much more than the knowledge of drawing. Medical illustrators must know their subject very well. This means that they must have a sound knowledge of human anatomy as well as a master of the terminology used in the medical profession. The knowledge of terminology or anatomy is not expected in natural sciences just because of the enormity of the field. A general knowledge of biology and human anatomy will be a great help, but is not a requirement. In medical illustration a lot of drawings are prepared from verbal description, whereas in natural sciences most illustrative material is drawn from the actual specimen. In medical illustration, the human body is the prime subject of interest, whereas diversity exists in natural sciences. Because of this, illustrators working in natural science research centers will encounter more structures unknown to them. They will encounter more puzzles and will have to struggle with lack of knowledge about those subjects. Their subjects will range from pictorial reconstructions needed for anthropology or paleontology through detailed studies of plants for botany as well as various forms of life for zoology, encompassing past, present and the future.

CHAPTER TWO

Areas of Specialization

ONE of the most controversial and interesting subjects for discussion is specialization, and with good reason. For the illustrator, especially the scientific illustrator, specialization is very desirable but at the same time can easily become dangerous. It is very desirable because it promotes familiarity with the technique of observation of the subjects drawn by the illustrator: familiarity with materials used during the rendering, familiarity with the techniques of drawings, and most importantly, familiarity with positioning or as it is referred to most often by scientists and illustrators, the procedure of setting up the specimens for the purpose of the drawing. It is desirable because the more we look at a particular segment of nature, the more we are able to see and the more and better we are able to understand.

With continual practice, drawing proficiency will be increased. With practice, an illustrator will draw better. The combination of experience gained by practice in observation and the expertise of drawing will improve the artist's capabilities as a scientific illustrator in his area of specialization. Through practice he will become better, but only better in observing certain types of subjects and only better in drawing when using techniques in which he had practiced. More specifically, the scientific illustrator will be better equipped to understand the demands placed upon him by the scientist or group of scientists conducting research within a particular area of science.

Why may specialization be dangerous? Why may it be dangerous especially at the time when the illustrator, after some years of experience, considers himself an expert? Specialization, any specialization, will place you in a certain frame of mind from which may be usually impossible to escape. Being accomplished and settled in a narrow area does promote self-assurance, which in turn can easily lead to the lack of self-criticism. An illustrator reaching such a point of self-assurance is extremely positive in his judgments; he has a strong feeling of being right; he believes that he knows all he needs to know. In many situtations this may be the truth, but it is better to leave a little room for improvement in the observation of subjects as well as improvement in the technical skills of drawing. We should remember not to blind ourselves by finding or looking for some kind of justifications at the moment when we should face the facts. Justifications are easy to find and easy to believe in.

The narrowness of specialization creates boundaries marked by our knowledge of drawing techniques. Limited knowledge of such techniques, when combined with the choice of subjects, does not allow us to be flexible enough. When situations arise that require us to change the subject and the technique, we may not be able to face the new challenge, hence producing mediocre illustration. When changing from a familiar technique of drawing practiced over a number of years and from a subject to which he or she is so accustomed, an illustrator will face numerous difficulties, which will have to be overcome in a very short period of time. Newly encountered problems will

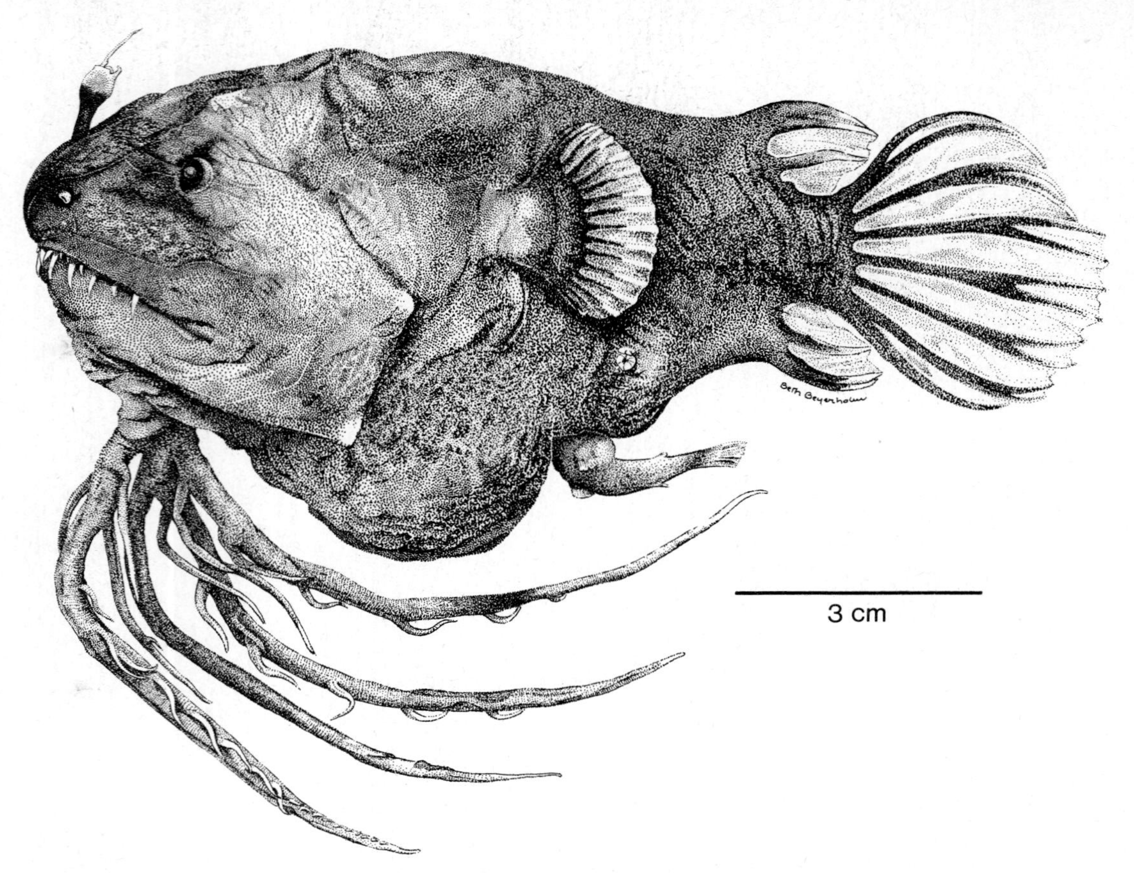

Figure 2-1. *Linophryne brevibrabata*. Pen and ink. Drawing by Beth Beyerholm, Institute of Cell Biology and Anatomy, University of Copenhagen and Danish National Museum (Antiquities), Copenhagen, Denmark. Published in *Steenstrupia*, Denmark, 1980. Drawn in 1980.

have to be solved during the sketching and rendering of the very first illustration. Such a simple-sounding project as drawing a stone implement may turn into a disaster. Proper usage of a crow-quill pen may become impossible, and the quality of the finished drawings will suffer (as well as the reputation of the illustrator). Limited specialization may be dangerous because it does not allow an artist to become a well-rounded craftsman; it will limit his capabilities. With technical limitations, a pronounced diversification is impossible, and it may be hard to make a reasonable living as an illustrator. Before entering any particular area of specialization, illustrators cannot forget that their attitude must be businesslike, making their approach more professional.

Nevertheless, it is hard to escape from specialization. When working, the scientific illustrator is bound to an institution and hence to the areas of research conducted by scientists within the institution. This means that if natural sciences predominate, the illustrator will be directed into a specialization in natural sciences; and when other branches like nuclear physics or medicine are the core, the illustrator will be forced to specialize in those areas. Again, within the general boundaries designated by the research, an illustrator can choose or be forced by circumstances into a specialized subdivision, or can remain faithful to the principle of diversification. When specializing in a narrow divisional area like anthropology and not in the enormity of natural sciences, an illustrator will discover very quickly the complexities of what once had been only a division and perhaps the specialization will become narrower. Reconstruction of pottery or drawing stone tools are highly sophisticated specializations by themselves.

By drawing soft tissue, skeletal structures, baskets, Indian clothing patterns, muscular reconstructions, stone tools, anatomical futures of *Homo sapiens*, pottery sherds, and holotypes of fishes as well as botanical specimens, an illustrator will be exposed to subjects requiring different techniques and different approaches. He or she will be in physical and visual contact with objects and, by working, will gain skills otherwise not obtainable, at the same time practicing techniques of drawings which would have re-

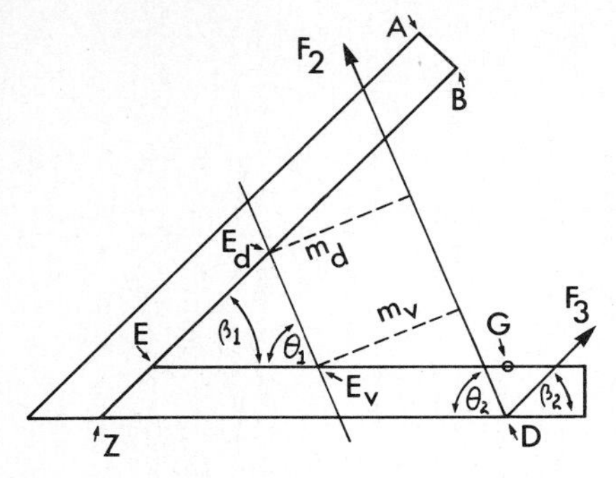

Figure 2-2. Changes in palate from primitive to advanced rhachitomes. Pen and ink on three-ply Strathmore drawing paper. Drawing by Dr. John R. Bolt, Field Museum of Natural History, Chicago. Published in John R. Bolt, "Evolution and Functional Interpretation of Some Suture Patterns in Paleozoic Labyrinthodont Amphibians and Other Lower Tetrapods," *Journal of Paleontology,* by Society of Economic Paleontologists and Mineralogists. Drawn before 1974.

mained foreign to him or her. As you can see, the decision regarding a broad specialization or a narrow one is a personal one, at times influenced by outside forces. Such a decision should not be treated lightly and should be reached only after some serious consideration.

Let's look at the very beginning of an artist's career. Such a career starts during the earliest years of study, during the initial introduction to drawing and painting. Usually for the art student the meaning of the word *specialization* is a bit different than for those who are already involved in the production of illustrations for consumer use resulting in personal income from such a production. An artist-to-be cannot forget that all artists, including himself, are involved in a business. Art is a practical enterprise, and the school years are the time designated for the necessary preparation for future competition. In school, all students have to choose some area of art as a specialization for that very practical reason. Specializations offered by various art schools are diverse, be they drawing, painting, printmaking, or scientific or medical illustrations. Concentrated effort does produce better results and prepares the student for quicker and stronger entry into the positions available.

In a few years of study it is impossible to gain knowledge in all aspects of art. All you can do is acquire the basics. Let's take a four-year curriculum as an example. Four years of study sounds very impressive and may appear to be a lot of time, but in actuality this does not allow for extensive practice. Let's examine those four years closely and compare some numbers—maybe then we'll understand the meaning of extracurricular work. Drawing is a base for all artists, therefore all art schools require that students take at least one drawing course during every semester of four-year study. It is a must for the student in order to fulfill requirements and receive appropriate credits. One drawing course every semester provides six hours of practice a week. As there are approximately eighteen weeks to one semester, it should be easy to calculate the total hours of practice during four years of study. There are 108 hours per semester, which, if multiplied by four years (eight semesters), provides 864 hours of total practice. Returning to reality, we should realize that 864 represents 17.6 weeks (at 7 hours a day) of total practice. This is less than one-half year. So, in four years of study, a minimum of one-half year is devoted to drawing. Naturally, all of this time does not include coffee breaks and all of the precious minutes wasted on small talk or other things. An obvious conclusion to reach is that in order to obtain the technical skills needed to produce a high-

Figure 2-3. *Amphidromus perversus,* snail's dissected genital system. Pen and ink on plate-finish bristol board. Drawing by Linnea M. Lahlum, Field Museum of Natural History, Chicago. Published in Alan Solem, "First Record of *Amphidromus* from Australia, with Anatomical Notes on Several Species," *Records of the Australian Museum,* by Australian Museum, Sidney. Drawn in 1982.

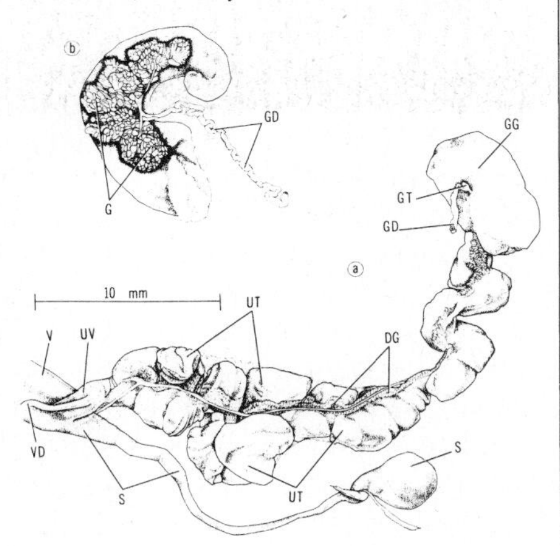

Figure 2-4. Woman's dress from Siaelland, (Zealand), 1850–1900. Tempera paint. Painting by Beth Beyerholm, Institute of Cell Biology and Anatomy, University of Copenhagen and Danish National Museum (Antiquities), Copenhagen. Published in *Folkedragter (Old Danish Costumes)* by Lademann, Copenhagen, Denmark. Painted in 1978.

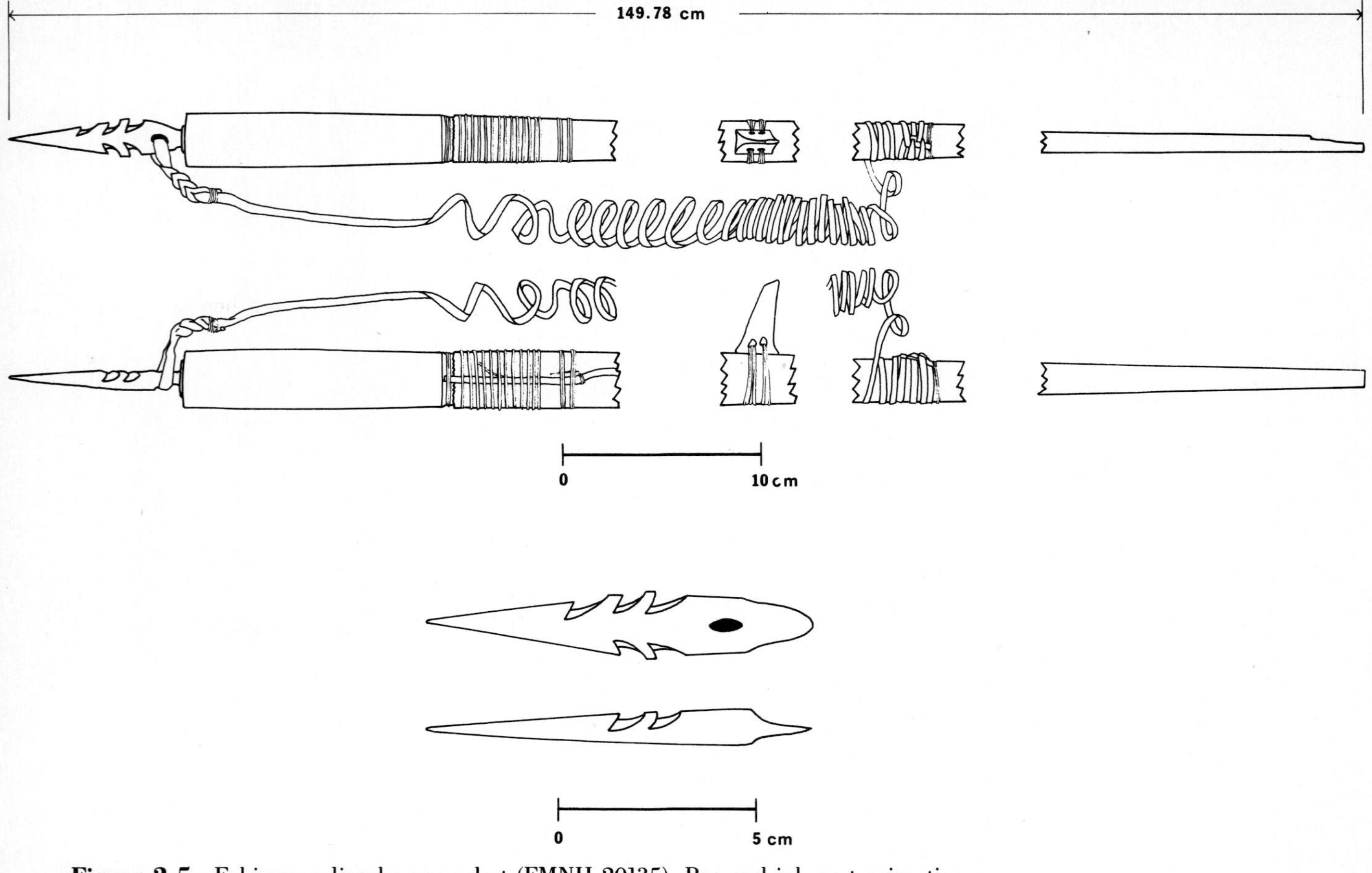

Figure 2-5. Eskimo sealing harpoon dart (FMNH 20135). Pen and ink on tracing tissue. Drawing by Zbigniew T. Jastrzębski, Field Museum of Natural History, Chicago. Published in James VanStone, "The Bruce Collection of Eskimo Material Culture from Kotzebue Sound, Alaska," *Fieldiana*, Anthropology, New Series No. 1, by Field Museum of Natural History, Chicago. Drawn in 1979.

quality illustration, the practice of drawing and painting cannot be confined to the classroom alone. A lot of personal time should be devoted to drawing in order to achieve a desired goal. The most important problem facing us is the impossibility to predict what technical skills may be needed for the future. Illustrators' initial decisions regarding specialization may not be correct and, by exclusion of certain techniques of drawing and painting, may not allow them to achieve proper results. At the time when they need certain skills, they will not have them.

It is fair to say that a large portion of scientific illustrators are specialists in their respective fields by accidents of nature. Such specialization happens because an illustrator, being busy with the production of illustrations for income, does not have the time to practice different techniques of drawing, especially new ones, and does not have the time to diversify into new subjects.

On the other hand, it is possible to plan ahead, to set goals, and to prepare for the ideal. Diversification into drawing new subjects does require the ability to adapt known techniques of observation and to make conscious effort in finding new approaches. To realize that an unknown subject is exactly as stated, an unknown subject, is a very important first step in the complicated process of observation and will help in difficult moments, but will not replace the time-consuming acquisition of experience. Usually, an illustrator does not have the luxury of time. Time will work against him, as an illustration will still have to be completed by a prescribed deadline.

Being employed as a scientific illustrator, an artist will have to prepare drawings for a very specific purpose and most often in a designated technique, a technique of drawing not left to his choice. Usually, the illustrator will be asked to employ one technique for a considerable period of time ranging from a few months to a lifetime. By starting to draw specific subjects, whether it be stone tools or insects, in only one technique, the newcomer slowly but surely will become a highly specialized illustrator by pure accident, forced into this specialization by the combination of monetary needs and availability of positions. When changing to new techniques and new subjects the illustrator may not be able to understand quickly the new demands placed upon him by the scientist. A new technique suddenly introduced may be foreign to him. That lack of knowledge combined with unfamiliar demands of observation can produce unpleasant results.

Not all scientific illustrators who specialize in certain techniques or in certain areas of natural sciences do so because this was their goal. Let's look at the most common specialization: black-and-white, pen-and-ink technique. It is a fact that black-and-white techniques, especially so-called line drawings, are in great demand. *Line illustration* does not necessarily refer to an outline drawing with formal values created by the changing densities of lines. A line illustration can be a delicately rendered image using contrasts and dots as the base for the presentation of values. It can visually imitate mezzotint or the graininess of autolithography. Such illustrations are considered line drawings because during the platemaking procedures the tone—an artificial introduction of dotting

Figure 2-6. Eskimo sealing harpoon (FMNH 20135). Pencil sketch (final sketch) on tracing tissue. Drawing by Zbigniew T. Jastrzębski, Field Museum of Natural History, Chicago.

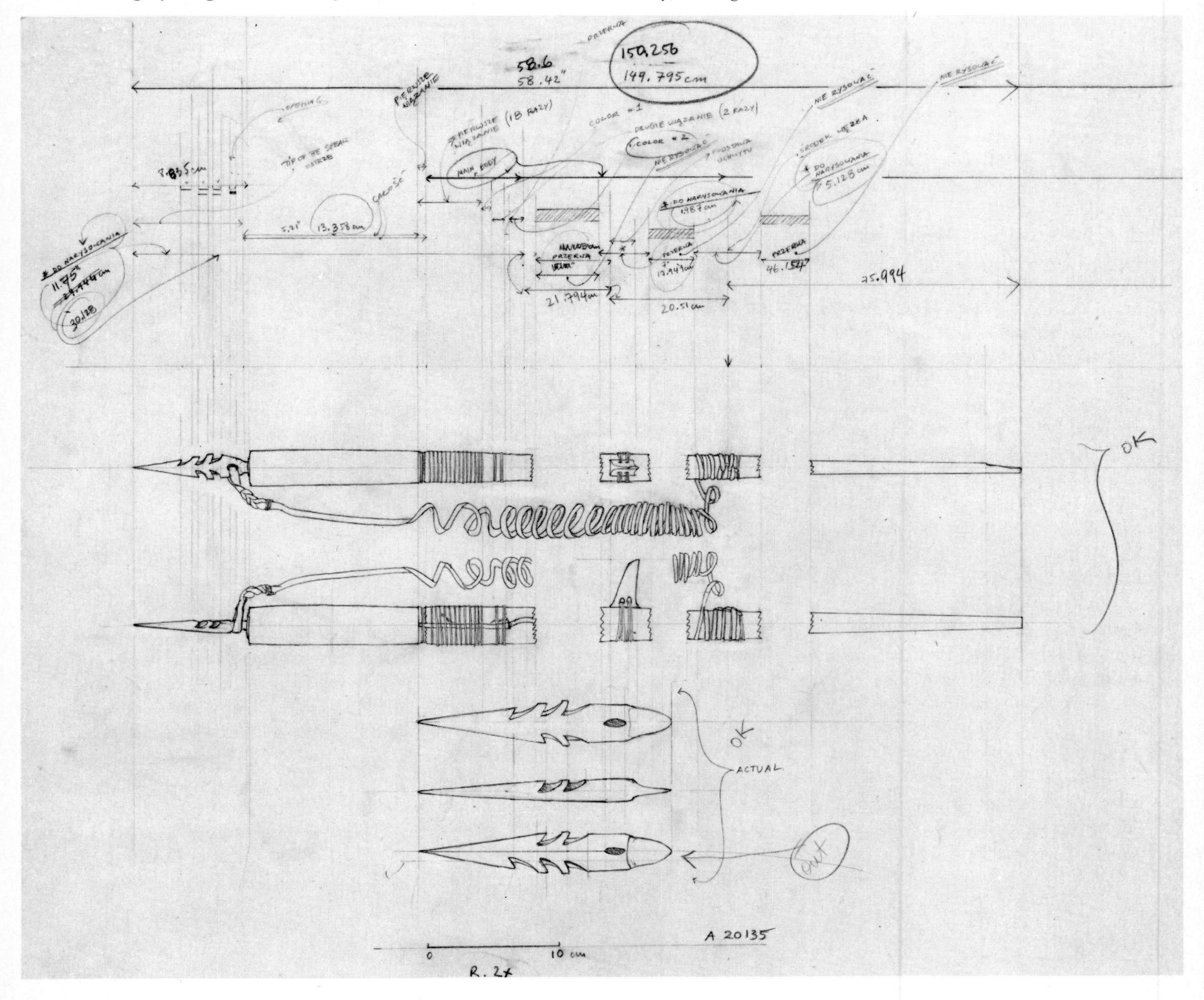

pattern onto a screen, preserving and changing into the dotting pattern the tonal values found in wash, continuous-tone pencil drawings, and other similar drawings—is *not introduced*. In other words, the illustration is photographed without the screen. Platemaking is simplified, resulting in cheaper production of printed matter. For those reasons it is very likely that an upcoming scientific illustrator will start with pen-and-ink technique, rendering illustration in stippling. Over the years the illustrator will become an expert, improving his or her renderings, and through the ritual of daily practice becoming a specialist in such technique. Researchers are known for their fondness for good-quality illustrations; they do appreciate an expert illustrator. After some years of experience, the clarity of information is combined with the elegance of stippled tone and the quality usage of different techniques. When a highly specialized illustrator receives a project involving a technique foreign to him, he will face an extremely challenging situation. By not practicing such a technique continually over a period of time, the outcome will be crude compared with drawings rendered by other artists who had practiced such technique. The illustrator's production will drop drastically, as he will be forced to spend more time per drawing than otherwise would be required. It takes more time to render a subject in an unfamiliar technique than it does when using familiar tools in a familiar way. The result may be unsatisfactory. The client may not be pleased with the drawing and may ask the illustrator to return to the technique with which the artist is familiar, thus reinforcing the specialization.

The very same applies to the subjects drawn over the years. The more familiarity with a set of problems, be it observational or technical, the better the outcome. The better the production, the more satisfied the customers. It is easier to be a specialist during the later years of work as an illustrator because of acquired knowledge related to the field one is working in. We have to acknowledge the fact that problems related to the narrowness of specialization do exist. First of all the market is limited. The illustrator is limiting himself to fewer customers, thereby cutting his potential income. His technical (and most of the time self-imposed) limitations will not allow him to diversify. It is obvious that not everybody will ask for pen-and-ink technique even if such technique is predominately used through the field of scientific illustration.

It is possible to attempt to plan for more general specialization in the areas of natural sciences and medical illustration by just narrowing the field to zoology, paleontology, botany, anatomy, physics, and so on. To do this, you should practice drawing and painting in all techniques in order to be acquainted with the variety of possibilities as well as try to understand and learn about as many subjects as possible pertinent to the choice of subject matter. Such an area will be broad, but it will still be a specialization. An illustrator in anthropology will be drawing a variety of Indian artifacts and clothing patterns, reconstructing pottery and pottery sherds, reconstructing musculature for the study of facial features, drawing stone tools, and many other tasks.

It should be obvious that the more you draw and the more you paint the better you become in the trade of illustrating scientific material. The more techniques are known to the illustrator, the better chance he has in obtaining a job, the more skilled he is, and the stronger he can stand up to the competition. This is an important problem worth serious consideration, but the internal strength of broad specialization can and sometimes does lead to disaster. In our specialized society, specialization has become a symbol of high quality. But high quality is a relative term usually determined by the person who hires the illustrator. In a great many cases, a well-rounded and capable illustrator who is represented by a variety of works on different subjects rendered in various techniques may not be regarded as a suitable artist for some very particular project. In such a situation, the versatility of the illustrator is held against him on the premise that if he can draw so many subjects in so many techniques, he will not necessarily be very good in any one technique. Such a premise may not be valid, but it is used by many, resulting in rejection of applicants. Unfortunately, it is beyond the scope of understanding as well as acceptance that if the artist draws well in a variety of techniques he is more likely to draw better and with more intelligence than one who is confined within definite boundaries. When organizing a portfolio used for presentation to specific customers, this should be kept in mind.

There is no perfect middle road. Personally, I think that versatility is a great plus for illustrators, as the market is not artificially closed by the artists themselves. Due to the different approaches, various and differentiated observations of subjects, and the knowledge of various tools of trade, the artist's improvement of skills is a natural process, at the same time creating necessary self-imposed constructive criticism of illustrators' works, resulting in self-assurance and helping to solve

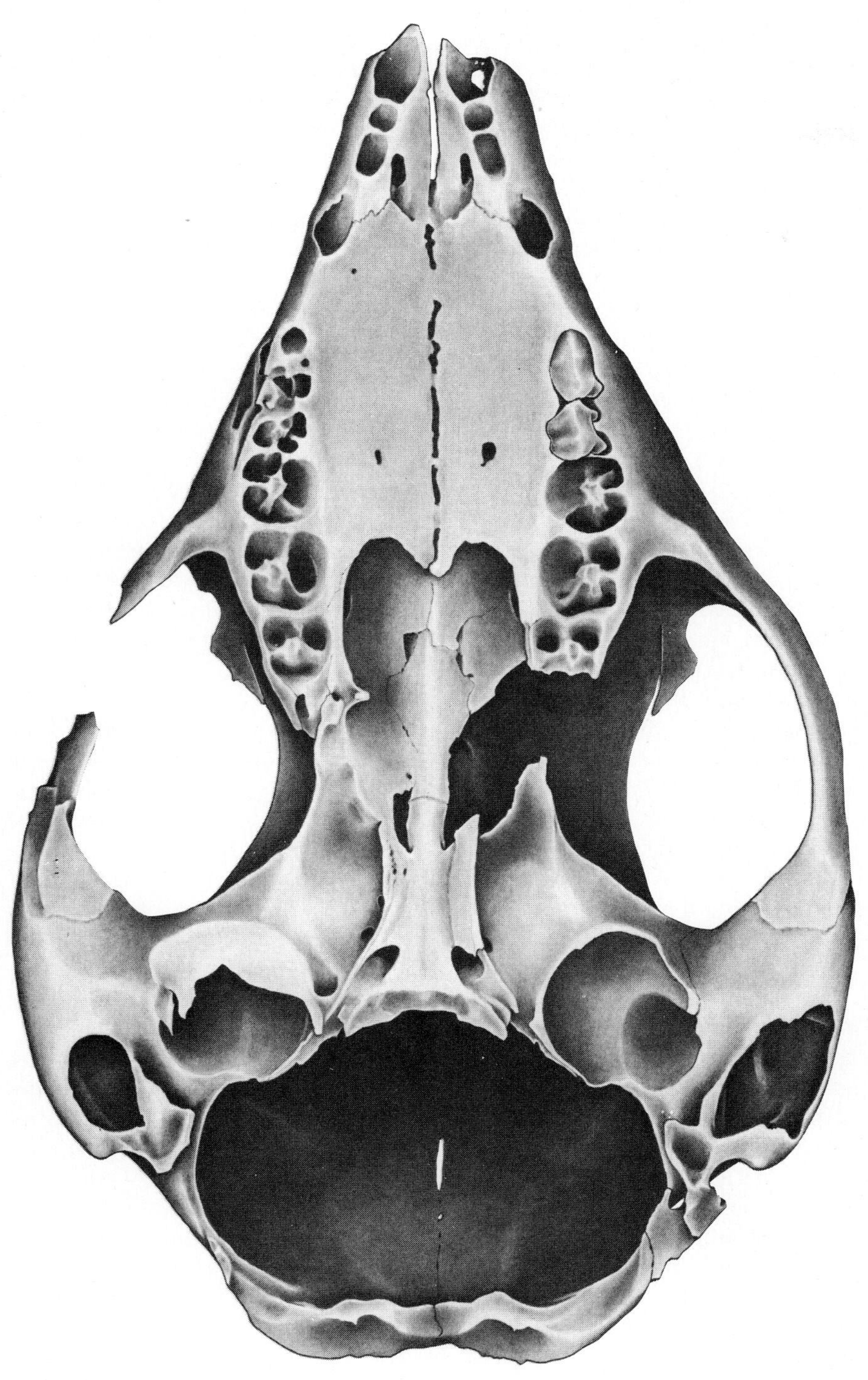

Figure 2-7. *Potorous platyops,* PM 4355, recent skull from surface of Webb's Cave, Western Australia, ventral view with unerupted right P^4 removed from crypt. Pencil on one-ply Strathmore drawing paper. Drawing by Zbigniew T. Jastrzębski, Field Museum of Natural History, Chicago. Published in William D. Turnbull, "The Mammalian Fauna of Madura Cave, Western Australia, Part VI," *Fieldiana,* Geology, New Series No. 14 (1984) by Field Museum of Natural History, Chicago. Drawn in 1982.

problems that may be completely new to the illustrator.

NATURAL SCIENCES

The term *natural sciences* refers to the field of study involving nature which, of course, is our civilization, past and present; all of the animal kingdoms; as well as matter on which we stand and from which we are constructed. It is a vast area of diversified research leading us to better understanding of our world and to better comprehension of the intricacies of its mechanisms. Within natural sciences the major areas of scientific inquiry are grouped in anthropology, biology, zoology, geology, and botany. Naturally, such major areas of research are subdivided and more precisely defined. For example, within geology we will find invertebrate paleontology, vertebrate paleontology, mineralogy, and structural geology; within zoology are insects, invertebrates, fishes, and mammals.

Illustrators working within such broad areas as natural sciences have to remember that each project in which they will be involved and each illustration rendered by them will be different. At the same time all of the subjects will have to be drawn in accordance with very rigid specifications dictated by the particular scientist conducting the research as well as, in many cases, internationally accepted codes of visual presentation applicable to the particular subject. From the scientific illustrator's point of view, the responsibility for the clear readability, technicalities of rendering, and presentation of finished product rests on his shoulders, the finished product being a pictorial representation of a particular specimen prepared for reproduction in a scientific journal.

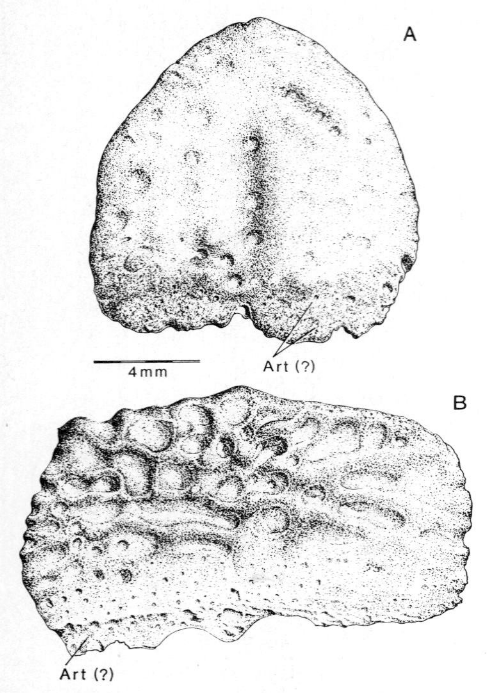

Figure 2-8. A, dorsal view of possible anterior member of armor series, FMNH UR 2403; B, dorsal view of large broken scute, FMNH UR 2404. Crow quill on Strathmore drawing paper. Drawing by Samuel Grove, Field Museum of Natural History, Chicago. Published in John R. Bolt, "Armor of Dissorophids (Amphibia: Labyrinthodontia): an Examination of its Taxonomic Use and Report of a New Occurrence," *Journal of Paleontology*, by Society of Economic Paleonotologists and Mineralogists and the Paleontological Society, Tulsa, Oklahoma. Drawn in 1974.

Illustrators specializing in depicting such a broad range of subjects should be very skilled in all technical aspects of drawing and painting. Why should so many techniques be mastered? Some subjects will be presented or rendered in a certain technique because that particular technique will help in the clarification of the subject. Such clarification is needed because drawings prepared by scientific illustrators are used to illustrate specific parts of scientific research and must present pertinent problems explicitly. An illustrator must understand that knowledge of drawing really means a knowledge of many techniques. An artificially created, personalized identity known as "style" definitely does not have a place in scientific illustration as it does within the world of fine arts. In scientific illustration technique should be substituted for style. Different drawing techniques, combined with properly selected materials and technical ability of the artist, can give the appearance of a very personalized style by which the artist will be recognized. Style, like handwriting, can be easily associated with a particular person. The reason for this is that individual muscular structure combined with the imprinting of our subconscious by input of information such as education and personal experience received during our lives results in certain characteristics visible in the drawn line. The same principle applies to drawing and painting. Without stressing an artificial style, through practice and improvement of technical abilities an illustrator will acquire certain characteristics which will be present in his work. His illustrations will be recognized as his and only his. In scientific illustration, the most important fact is the information presented in the form of pictorial representation to other scientists. For this reason, the technique and choice of materials, such as the proper paper combined with an appropriate pencil, should be made consciously and not by accident. Only then will the illustration be acceptable, not only from a scientific point of view but also from the visual, satisfying both the scientist and the artist. The reason for mastering all possible techniques is an important one. From the artist's point of view there is the idea of market: The more you draw the better you will become, thereby conquering larger portions of it. From the scientist's point of view, a variety of techniques will offer more possibilities for presenting his data. In such a large area as natural science that is important.

Natural sciences cover a variety of subjects and a variety of researches. The available market is very large and very versatile. For example, in the research institution where natural sciences are the primary object of study, the subjects presented to the illustrator for drawing will vary, ranging from anatomical studies of muscular tissues called soft tissues through the hard parts of skeletal systems, from artifacts of past civilizations like stone tools or pottery through recent artifacts produced by the nineteenth century tribes from all over the world. The subjects will range from detailed studies of pollen grains to colorful representations of large plants. For the anthropologist, a pictorial reconstruction of pottery, the deciphering of weaving patterns, visual presentation of designs found on various artifacts such as pottery sherds or seal engravings, or deciphering clothing patterns are all of great importance.

The subjects presented for drawing by archaeologists will depend on the scientist's area of research. When conducting studies involving inscriptions found on walls of ancient buildings, the artist will be asked to climb scaffolds in order to prepare a copy of hieroglyphs. An illustrator must combine patience with precision as his or her work will form a base for deciphering and further linguistic studies. Such a project is an exciting challenge, as the process of copying involves not only inscriptions but all the traces of coloration as well as any possible markings found on the surface. This will tell us how artists from thousands of years ago worked, what tools they used, and how their society was structured, thus shedding light on the past.

Geology is another interesting segment of natural science. Actually, geology consists of two large basic specialties. First is paleontology, within which studies of past life are conducted, studies involving the grand reptiles as well as the beginnings of life on our planet. The other branch of geology is the concern with physical and chemical properties and compositions of the crust of the earth and with the laws governing all that material. With our efforts to reach the stars, all of the planets of our solar system must be included. Studies of meteorites and the structural qualities of the moon are a part of this branch of geology. Such studies are based on actual samples of matter delivered to us from space by astronauts or by nature, as in the case of meteorites. The paleontologist bases his research on fossil material embedded in the crust of the earth. Paleontology is branched into various subdivisions such as fossil invertebrates, fossil mammals, fossil reptiles, or fossil plants. Illustrators participating in any of those researches will look at and draw parts of past life, handling specimens millions of years old. Sometimes reconstruction of skeletal structure will be needed; under the direction of the scientist or team of scientists, a drawing will have to be composed based on fossil remains as well as on the verbal explanation given by the scientist. Thinking and an analytical approach combined with the ability to communicate are all necessary.

Life of the present day such as fishes, birds, insects, and invertebrates are grouped under zoology. There are many unknown forms of life existing which are collected, studied, and described. All newly discovered species will require il-

Figure 2-9. *Chylonycteris macleayi*. Wash on bristol board. Drawing by E.S.T. (name unknown), Chicago Museum of Natural History (Field Museum of Natural History) Chicago. *Sea Mammals of Middle America and the West Indies* by Chicago Museum of Natural History (Field Museum of Natural History), Chicago, 1904. Drawn between 1900 and 1904.

lustration for taxonomical purposes (taxonomy being description based on very careful study for the purpose of classification). To many of us it may appear that at present we humans know all the secrets of nature, but it is not so. New, never-before-seen and described subjects such as fishes are brought from expeditions constantly. Detailed drawings of such creatures are necessary because photography cannot always present the subject accurately. Mechanical devices such as cameras cannot distinguish, for example, between a reflection of light and the real shape of the object. The majority of zoological illustrations are rendered in black and white, usually in pen and ink. Many are also drawn in continuous tone and in color whenever delicacy of tonal changes or actual color is of crucial value to the researcher or whenever budget will allow for halftone or color reproduction.

Insects offer the widest area of unknown. We can surely say that by numbers, insects are in the majority. As they are usually greatly specialized—that is, adapted to a special kind of living in accordance with an ecological system—insects offer great diversity. By being very small, they do present a challenge to the illustrator. Most have to be viewed through very powerful microscopes equipped with a camera lucida. The camera lucida is actually a very old tool used by illustrators

Figure 2-10. Acari: *Allothyrus australesiae*. Male, right leg II (40 ×), trochanter through genu, posterolateral aspect, genu II anterolateral aspect, (Allothyrid mite). Pen and ink on tracing tissue (Rapidograph® technical pens), drawn using various pen points ranging from number 000–3, and French curve. Drawing by Dr. John B. Kethley, Field Museum of Natural History, Chicago. To be published. Drawn in 1978.

for tracing. Since Renaissance times the device known as camera obscura has been used for precise copying of subjects. Such a device allows the illustrator to be very precise as well as fast in producing the desired illustration.

The parasites living on the insects, when viewed in magnification of hundreds of times, appear as strange spaceships. Usually the scientist will draw an initial sketch explaining the particular problem to be illustrated. With such help and with constant guidance, the drawing will slowly emerge, starting with a series of sketches of the parts of the insect which are later combined into a complete illustration. Because of the very high magnification required for viewing very small subjects, the depth of field presents a problem. When the subject cannot be viewed as a whole, then it is drawn in a similar fashion to the way mountains are drawn on a map. A layer of sketches represented only by outline gives a visual presentation of this structure at that particular depth of field. Layer upon layer of structures will represent the complete and accurate portrait of a mite.

Botanical illustration is a world in itself. The fascinating forms of ferns will never be forgotten by anyone who has seen and drawn such structures. The intricacies of flowers always present an unquestionable challenge.

In natural science the smallest living forms of life as well as gigantic mastodons come back to life in drawings. The reconstruction of the unknown waits for the scientific illustrator. It is a combination of learning about our world and constant improvements on the skills of drawing.

The variety of techniques is almost as wide as the variety of subjects. Black-and-white illustrations are rendered in pencil, using different types of paper. Pen-and-ink lines are applied with crow quill or technical pens. The technique of stippling, creating an illusion of tone by the application of dots, is an extension of the richness of tones of the nineteenth century stone lithography. Tools such as stomps, pencil erasers, chamois cloth, and cork can and will produce the most delicate of halftones if properly used. The color may be a combination of a variety of techniques such as watercolors, acrylics, or gouache needed for depicting coloration patterns on birds, mammals, insects, or shells. Every specimen presented to the illustrator is different. With that many different subjects, the work of the scientific illustrator is extremely interesting; it is the life of a discoverer.

BOTANY

The depth of involvement in a particular specialization depends on your interests and your personal needs. This in turn leads us to the type of institution conducting research. It is possible to be involved in this field through an institution where botany is the only subject of study, a good example being the United States Department of Agriculture, where a staff illustrator will be involved in the production of botanical drawings only. At present, just as in the late nineteenth century, the need for skillful artists exists. The only limitation is artists' technical ability to produce clear, understandable, as well as beautiful illustrations. Matching the quality of line presented in *Illustrations of North American Grasses*, published in 1891 by the United States Department of Agriculture, may be a bit difficult. *Florae Insulorum Maris Pacifici* by Drake del Castillo, published in Paris by Masson in 1886, is another good example of illustrations drawn by a skillful hand. Even while admiring past commercial achievements—the drawing and platemaking processes involving hand engraving and printing—we cannot overlook present-day accomplishments such as illustrations published in *Vascular Plants of the*

Pacific Northwest by Hitchcock, Cronquist, Ownbey and Thompson, published in 1969 by the University of Washington Press.

Examples multiply. While it is true that the charm of hand-retouched (actually painted over printed tone) illustration found in volumes of *Botanical Cabinet*, published by a London-based publishing conglomerate led by Conrad Loddiges in 1822, is not to be found in present-day publications, it is also true that many color illustrations currently reproduced are more accurate if not more beautiful.

Besides government-sponsored research facilities, other organizations—whether commercially involved in the private sector of industry or part of universities or nonprofit institutions—offer opportunities to the botanical illustrator. Naturally, in order to be completely immersed in the subject, a sound knowledge of botanical terminology would be of great help. Such knowledge would be useful in understanding the subject as well as during interactions with scientists. At the same time, it is not necessary to become a botanist, though many illustrators have learned enough to become competent researchers. This also holds true for scientists who through practice, without studying drawing or painting, became very good artists. Scientists will always supply information, if necessary, in plain, easily understandable language, explaining what and how details should be handled and illustrated. With time, an inexperienced artist will learn.

Among the techniques used for rendering, the black-and-white ink drawing is most prominent, because virtually all illustrations produced for research are printed in black and white. That means RAPIDOGRAPH® technical pen or crow-quill pen used on any white surface—Strathmore paper, boards of all kinds, tracing paper, or drafting film. Tonal changes will be achieved by the thickness and density of line or by stipple. Pen-and-ink, black-and-white illustration is easily reproduced, retaining the most readability per cost of production of printing plate. A knowledge of painting will be required during preparation of field guides and in some cases involving pure research. Most scientists have their own preferences regarding technique of drawing or painting dictated by their habits or the feasibility of adaptation of such technique to the particular subject as well as to the publication. Drawings are prepared from actual specimens or small sections of specimen. If the subject is too small for viewing with the naked eye, it will be viewed through a microscope. Naturally, proper care must be executed; specimens entrusted to you should not be injured, broken, or destroyed. In some cases illustrations will be based on verbal description and scientists' notes, which is quite challenging for both the scientist and the illustrator. For the illustrator, the challenge is comparable to solving a puzzle, as he or she does not know how that particular plant or a section of it really looks. During such a project, a number of sketches will be prepared and the work, the depiction of the actual subject, will be judged by the scientist as the work progresses, with necessary changes implemented to improve and enhance the illustration until the finished product represents the subject to the scientist's and artist's satisfaction.

Many subjects received by the illustrator will be small, such as a section of a root, a cross section of a leaf, a grain of pollen, or a flower. The microscope is the essential tool. Under the microscope the form and exquisite beauty of nature are visible. The world of vegetation abounds in unsolved, undescribed mysteries. Under the microscope,

Figure 2-11. *Elaphoglossum petiolatum*, margin and glandular abaxial surface (12 ×). Pen and ink on mylar drafting film. Drawing by Zorica G. Dabich, Field Museum of Natural History, Chicago. Published in Robert G. Stoltze, "Ferns and Fern Allies of Guatemala, Part II, Polypediacae," *Fieldiana*, New Series No. 6, 1981, by Field Museum of Natural History, Chicago. Drawn in 1980.

Figure 2-12. *Undaria pinnatifida*. Watercolor. Painting by Feng Ming-hua, Institute of Oceanology, Chinese Academy of Sciences, Quingdao, People's Republic of China. Unpublished.

Figure 2-13. *Thunbergia grandiflora*. Watercolor. Painting by Zhongyuan Feng, South China Institute of Botany, Chinese Academy of Sciences, Guangzhou, People's Republic of China.

an intimate view of the unseen and unnoticed is revealed. As in all major areas of sciences, in botany the specific research will call for a very specialized type of drawing. It is possible to specialize in a technique as well as in a subject. Wherever research is conducted, the results will be published and the need for a botanical illustrator will exist.

It should be remembered that there is a difference between the book publisher's interest in botanical subjects and scientifically prepared drawings for the sole purpose of communicating that very specific data to the other scientists. Botanical illustrations prepared for a scientific purpose must be drawn precisely, the subject must be observed precisely, and the result must be coherently presented, explaining clearly very specific problems pertinent to the particular research. The first is similar in presentation to so-called editorial illustration. Editorial illustration is depiction of a botanical subject, drawn or painted for the primary purpose of interesting the viewer in buying the book; and secondarily, and in many cases not even that much, to inform him about complexities of structure of that subject. Many publications have very well executed illustrations attractively painted and printed, but actually of little or no scientific value. Botanical illustration, being functionally descriptive and explanatory, presents the whole subject as well as numerous details appropriately enlarged for easy viewing by the reader. For example, an illustration of a tomato plant would present the whole plant with the roots, flowers, and fruit as well as detailed enlargements of seeds and an explanatory cross section of flower. In such an illustration the information is of primary importance; the visual attraction achieved by the quality of the drawing or painting is secondary but highly desirable. The scientist, as initiator and recipient of the illustration, considers the image as an explanatory supplement to his text and not as a primary object of admiration. Naturally, if such an illustration is rendered by a skillful artist, it will be very attractive. As we look at an example of descriptive illustration, be it in color or black and white, we will quickly notice that the subject is arranged in such a manner that the main structure of the subject—the body of the plant with its proportions—is clearly visible and occupies the main area of the plate. The so-called plate usually occupies the whole area designated for the printed page, making the plate a full-page illustration.

The structural elements of the plant, the stems, and the leaves are presented in such a way that it is easy to see how that plant is built, how those segments emerge from each other; the angle of the curvature of the leaves and stems will not be questioned. Drawings of the leaves will have all pertinent information such as coloration, texture, and the veins properly depicted. The front and back of leaves, usually different from each other, are presented in separate illustrations by drawing the leaf as if it were flat.

All separate illustrations pertaining to one main subject are montaged into one plate, in a coherent presentation of information. When different enlargements are necessary for the depiction of separate sections of the plant, the appropriate information regarding the size of the subjects will be presented in the form of a scale and placed near the respective drawings. This is achieved by drawing a pen-and-ink line of a certain length proportional to the enlargement and designated with metric numbers. As long as the readability and clarity of the information is not obscured, such a composite layout can contain a large quantity of information. As I have said, in scientific publications the text is of primary importance, and as much as the illustrator would like to produce visually pleasing drawings or paintings, numerous restrictions will be placed upon him. First of all, most illustrations must be rendered in black and white for ease of presentation of a clearly read illustration and because of the lesser cost of reproduction of that type of illustration.

Most of the drawings will be rendered from *dry specimens*. The dry specimen is a whole plant or part of a plant specially preserved, prepared, and mounted on the herbarium sheet. Prepared specimens may hinder the illustrator in some way because they are glued to the paper or board. Being flat, some parts of the subject will be obscured by overlapping. A dry specimen

Figure 2-14. *Sterculina nobilis*. Watercolor. Painting by Zhongyuan Feng, South China Institute of Botany, Chinese Academy of Sciences, Quingdao, People's Republic of China.

must be handled very carefully because it is very easy to chip, break, or crack parts of it, damaging the plant. All of the specimens are carefully catalogued and numbered so scientists will have easy access to the collection. The illustrator should note the number visible on the herbarium sheet and make a habit of writing it next to the sketch as well as next to the finished illustration. Such a procedure reassures all involved parties as to the correct placement of a drawing in the appropriate plate. In a number of situations specimens may be a size not easily placed under the microscope. Camera lucida attached to the microscope cannot be used and outlines of the subject cannot be traced.

If a deadline prohibits the old-fashioned, time-consuming technique or triangulation method, the Xerox copier can be quite satisfactory. The subject is placed on the Xerox machine and an image can be instantly obtained in the form of a print. If it is possible to adjust the tone of the Xerox machine, the image will be very satisfactory, showing the outlines of separate leaves as well as the veins in the leaves. As the obtained image will be the actual size of the specimen, reduction of the following sketch will be the next logical step. The final drawing must not be cumbersome for rendering, thus maintaining the speed of production and preserving the accuracy of the drawing.

A collection of botanical materials may have some of the specimens preserved in various liquid solutions. Such solutions may vary chemically, but most will have formaldehyde as a basic ingredient. Usually formaldehyde will be mixed with ethyl alcohol, with the addition of a compound to neturalize the strong smell of formaldehyde. Such specimens will be drawn while submerged in the solution or else they will dry and become damaged.

Other small specimens may be immersed during drawing in a few drops of water, preparing and softening the tissues. When moisture is kept at an appropriate level the subject will not dry out under strong illuminating lights. A specimen that was dry-led and then submerged in liquid will unfold itself, allowing better observation and, if necessary, dissection. The drawing of an "unfolded" specimen will progress easier and with less damage to the specimen.

Do you want to specialize in botany? Love of plants is not enough. Decisions should be based on technical ability as well as personal interest.

FISH

Drawing fish is a separate specialization within the large area of biology, serving ichthyologists' taxonomical needs. It is a good example of specialization that combines one technique of drawing, pen and ink, with one subject, fish. Very rarely will pencil, watercolor, acrylic, or carbon dust be used. The intricacies of the subject combined with standardization of presentation of information do not permit other techniques. Readability is of prime importance. An illustrator will have plenty of material to draw from. Specimens will come from all parts of the world, from the depths of the sea and from the rivers, some monstrous in appearance and others plain looking, both large and small. The illustrator should be prepared to handle carefully subjects which are no bigger than ten millimeters. Ten millimeters equals about three eighths of an inch. Some fish have a multitude of scales and others are completely without them, but covered with a complicated coloration pattern. Most are previously undescribed and unclassified. Because of this, the illustrations are of great importance to the scientist as supportive material and to helping him present the subject clearly, helping him to communicate all necessary data needed for classification of the newly described creature.

In order to present the subject correctly, the illustrator has to comply with rigid standards of visual presentation. Fish are always drawn from the left side (the side to the right when looking directly at the head of the fish). Fins are stretched slightly to allow for clear depiction of spines and rays. If scales are present on the body of the specimen they all must be counted and appropriately drawn. Breathing pores should be noted and, if necessary, certain features underlined more strongly than in nature, making them obvious to the reader. Because of clear presentation of minute details such as the segmentation of rays, scar tissue, or the structures composing the lateral line, such

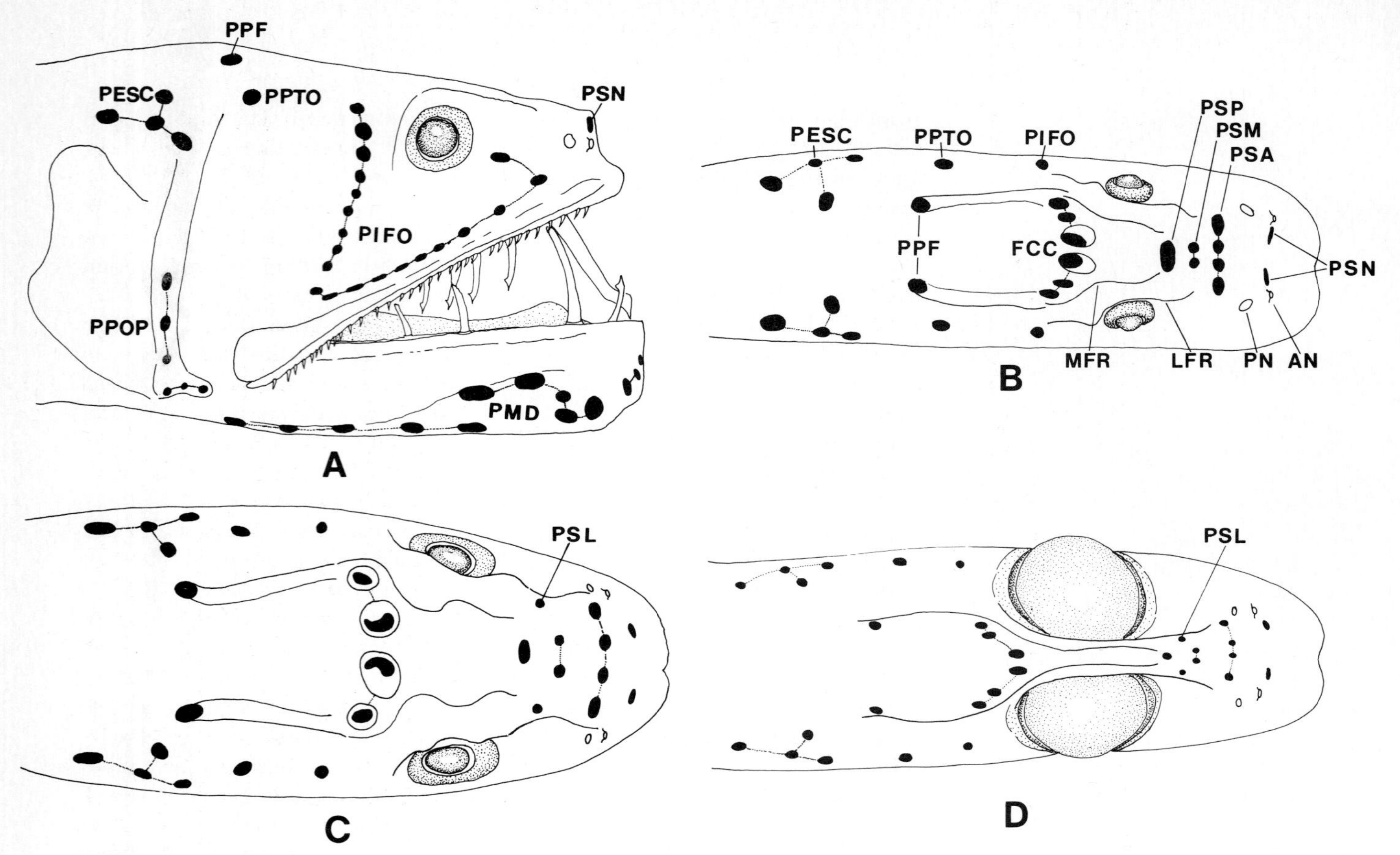

Figure 2-15. Cephalic laterosensory pores in *Odontostomopys normalops*, SIO 72-316. Pen and ink on tracing tissue. Drawing by Dr. Robert Karl Johnson. Field Museum of Natural History, Chicago. Published in Robert Karl Johnson, "Fishes of the Families Evermannellidae and Scoperlarchidae: Systemmatics, Morphology, Interrelationships, and Zoogeography," *Fieldiana*, Zoology, New Series No. 12, by Field Museum of Natural History, Chicago. Drawn in 1980.

drawings are much more readable than a photograph. A photograph will present whatever the lens transmits upon the film; a camera cannot distinguish between reflections and an unwanted feature, nor can it find the actual edges of the scales on the body of the fish. A photograph may not be able to capture the coloration pattern of the subject, because in a collection specimens are kept submerged in ethyl alcohol, which helps to preserve the subjects but at the same time damages the coloration. Through very careful observation, an illustrator, using his ability for analytical thinking and knowing from scientists what features to present, is able to clarify the necessary information. The illustrator is able, through sketching and meetings with the scientist, to untangle a specimen which may have been mutilated by dissection or trawling and to present visually coherent and truthful illustration. Such a specimen, referred to as a holotype, is selected by the scientist very carefully. It has to represent the features of the family to which that particular fish belongs and forms the base for classification and naming of that group.

As for taxonomic purpose, all features have to be drawn as they are on the subject. Some details, being small, will dictate the final size of the rendering as well as the enlargement. Being confined to the tools of his trade, the illustrator needs a reasonable space to work within, to move the tip of a pen on the surface of the paper without obliterating previously drawn features.

How does the process of drawing start? After receiving the specimen from the scientist and the initial instructions, the illustrator has to set up the subject. The fish, no matter how small, cannot be damaged. Pins cannot be inserted through the tissue, and nothing can be ripped off the specimen. The subject must be submerged in the ethyl alcohol during the drawing in order to protect the tissue from any damage caused by drying. If the illustrator happens to be allergic to such a chemical compound, the subject may be removed from the ethyl alcohol and for a short duration kept in water, to be submerged again later in the solution.

After setting up the specimen, the sketching procedure can start. Preparation of the sketch, actually a number of sketches, is another matter. The sketches must be correct, as only through correct sketches will the final rendering be achieved. The illustrator should judge the initial size of the preparatory and final sketches in accordance with his observation of the subject. Large sketches can be reduced later by photostating or by tracing, using an enlarging-reducing camera to reach a manageable size for rendering. The rendering cannot be too big nor too small to achieve properly the final outcome, which is a printed image of the drawing. If the finished illustration is too large in relation to the designated size of reproduction and is not rendered accordingly, after the reduction many areas will tend to fill up, resulting in an unplanned and unpleasant appearance caused by darkening, spot lines, and loss of clarity. By the same token, a drawing cannot be too small. When an illustration is not drawn in the proper proportion to the final reduction, the majority of the details may be lost. For absolute accuracy, the specimen must be measured by the illustrator so the final sketch—and most importantly, the rendering—will correspond with the scientist's numerical data. For this reason as well as others, such as an illustrator's inexperience with the subject, a very close intraction with the scientist is a must. The scientist should and will direct the illustrator as to what should be drawn and how it should be presented.

As photographs do not show clearly the majority of important information due to the difficulty in photographing preserved fishes, they are not very useful during the preparation of the illustration. Nevertheless, sometimes a photograph of the specimen can be used for the purpose of establishing general proportions of the subject. Use of a photograph will not speed up production as first it must be taken, then developed and printed. Usually an illustrator is not equipped to do such work himself, so it must be delegated to somebody else. During such a process the intermediating person must receive appropriate directions. Time is taken away from the illustrator, who must explain and then wait for the image to be printed, hopeful that the correct proportion of the enlargement will be maintained, and that the specimen will not be damaged in the process. The proper enlargement is important, as it must coincide with the detailed sketches prepared with the use of the microscope. If it is possible to sketch a given subject within a few working days, there is no need for a photograph. The final sketch should contain all necessary data presented in such a way that meaningful meetings between illustrator and the scientist can take place. All scales present on the surface of the subject have to be drawn in a clear fashion, and the fins properly labeled with visibly presented and numbered rays and spines.

The procedure of drawing the scales covering the body of the fish is very involved and requires a lot of patience. All scales covering the fish, with the exception of the fins, usually have to be counted and drawn in their respective positions. Scar tissue will be represented as it appears on the subject. The scales are not reconstructed or substituted. Catfish are the only ones without scales, but usually they have complicated coloration. The problem lies not just in counting separate scales, but in seeing where they are. Not all subjects have easily visible scales—as a matter of fact, scales on most fish will be rather transparent, therefore not easy to find and delineate. In such a situation where the scales are not easily visible, each scale must be observed separately, lifted up by hand with a teasing needle in order to determine its truthful position and real shape. As you can see, during the preparation of the first sketches, the final sketch, and the rendering of the illustration there will be plenty of examination awaiting the illustrator—plenty of poking, counting, and thinking all done without injury, without the slightest damage to the specimen. Naturally, all of the spines and the rays in the fins must be counted also, with their length as well as width and position carefully noted. Coloration present on the subject, sometimes appearing as a slight change of value, must be noted and the proper darkness discussed with the scientist. It is obvious that when the subject, a fish, is very small, enlargement will be necessary; but when drawing a large specimen, an appropriate reduction is a must. Having modern equipment—that is, a good microscope equipped with camera lucida—helps, but if such is not available an illustrator must use the grid method during all preparatory stages. The camera lucida will allow for very precise measurements and high-quality tracing, but nevertheless, if the subject is larger than the opening of the lens when looking through the microscope, it is a good idea to measure the subject by hand just to make sure that proportions are correct. As you know, the lens of a microscope will distort the subject when the subject is viewed near the edges of the lens.

Beside taxonomical drawings, other structures of fish anatomy will be drawn, but none are as involving nor as rewarding visually. Positioning, studying, and sketching the specimen is very tedious work. The

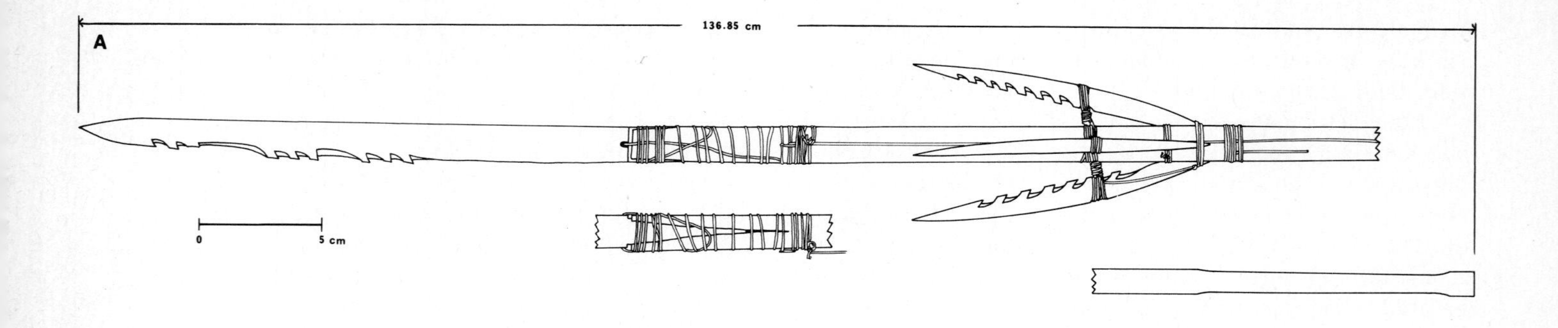

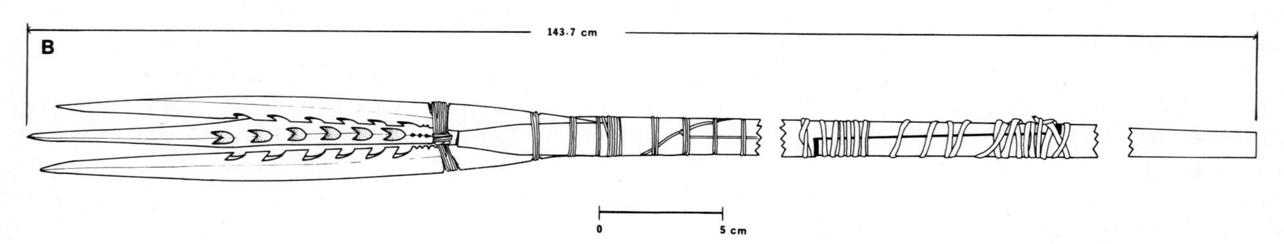

Figure 2-16. A—bird spear (20129); B—fish spear (20130), Eskimo. Pen and ink. Drawing by Zbigniew T. Jastrzębski, Field Museum of Natural History, Chicago. Published in James W. VanStone, "The Bruce Collection of Eskimo Material Culture from Kotzebue Sound, Alaska," *Fieldiana,* Anthropology, New Series No. 1, 1980, by Field Museum of Natural History, Chicago. Drawn in 1979.

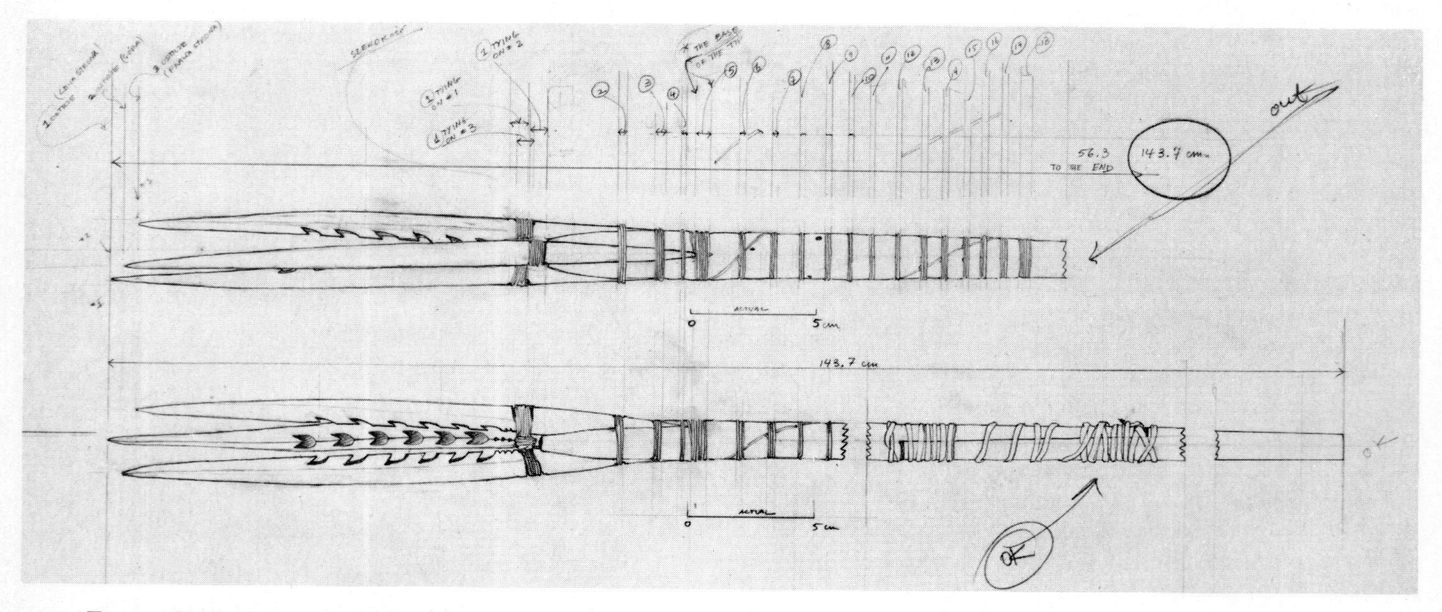

Figure 2-17. Bird spear (20129), final sketch. Pencil on tracing tissue. Drawing by Zbigniew T. Jastrzębski, Field Museum of Natural History, Chicago. Drawn in 1979.

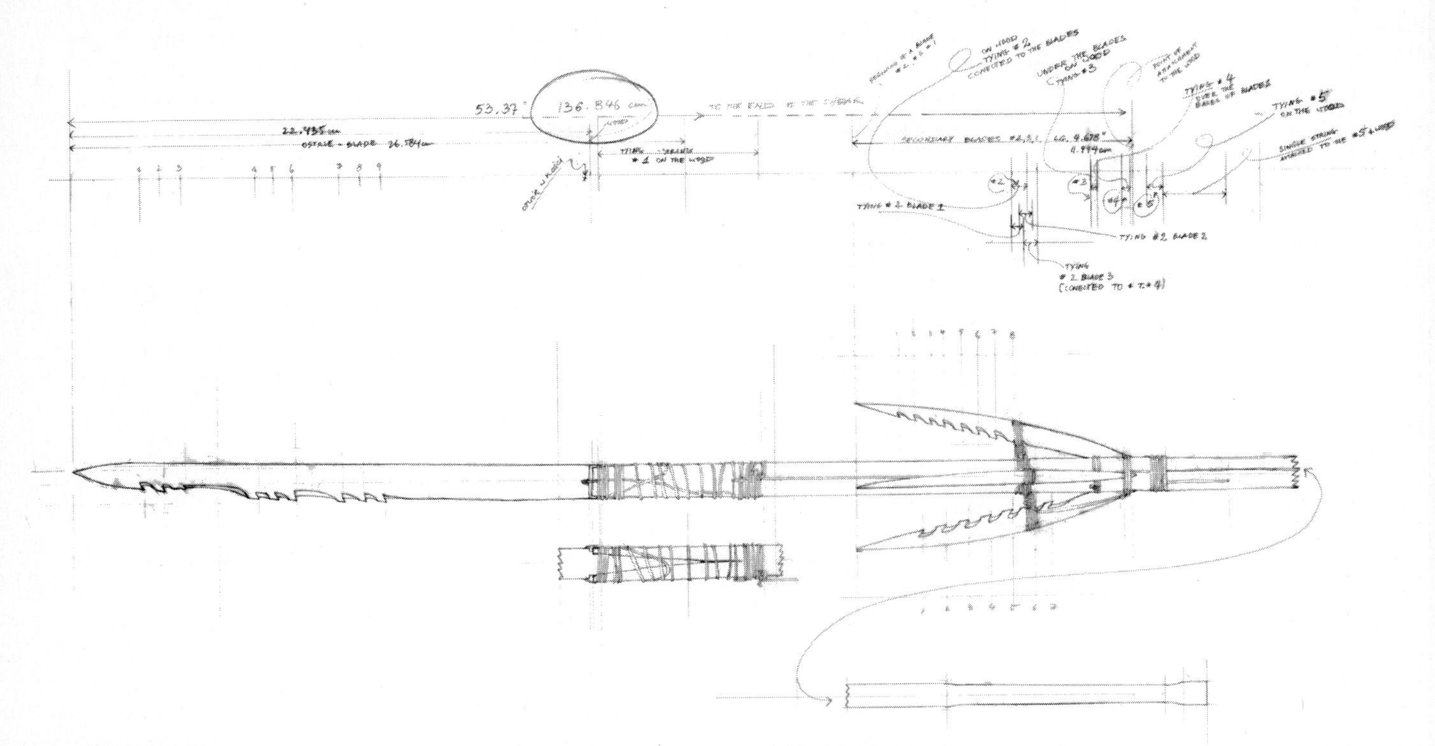

Figure 2-18. Fish spear (20130), final sketch. Pencil on tracing tissue. Drawing by Zbigniew T. Jastrzębski, Field Museum of Natural History, Chicago. Drawn in 1979.

knowledge of terminology is helpful, but not necessary. More important is logical thinking and the power of observation combined with knowledge of drawing.

ANTHROPOLOGY

The range of subjects encountered in anthroplogy is limitless, bound only by the imagination—from the remains of prehistoric cultures to the ethnographic materials produced by the present-day hunting and gathering tribes, from the participation in studies of human biogeography by tracing migration patterns of early tribes to copying and reconstructing ancient inscriptions, decorative patterns, and designs. Many times archaeologists need precisely drawn maps in order to present the location of an excavation as well as the plan of excavation itself. They will need correctly mapped positions of the artifacts in which they are interested. Such may be human remains, pottery, and other manmade objects or large constructions built from stone and earth. Charts and diagrams as well as maps are part of an illustrator's work. The artist's knowledge is enhanced by combining the daily practice of drawing with touching, looking at, and analyzing artifacts not easily accessible to the average person. Besides the good feeling resulting from personal accomplishment, the excitement of participation in the discovery of a lost civilization such as the very recently discovered irrigation system of enormous size and technical complexity in Peru cannot be matched by anything. In this example, modern technology helped. Photographs taken from the satellites circling the earth were a major tool in pinpointing structures stretching hundreds of miles across the land. Mapping and illustrating such immense and complicated remains had to involve specialists from different fields. The final drawings, the drawings for publication, were prepared by scientific illustrators.

Most of our knowledge of the past was delivered to us in a form of small, broken pieces of various kinds—pottery, parts of stone tools, bits and pieces of woven material. The garbage dumps of yesteryear supply the information regarding the daily habits of the people who lived there. Those bits and pieces help us to understand not only the past but the present. By studying the remains, we become more positive about our knowledge of where we came from and how we arrived to our present state of civilization.

Each small subject submitted for illustration requires a different approach and may require a different drawing technique as well as certain imposed codes of presentation. A good example of the differentiation of codes prevalent in anthropology is the position of a cross section of a pottery sherd. In the United States the cross section will be placed on the right side of the frontal view of the same sherd, while in Great Britain and Europe it will be on the left side.

Drawing pottery sherds calls for concentration. An illustrator will have to draw a great many pottery sherds. Listening to the scientist and preparing the drawings under his direction can make a specialist out of the artist. The sense of observation developed during such work is of great value to the researcher but it is primarily of great value to the artist himself. Pottery sherds are usually presented from the frontal view, which is the outside wall of the vessel. Such a direct view enables us to visualize all parts of the design preserved on the sherd. The second drawing is a representation of the cross section of the specimen. While the frontal view gives information about the shape of the design, the cross section will tell all about curvature of the wall as well as the thickness of the various sections of the sherd. Sometimes a reconstruction of the design found on the surface of the pottery sherd is needed. Such involving study results in a hypothetical presentation of that design. Reconstructions in general require patterns a very methodical approach. In anthropology, reconstructions—that is, pictorial reconstructions of the specimen based on remains, whether of an architectural structure or a small artifact—present complete information about the object.

A descriptive drawing of objects such as spears, harpoons, maces, figurines, clothing, fishing nets, or buttons are usually drawn diagrammatically in black and white. A pen-and-ink technique using technical pens such as RAPIDOGRAPH® technical pen or crow quill are used. Black-and-white drawings predominate in the field because of their superb clarity in projection of the information. Such illustrations may be stippled, while some will be a combination of dots and lines and others will be drawn in line only. Speaking of line, it is important to note the technique used for drawing stone tools. Stone implements, as the name indicates, are tools or stones left over from the process of preparation and production of stone tools. Illustration of stone implements is a difficult process. The only information contained in the drawing is the direction of a cleavage and delineation of chips or flakes present on the surface. The light must appear as if it is coming from the upper left side. As a matter of fact the right side, the side opposite to the direction of light, is and must be slightly back lit. The *Bulletin de la Société Prehistorique Francaise*, volume 69, has the best stone tool illustrations by P. Laurent. Why is it such a difficult process to draw a piece of stone? The scientist is interested not just in the shape of the implement, but also in the linear direction left by the process of flaking, so the illustrator has to struggle with extracting proper information by finding the direction of breakage. Such information gives the scientist a base for deducting how the stone tool or tools from that particular site were made. This information will give him insight into the technology and structure of that civilization. The object itself is drawn using crow-quill technique. A crow-quill pen allows an illustrator to control the thickness of the line through application of pressure on the tip of the pen. Technical pens will produce continuous and very uniform lines, not suitable for representation of information contained on the subject. The quality of the line and the changeability of the width of the line help the artist to present three-dimensionality of the object while at the same time allowing him to remain faithful in the presentation of information.

Artifacts such as harpoons, baskets, maces, and weaving patterns are drawn in outline in order to clearly present pertinent information. Figurines as well as other similar objects may have to be rendered with a combination of line and tone, either stippled or continuous, to underline the three-dimensionality of the subject. As exemplified by the extreme rigorousness of the procedure used during stone-tool drawing, the direction of the light illuminating a given artifact is imposed by prevailing convention or by an individual scientist. Shadows are excluded from drawings presenting definite information. Papers used for drawing will vary according to the project. The same brand or type of surface will not always be available to draw on. An illustrator should be prepared to draw on virtually all types of surfaces—illustration board, Strathmore paper, tracing paper, or drafting film. Each will handle differently. Many times the paper will be supplied by the scientist or he will specify a certain type, Strathmore, tracing tissue, or drafting film being most common. The pen-and-ink technique is not always a rule of thumb. Rubbings are commonly used to obtain the information. A shallow relief can be copied very precisely through a variety of rubbing techniques. A decorative design or an inscription will be presented clearly when covered with a thin paper and rubbed with some graphite. Many times such rubbings provide a final illustrative material and are used for reproduction. At the same time, not all illustrations are in black and white. Color is used at certain times. It is used especially during the designations of coloration of decorations remaining on architectural structures.

Some problems encountered in hieroglyphs are similar to problems

Figure 2-19. A typical wooden plant pillar building from the sixteenth century, Zürichbergstrasse 59, Zürich, Fluntern, Switzerland. Illustration executed from plans and surveys is a reconstruction of a typical construction type from the sixteenth century. Pen and ink. Drawing by Christoph Frey, Archaeological Department of the City of Zürich, Switzerland. Published in J. Hanser, et al., "Das neue Bild des atten Zürich" 1983. Drawn in 1982.

found when drawing clothing patterns. In both cases the illustrator cannot injure the specimen. In hieroglyphs, all of the data—that is, all of the pictorial characters, the leftover particles of paint, and in some cases all of the scratches present on the surface of the subject—are copied. Only after the scientist has made his decision can the work begin on the final stage of illustration. When drawing clothing patterns, the intricacies of scientific illustration are a surprise. Tracing the outlines of the garment is impossible, as all the seams have to be found without destroying the garment. Each separate piece of material has to be measured and presented correctly. Deduction of clothing patterns is a puzzle which can be solved if a basic linear measurement is established. Only through very careful measuring can a pattern be sketched, corrected, and sketched again. When a drawing is rendered it presents a coherent and easy-to-read illustration. Another subject requiring logical solution is a representation of weaving patterns of basketry, cloth, or Indian quill works. Patience and thorough understanding of the subject are musts.

Drafting techniques are often used for architectural renderings. An architectural rendering of North American Indian dwellings serves a double purpose. First, such illustration preserves important information about the structure, allowing the scientist and his team of conservators to preserve wooden segments. Second, it serves as material for careful study and is used as an illustrative material in the publication of the research. Objects such as garments, totem poles, or houses are large and cannot be placed under the microscope. On the other hand, small chips of stone, fragments of pottery, or quill works must be viewed through the microscope; the drawing will be an enlargement of the subject. Within anthropology it is possible to find very well-defined fields of subspecialization. Some artists draw only stone tools; others feel more at ease with architectural renderings; while some others may find pottery reconstruction more of a challenge. Personal interests and technical ability will dictate the direction taken by the illustrator. It is possible to deduce the extent of knowledge of engineering of past civilizations from their garbage dumps. Fragments from the past supply us with a wealth of information, whether it be in the form that puzzles us today or in the clear writing preserved on stone or clay tablets. Broken pieces of pottery can offer information similar to that contained in large architectural remains. To illustrate his or her research, an anthropologist will need a logically thinking illustrator, because in some cases the illustrator may be the one who will notice something unusual or of great value.

RECONSTRUCTION IN ANTHROPOLOGY

When looking through illustrations in *The Long House Mesa Verde National Park* by George Cattanach, *The Excavations at Kotosh, Peru* by Izumi and Terada or *Journal of Saudi Arabian Archaeology*, we can appreciate the quantity and quality of presented information pertaining to architectural structures. Pictorial reconstruction of such large architectural complexes as irrigation and road systems, palaces, temples, or whole cities are large and complicated projects, requiring from well-versed illustrators a knowledge of drafting and some engineering background. Such an illustrator, besides having the ability to draw and sketch freehand, must be a surveyor and draftsman. A freehand drawing may present a three-dimensional hypothetical structure of the temple but will not explain its structural intricacies. Production of such pictorial representations goes slowly and through many stages, usually being a cooperative effort. Because the illustrator must know what to present in the finished rendering, he must receive the proper information from the researchers in the form of drawn, written, and verbal instructions. The plan of the structures in question as well as a topographic representation of the surrounding landscape are drawn very precisely; as sources, photography from the air is intermixed with measurements taken on the ground. Basically, a series of detailed sketches organized and redrawn in the office is the outcome of the fieldwork.

During excavations of a smaller site, every centimeter is measured and positions of all findings, whether human remains or artifacts, are carefully noted. Rendering illustrative materials presenting various levels of excavations or pictorially reconstructing layouts of excavated structures call for skills not easily attainable to studio drawing classes. Knowledge gained in school should be complemented by experience gained during work on actual projects. Excavations usually yield quantities of broken pottery, and pot sherds form the basis for reconstruction of whole vessels. During such reconstruction the illustrator must carefully measure the thickness of walls of all available sherds. Such measurements will be of help during deduction of the shape of the vessel. Walls of handmade pottery tend to be thinner toward the top and thicker at the lower section. Otherwise, freshly made pottery will collapse under the weight of clay not having sufficient structural support. When some kind of design remains on the sherds' surface, it must be noted through sketching for the very same purpose. The section of the rim will give the angle and vertical curvature of the upper part of the subject, and will help in deduction of the whole vessel. Reconstructing a design as well as presenting a cross section of the structure are part of such a subject. The anthropologist and il-

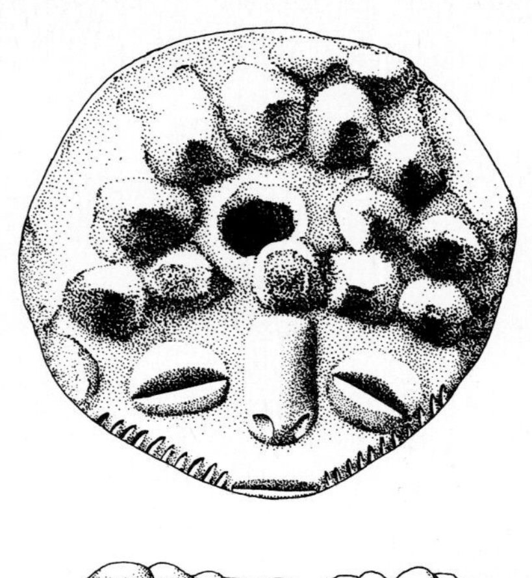

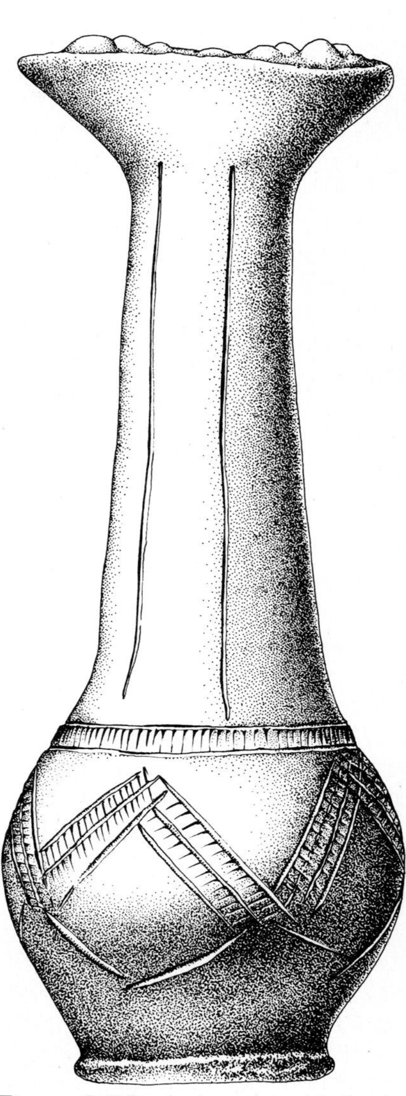

Figure 2-20. Anthropomorphic bottle from Sofwe (classic Kisalian), Upemba rift, Shaba Zaire. Pen and ink, crow quill. Drawing by Yvette Baele, Musée Royal de L'Afrique Centrale, Belgium. Published in F. Van Noten, *The Archeology of Central Africa*, by Institut Musées Nationaux du Zaire, Belgian Center for Excavations in Africa. Graz, Austria. Drawn in 1982.

lustrator together discuss and finally render the puzzle. The result is a presentation of complete information about that particular vessel in the form of a black-and-white illustration. All the broken pieces are depicted in their original positions, and the shape, the thickness of the walls, as well as the curvature of the lip are all shown.

Techniques of drawing will be chosen in accordance with the objective. Virtually any technique, whether black and white or color, can be used during pictorial reconstruction. For example, if a researcher concentrates on some aspects of fresco painting, few color illustrations will have to be prepared. The images will have to be copies made with great care and exactness. All chips of paint and plaster will have to be measured and appropriate notes taken depicting coloration and position. Perhaps some chosen examples of the composition will be reconstructed through color illustration.

Extreme precision will be particularly needed during epigraphical research. Epigraphy is an area of archaeology devoted to studies of old inscriptions. Photography is rarely used for recording carved, engraved, chiseled, or painted scripts found on walls of tombs. Forces of weathering have usually destroyed images so the eyes of the illustrator are better equipped to detect traces of images and variables of color than a camera. In some cases, parts of text copies have to be reconstructed pictorially. Old scripts are composed of drawings symbolizing the action and therefore the meaning of an abstract thought. When the final objective of illustration is not immediate publication but continuation of research, pencil will be used to prepare a series of sketches depicting variables found in the carved original. Such sketches should have the quality of a finished illustration: drawn in pencil for reference, but clean, properly rendered, presenting more than a pen-and-ink final rendering. Such visual material is needed for preservation of data for future studies and, being flexible, allows for changes as research progresses. For such recording, use of tracing paper will be more beneficial than opaque paper, because transferring the drawn image can be done without difficulty through blueprinting. If needed, copies can be easily made and changes implemented.

Anthropology is based on collections of manmade materials which, when displayed in an art gallery, would be regarded by the viewer as fine art. Fabrics, rubbings, sculptures, paintings, and artifacts such as jewelry or weaponry or ceramics are all part of the material regarded by the anthropologist, ethnographer, or archaeologist as scientific specimens upon which their research is based. The conservator will keep the objects in proper shape and in many cases will reconstruct some of the specimens back to their original three-dimensional state. The conservator will work *on* the actual object, while the scientific illustrator will work *from* the object itself, his concern being the illustration of needed material. Conservation is not part of scientific illustration; it is a completely separate field dealing with preservation and restoration. A good knowledge of chemistry as well as knowledge of various painting, weaving, and sculpting techniques are needed for those who are entering the field of conservation. Nevertheless, in some situations a team of anthropologists, conservators, and illustrators will cooperate in order to produce the final product. What type of project may involve that many specialists? The excavation of the ruins of Herculaneum and Pompeii are good examples. In such a project architects, surveyors, sculptors, conservators, scientific illustrators, and anthropologists all worked together. Drawings had to be prepared for publication of the research and all excavated material had to be preserved. In such a case, just as during pictorial reconstruction of an individual vessel, drawings of separate structures as well as topographical maps and planes had to be based on the actual measurements of the actual objects. Some material had to be pictorially reconstructed before preparators were able to do the same in three dimensions. Besides pots, buildings, and irrigation canals, there are numerous smaller artifacts. The illustrations found in *Pre-Columbian Shell Engravings from Spiro* by Phillips and Brown should serve as good examples of innovative as well as constructive reconstructions of designs. Rubbings as well as pen-and-ink live drawings present these difficult subjects properly. It is important to be aware that in addition to drawing techniques, the knowledge of sculpture and casting techniques plays an important role in anthropology. Knowledge of human anatomy is essential during the processes of reconstruction of facial features based on skeletal material. Reconstruction of the anatomical features, the musculature, and the facial tissues is a long, involved process. The skeletal remains will supply needed information regarding the placement of muscles as well as the thickness of tissues. After a plaster cast is made from the actual specimen, layers of Plasticine are applied to its surface to stimulate muscular structure. Further modeling dictated by the underlying muscular structure will present the reconstructed facial tissue. From such prepared, sculpted reconstruction a plaster cast is made from which the deduced face is presented in permanent form by positive casting, ready for the scientist to study. Such tedious reconstruction enables

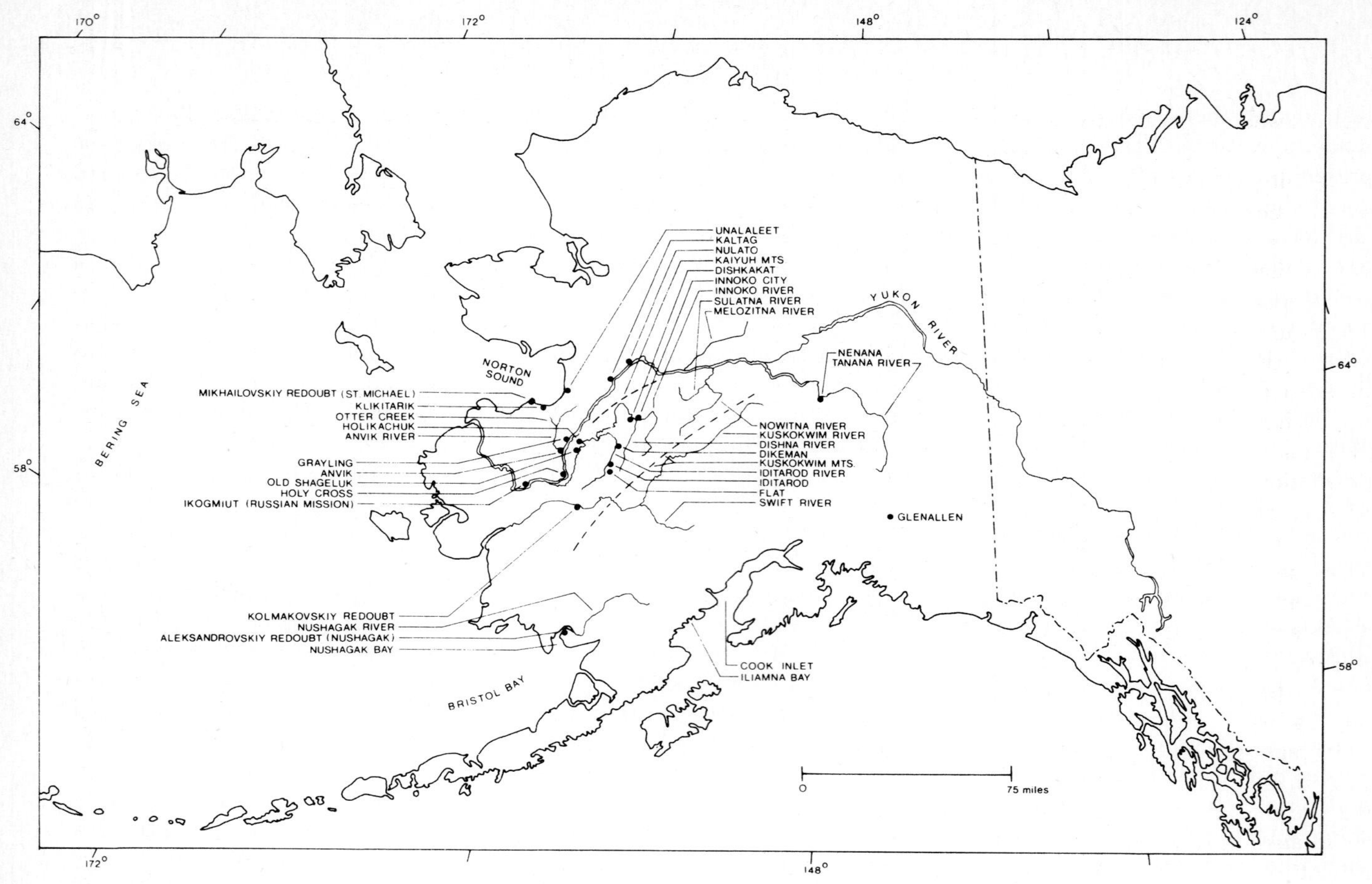

Figure 2-21. Map of Alaska. Pen and ink on tracing tissue. Drawing by Zbigniew T. Jastrzębski, Field Museum of Natural History, Chicago. Published in James VanStone, "Ingalic Contact Ecology: An Ethnohistory of the Lower–Middle Yukon, 1790–1935," *Fieldiana*, Anthropology 71 (1979), by Field Museum of Natural History, Chicago. Drawn in 1978.

the anthropologist to study and compare the features of humans living thousands of years ago. It enables researchers to compare the humans from past ages with the present-day *Homo erectus*, revealing cultural differences as well as behavioral patterns.

CHARTS, DIAGRAMS, AND MAPS

There is a deadline in three days. The map hanging on the wall can be measured in feet. The project calls for copying the map with selected features clearly designated, reducing the traced image to a size that can be worked on at the drawing table, and using Letraset transfer letters to designate over 100 names of cities, settlements, and rivers, some of them containing as much as 30 characters. So far, the only problem has been speed of production. But when the map is to be reduced to five by six inches in its final printed form, the artist does face an interesting challenge. All designations should be legible; all should be readable. Look at the map of Alaska published in *Ingalik Contact Ecology* by James W. VanStone to find an example. Such a diagrammatical map cannot be compared with the beauty found in the clarification of stratigraphy or the professionalism of rendering present in the *Atlas of Mars* published by NASA. Informational context and type of research will always separate both kinds of illustrations. Diagrams and charts? Chapter Eighteen of *Graphics for Engineers* by Hoelsher, Springer, and Dobrovolny starts with an excellent definition: "The purpose of charts and diagrams is to present facts and their significance in a more easily interpreted form than could be done with words or tabular data." Charts, diagrams, and maps play an important role in communicating a wide variety of scientific data. Specialized publications such as *Marine Geology, International Journal of Marine Geology, Geochemistry and Geophysics*, or everyday, easy-to-reach *Scientific American* will always present us with good examples.

Drawing charts and diagrams requires good knowledge of drafting tools as well as an understanding of graphics. Composition, simplicity, readability, and overall elegance of presentation are as important as the exactness of representation of the information. It is true that during the preparation of illustrations of this nature, the objects to be represented diagrammatically are not given to the artist, and he cannot touch, handle, or look at them as easily as in other areas of scientific illustration. But this personal pleasure is of secondary importance. Charts, maps, and diagrams are of immense value to the scientist and represent the essence of months if not years of study. As to be expected, line is predominately used during preparation. In many instances use of color or tones is encouraged if not expected or required. As always, it all depends on the area of research, the type of information to be presented, and the budget designated for publishing. Besides copying a scientist's sketch or a computer printout in ink for a reproduction, in many instances the illustrator is presented with the problem to be solved. Among such are diagrammatical representations of anatomical features, graphic depictions of a biogeographical nature for anthropologists, stratigraphy and morphology of rock formations so important in geology, or graphic presentations of the insides of nuclear reactors. When working for NASA, outer space is the limit. During preparation of maps based on the photographs taken from satellites and space probes, the illustrator is facing a double problem: The presentation must be extremely precise, but the artist must first use his or her puzzle-solving skill to achieve this end. When working with scanning electron-microscope photographs, the illustrator will focus on the mysteries of inner space. Bits and pieces of information, in the form of photographs similar to those that come from satellites, will have to be arranged, retouched, and underlined so the finished product will be clearly readable to the viewer. As always in our profession, the illustrator will have to work under the direction of the scientist or the team of scientists conducting the research.

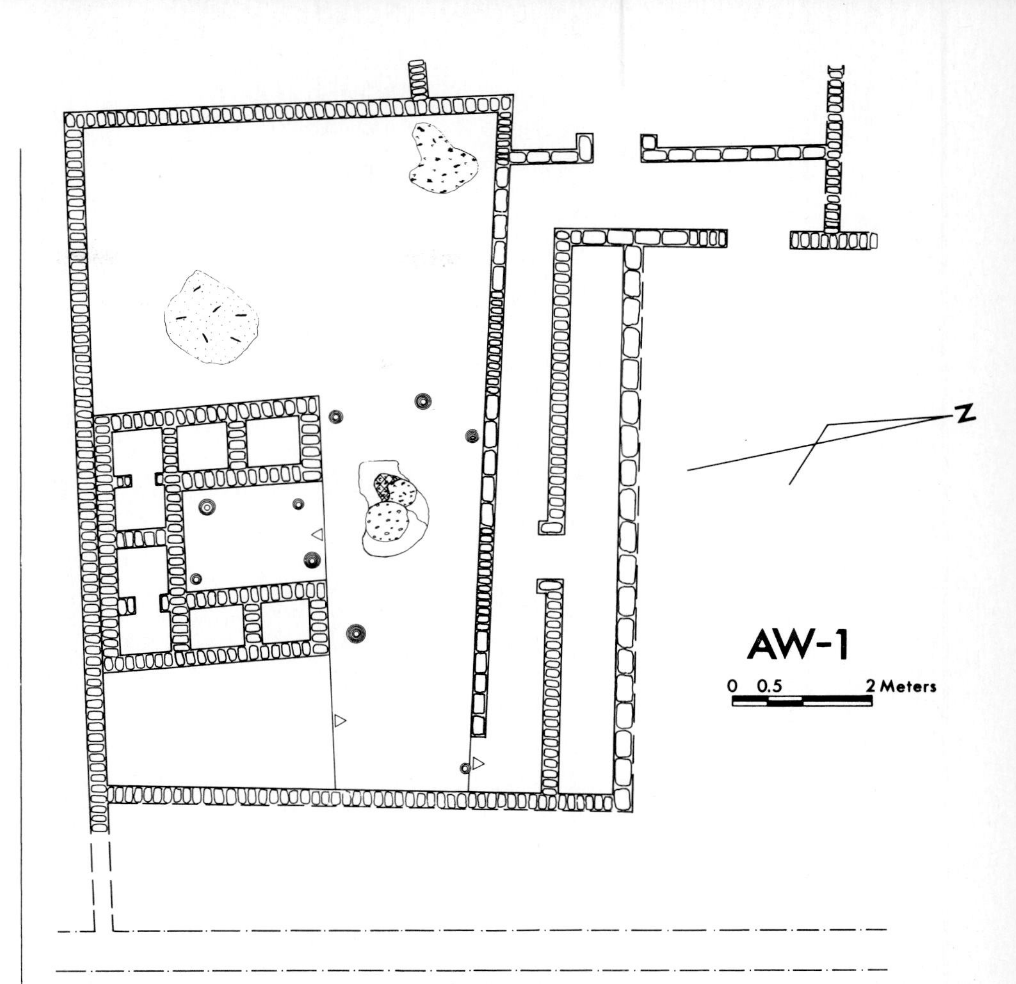

Figure 2-22. Architectural plan, administrative architecture, Chan Chan, Peru. Pen and ink, drafting film. Drawing by unknown artist, Peru. Published in A.M.U. Klymyshyn, *Intermediate Architecture: Chan Chan, Peru*, Ph.D. dissertation, Harvard University, Cambridge, Massachusetts, 1976. Drawn in 1970.

When drawing a chart from a scientist's sketch, the artist may think that a quick tracing will suffice, but this is rarely so. Most charts are so precise and exact that a millimeter of variation will make a big difference. Charts, diagrams, and maps require as much precision as the rendering of anatomical features. Obviously, such drawings present a technical challenge. Visual aspects cannot be forgotten; a thinking mind will find numerous ways to present the material in an interesting and well-organized manner. Good examples of such presentations are the charts needed for demograhical studies. Studies of ever-changing populations are continually being conducted. Population changes within human and animal kingdoms are monitored, and the evaluations of the impact of population densities on particular ecological systems are presented through charts and diagrams. Illustrations for such studies are sometimes rendered in color, many being extremely imaginative and beautiful without losing their scientific value. Various products avail-

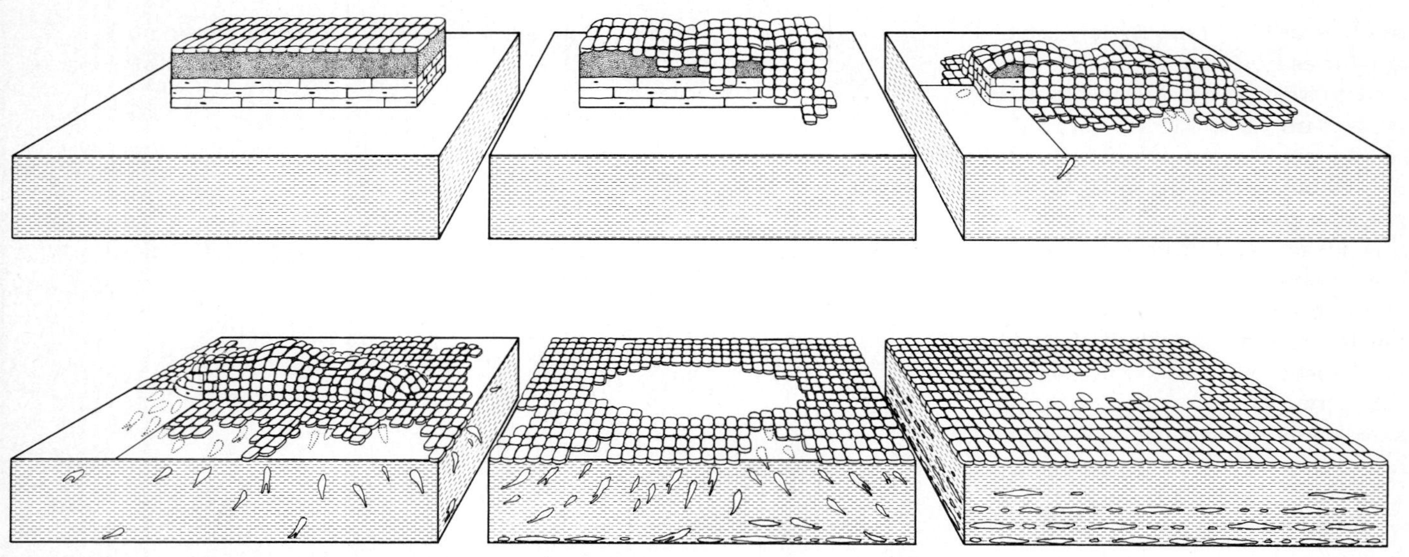

Figure 2-23. *Gallus domesticus*, growth of heart tissue explants on collagen gels. Pen and ink. Drawing by Stephen L. Sickerman, Texas Tech University Health Science Center, Lubbock, Texas. Published in Roger R. Markwald et al., "Use of Collagen Gel Cultures to Study Heart Development: Proteoglycan and Glycoprotein Interactions During the Formation of Endocardial Cushion Tissue," 42nd Symposium of the Society for Developmental Biology, by National Institutes of Health (NIH), New York. Drawn in 1983.

able on the market—such as instant tones or instant letters as well as a variety of colors that can be quickly transferred or applied to the surface of the paper by pressure—assure the illustrator that the finished product will be cleanly prepared and camera-ready. The scientific illustrator should be familiar with available materials and be aware of the possibilities offered by them.

Examples of subjects presented through charts, graphs, or maps are endless. All of the branches and subdivisions of sciences need skillful illustrators who can transform the information given to them by scientists into clear, clean, readable illustrations, so the reader of scientific publications can see and compare the data presented by works and numbers in a cohesive representation of a deduction, theory, or explanation. Statistical analysis pertaining to biochemical changes in the composition of the earth can be represented by charts, graphs, and maps, as can studies conducted on manmade objects and the presentation of worldwide distribution of a given artifact. Selected examples of designs found on pottery sherds can be presented in the form of a chart or diagram. Conclusions regarding the dating of various excavated strata, variations in morphology of a living organism, or a hypothetical bombing pattern can be depicted in a form of chart or diagram. Not all diagrams or charts are flat; some are drawn as three-dimensional renderings. The size of the objects may be easier to grasp by the reader if such are presented in a form of a three-dimensional diagram. The cross section of a complicated structure or the comparative analysis of the chemical composition of meteorites may best be understood in the form of a chart or a diagram.

Is this art difficult? Yes, some graphs can be very difficult, requiring great precision from the illustrator. By precision I mean the differences created by the width of a line, perhaps the difference of a millimeter. Such close tolerances are very important when particulars rather than general information are being presented. A diagrammatical representation of the sperm of an African beetle can result in considerable technical difficulties. The length-to-width ratio of such a subject, combined with the precisely counted twists found on the main portion of the body presented at the angle particular to that subject, will not allow for any mistakes. When considering the reproduction, which involves great reduction of the original illustration, the question of final readability can be raised. Nevertheless, the illustration must be correct, reproducing clearly during printing. What about your knowledge of the subject to be drawn? A diagrammatical representation of a "dig" containing numerous graves in which various skeletal remains were found at different levels requires good understanding of human anatomy. Such a presentation should include all artifacts found in the excavated grave, with skeletal remains presenting a difficult problem. All of the objects should be depicted in their original places, with skeletal parts overlapping most of the time. Adding to the difficulty, the remains are sometimes found on top of each other. Here, the problem of accuracy is combined with presentation of depth. In order to present the depth of field, a skillful application of thickness of the line is necessary.

This is archaeology, but what about fields like chemistry, physics, medicine, and engineering? Many if not most charts or diagrams are originally drawn by the computer. A scientist translates his findings into computer language and then, by punching a couple of buttons, creates a visual image. Unfortunately for the scientist, most of such drawings are not fit for reproduction and must be redrawn by illustrators. The reason for this is that machines use felt-tip pens for drawing, and the flow of ink from the pen to the paper is not perfect, resulting in a weak image that is not good for clear, readable reproduction.

Where are the beautiful maps of the nineteenth century? They are precisely structured complexities, beautifully rendered, painted, and printed. I have no doubt that at one time or another you have seen a well-drawn map and probably have commented on its beauty. A map does not have to look dull—a lot depends on the artist's ability of rendering, the cleanness of presentation, and the readability of the final product. Let us not forget that preparation of such illustrative material requires great technical skill.

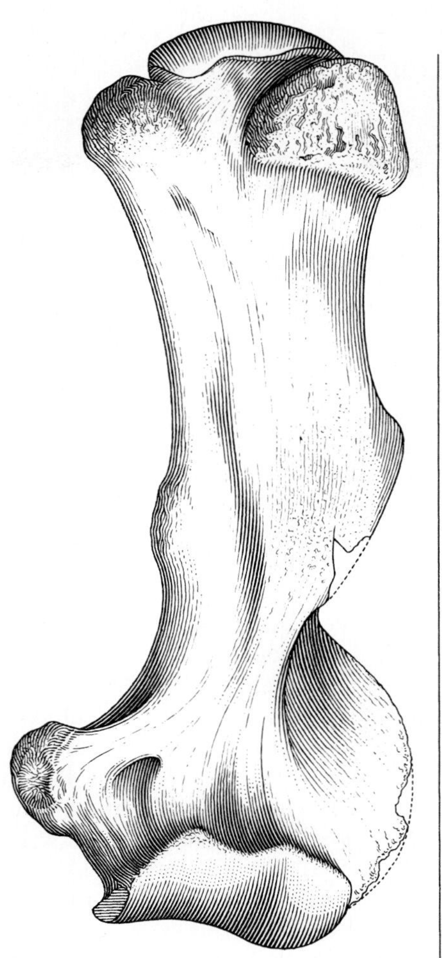

Figure 2-24. *Titanoides faberi*, left humerus, P14902. Crow-quill pen on two-ply Strathmore paper. Drawing by Carl F. Gronemann, Chicago Museum of Natural History (Field Museum of Natural History), Chicago. Published in Bryan Patterson, "A Contribution to Osteology of *Titanoides*," *Proceedings of American Philosophical Society* LXXII, 2 (1934) by the American Philosophical Society, Philadelphia. Drawn before 1934.

From simple, plain, but informative maps through complicated topographic delineations of surface configurations, cartography is a specialization in itself, offering numerous sub-specializations for the artist. Fields such as engineering, the military, or geography present opportunities for interesting work. As for natural sciences, a general knowledge of surveying methods and drafting will be of great help to the scientific illustrator.

BLACK–AND–WHITE ILLUSTRATION

Illustrative presentations of research material in black and white, tone or line, are the most popular and most widely used in all sciences. Virtually all scientific publications and all scientists will prefer such drawings because of the tangibility of the final result, the printed information. Specialization in black and white is not just knowledge of many drawing techniques. It is much more than this, being also a specialization in a particular branch of science. It does require from the illustrator a well-defined understanding of the application of a particular drawing technique to the particular subject. The technique of drawing will change in accordance with the subject, but most of the time is predetermined by the given branch of science, the particular research, and the need for explanatory presentation of particular material. Naturally, both familiarity with materials and the ability to draw as well as think logically are important. The illustrator should be familiar with as many as possible surfaces and tools used in the process of drawing: transparent, semitransparent, or opaque papers; illustration board; drafting film; acetates; tracing tissues. The artist should also have a good knowledge of pencils, pens, inks, litho crayons, carbon dust, charcoal sticks, and brushes. Brushes are needed for some black-and-white renderings, as well as to keep unwanted particles of dust off the drawing or for direct application of carbon, charcoal, or pencil. Familiarity with production tech-

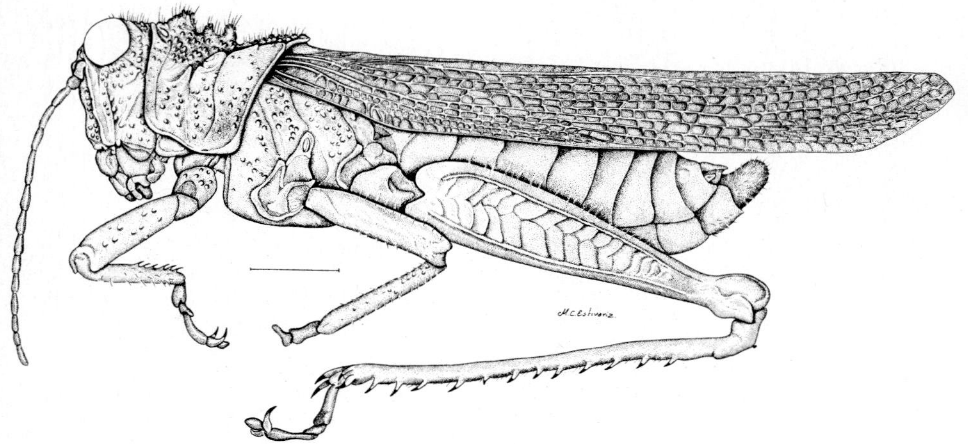

Figure 2-25. *Cralazella quadrata* ♂ lateral view. Pen and ink on vellum. Drawing by Maria Cristina Estivariz, Museo de la Plata, Buenos Aires, Argentina. Published in R.A. Ronderos, *La Familia Ommexechidae,* Acrida by Association d'Acridologie, France. Drawn in 1979.

niques is also necessary, including knowledge of techniques and tools needed to prepare the illustration for reproduction. Tone and line separation, type specification, paste up, and keylining are all important. Various combinations of tools when applied to different surfaces in different manners will produce results that should be consciously controlled by the illustrator. Carbon dust on Ross board will behave differently than when applied to drafting film or Strathmore paper. An open-minded illustrator will be acquainted with materials not necessarily associated with drawing. For example, a material known as Rubylith, a sheet of thin plastic covered with another thin layer of peelable red plastic, can be very helpful in bringing diagrammatic pen-and-ink drawings away from the white background, adding a much-needed differentiation of contrast. Knowledge of production materials will let the artist effectively control the final image. When brush and ink are used on drafting film, a very good surface will be provided for scratching, thereby allowing the illustrator to use the same technique as on scratch board. There are as many technical possibilities within black-and-white mediums as there are subjects to draw.

In all areas of science, whether it be natural, applied, or medical, illustrators specialize within one or two techniques. It is good to know various techniques, but by the simple fact of drawing similar subjects an illustrator is confined to a narrow range of techniques. It is possible to pick a technique that you like the best or in which you are most proficient, but this will dictate the subject that will be given to you to draw. It may be a technique that you favor combined with a subject you may not like, or it may be a combination of both a subject you like and a technique you know. A good example of successfully combining the technique with the subject is represented by the work of French artist P. Laurent, who excels in crow-quill technique, confining himself to drawing stone tools and implements. In many cases such specialization occurs by accident. An artist with limited knowledge of techniques will start on one project and over the years will improve his drawing to a great extent. Such an illustrator will become a specialist who excels in one technique as well as in understanding one particular subject. On the other hand, an artist with the capabilities and sound knowledge of various techniques may start on a certain demanding subject, such as the anatomical structures of mites, and will become fascinated with the subject to such an extent that he or she will remain with the same scientist and the same technique for years, working on the same subject. By doing this, such an illustrator will become a specialist in his área, combining a very specific technique with the subject. The knowledge of the subjects to be drawn greatly helps the illustrator in the visual presentation of a particular problem.

Naturally, it is easier for those who know more than one or two techniques to find desirable work, but most scientific illustrators remain faithful to the subject as well as the area of science by boundaries dictated to them for their technical development. As for techniques, black and white does not mean just outlines and flat images. Drawings in pencil, carbon dust, or charcoal, carefully shaded with a multitude of delicate values, as well as pen-and-ink, stippled renderings are all black-and-white illustrations. Stippling, creating tone by making dots, can produce similar tonal differentiations. Stippling is a handmade effect that can be compared to photographs reproduced in newspapers. Illustrations which are stippled, like simple outline drawings, are considered line drawings by the platemakers and therefore do not require screen photo application. Such illustrations are much cheaper to reproduce because of simplified platemaking procedures. Line drawings are very desirable because of lower production costs, and are very effective in presentation of the subject. Continuous-tone renderings are referred to as tone drawings. They require translation of the image into a dotting pattern, achieved by photographing the illustration through a screen in order to be transposed onto a printing plate without losing delicately rendered tones.

Figure 2-26. *Abystoma* spp. (composite of four species); tissue matrices of the stump-regenerative complex at completion of the initiation phase of limb regeneration (30 days postamputation) in adult land phase salamander. Pen and ink. Drawing by Stephen L. Sickerman, Texas Tech University Health Sciences Center, Lubbock, Texas. Published in Henry E. Young, "Temporal Examination of Glycoconjugates During the Initiation Phase of Limb Regeneration in Adult Abystoma," Department of Anatomy, Texas Tech University Health Science Center, Lubbock, Texas. Drawn in 1983.

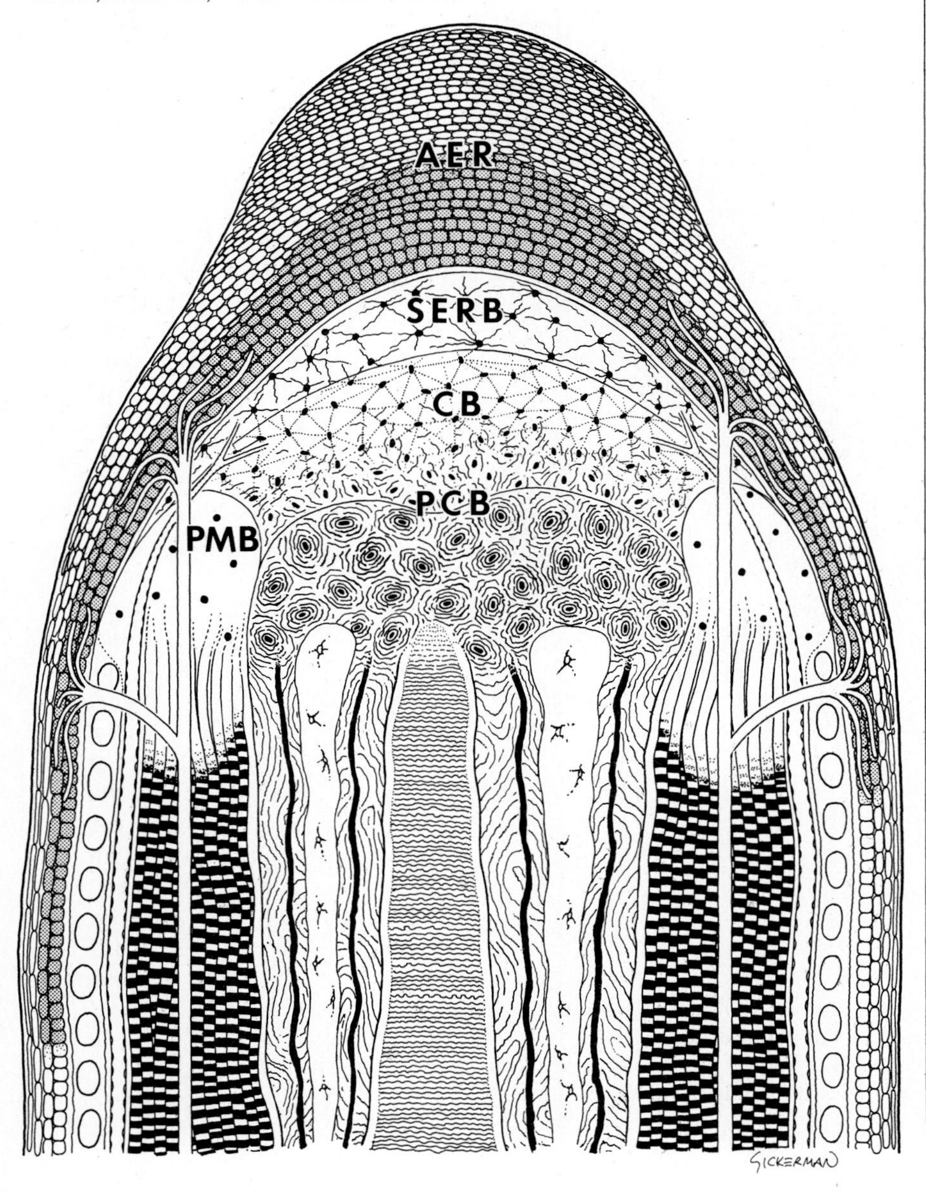

Subjects for black-and-white illustration range from simple charts and diagrams through maps and very complicated studies of various forms of tissues, skeletal components, insects, and plants, as well as enlargements of sections of living matter. Artifacts of the past and present are also included, such as pottery, shell engravings, tools, clothing, decorations, buttons, fishing nets, and weaponry. Plants reconstructed from verbal description will compete with reconstructions of early forms of life. Really, it is impossible to mention all potential subjects. Within the area of medical illustration techniques such as carbon dust, scratch board, pencil, and pen and ink are used to illustrate the anatomy of *Homo sapiens*. Drawings of operating techniques and other various supportive visual materials are of great importance to the medical community and to students of medicine. Research conducted in the medical field needs as many varieties of illustrated subjects as are found in natural sciences. An illustrator specializing in black and white cannot possibly be an expert in all subjects, nor can he or she excel in all drawing techniques. Certain subjects and techniques will have to be chosen, either deliberately or by accident.

BIOLOGY

Biology is the study of life, from the Greek *bios*, meaning *life*, and *logikos*, meaning reason. Biology is reasoning about life. All forms of life are under scientists' scrutiny creating a large field full of diversified research calling for specific skills from the illustrator. The whole field of biology is so large and so diver-

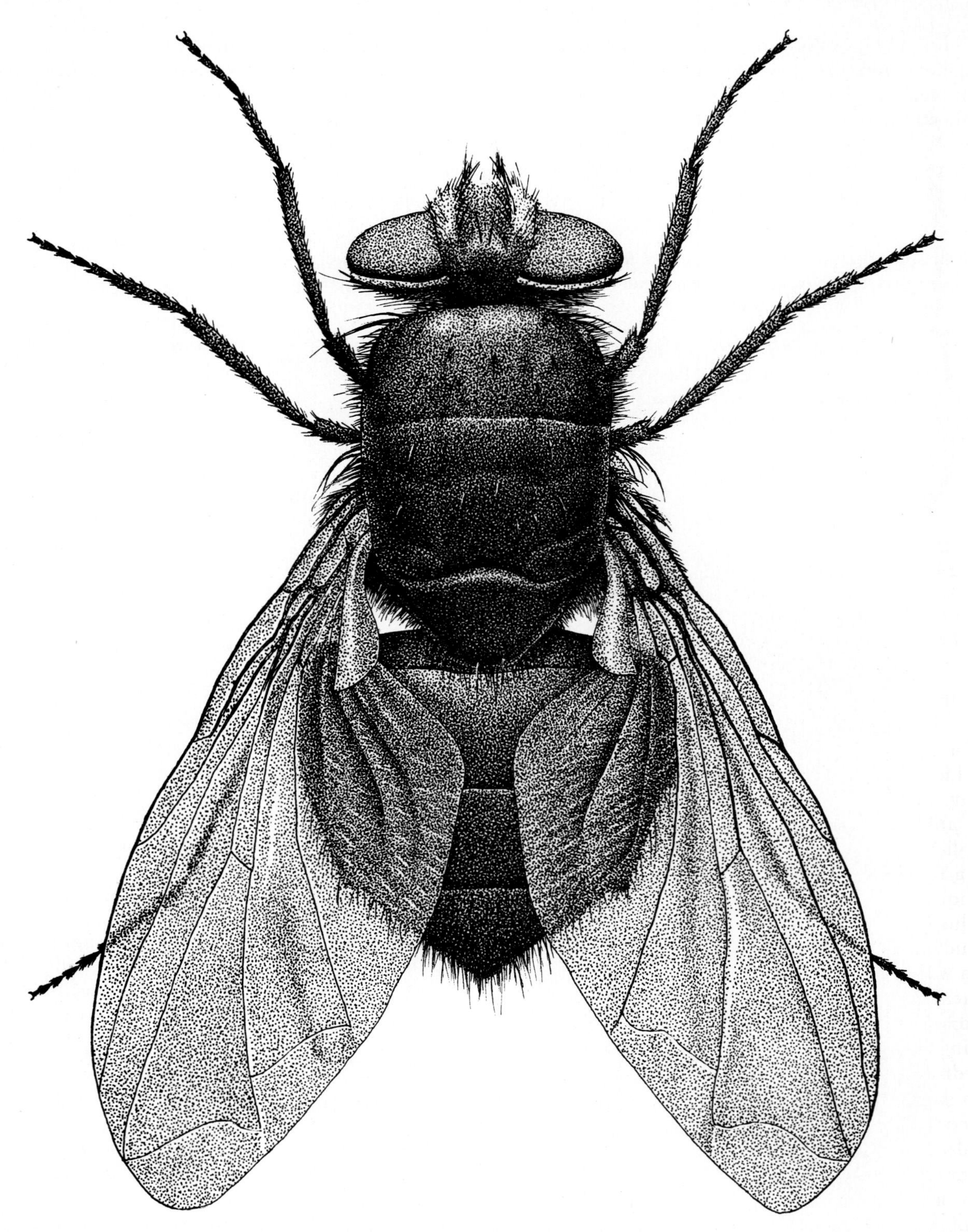

Figure 2-27. *Calliphora erythrocephalus,* Belgian fauna (Insecta, Diptera). Pen and China ink. Drawing by Auguste Van Der Kelen, Institut Royal des Sciences Naturelles de Belgique, Brussels, Belgium. To be published in *De Nieuwe Natuurkalender* and in *Le nouveau Calendrier de la Nature,* edited by the Patrimoine de l'Institut Royal des Sciences Naturelles de Belgique, Belgium. Drawn in 1980.

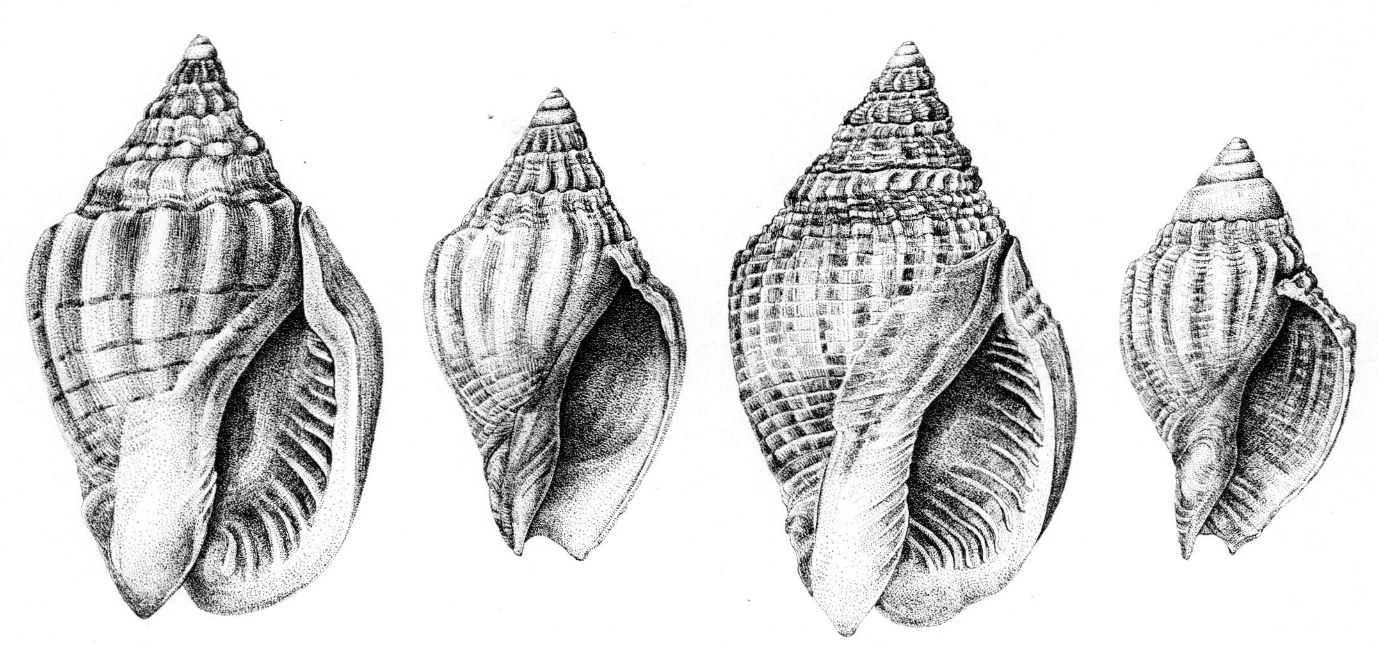

Figure 2-28. Shells of Nassariidae. Rotring (technical) pen and China ink, stipples only. Drawing by Jacqueline Van Melderen, Institut Royal des Sciences Naturelles de Belgique, Brussels, Belgium. Published in W. Adam and J. Knudsen, "Révision des Nassariidae de l'Afrique occidentale (Mollusca, Gastropoda, Prosobranchia)," *Bulletin de l'Institut Royal des Sciences Naturelles de Belgique*, 1984.

sified that a scientific illustrator is forced to specialize.

For example, illustrations required for entomology, the branch of biology devoted to the study of insects, are complicated presentations controlled by very rigid standards, calling for skills of observation and techniques of drawing different from the illustrations needed for ecological studies. The anatomy of invertebrates will impose a different set of requirements upon the illustrator than drawings of vertebrate anatomy. Drawing shells cannot be compared with drawing bones of mammals. When asking yourself about the importance of specialization, you may ask also about your knowledge of anatomy. Is it necessary, and if so, which branch of anatomy? It is impossible to be knowledgeable in all anatomical systems of living matter, but for the illustrator specializing in medical sciences, a good knowledge of human anatomy is needed. It is also very helpful when an illustrator plans to work within a narrow area of biology such as endocrinology, where he or she will prepare drawings pertinent to research on growth control, development or reproduction of members of animal or human kingdoms. At the same time, when illustrating actions of hormones, such knowledge may be altogether unnecessary. Entomology will lead you into ecology, as insects are an integral part of ecological systems through their direct influence on vegetation, linking agriculture to our personal well-being. The scientific illustrator can also devote his skills in biology to taxonomical studies of various members of fauna and flora: fishes, insects, plants of today as well as creatures preserved from the past in the form of fossils. An illustrator can be fascinated by the complexities found in pictorial presentations of the structural characteristics and useful influence of microorganisms such as molds or yeasts upon our existence. Even within such seemingly narrow area it is possible to diversify. Microbiology is subdivided into structural and functional areas; both require capable and intelligent illustrators.

As you can see, biological subjects range from picturesque taxonomical descriptions of presently living forms through soft tissue and skeletal anatomy, charts and diagrams, studies of microorganisms, the intricacies of flower structures, and surgical procedures. Let's look at an entomological publication issued by Rijksmuseum von Natuurlijke Historie in Holland, entitled *Tijdschrift Voor Entomologie,* for typical entomological illustrations. The quantity of illustrations should tell you something about the need for skilled scientific illustrators. Most illustrations are black-and-white line drawings that carefully explain the complexities of the subject. As you remember, line drawing is much cheaper to print than halftone illustration. But when necessary, such extra expense can be accommodated in the publication budget, allowing the illustrator to present the subject not only with scientific accuracy but also with the beauty of rendering.

Biology offers more subjects to draw because it encompasses a multitude of areas in science. The range of subjects and their impact on illustrators' perceptions of nature will have lifelong implications. Anatomical illustrations will compete with complicated explanatory presentations of molecular structures. The pictorial reconstruction of long-gone forms of life, so important for paleontological studies, will provide an everlasting challenge for the illustrator. Virtually all natural sciences are a part of biology. Knowing this, we can see how broad and complicated a specialization in biology can be. For this reason an illustrator must choose, making a decision based on personal interests. In order to do this it would be very advisable for the artist to take few courses in biology. This would help to broaden the horizons and would prepare the illustrator for the proper decision. A biological illustrator will always find himself specializing in certain areas of biological science. He or she can specialize by subject, technique of drawing, and within boundaries set by the structure of the institute conducting research.

Virtually all known techniques of drawing and painting can be used effectively, providing that the technique is not too extraordinarily time-consuming and thereby too costly. For this reason, oil painting is not very suitable. Black-and-white illustrations will receive priority. The most favorable techniques are pen-and-ink line drawings, pen-and-ink line and stipple, pencil in various forms, carbon dust, washes on various surfaces, colored pencil, watercolor, tempera, acrylic, and combinations of these—really, any technique that can produce tangible results in a relatively short time, allowing for clear, easy-to-read reproduction. In the *Quarterly Review of Biology,* September 1975, the paper titled "The Adaptiveness of Social Wasp Nest Architecture" has a few well-produced well-printed illustrations. When we judge the quality of illustration, we base our responses not on the original but on greatly reduced prints appearing in publications. It is very important to remember that if an illustration is greatly reduced, the possibility of presenting a clearly understand able image is also diminished. Look through the most accessible scientific journals, such as *Science* or *Scientific American*. In both you will easily find good examples of biological illustration, ranging from presentation of anatomical structures executed in pen and ink or pencil to reconstructions of structures of the brains of animals extinct for millions of years. Go to the library and look for yourself. You will quickly be able to judge the skill of illustrators by viewing a cross section of subjects and techniques.

You also have to remember that when working with small subjects you will use very sophisticated tools—microscopes of various kinds as well as tracing devices. Learn something about optics, learn how to operate a microscope, and when using such a tool, be very careful not to damage the specimen when focusing. Without the specimen you cannot produce the required drawing, nor can the scientist conduct his research.

Future scientific illustration will no doubt be carried into space. The new frontier has been established and it is open to artists, especially to those who like to cooperate with science, who enjoy the thrill of discovery, and who are pleased to be part of a research team combining a love for drawing with inquisitiveness into the nature of matter. While here on earth the scientific illustrator finds the unknown within easy reach and without participating in adventurous expeditions, soon outside the earth there will be a need for the same—but it will not be as easily reached. The human eye combined with that powerful self-motivated computer, the brain, comprise the best scientific instrument available, helping us to advance our knowledge.

CHAPTER THREE

Tools: Their Description and Usage

THE scientific illustrator encounters a variety of tools—tools needed for measuring, viewing, poking, drawing, and painting. The selection depends on the field of specialization and availability of tools as well as the complexities of the project. Special ruling pens, planimeters and Ellipsographs are in actuality simple-to-use devices, but are rarely used in natural sciences. An illustrator working for an industrial research area or a cartographer will have greater need for such specialized instruments than an artist preparing medical illustrations. When considering the need for a microscope as a viewing tool, the opposite will be true. Tools used by various scientific illustrators can range from simple devices all the way through complicated and expensive machinery. Neither the structural intricacies of the tool nor its price will not have any bearing on the quality of the finished product. High quality can only be accomplished by an illustrator's conscious control of his hands, his ability to observe, his patience, the proper combination of drawing or painting tools for the project, and the proper choice of papers. Cleanliness is an essential ingredient. Keeping all tools in operational condition will require some work, but it is well worthwhile to invest some time in such routine procedures. Do not expect a mistreated brush to work properly, and do not expect a technical pen full of dried-up ink to be of any use. All available drawing, measuring, tracing, or painting tools should be clean. Drafting machines or microscopes will not work properly when handled roughly. Once a microscope is dropped to the floor from any height it will have cracked or misaligned optical components, causing a lot of expense in repairs and a lot of pain during observation and drawing. Learn how to operate and find out what to expect from any tools before using them. If the only chance to learn is through "on-the-job experience," then be cautious and think fast. Remember that regardless of the complexities of optical instruments, advertised technological achievements, and photographs of instruments in catalogs, you should not expect miracles. The illustrator is the only individual who, with the help of those tools, can produce the drawing. If you are interested in learning more then ask, buy, or acquire catalogs from art-supply stores. Such catalogs will have a listing of all drawing, painting, and drafting equipment. Write to manufacturers of microscopes and ask for information. Write to producers of papers, inks, drafting films, pens, and do the same. In most instances you will receive an answer because you, the illustrator, are a potential buyer of the tools produced by those companies. In the field of scientific illustration all tools will fall into eight general categories:

1. Tools for positioning specimens
2. Tools used for observation
3. Tools used for measuring
4. Tools used for tracing
5. Tools used for drawing and painting
6. Drawing surfaces
7. Tools used during preparation of the illustration for printing
8. General-purpose tools

Lots of tools will be used for a variety of purposes. A simple X-Acto® knife can be used to position a specimen, cut paper, sharpen pencils, or stab your own finger. Similarly, a viewing camera can be used for observation of certain specimens, for tracing an image directly from the specimen, tracing the sketch, or even for photostatting, when used with some skill. Interchangeability of tools will depend on your own ability to think. Some of the tools will be applicable to a variety of tasks, while others will remain as specialized contraptions used only for a definite purpose. Various areas of research will require respective drawing, painting, or plotting tools. Obviously, a drafting machine will not be used in rough terrain but rather in more stable situations, just as a sophisticated microscope will not be taken to the field, but specimens will be brought to the illustrator, who should be located in a reasonably dustproof enclosure such as the research institute.

TOOLS FOR POSITIONING SPECIMENS

Depending on the size of the specimen, mode of observation, and technique of transferring the image of the object during initial preparation steps to the tracing tissue, supportive devices will range from small pieces of glass to large dissecting trays to the tabletop. Similarly, "poking" tools used to position the specimen or discover its structure or surface configuration will range from teasing needles to your own fingers.

GLASSWARE The petri dish forms the basic line of tools for positioning relatively small skeletal or soft-tissue specimens. The petri dish is a round, completely transparent container with short vertical walls. It comes in a variety of sizes for everyone's convenience. Usually petri dishes are used by scientists for other purposes, such as growing cultures, but scientific illustrators often use them to position specimens. Because of the transparency of glass, such a dish is an excellent container, allowing light directed from below the dish to travel through a semitransparent or transparent specimen. Because of the restricted diameter of the dishes, only relatively small specimens can be positioned in the liquid used to preserve soft-tissue specimens. As you know, soft-tissue specimens without the protection of liquid will be badly damaged by hot microscope lights or just by drying out.

Positioning procedures will vary, depending on the type of specimen. Some will be placed on a small sheet of gauze when submerged in glycerine, while others will be pinned down by thin metal pins. In order to pin a specimen down while using a glass dish, some kind of material easily attachable to glass must be used. Plasticine is easy to both find and use. To start, the dish has to be completely dry, assuring good contact between the Plasticine and the glass and the glass walls of the petri dish. Thin metal needles inserted horizontally into the Plasticine will hold the specimen pinned down to the bottom of the container or between the needles. Before placing a soft-tissue specimen in the container, the preserving liquid will have to be poured into the dish. Restrain yourself from holding and handling such specimens without protective liquid, as it is extremely easy to damage the tissue. The specimen should be positioned according to the scientist's specifications. This may take some doing, because none of the pins can obstruct important parts of the specimen, and some pins may have to be bent in various ways to hold the specimen or to stretch the tissue. You must be extra careful when handling glycerine so as not to get that greasy substance onto the microscope, the drawing table, or the drawing. Instead of pouring liquid from a large container into the petri dish, try to use a squeezable plastic bottle, a rubber bulb with a bit of plastic hose, or a pipette.

Skeletal material is positioned in a dry glass dish. It will always rest on a small mound of Plasticine or some kind of similar material. Depending on the size of the specimen and the preparation for storage and study, the specimen can either be placed directly on the Plasticine or supported by a metal pin which will be inserted into the small mound of Plasticine. As proper positioning and handling of the specimen are essential, usually small skeletal subjects are glued to the tip of the metal pin. Similarly, dried material such as insects are stored by being pierced with a very thin metal pin. The pin helps during the handling of the specimen because the human hand does not touch the subject. Also, it is easier to turn the specimen around during positioning. Vertical and horizontal motion are controlled by simply sticking the pin into the Plasticine at the correct angle. For dry specimens, a petri dish is not necessary, although under laboratory conditions such a dish is more easily accessible than other materials.

Besides the glass container used for positioning specimens, you will need something to hold necessary preserving liquids. Any glass jar will be good for this purpose. It is better to use glass instead of plastic jars because some plastics may be adversely affected by preserving liquids. If you know that the plastic container will not be dissolved by ethyl alcohol or a mixture of formaldehyde and ethyl alcohol, then feel free to use it.

FORCEPS Various forms of tweezers made from metal are used to hold specimens while transferring the subject from one container to another. Forceps are usually used with soft tissue or live specimens. They come in a variety of shapes and finishes. Some are small and have extremely fine and sharp tips, while others are large and bulky. The size of the specimen will dictate the size of the forceps. Extra-long forceps are used for removing specimens from the bottom of glass containers. Fingers will not fit into most jars, and even if it were possible to reach a specimen with your hand, the preserving liquid would spill all over. Forceps must be treated as an extension of your arm and should be handled properly. It is easy to damage part of the tissue when the specimen is squeezed too tightly, but the specimen will slip out of the forceps' hold if not grasped tightly enough. Do your best not to stab the subject, but gently grab it during the transferring process.

Skeletal material will easily slip out from the forceps' hold, so if it is necessary to pick up a small piece of bone, use another variety of forceps called *triceps*. Triceps have three thin, hooked metal pins activated by a built-in spring and enclosed in a housing similar to a technical pencil. When the button located on the top of the housing is pressed, the three prongs will extend downward. The specimen is grabbed from three sides at once and moved upward when pressure on the spring-release mechanism is slowly lessened. Although triceps are much safer than forceps, be careful at all times, as small, hard subjects do have a tendency to jump away. If this happens, you may have a very hard time finding a lost specimen. Triceps are not very good for removal of soft-tissue material because the three-hooked metal pins will pierce the subject, injuring or damaging it badly.

STAGES, MICROSCOPE Stages are small, fixed, manually or mechanically operated platforms mounted under the objective of a microscope. All specimens must be placed on some kind of flat surface for observation through the microscope. Again depending on the size of the specimen and availability of equipment, an appropriate stage should be used. The most common stage is one formed by the base of the microscope, which has a round opening cut through it. Inside the opening there is a stage plate, a round piece of glass or plastic. Sometimes one side of the plastic stage plate is coated black and the opposite side white. The differentiation of colors gives the viewer two possible backgrounds on which to observe the specimen. The stage plate is removable, at times secured by a small screw. It is used for the incident illumination of the specimen. The subject does not have to be positioned directly on the stage plate, as it can be secured in a petri dish or other flat surface and then placed under the objective of the microscope. For transmitted illumination, frosted glass is best. The frosted glass stage plate is located in a slightly different kind of microscope base from the black-and-white stage plate. The base is usually taller, allowing for a source of light to be mounted under the base. The light is directed upward through a transmitted-light stage plate by a mirror.

The gliding stage and cup stage have one thing in common. Both can be rotated horizontally, with the cup stage allowing for a certain degree of tilting. The gliding stage is a round, thicker object, with the stage plate essentially performing the same functions as the incident or transmitted simple stage plate, but allowing for horizontal rotation of the viewed subject. It can be mounted in especially designed microscope bases and will be available only if the manufacturer of the microscope you are using has such accommodations available. The cup stage is only for incident light, as it cannot transmit any, being completely opaque. The cup stage in actuality is a half sphere. With its bottom round and upper section flat, it can be rotated as well as tilted. A cup stage is unfortunately not firmly attached to the base of the microscope, presenting a rather dangerous situation. The specimen can easily fall off if it is not secured well, and the whole stage can jump out of place during rotation. If by any chance you need such a device, make it yourself by cutting a wooden ball in half and painting it black or white depending on the background you need. Building a stand should not be difficult either.

A mechanical stage is a sophisticated platform firmly attached to the metal rod that holds the microscope. It can be operated by hand or by electric motor. A mechanical stage is used for observation of very small specimens mounted on glass slides. The stage has a device for holding glass slides securely and for moving the specimen in any horizontal direction. Because subjects are very small and therefore, for any meaningful observation, magnification must be large, the mechanical stage has two turning knobs allowing for back-and-forth as well as left-to-right horizontal adjustment. With large magnification, the distance between the ocular and the specimen mounted on the glass slide is minuscule. Extra caution has to be exercised while focusing the microscope, as it is easy to jam the objective of the microscope into the glass slides. Compound microscopes have a dual system of focusing: One is calibered for general relatively small magnification viewing, suitable for finding the subject and positioning it under the objective, while the other is calibered differently, allowing the observer to focus on the desired feature of the specimen without immediately endangering either the microscope or the glass slide.

TRAYS A flat, traylike container of any type should be used when the specimen requires protection of preserving liquid and needs to be positioned for drawing. The dissecting tray or dissecting pan, as it is sometimes called, will allow for pinning a large specimen in the prescribed position. The bottom of the dissecting pan is covered with a heavy layer of black wax, allowing inserted metal pins to stay in place. Usually the wax covering in the dissecting tray will not crack, nor will it separate from the walls or from the bottom. In a homemade dissecting tray, wax adherence will not be as strong and there is a possibility that the whole sheet of wax will begin to float and wobble in the liquid. Naturally, it is difficult although not impossible to draw under such circumstances. In order to correct the situation, the specimen should be removed, the liquid poured out, the tray dried, and without burning your own fingers, the wax should be melted back into stable condition. Be very careful when melting wax, as it is hot and severe burns can result if the hot wax gets on your skin. If you can, avoid such experiments and obtain factory-produced dissecting trays. The only way to illuminate the specimen is from above, with incident light. If you do wish to illuminate the specimen from below, build or buy a plastic or glass container and line the bottom with transparent polyurethane plastic. Light will be able to travel through the specimen; the plastic will allow for insertion of pins. To make things better for yourself, install a movable mirror underneath. Such a mirror will allow for transformation of regular incident illumination into the transmitted system. The light from above can easily bounce off the mirror located under the clear plastic or glass container and directed at various angles through the specimen. Whenever using trays, tight covers will be necessary. You will have to make a cover for the container. Usually a sheet of plastic will suffice, especially when clips are used to hold a sheet of Plexiglas to the edges of the dissecting tray. For custom-made trays, custom-designed tops will have to be made. Covers will prevent evaporation of the liquid during overnight storage and work breaks. A Plexiglas cover can be used as a base for a grid. Horizontal and vertical lines drawn in black ink on top of the cover will help you judge the proportions of a large subject.

NEEDLES, PINNING Stainless steel needles of various lengths and widths are used to pin a specimen in the desired position. Depending on three-dimensionality and the size of the specimen, appropriate needles should be chosen. During the process of pinning, make sure the specimen is not injured. If this happens, inform the scientist immediately. In all probability pins will have to be bent in order to hold down sections of the specimen. A pair of pliers will be of some help. Pins will be used for holding soft-tissue specimens submerged in preserving liquids. For this reason, avoid pins with heads. If that's what you have, then cut the heads off. Heads of pins obstruct the view and, when sticking out of the liquid, reflect in the solution, interfering with observation and therefore with drawing. For best results, have all pins submerged in the liquid, thereby avoiding unnecessary reflections and obstructions. Do not use pins coated with black paint or pins with gold-colored heads. The color, because it is separate from the metal structure of the pin, will eventually dissolve in the liquid and the color will flake off, possibly leading to misjudgment of a specimen's features and coloration markings.

NEEDLES, TEASING Teasing needles are sharp, thin needles mounted in wooden or metal handles and used for touching various structures of the specimen. During observation and drawing, a need to check certain things will arise. Our fingers are too large and cannot be used efficiently, while thin needles can reach into almost any place. For checking features as well as for moving small specimens around, teasing needles are indispensable. Because of their sharpness, certain caution must be exercised, especially when working with a microscope. When larger magnification of a small subject is necessary, the depth of field decreases and our judgment of distances is profoundly affected. Be very careful not to scratch the lens of the microscope with the teasing needle. It is worthwhile to exercise patience and avoid serious damage to the expensive tool. Place the tip of the needle under the objective and then look through the oculars. Move the teasing needle horizontally until the motion is visible. Once the motion of the teasing needle can be detected, slowly bring the tip of the needle into focus by lowering it. Teasing needles are produced in two varieties: straight and with the tip bent at approximately a 45-degree angle. For very small specimens you will have to sand the needle down to reduce its diameter. In order to have a slight hook at the tip, you will have to bend the very tip of the needle a bit. Such a very small hook may be helpful in pulling off tissue covering a structure to be drawn or when checking the forms of transparent scales of a fish. Teasing needles are used during observation and drawing of small specimens, provided that subject will not be scratched and damaged by their sharp tips.

TOOLS FOR OBSERVATION

The most important of all tools used during observation of any specimen are your own eyes. This does not mean that the most perfect vision is the best. Observational skill depends on personal ability to investigate, to compare, and to be interested in the work you do. Additional help will come from magnifiers of various types as well as from illuminators. Any magnifier, such as a hand-held enlarging lens, and any illuminator, be it a desk lamp or a flashlight, will accomplish the basic task. As a matter of fact, in many situations more advanced viewing and illuminating tools can become a burden. A microscope may enlarge a particular subject too much, prohibiting the viewer from grasping the form of the whole subject. A very bright light can affect the viewer's understanding of the depth of field. During observation the scientific illustrator becomes an investigator whose primary function is to decipher the unknown, to check and recheck information received from the scientist as well as his own line of reasoning. Nothing can be taken for granted and nothing can be disregarded.

FIBER OPTIC COLD LIGHT SOURCE This is a source of light that emits so-called cold light used for illumination of specimens. Light is an energy and as such emits heat. Regular light bulbs are hot and dissipate light in various directions at the same time. Such a situation endangers the specimen unless certain precautions are taken. The fiber optic cold light source consists of a transformer and tubes made of perfectly aligned fibers of glass. The transformer is a primary source of light, while fibers of glass are the light conductors. A high-intensity light bulb placed inside the transformer emits light and therefore most of the heat away from the specimen, but nevertheless the object is perfectly illuminated. The density of illumination is controlled by an iris diaphragm located at the place where tubes containing the fibers of glass are inserted into the transformer. Usually at the same place there is an opening for color filters. The filters, made of heat-resistant glass, will affect the color of emitted light and can be helpful during observation of some specimens. The tubes containing fibers of glass are flexible, and therefore can be bent easily, directing the light toward the proper area. A focusing lens is attached at the end of those tubes. With the help of the focusing lens it is possible to attain extremely high illumination of the subject. Because the light that comes from the end of fiber optic tubes is cold, neither the observed specimen nor your fingers will be burned. There are numerous designs of cold-light systems available on the market. Some will have one fiber optic tube, while others are more diversified, having up to three separate light-transmitting fiber bundles. Having more than one tube allows the illustrator to illuminate a specimen from various directions at the same time. The flexible tubes can be attached to the microscope with special attachments, while metal-enclosed gooseneck tubes can be left sticking out of the transformer, the gooseneck light guides bent into the desired position. For the illustrator, the biggest problem will be the vibration caused by the cooling fan in the transformer. When such a unit is placed on the drawing desk, the vibrations caused by the cooling fan will cause the small specimen placed under the microscope to move around. Volpi Intralux Cold Light Source is the only one that does not produce vibration. Naturally, placing a vibrating condensor next to the drawing desk will solve the problem, but unfortunately the length of the light guides prohibits such a simple solution in most of the cases.

INCIDENT-LIGHT STAND The incident-light-stand is the base of the microscope, usually with a black-and-white stage plate centered under the objective and used for positioning specimens. Illumination must be from above. Most microscopes have an incident-light base routinely attached to the stand holding the main body of microscope. Whenever a specimen is illuminated from above, the most practical and most stable base will be the drawing table. In order to attain such a luxuriously large surface, a carefully chosen microscope must be used. Most bases are nonremovable, serving two functions at the same time by supporting the microscope in an upright position and allowing some space for the specimen. In scientific illustration flexibility is advisable unless the artist is specializing in drawing very small subjects. It is quite difficult to place a large specimen on top of the standard base of the microscope. As a

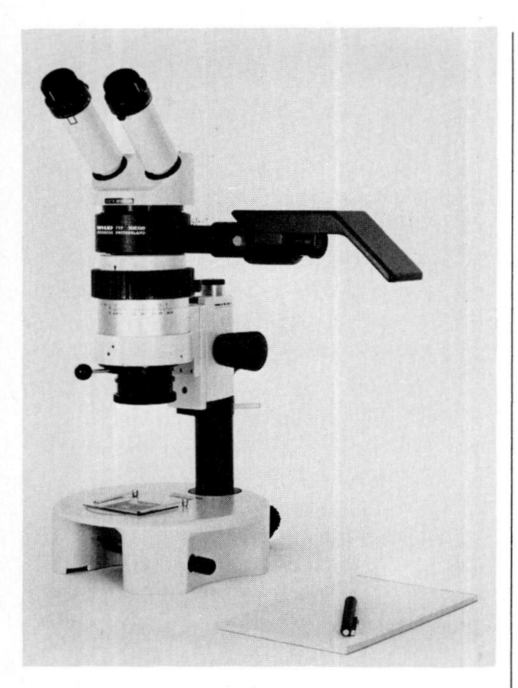

Figure 3-1. Wild Heerbrugg M 7 A zoom stereomicroscope with camera lucida (drawing tube) attachment.

matter of fact, in some situations it is impossible. For that reason it would be good to have the microscope attached to the edge of the drawing table rather than to a small base. Special swing-arm mounts are provided by manufacturers for that purpose. In such a situation the tabletop will become the incident-light stand.

TRANSMITTED-LIGHT STAND This is the base of the microscope with a built-in light source and a mirror used for illuminating specimens from below. The light bouncing from the adjustable mirror through the subject allows for examination and drawing of semitransparent or transparent structures. A transmitted-light stand usually is good for bright-field illumination. More sophisticated stands have combined the bright field with dark-field illumination. Bright-field illumination lets the light travel through the specimen directly into the objective of the microscope, making the subject appear too dark or darkish, with a brightly illuminated background. Dark-field illumination is the opposite, as the light coming from below does not enter the objective directly but first bounces off the specimen. The structural irregularities of the subject cause the light to refract and diffract before entering the microscope's viewing system, causing the specimen to appear bright against a black background. The adjusting mechanism, located on the side of the base, makes the selection easy. It is possible to build a simple stand with a fixed mirror in an inclined position of about 30 degrees simulating the bright and dark-field effect. The result will not be exactly the same but in many situations will be very adequate. Such a contraption can be a sophisticated stand with light directed horizontally at the mirror or it can be a very simple box. It must have a sturdy, transparent, flat surface for placing specimens on. When incident light is used, the mirror will reflect the light coming from above, directing the path through the specimen. Simple horizontal rotation will produce the change of illumination from bright to dark field.

MICROSCOPES A microscope is a simple or complex magnifier used for observation of specimens. It is made of a combination of various lenses and prisms enclosed into some kind of housing. In order to choose the correct viewing tool, an illustrator must consider the size of the subjects and the required magnification for drawing. Manufacturers such as Wild Heerbrugg, American Optical, Zeiss, and others offer a wide variety of microscopes to choose from. Some of the microscopes are capable of very high magnifications, although for general work any magnification over 50× is a deterrent rather than a help. Some microscopes are built of interchangeable parts, while others cannot accept any structural additions. In all types the light will enter the objective, passing through a combination of lenses, with its path bent by prisms and directed to oculars, entering the viewer's eye through the eyepieces. Magnification depends on the selection of the appropriate lens in the objective as well as in the eyepieces. Specialized microscopes for photography can be operated manually or automatically. Camera lucida can be attached to the stereoscopic microscope by adding that section into the general structure. For monocular microscopes or for those which do not include the drawing tube in their main body a "clamp-on-the-ocular" small version of the same exists.

Microscopes are not difficult to operate. Every microscope must have a light illuminating the specimen, as without the light not much can be seen. The operation of a typical stereomicroscope consists of the following steps:

1. The specimen is placed under the objective of the microscope.
2. Magnification changer is moved to one of its lowest positions.
3. The transformer must be activated, causing electricity to flow to the lights illuminating the specimen. The strength of the light is regulated by a knob located on the transformer.
4. Light must be directed on the specimen.
5. The interpupillary distance has to be set. This is accomplished by moving both oculars simultaneously in a horizontal direction until, when looking with both eyes, one field of view is visible.
6. Viewer must use both eyes while looking through a stereoscopic microscope.
7. Using the focusing knob, the microscope is focused on the specimen. Turning the focusing knob will make the microscope move up and down until the

image is found and clearly visible.

8. The dioptric difference must be adjusted by rotating the adjustable eyepiece while the viewer's other eye is closed. Upon achieving desired sharpness, open both eyes and refocus the microscope again.
9. Now the viewer is in the position to observe the specimen.

Whenever magnification is changed drastically, the dioptric difference should be readjusted again. Having a drawing tube as an integral part of the microscope, the illustrator should know that such addition does change the total magnification of the specimen. In order to properly operate the camera lucida built into the microscope, additional steps must be taken:

1. When the microscope is properly adjusted and focused, the camera lucida's engaging mechanism must be activated. At times this is a simple knob which has to be pulled away from the extended arm of the camera lucida or some kind of switch. The pathway to the light is open and while looking through the microscope it may be possible to see the surface of the table overlapping the specimen.
2. Drawing paper has to be placed under the mirror of the camera lucida.
3. Density of the light illuminating the specimen should be diminished considerably.
4. The area on the table where the drawing paper was placed has to be illuminated considerably.
5. While looking through the microscope, the tip of a pencil must be moved across the area directly under the camera lucida's mirror.
6. When a faint image of the pencil is located, a line must be drawn on the surface of the paper.
7. Using the focusing ring located at the end of the extension tube, the camera lucida has to be focused on the drawn line.
8. When the drawn line is visible as a clear image, it is time to find out the truthful enlargement factor for the specimen under observation.

The microscope has all magnifications designated on the objective, but while this may be correct for viewing, the magnification used for drawing will be in actuality smaller. In order to be correct, take a small metric ruler and very carefully place it next to the specimen. While looking through the microscope, with the magnification factor on the microscope adjusted to the specifications given by the scientist, trace the visible distance between millimeters printed on the ruler. Make sure that the distance is correctly traced. Remove the ruler and measure the traced distance with the same ruler. When the length of the traced distance is divided by the actual distance, the truthful magnification factor will be obtained. If a more precise measuring device than a simple ruler is on hand, use it, although it will not be necessary to obtain greater accuracy. When studying three-dimensional subjects, the depth of field will interfere with such a process. In order to alleviate discrepancies, take an average or simply place the ruler at the midpoint of the vertical dimension of the specimen. I assume that when the need arises the lens on the objective as well as the eyepieces will be cleaned, using only lens tissue. Be very gentle when doing this, and of course be sure to use an appropriate material to clean such a vital and expensive viewing tool.

VIEWING CAMERAS A viewing camera is a large camera with movable bellows and a movable bed used for viewing or tracing three-dimensional objects as well as two-dimensional artwork. Some viewing cameras are adaptable to photography or Photostats. Design and construction vary slightly depending on the purpose of usage. In scientific illustration such a camera is utilized for observing and tracing large objects because of the built-in possibility for reducing the image. The ground glass located on the back of the camera serves as the surface upon which tracing tissue is affixed and the image of the subject drawn. Illumination for proper viewing and tracing is provided by lamps located at the edge of the movable bed. The proportions between the lens mounted on the movable bellows and the stationary ground glass as well as between the bed and the lens will give, respectively, reduction or enlargement of the viewed subject. Viewing cameras are simple to operate. Two separate cranking mechanisms, one for the bellows and another for the bed, are provided. In order to attain proper reduction of the subject, the desired length or width should be marked on the tracing tissue and by using both cranking mechanisms the image of the subject will be brought to the previously marked designations. Naturally the lens cup must be removed and the proper f-stop adjusted before proceeding with operation. The only problem encountered while using a viewing camera lies in the fact that horizontally constructed viewers do not have proper support for specimens. Large specimens will have a tendency to fall off the bed, leaving the illustrator and scientist in great distress. Ingenuity on the part of the illustrator will help to preserve the specimen intact.

TOOLS FOR MEASURING

Measuring can be accomplished with modest tools as well as with the most sophisticated ones. Such range is dictated by the size of the specimen. It is impossible to correctly measure a two-millimeter-long subject with a ruler, but on the other hand, a measuring tape will be quite sufficient when clothing artifacts are being studied. By the same token, a hand-held measuring tape will not be sufficient when an archaeological survey of very large structures is undertaken. Whenever measurements of distance between two points of interest separated by kilometers are to be taken, optical measuring devices such as theodolite or transit will be used. With those measuring devices it is possible to establish distance as well as vertical measurements of structures that are large and far from each other. For mapmaking, special electronic measuring tools are used that allow for great precision as well as speed. As you can see, compasses used in drafting, a simple ruler, and measuring tape or calipers can and will be used by illustrators. Each separate project will require specialized tools if measurements are to be accurate enough for the truthful representation of the subject. Essential measuring devices are few, with the proportional divider and metric ruler being the most important.

CALIPERS Calipers are measuring devices used for obtaining an accurate reading of thickness of the walls of pottery sherds, the height or length of various specimens, the diameter of cylindrical objects, or any general measurement. Some calipers are highly complex tools capable of extremely precise measurements, while others are very simple, giving only a general idea of measured distance. Every caliper will have two prongs protruding from the main shaft. The prongs are of importance because their shape will dictate usage. Pottery calipers are among the simplest, with two inwardly bent prongs. Pottery calipers are used to measure the thickness of the clay as well as the outside diameter of the vessel. Various sizes are produced, some quite small. On some the tips of prongs are bent outward rather than inward. Such calipers are very useful when the measurement of the inside of a difficult-to-reach area is needed. It is possible when the caliper is closed to insert it into an enclosure and, with the outwardly bent tips, measure the inside distance between the walls of the vessel. The Vernier caliper is constructed differently, having horizontally sliding jaws. Measurements are taken between outside fixed and stable and inside movable sliding jaws. Calibration can be in the metric or English system, with some Vernier calipers having both. It is the scientist's pocket tool. Virtually all larger specimens are measured with Vernier calipers. The length, width, diameter, and distances between structures on the specimen can be obtained within 0.1 millimeter precision. The micrometer caliper, capable of 0.01-millimeter measurements, is operated by rotation of the arm rather than sliding. Because the fixed and movable ends of the caliper do not have sharp protruding points, but are round and flat, allowing the instrument to close precisely, it is used primarily to measure the diameter of objects. When using calipers, an illustrator must be careful not to damage the specimens. Closing the caliper too hard will chip the pottery sherd and can crack a stone tool.

DIVIDERS These are two-pronged instruments similar to a compass, but having sharp needles at the ends of both arms. Dividers are used for horizontal and vertical measurements. Calibration is not marked, therefore distance, although precisely measured, will not be known until a metric ruler or other object with a clearly marked scale is matched against the divider. Dividers are the most commonly used tools, replaced only by tracing contraptions such as the camera lucida. While measuring horizontal distances, the tool must be held parallel to the subject and the illustrator's head must be positioned directly over the specimen. When the specimen is not directly aligned with the line of vision, distortion of the image will follow. With dividers it is possible to reduce or enlarge measured distances by dividing or multiplying obtained data. The pointers are a tool useful for establishing proportions between points of interest, but without the speed and precision given by proportional dividers.

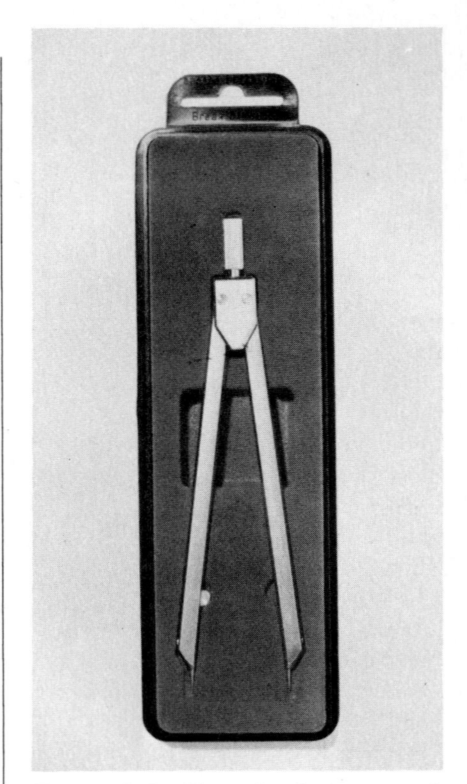

Figure 3-2. Plane dividers, 5.7 inch with up to 7-inch opening. Replaceable alignment gears. Reproduced by permission of KOH-I-NOOR RAPIDOGRAPH, INC.

PROPORTIONAL DIVIDERS These are measuring tools capable of transferring measured distance into proportionally reduced or enlarged linear value. They are constructed from two independent, flat but narrow strips of material with sharp, pointed needles attached to the opposite long, narrow space surrounded by the frame. Both sections are joined by the retaining device inserted into the long, narrow opening. A horizontal calibrating line is etched on one of the flat parts of the retaining screws. When a proportional divider is joined together and closed, that calibrated line can be matched with other similar lines etched on one of the flat sections. The proportions are established when the tool is closed and the retaining device is moved along the narrow open space. Naturally lines designating the proportion must be completely aligned with the line on the retaining device. Once alignment is made, the retaining screw is tightened and the proportional divider is ready for use. In order to measure any distance, the divider must be open, forming the letter *X* with arms moving crosswise, easily adaptable for measuring any distance. Because of calibrations of proportions, the letter *X* will have one end narrower and the other wider. Measurements taken with the narrower end will be enlarged on the opposite side of the tool, while anything measured with the more open side of the divider will be automatically reduced by a designated factor marked on the divider. A proportional divider is one of the most essential tools for the scientific illustrator. Depending on how it is calibrated, it makes possible quick reading of ratios in linear form, circles, planes, and solids. In some proportional dividers the rock movement of the retaining screw with the horizontal calibration line will help to attain extremely precise alignment of that calibration line with marked ratios, while in others the sliding movement will be controlled by your own fingertips.

METRIC RULERS The metric measuring system is used by most of the people on earth. Because it is simple, precise, and easy to understand, scientists have adopted the metric system to their needs. For that reason alone, an illustrator should have one metric ruler. The length of the ruler is of no importance, because an illustrator's primary objective will be to obtain a designation of proportion. Such designation is called the *scale*. All scales must be measured in metric and marked respectively. The scale does state that if the length of the drawn line is compared with the drawing of the subject, the viewer will know the exact size of the actual object, because the scale is drawn in proportion to the enlarging or reducing factor used during preparation of the illustration.

Figure 3-3. Proportional divider, 7.9 inch. For division of straight lines into ¾-inch to 10-inch sections, and circles into 3 to 20 sections. Movement of slider is accomplished by a toothed-wheel straightening device. Reproduced by permission of KOH-I-NOOR RAPIDOGRAPH, INC.

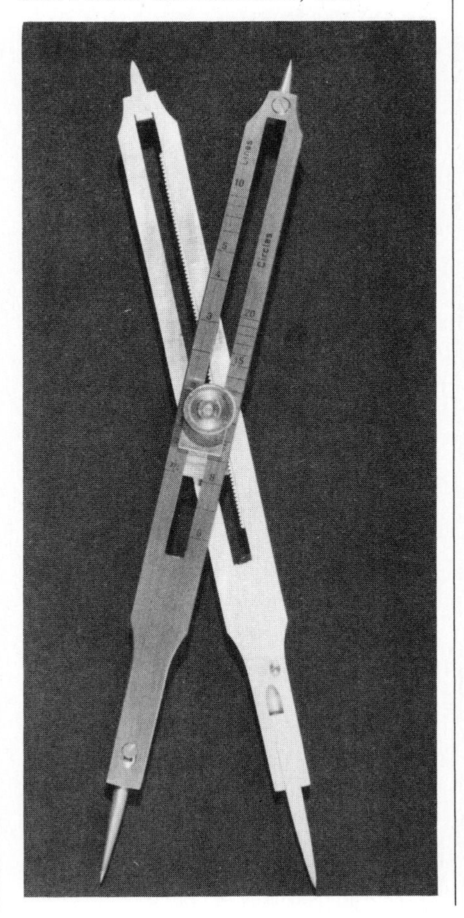

STRING Thin string is an excellent tool when grid coordinates have to be marked on the specimen to be measured. As long as right angles are precisely established between horizontally and vertically stretched pieces of string, proportions can be measured with greater ease and greater definition. Naturally, the size of the specimen will dictate use of the string. Large specimens are very difficult to trace. The camera lucida may not be of any use, especially if it is mounted on the microscope. Large specimens usually are bulky, creating problems when placed on the bed of the viewing camera. The depth of field is affected by optical components that produce proportional discrepancies during tracing. A grid made of stretched string over the specimen solves this problem. In order to speed up the transfer of the image to the paper, a replica of the grid produced by the stretched string should be drawn. When proportions of distances between vertically and horizontally stretched string are reduced on the paper, the reduction of the image will follow. In order to accomplish correctly the production of the grid, some kind of ruler should be used. Designation on the ruler is of no importance because it is the proportions between drawn lines and the stretched string that matter. As a matter of fact, it is possible to make a ruler uniformly marking some kind of distances on its surface. You must make sure that the grid produced by the stretched string will stay in its designated place for a considerable length of time and that it will stay properly stretched. In order to measure the specimen dividers should be used, as a grid by itself will only help in such processes.

TOOLS FOR TRACING

Tracing can result from direct, "face-to-face" confrontation with the specimens or from cleaning up final sketches before actual rendering. The outcome of tracing may be the final product, the finished illustration, or it can be a reduced or enlarged sketch of the subject under scrutiny. It is possible to produce the final version of an illustration using camera lucida without preliminary sketches or measurements. Tracing saves a lot of time and therefore speeds up production. Tracing tissue through which the image is visible is the most commonly used material during preparatory stages and sometimes during rendering. Transferring the approved image to another drawing surface is not necessary. Among tracing tools, graph paper is the most important. Graph paper can be bought or made by the illustrator. As long as proportions are properly matched, enlargement or reduction of the image will be correctly accomplished. Graph paper placed under tracing tissue will serve the same purpose as a viewing camera. Nevertheless, whenever a microscope is used, the drawing tube will be the necessary tool for tracing of the image.

CAMERA LUCIDA The camera lucida is an instrument used as an aid during the process of tracing an actual specimen onto the paper. Among the varieties of camera lucida the most prominent is a device adapted to the microscope and based on the fact that a prism bends the path of light. All that is necessary to accomplish the desired effect is a 45-degree angled prism. When the viewer's eye is correctly positioned over the prism, looking down through its edge to the paper below, it is possible to see two images at once: the image of the object positioned at front of the prism and the paper underneath the prism. Because both objects are visible at the same time, the object in front of the prism appears superimposed on the paper, allowing for tracing of the object's contours. The principle is simple, but a sophisticated construction combining mirrors, focusing lens, and other optical components is necessary when the camera lucida is an integral part of the microscope. It is possible to obtain table models, clamp-on-the-eyepiece versions, and toys built with only a see-through mirror. The choice of the tool depends on the project. Naturally, the more precisely constructed camera lucida, the better it will help you to accomplish more correct tracing of the subject. Some drawing tubes have to be used in conjunction with a microscope while others are freestanding models. As a matter of fact the clamp-on-the-eyepiece type has two principal variations. One allows the viewer to look through the microscope while tracing, while the other projects the image onto the table in a similar fashion to what a pantograph will do. The image can then be traced without looking through the microscope.

LIGHT BOX As the name implies, a light box is a metal, wooden, or plastic box with some source of light inside and the top made of a semitransparent, sturdy sheet of white plastic, usually covered with a sheet of glass. It is one of the most important tools used for tracing and organizing a sketch or for transferring a final sketch to the drawing surface. Light boxes, ranging in size from small portable ones to large tables, are available in any art-supply store. The tool is not hard to make, as all you need is some kind of box, a light fixture to mount fluorescent tubes, an on-and-off switch, and a white semitransparent cover made of Plexiglas. With a light box, guesswork will be eliminated and every sketch can be traced onto the drawing paper with great precision. A sketch will be visible through overlaying paper when the light inside the box is turned on. At times, a light box can be used for observation of a specimen. Being able to illuminate a semitransparent specimen from below may help you in observation and counting of its internal and external structures.

MICROPROJECTOR A microprojector is a microscope turned upside down or mounted in a horizontal position with a stage carrier for slides and a strong light that comes from above the specimen and is capable of projecting an image on the table or wall. Some microprojectors can be used for observation and tracing of live specimens. The instrument can be focused, and various combinations of lenses allow for certain enlargement of the specimen. The horizontal models will have a greater capacity for enlargement because the distance from the objective to the wall is not fixed but flexible. Some table models may have a mirror allowing for projecting the image on the wall. After the specimen is placed on the stage and its image projected it may be beneficial at times to work in a darkened room because the projected image will be more visible. When three-dimensional specimens are placed on the carrier stage, a struggle with the depth of field will follow.

PANTOGRAPH This is a device made of two X-shaped structures similar to a proportional divider. It is used for enlarging and reducing drawings. Ratios for enlarging and reducing are clearly marked, making on-the-surface adjustments easy. A pantograph can be obtained from any large art-supply store. With over twenty ratios to choose from, it will be easy for the illustrator to find the proper one for his needs. First, a correct proportional ratio has to be established and set on the instrument, then the sketch must be placed near a new sheet of paper or tracing tissue. By placing the arm with a suction cup or a securing device on the drawing table, the middle arm and the opposite arm from the suction cup will follow the sketch, reducing or enlarging it, depending on which of those arms is used for "tracing" over the already drawn image. A sturdy hand is necessary for best results, but even if the resulting drawing appears correct, retracing or redrawing should follow, to clean up small inconsistencies and eliminate shaky lines. Although the pantograph was invented and produced for drafting purposes, it is a very versatile tool capable of reproducing a complicated image quickly and clearly. When a viewing camera or photostatting is not available, a pantograph is the only sure way to reduce or enlarge a drawn image.

PROJECTOR A projector is a box with a lens, mirror, and source of light, usually used for projecting sketches or drawings on the wall. It is a simple device but very useful when a large-size original drawing must be produced. A projector will only enlarge. It will not reduce. For that reason it can be used as an observational tool when small, solid, objects are to be studied. Depth of field will again be a problem if the subject under study is very three-dimensional. The image will be projected on the wall and can be easily traced. The clarity of the projected image will depend on the type of specimen. Because of the heat produced by the light inside the projector, be sure that only those specimens that can withstand high temperature without damage are chosen for observation and subsequent tracing. Be sure that none of the specimens are damaged by you or by any of the devices you use for observation and transposition of the image.

TISSUES FOR RUBBINGS When deciphering archaeological and paleontological specimens, at times the only way to obtain a good image of actual structures is through rubbing. A carefully prepared rubbing can reveal more information about the structures than the eye can catch. The best tissues are thin and delicate, such as Kleenex or toilet tissue. After placing a piece of tissue over the specimen, use a finger dipped in ground graphite as a smudging medium to gently rub across the surface of the specimen. Do not expect all structures to appear instantly, as some may be too small to catch smeared graphite the first time. Do not smear anything on the surface of the specimen, and do not use any wet medium, as wetness can destroy the specimen. Water will dissolve clay and will dissolve some of the natural-earth materials that bind parts of the fossil. Rub only on the surface of the tissue covering the whole or a section of the specimen. Repeat the process a number of times to obtain as much information as possible. Tissues can be sprayed gently with fixative, but be careful, because graphite placed on the surface of thin tissue can easily be displaced by too much fixative.

Again make sure when spraying tissue that you remove it from the surface of the specimen and place it on clean illustration board. It is possible to obtain transparencies of rubbings by laminating separate pieces of tissue from both sides. Use thin, glossy laminating tissue and place the rubbing between two pieces of laminating material. Make sure that the adhesive, dull side is facing the rubbing, otherwise a disaster will follow. Use masking tape to secure the laminating tissue to the illustration board before applying heat, otherwise it will curl. A drymounting press or regular iron can be used, providing that a protective sheet of tracing tissue is placed between the source of heat and the laminating material. Transparencies obtained in such a way can be scrutinized on the light box, projected on the wall, or set into slides printed on a blueprinter and used as illustration for lectures. For the illustrator, rubbings made into transparencies are of great importance when designs on pottery sherds or structural complexities of a fossil are hard to decipher. When numerous rubbings are taken from the same area, made into transparencies, and overlapped on the light box, the prominent structures will become more visible. An added benefit is the fact that such transparencies are virtually indestructible.

TOOLS FOR DRAWING AND PAINTING

It is not possible to draw and paint well using poor-quality tools. A bad brush will affect the outcome. The wrong combination of drawing or painting surface with improper paint or pencil will make it impossible to finish an illustration professionally.

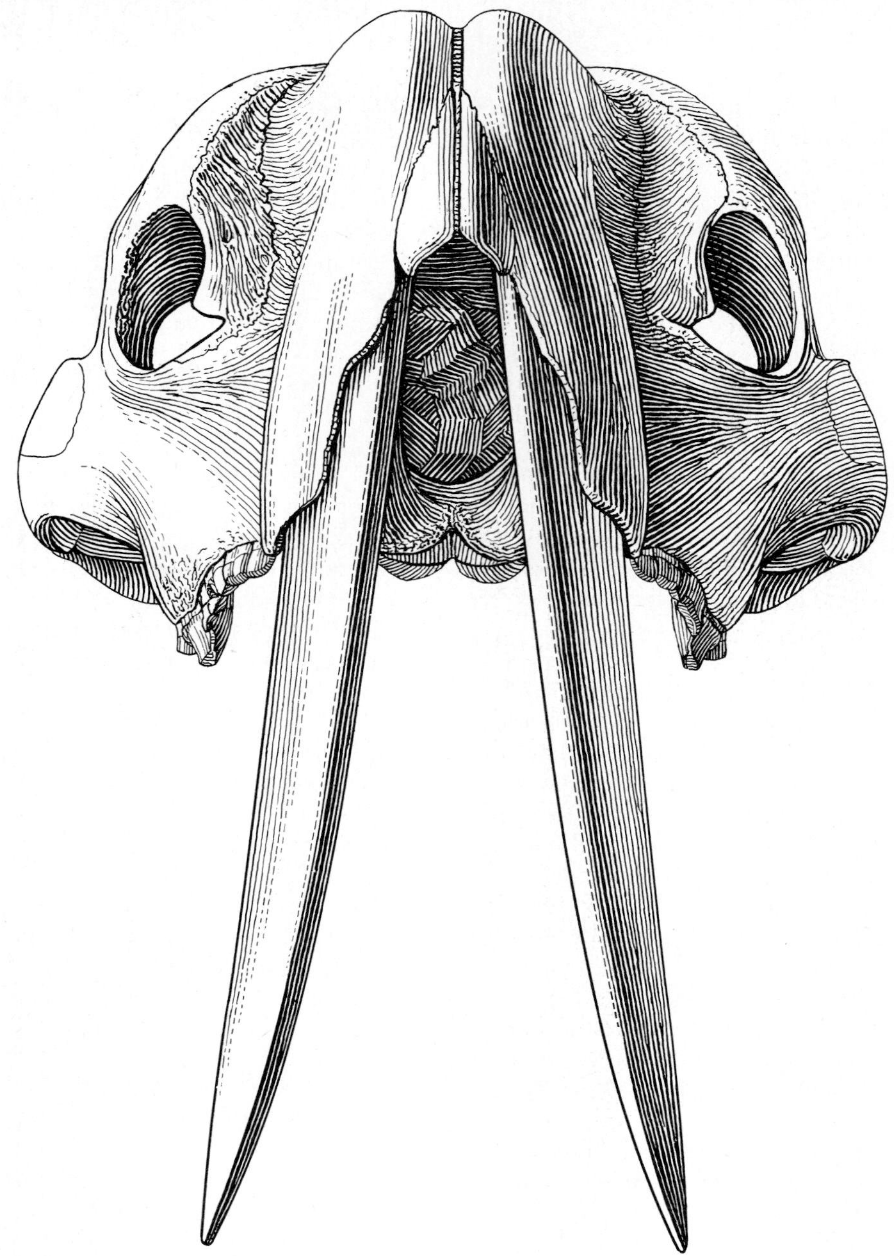

Figure 3-4. *Thylacosmilus atrox,* holotype, (sabre-toothed marsupial). Crow-quill pen on two-ply Strathmore paper. Drawing by Sidney Prentice, Chicago Museum of Natural History (Field Museum of Natural History), Chicago. Published in Elmer S. Riggs, *Transactions of the American Philosophical Society,* New Series, Vol. XXIV, 1934, by The American Philosophical Society, Philadelphia. Drawn in 1931.

Your choice of materials and choice of technique should match the subject. Being able to draw and paint is not enough, as you also have to learn how drawing and painting tools behave. Such expertise can be attained only by practice. In scientific illustration the choice of tools and materials is limited. It is obvious that although oil painting is a very versatile medium, it takes longer to dry then tempera or watercolors. The choice of drawing tools will be dictated by the type of drawings you are expected to produce. Is it possible to do drafting freehand? You know that it is not, therefore a drafting machine will become an everyday companion. The area of research of the institute you may work for will also dictate certain tools of the trade. Because it is impossible to predict in what area we will be working and what type of tools are waiting for us in the future, it is good to have at least a general idea about basic and major aspects of most useful drawing and painting supplies.

AIRBRUSH The airbrush is an instrument that allows the illustrator to spray paint in a more or less controlled fashion. It is built in such a way that a flow of air produced by a compressor will push liquid paint in a preadjusted way through the nozzle of the metal tip, which is equipped with a needle. A container, small or large, is attached to the metal tip of the instrument. In some models the flow of air and the flow of paint can be adjusted and controlled simultaneously while in others two separate controls direct the flow. Because of the two different constrictions, one type is called double-action while the other is termed a single-action airbrush. The difference does not end here, as a variety of sizes of opening are available. The nozzle is connected to the air compressor, with a hose providing flexibility of operation. Airbrushing requires a steady hand as well as preparatory illustration procedures before paint is applied. A variety of materials are used as stencils, depending on the result desired. An airbrush is an excellent tool for producing smooth tones on an illustration as well as overlays for already drawn or painted illustrations. Watercolors or inks can also be used safely. Naturally, paint should not be left in the tool to dry, as an airbrush must be cleaned after each use if its future operation is to be satisfactory.

BRUSHES The brush is the most important of all painting tools. Always be sure that the brush you are buying is of good quality, with a tip that ends in a fine point. The most expensive brushes are not necessarily the best. Examine every brush separately, holding it to the light to see if the hairs are clustered correctly. In scientific illustration only soft-hair brushes will be of use to you. Any so-called oil paint brush which looks like a toothbrush or a broom will not do. The need for delicacy in paint application prohibits the use of such monstrous tools. The size of the brush will depend on the size of the illustration or the usage of the brush. A retouching brush will most likely be of small size, although any good-quality brush will do the job. For general use a number 6 brush is sufficient as long as its tip ends in a point. Very small brushes do not hold much paint, making it difficult to draw long, thin lines, while too large a brush may have too thick a tip, interfering in very detailed rendering. Brushes will be used during drawing as well as painting. A good-quality, soft-hair brush must be used whenever small dust particles are to be removed from the surface of the drawing. Harsh hair will smear gently placed graphite dust, creating smudges. Round-tip as well as flat-tip brushes will be used for application of carbon dust to any drawing surface. The need for a gentle touch will be obvious to you once use of this tool has been experienced.

COMPASSES A compass is a two-pronged metal tool with a sharp needle on the end of one arm and graphite or a pen on the other. Obviously, it is good for drawing circles. Some compasses are able to span a great distance. A beam compass is capable of producing a circle of approximately 40 inches or more. A circle of such diameter will very rarely if ever be a part of your drawing; for that reason, when acquiring a compass settle for a so-called bow compass, and be sure to have a pen included in the package. Another variety of compasses will allow for attachment of a technical pen. Such compasses are easier to handle when ink lines are used because the flow of ink is controlled with greater precision by the mechanism of the technical pen.

CROW-QUILL PENS These are metal pens capable of producing a line of variable thickness. The thickness of the drawn line is controlled by the amount of pressure exerted on the pen by the hand of the illustrator. A great variety of such pens exists on the market, with some having a permanent ink container attached to the drawing nib. Graphos pens are an example of this type. The crow-quill pen should have a pointed tip rather than being a rounded or squared. The roundness or squareness of the tip of the metal pen dictates the basic thickness of the line. Pens used for calligraphy are called lettering pens, while those that have a rounded tip are at times referred to as writing pens. Pens produced for calligraphy are not suitable for drawing purposes because the thickness of the line produced by the pen cannot be controlled easily. The charm of the crow-quill pen lies in the fact that variations of tone are achieved by the density as well as the thickness of the drawn lines. The size of the pens varies as much as the metal alloy from which those pens are made. The stiffness of the pen and its resistance to pressure must coincide with the illustrator's muscular structure, as otherwise too much stress will be exerted on the hand during control of the pen. In other words, a heavy-handed artist will need a stiffer metal alloy than the light-handed illustrator. When hands get tired, craftsmanship suffers. It may be possible not to notice cramping muscles until it is too late. The first sign is an uncomfortable feeling. When using drawing tools other than the crow-quill pen, such a feeling can be dismissed without unpleasant results to the finished product, although you should feel as comfortable as possible at all times while drawing or painting. Make sure that you wash new crow-quill metal pens well before beginning to draw. The outside as well as the inside of a new pen is covered with a thin layer of grease. If this layer of grease is not removed, the ink will not flow, but will build up on the upper section of the pen, prompting you to lose patience and shake the pen. Shaking a crow-quill pen will produce a blob of ink, and naturally you do not want to do this, especially on the drawing's surface. If a pen is stubbornly refusing to draw, it can be prompted to work if the tip is dipped in a small puddle of ink. At times such action will be necessary regardless of the age of the pen. The crow-quill pen should glide with ease over the drawing surface without casting surface irregularities. Any pen that catches the surface while drawing a line should be discarded.

DRAFTING TABLES A drafting table is one with a flat rectangular surface used for drawing. The most important feature of the drafting table is the flatness of the top and the stability of construction. As the name implies, the table is produced for drafting purposes, therefore the top of the table can be tilted to any position comfortable for the illustrator. Some of the tables have pneumatic adjustments activated by various levers allowing the top to be raised or tilted. For drafting purposes this is fine, but when natural-science illustration is being done make sure that the specimen, microscope, illustration, and drawing tools will not slide off the tilted surface of the table. Avoid flimsy constructions—it is better to have a good-size regular, flat, nonadjustable table than a shaky "professional" drafting table. Even a couple of wooden horses with a sheet of plywood across them will give more stability than some of the expensive contraptions. An old-fashioned wooden drafting table is best for the scientific illustrator. Its stability and the level of the top are controlled by locking screws, while at the same time simplicity of construction and relative lightness of the table will allow for guide adaptation to any situation. When you use a drafting machine, the front edge of the table will suffer terribly from the small wheel that supports the vertical arm of the machine. In order to alleviate this problem, build a metal edge into the table. The supporting wheel will have a sturdy surface for its horizontal travel and the metal will protect the edge, not letting the wood chip off. Some of the more sophisticated and complicated drafting tables have reversible tops. One side serves as a drawing surface while the other is a light box, allowing the the illustrator to switch tops according to his or her needs.

DRAFTING MACHINES This is a sophisticated version of the regular T square. A professional drafting machine is attached to the table at the opposite side from the illustrator. It consists of a horizontal bar clamped onto the top of the table and a vertical arm attached to a horizontal bar. The vertical arm resting on the supporting wheel travels across the table, while the horizontal bar remains stationary. The drafting head, containing vertical and horizontal rulers as well as the necessary digital designations, is attached to the vertical arm and travels in a vertical direction. Such a combination allows for up-and-down as well as side-to-side movements, while the drafting head controls circular rotation of the L-shaped attached rulers used for support of the pencil or pen during drawing. With a drafting machine it is possible to choose any angle, any horizontal or vertical position of the drafting head, and lock the desired position in place before proceeding with drawing. Rulers attached to the drafting head have separate adjustments for calibration of the right angle formed by the vertical and horizontal rulers. Such a machine is not needed unless serious drafting is being done or you are drawing a highly precise illustration involving a perspective, a cross section of very complicated fossil, or an explanatory presentation of an engine. Drafting machines are produced in various sizes ranging from very portable through very large models.

ERASERS A good friend of the illustrator, erasers come in various shapes and degrees of hardness. You need a variety of erasers, some for cleaning up and others for smudging. Depending on the materials you use for drawing, erasers that can clean ink, pencil, or rubber cement should be handy. Pencil-type erasers such as those used by secretaries to erase mistakes during typing are excellent for continuous-tone drawings. Stenotrace number 1507 is perfect for smudging graphite, while number 1607, being much coarser, will help you to erase fine, thin lines from already-rendered areas. For best results, the tip of any pencil-type eraser used for such purpose should be cut at a 45-degree angle; the tip of the coarser eraser must be kept clean as otherwise smudges will result. Faber Castell produces similar general-purpose pencil erasers. Gum enclosed in wood gives better stability during use, helping to control the pressure exerted on the eraser. Other pencil-type erasers are a little too flexible and often give under pressure, prompting the illustrator to apply too much force, thereby losing precision. The kneaded eraser is also essential. Being flexible, it allows for shaping small cones with very fine points, helping to remove very minute particles of graphite with a gentle touch. The kneaded eraser is best when a bit of graphite powder is imbedded into its structure to separate the gum base; this makes shaping of the eraser easier and increases stickiness of the gum. Art gum—a large, squashy and bulky eraser—crumbles easily when moved across a drawing surface, removing smudges without great injury to the surface, while all other types of erasers will in one way or another destroy the surface of the paper. Technical erasers used on drafting film are produced to be used in eraser holders or as a block. Although it is possible to wash ink from the drafting film using Q-tips and alcohol, such a procedure may endanger the drawing. In some situations, because of congested lines there is no space to wash certain areas safely. An eraser is the answer in such emergencies. Besides hand-operated old-fashioned erasers, an electric eraser can be quite handy when a large area of the drawing has to be eradicated. An electric eraser is a small hand-held machine with a motor rotating a stick of gum at considerable speed. It should be used only if the drawing surface is thick, as it will remove the upper layer of the paper. It is easy to make a hole in a drawing when not being cautious. Before you use an electric eraser, the drawing should be secured well to the table. The paper will move easily when pressure is applied to the rotating eraser.

GRAPHITE Powdered graphite should be obtained from the most reliable source—the pencil. For continuous-tone drawings it must be ground to a powder and must be of smearing quality. Try various pencils to find out which smear the best. The majority of graphite sticks and graphite contents in pencils have limited smearing capabilities, and from their produced lines it is impossible to tell how well the graphite will smudge. Try to collect a reasonable quantity of graphite on the surface of the illustration board and smear the powder using stomps and the eraser. Use a regular Venus drawing pencil to start. After making a choice, be sure that grinding will not leave sharp particles nor produce extremely fine powder. Ground graphite must have some adhesive qualities to be of use during drawing.

FRENCH CURVES These are flat pieces of plastic curved in various ways used as guides for the tip of the pencil or pen during drawing. French curves come in a great variety of shapes and sizes, packaged in sets or sold separately. The shapes are designed with numbers. Curve number FC 341 is most useful for general work. A scientific illustrator who specializes in natural sciences will rarely need other types of curves. At times, depending on the shapes of curves, names such as Copenhagen ship curves, arc curves, aircraft curves, or Flexi curves are applied. With the exception of Flexi curve, they all have a permanently molded structure. Flexi curve (a flexible ruler) is made of rubberlike plastic reinforced inside with wire, allowing the illustrator to bend the rubberlike structure into various curved shapes. For best results use plastic curves, as a plastic edge will give more stability and smoothness during drawing. All French curves can smear a freshly drawn ink line. Whenever ink is used, be sure that the tip of the pen is pointing slightly away from the guiding edge, otherwise ink may slip under the curve and will be smeared easily. When drawing, do your best to draw a continuous line rather than drawing sections and trying to join them later. It is difficult to accomplish such a task. Joining sections of drawn line usually results in mistakes such as a thicker line in the place where two lines meet. It is easy to bypass completely that crucial point and have two lines that do not meet at all.

Templates are similar to French curves in application to the drawing. They are flat square or rectangular sheets of plastic with definite shapes cut through the plastic which are used for quick transfer of those shapes to the drawing surface. Again, there is a great variety of cut shapes. The templates are organized into categories according to the shapes, such as circles, squares, letters, symbols, triangles, or ellipses cut into their basic structure.

INKS All black liquid inks will be used primarily for drawing. The selection is formidable, but you do not need a great variety of brands or types. Rarely will white or colored inks be used, although when preparing charts or maps, colored inks may be a routine rather than an exception. Ink should be properly matched with the drawing surface. Manufacturers provide explanations or recommendations, which are clearly printed on the bottle. For general use, Pelikan number 17 black ink or Artpen India ink produced by Koh-I-Noor will be sufficient. For special-purpose effects or drawing surfaces that do not easily accept water-soluble inks, consult art-supply catalogs. Besides drawing, you may want to use ink for line and tone separation. Grumbacher Patent Black, number 1471–2, will adhere easily to acetate. Such ink should be applied with a brush rather than a pen, because it will not flow through a technical pen. After an application of Patent Black masking ink, rub dried ink with a bit of Vaseline, spreading on a thin coat of the greasy substance with a soft rag in order to protect the surface. Patent Black is a sticky ink and when not protected with a thin layer of Vaseline may stick to other papers.

PAINTS Paints are composed of ground-up pigment mixed with various bases and used for painting. A variety of brands and types of paints are available. Before starting any project, consider the time involved in paint drying processes. In scientific illustrations watercolor or tempera is the most likely medium, although acrylics, oils, or lacquers may be used. When buying water-base paints, consider the porosity of the surface to which the paint will be applied. Cheap watercolors will have coarsely ground-up pigment, allowing for on-the-surface application of the mixture, while more expensive and very finely ground pigments will have a tendency to penetrate the surface of the paper. There is nothing wrong with using the best watercolors or tempera paints, assuming that each painting is well preplanned and the illustrator will not make many mistakes. Unfortunately, because the painting process is based on making mistakes, noticing and correcting them, finely ground pigments will interfere with the removal of wrongly placed tones. Coarsely ground pigments can be washed out with relative ease from the surface of the paper, while better-quality water-base paints will remain imbedded in the fibers of the paper. The choice of the type of paint is a very personal one; it is closely related to the materials used for a painting surface as well as the technique of application of the paint. To start, use any cheap watercolors, being prepared for controlled opaque application of the pigment rather than impulsive transparent presentation. The bibliography will provide descriptions of a variety of painting techniques, tools, and surfaces.

PALETTE A palette is a place to mix paint. It can be made of any suitable material conforming to the type of paint in use as well as to the tools of application of mixed paint. A large sheet of paper will be needed when trying an airbrush, while a sturdy piece of wood will be more appropriate for oils. For water-base paints, avoid plastics. White plastic will interfere in proper judgment of the proportions existing between water and the pigment. Plastic will make water bead up, not allowing you to realize the viscosity of the mixture, and by this causing a loss of control over the painting. Without knowing for sure the proportion existing between water and pigment, it is impossible not to make lots of misjudgments. For that reason, when painting with water-base pigments, use white porous cardboard. The whiteness of such a surface will let you know the color of the mixture, while porosity will allow for proper judgment of tone.

PENCILS Pencils are used during drawing, sketching, painting, or marking up illustrations for production. Just like paints, the pencil must match the drawing surface as well as the chosen technique. Drafting calls for a different type of pencil from continuous-tone rendering. Smearing of graphite will be a deterrent during drafting, while it is a must for the smooth rendering of any tone. You know that the addition of a greasy substance to pressed graphite will result in a diminished capacity for smearing. The Venus brand of pencils is excellent for application of continuous tone, while Eberhard Faber Ebony is not. Before deciding on any brand, try its qualities on an appropriate drawing surface. Koh-I-Noor carbon pencils are the best for carbon dust on mylar drafting-film drawing, allowing for a much greater range of tones than Wolff carbon pencils. When coquille board is used as a drawing surface, the Wolff product, being harder than Koh-I-Noor, will produce better results. Technical pencils are actually holders of thin lead inserts. Two standard sizes, 0.3 mm, and 0.5 mm, are available. The hardness of leads for those technical pencils ranges from 2H to 2B, with the softest lead being completely sufficient for achieving the darkest of all tones. It will be useful to have a couple of Pentel technical pencils with Pentel leads during the production of most illustrative materials. So-called "regular" pencils should not be discarded in favor of technical pencils, and as long as tips are kept sharpened, results will be satisfactory. Softer lead will produce a thicker line, while harder lead will do the opposite. During sketching and drawing procedures hard leads are necessary, but do not go to extremes. Leads such as 7H will rip the paper or damage the surface before a produced line will be visible. For general use you need a range of hardnesses starting with 2H and ending with 4B. The softest 6B pencils are much too soft or specialized detailed drawing and can be used only as a supply of powdered graphite. During painting, Caran D'Ache or Mongol water-soluble colored pencils will enhance the illustration. Please remember that choice of tools is closely correlated with the subject. A very detailed color rendering will be accomplished better with brush than with water-soluble pencils. During preparation of an illustration for production, a blue nonreproducing pencil should be used to draw basic boundaries of the layout as well as any other designations helpful for positioning separate sections of the plate. Blue will not be noticed when transferred to high-contrast film through the eye of the camera. Actually, any blue pencil will be fine as long as it is not too dark and the needed line is drawn lightly.

REDUCING GLASS This is a hand-held lens capable of reducing a drawn image. A reducing glass will enable you to judge the tonal range of drawing during production or after the finish. It will allow you to see a smaller image of the painting or drawing, helping you to estimate the final appearance of the illustration. It serves the same function as a mirror except it is more handy and portable. Holding a reducing lens in your outstretched hand, look through the lens at a drawing, trying to notice unwelcome congestions or unpleasant disappearance of tone. Remember that the illustration will be greatly reduced when printed, and try to establish the proper reduction of the drawn image and then make your judgment.

RULING PENS Ruling pens are metal-tip pens without a container for ink; they are used for drawing straight lines with the help of a ruler. They are a basic drafting tool needed whenever drafting or drawing charts is to be accomplished properly. Although they are slowly being replaced by technical pens, ruling pens are still an essential drawing tool.

RULERS Everybody knows what rulers are, but high quality rulers are seldom found. The most essential part of the ruler is the edge. A stainless steel straight edge is the best for drawing straight lines. The smoothness of the edge of this ruler cannot be compared with any other. The reason for such a high-quality product is the fact that a stainless steel straight edge does not have any measuring designations etched into its surface. Such designations make the edge of a ruler rough and not really suitable for drawing. A metal ruler is mostly for the professional illustrator. It will be used for measuring, cutting, and drawing. Among rulers with measuring designations, the Gabel 18-inch metal "The Publisher" ruler is most professional. It offers sizes for points, agates, picas, and inches. Naturally, you will also need some kind of metric ruler, but unfortunately professional metric rulers are not produced in the United States.

STOMPS Stomps are tubes of paper that are rolled, pointed at both ends, and well packed; they are used for smudging powdered graphite, carbon dust, or charcoal during drawing. This essential tool comes in a variety of sizes designated by numbers. Stomps must have a very smooth surface, as any roughness will produce accidental irregularities, uncontrolled by the illustrator, within tonal areas. During use, the tips of the stomps will become softer and fuzzy, making precise placement of tone difficult. Naturally, the chosen size will be in appropriate proportion to the size of the drawing. You know that a large number 10 stomp will not be applicable for smaller areas, while the smallest in diameter, number 1, will be too small for larger areas. The sides of the conelike tips should be used during placement of general tone, while the tips are applicable to a particular area. In order to place graphite onto the tip or the end of the stomp, a generous quantity of the powdered substance should be readily available. Taking the stomp in the hand, the end must be rolled into the graphite. Imbedding graphite particles onto the surface of the stomp will be achieved when some pressure is applied during this simple operation. Tips of the stomps can be resharpened by using fine sandpaper and rubbing the pointed end on its surface while rotating the tool. Do not confuse stomps with Tortillons. Tortillons are quite similar and serve the same function, but are cruder, with the edges of the rolled-up paper clearly visible. Tortillons are not suitable for good-quality continuous-tone rendering. The edges of the rolled paper will produce parallel lines rather than a continuous surface. The darkness of the tone is regulated by the quantity of graphite present on the end of the stomp as well as by pressure. A delicate touch will result in delicate tone. For the most subtle of tones, use Q-tips or a bit of cotton attached to a thin wooden stick. Cotton will help you to control the pressure exerted on the tool because it will "give in" more readily than a well-packed stomp. Do not expect that a stomp or a bit of cotton will produce smooth tones instantly. In all situations you, the illustrator, will have to go back into placed tone and remove darker spots with a kneaded eraser or add more tone using the very tip of the stomp or the pencil.

STYLUS A stylus is a tool similar to a pencil, with thin round metal protruding from one side, ending with a small round metal ball. It is used for holding down paper during pen-and-ink rendering to prevent unwanted "bounce" of the drawing surface. A stylus can easily be made of wood. The tip must be rounded and smooth, allowing for gentleness when touching and holding the drawing surface. Besides holding down papers, the stylus is used for rubbing instant-type transfer letters, causing friction and producing enough heat to melt underlying wax, which causes the printed letter to stick on the desired surface.

TECHNICAL PENS Technical pens are sophisticated drafting and drawing pens with a controlled ink flow consisting of three basic parts: the point, ink container, and holder. The construction of technical pens will differ slightly depending on manufacturer. Koh-I-Noor leads the rest in quality. Technical pens are designed to produce a uniform straight line with the help of a ruler. When drawing freehanded, the pen can be held at about a 45-degree angle for the best stipples. Koh-I-Noor is the only manufacturer with a pen that can maintain uniformity and stability of line when held at an angle in relation to the drawing surface. Such pens have interchangeable points ranging from number 7, which produces a very

Figure 3-5. Technical pens. Rapidograph® technical pens are provided with steel or jewel points producing metric line widths for engineering, architectural, or other uses (stipple, line in scientific illustration) application. Those metric lines are compatible with standard American drawing sizes and microfilm practices. Line widths are clearly distinguishable and are in geometric progression of the square root of two to facilitate the "half-scale blow back" technique common in microfilming. Reproduced by permision of KOH-I-NOOR RAPIDOGRAPH, INC.

wide 2.00 mm line, to the extremely fine number 6×0, capable of a 0.1-mm line. For our purpose, sizes number 3×0, 2×0, and 1 are most suitable. When preparing a technical pen for drawing, do the following:

1. Learn the construction of the pen by taking it apart, point included, and putting it back together. Be careful, as 50 percent of artists are heavy-handed.
2. Pay special attention to the point in the pen. The point is the conveyor of ink and must work properly in order to achieve a continuous flow of ink. Do not tighten the point too tightly or leave too much looseness. A loose point will wobble,

Figure 3-6. Dual-designated (standard metric) drawing points for manual pen. This is the heart of the pen, the most fragile part, which must be properly taken care of. Reproduced by permission of KOH-I-NOOR RAPIDOGRAPH, INC.

interfering in drawing. Make sure that when the pen is shaken you hear a "click, click" noise. Such a noise indicates that the internal part of the point is loose and not frozen in one place. Never bend the wire attached to that part. Without that thin wire, ink will flow continuously through the opening of the pen tip.

3. When everything is correct, fill the ink container with ink. Be sure that it is not overfilled.
4. Take a piece of paper towel and wrap it around the tip of the pen before the ink container is placed back onto the holder. Ink may flow through the front of the pen when refilling it because of internal pressure.

The heart of the technical pen is the point, referred to as the nib. This consists of an enclosure in which a delicate metal pin attached to a small cylindrical weight is placed. The pin is locked in place by a security device consisting of a plastic cup inserted into or screwed on the back portion of the point. Older pens may have a *U*-shaped metal pin. Ink cannot be kept in the pen for long periods of time. Once the point is frozen by dried-up ink, it may be hard or impossible to remove it from the pen. It must be soaked in water for a considerable time so the ink within the point and around the holder into which the point is inserted will dissolve. Adding alcohol to the water will speed up that process. It is possible to soak the whole pen in a stronger solution of alcohol and water, but be careful: Some plastic holders can dissolve, making the pen not fit for use. In order to remove the inside part of the point, remove that security device. Make sure that you will not break anything. Hold the point vertically, and then remove the cup or the *U*-shaped pin. Then, grasping the plastic cylindrical object securely with your fingers, move it up. Do not drop that part. The thin wire attached to the small cylindrical plastic weight can be damaged very easily. Once that wire is bent, the point will have to be replaced, as it would take considerable patience and experience to straighten it. Straightening is possible, but remember that if you do not succeed, the only result may be the waste of a couple of hours. In order to place

Figure 3-7. Proper way of holding Rapidograph® pen with point larger than 0 during production of line illustration. Pens with smaller points than 1 can be held at approximately a 30–40-degree angle. Such an angular position facilitates faster production of dots (stipples) without loss of control. Reproduced by permission of KOH-I-NOOR RAPIDOGRAPH, INC.

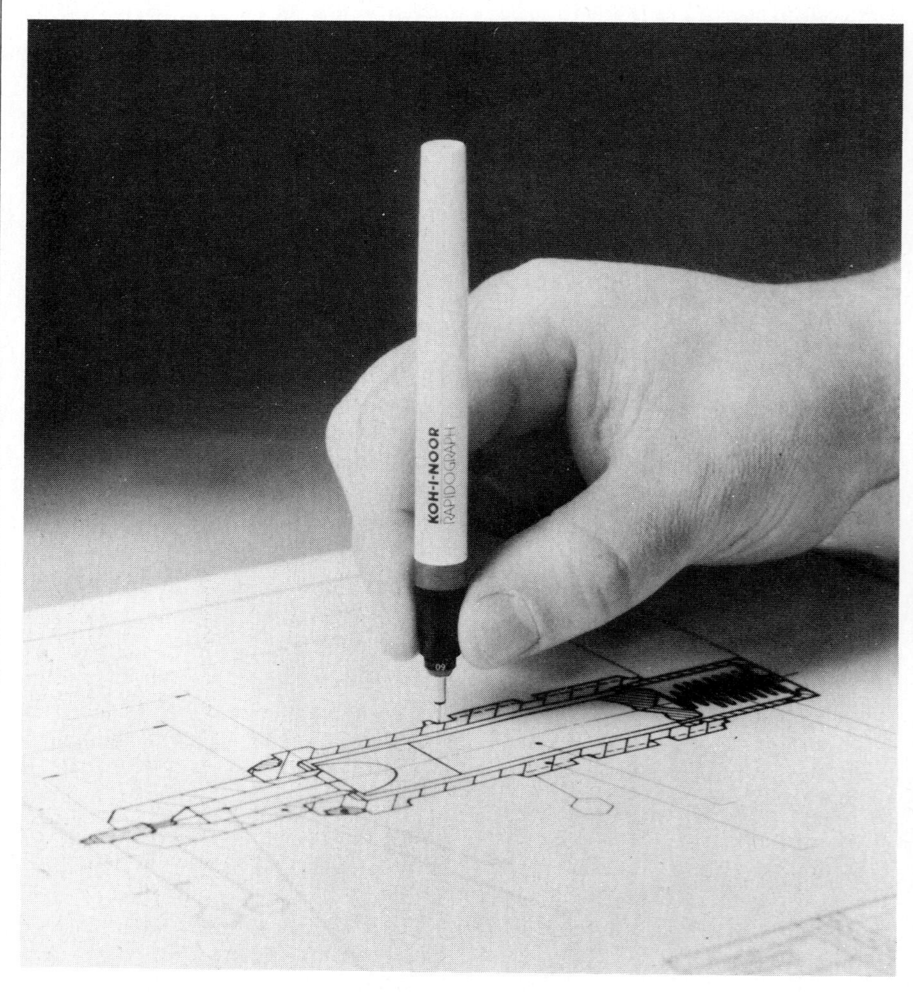

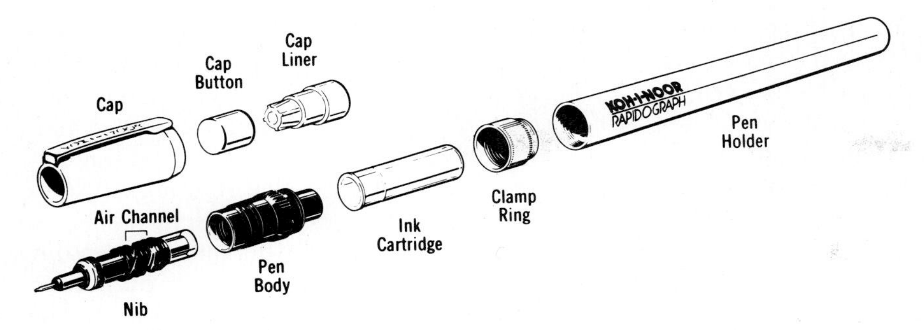

Figure 3-8. Reduced view of Rapidograph® technical pen. Reproduced by permission of KOH-I-NOOR RAPIDOGRAPH, INC.

wire back into the enclosure, gently lower the wire without applying any force. Once the wire is partially in the enclosure, tilt the point slightly and tap the side a couple of times. Slight tapping will make the wire fall into its designated space. If you are unsuccessful the first time, repeat the process. The wire will eventually fall into the opening. Be sure the security cup is placed over the end of the point. If that cup is missing, the tip will move too much inside the pen, allowing for a greater flow of the ink through the point and resulting in blobs and ink drops. For best results, clean the pen every three to four days. Such a procedure will preserve your drawing tool for years. When choosing ink, be sure that the thickness of the ink will not affect the drawn line. Very thick ink will not flow easily through the narrow opening, while diluted ink may result in a light line visible to the eye but not feasible for reproduction.

DRAWING SURFACES

The quality of the chosen surface for drawing or painting will affect the final outcome. More important still is the proper match of surface with technique and subject. The availability of materials as well as instructions accompanying the project will affect your personal choice. What is good or what is the best has been debated by artists for a long time. Be sure that whatever is chosen will help the project and the production of the illustration. When thinking about the future, do not imagine yourself as the one who has made the most important statement of present times but rather consider the quality of reproduction. Leave the rest to fate. Naturally 100 percent rag papers will have a better chance to survive than not-so-pure and not-so-classical surfaces; nevertheless, use of such papers may not be feasible for a variety of reasons. Remember that not everything which is expensive is the best. Cheap tracing paper will handle ink better than expensive vellum just because it is of better quality and does not have graphite repellent built into its surface. On the other hand, good-quality drafting film will cost you money. Make yourself a sampler of papers, carefully indicating manufacturer, type of paper, and other pertinent information, and in your spare time test the surface with drawing tools to see and to learn. Papers or other drawing surfaces can be affected by improper storage or humidity and will vary from time to time. This means that Strathmore drawing one-ply, smooth-surface paper may not handle graphite in a uniform way. At one time the graphite will smear on its surface with great ease, while a new batch of the same paper will not allow for the same. Never—and that means *never*—roll papers for transportation or storage. There is nothing worse than a damaged surface. A crease, break, or scratch on the surface of the paper will control the outcome of your drawing. Such damage will attract more graphite than undamaged areas, forcing you to change tone during drawing to conceal that dark spot that so suddenly appeared from nowhere. Make sure that the salesperson will place the paper flat, with cardboard backing for support, before wrapping the whole package. When a sheet of paper is too large, ask the clerk to cut the sheet into smaller convenient sections. Insist on proper packaging because your drawing, your customer, and your own pocket are at stake. When buying paper, be sure you get the type of paper you asked for. It is unfortunate but nevertheless true that lots of sales personnel will not know or care what they are selling to you. Check before buying, as there is a world of difference between such similar-appearing papers like Strathmore drawing or Strathmore bristol. In order to know for sure how to recognize papers, write to manufacturers and ask for samples as well as descriptions and specifications.

What should you look for when acquiring papers, drafting films, or acetates? Basically check for an undamaged surface, grease spots, and other people's fingerprints. You should not have to pay for all of this, as you can inflict enough unwanted marks on the drawing surface for free.

ACETATES Acetates are transparent or frosted sheets or rolls of plastic used for a drawing surface during preparation of overlays, as well as for tone and line separation. Acetates differ considerably in quality but most importantly, they differ in thickness. During production of a scientific illustration, the flexibility of very thin acetate is a deterrent rather than a help. Maps, charts, and overlays consisting of line work are the best suitable subjects for the use of acetates. It is possible to draw on any acetate provided that its surface has been carefully washed with soap and a very soft rag. The surface cannot be scratched, yet the thin layer of grease must be removed before ink will adhere to the surface. When untreated acetate is used for drawing, the India ink will not adhere well—it will tend to crack and peel off as well as being easily removed by friction. Specially produced inks will adhere strongly to the surface of untreated, or as it is sometimes called, unprepared acetate, although technical pens with small-size points may not be the best tools to use for drawing. Thick ink will not flow evenly through the narrow channel of the small point and may clog the pen permanently. For drawing it is possible to thin such thick ink with water, but be cautious and check the density of blackness before starting rendering. For best results use drafting film rather than acetate as a drawing surface. All instant transfer-type or instant-transfer tone will adhere very well to the surface of acetates, helping you with overlays. When necessary, instant-transfers type can easily be removed using masking tape rather than an X-Acto knife or eraser. Masking tape will pick up unwanted letters and other designations without damaging the surface. When tracing a large surface full of images, such as inscriptions carved into a stone wall, use 0.0015-gauge prepared acetate in rolls. The thinness of the drawing surface will give a certain flexibility that allows for proper placement of the acetate sheet over the subject. The prepared surface of the acetate will take ink quite well. Drafting film is not suitable for epigraphic work because it is not transparent enough and as such, will not allow the illustrator to see and therefore record all details of carved inscriptions. Acetates are attached to the illustration board with masking tape, although it is possible to dry-mount clear acetate permanently. In rare situations clear acetate can be affixed to the surface of illustration board, using spray adhesive and white removable dry-mounting tissue.

DRAFTING FILMS Drafting films are similar to acetate but have a semitransparent polyester drawing surface. Drafting films are manufactured with many variations in the quality of the surface. The best drafting films are frosted on both sides of the sheet; therefore it is relatively easy to separate a good drawing surface from one that will give you problems. Drafting films are very stable and will not tear easily. Cutting is the only way to trim excess material. When choosing drafting film as a drawing surface, be sure you are aware of the purpose of the film. Some are produced to accept pencil, while others are for ink. Whenever you are not sure what to use, write to the manufacturers for specifications. The best drafting film for ink drawing is Cronaflex U-C Tracing Film, produced by DuPont. The film comes in two thicknesses for your convenience. You must remember that drafting films are categorized into numerous groups and produced for other purposes than just drawing, such as cartographic reproductions or high-contrast, high-speed camera-produced negatives. When used for drawing purposes, drafting film speeds up the production of illustration. The process of transferring a sketch to a drawing surface is omitted because the sketch is visible through the film. Mistakes can be removed by washing the surface with alcohol or water or by scratching. A sharp X-Acto blade will remove the ink from the film's surface. Excessive damage to the surface will expose the underlying, unfrosted main body of the film, interfering with the adhesion of ink. Any pen can be used for drawing. A drawn ink line will be slightly thicker than a similar line placed onto the surface of paper or illustration board. Whenever using pens, especially technical pens, do not press into the surface of the drafting film too strongly, because the tip of the pen may scratch the surface, removing small bits of material. Such particles will cause uneven lines and ink smears. Remember that the surface is washable and the drawing ink acts in the same way as any other liquid, such as water, upon the upper layer of the film. For that reason restrain yourself from drawing overly wet lines and wait until the surface is dry. Whenever a mistake or a fingerprint is washed out, be sure that section of drafting film is completely dry before drawing, otherwise serious damage will occur. During drawing, small particles from the surface are removed by the tip of the pen, so technical pens will have a tendency to clog more often. Have a paper towel or soft rag near you and wipe the tip of the pen during drawing. Besides pen and ink, other drawing techniques can be very effectively used: carbon dust combined with pen-and-ink lines, graphite dust applied with stomps or with pencil, or a wash using a brush.

PAPERS FOR DRAWING The choice of paper will be dictated by the technique of rendering. Pencil, pen and ink, or wash will all require different surfaces. In scientific illustration, three basic types will be of use: paper for continuous-tone rendering, usually one or two ply with a smooth surface; sturdy board for wash drawings, also with a smooth surface; and heavily textured by relatively thin board used for imitation of stippling effect. The smoothness of the drawing surface is of great importance. Without interference from texture, rendering of details can be achieved with more control. Strathmore drawing paper is most commonly used for pencil, tone illustration. Make sure you dif-

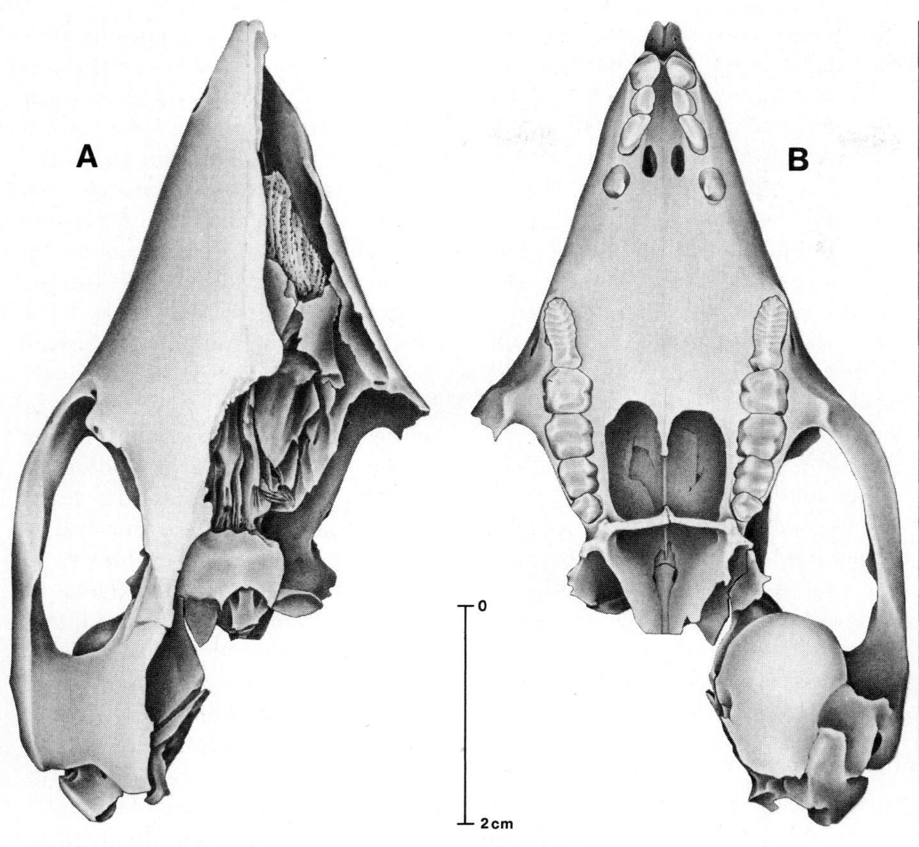

Figure 3-9. *Bettongia penicillata*, recent, FMNH 35325, skull in dorsal and ventral view. Pencil on Strathmore drawing paper. Drawing by Zbigniew T. Jastrzębski, Field Museum of Natural History, Chicago. Published in William D. Turnbull, "The Mammalian Fauna of Madura Cave, Western Australia, Part VI," *Fieldiana*, Geology, New Series No. 14, (1984) by Field Museum of Natural History, Chicago. Drawn in 1983.

ferentiate betwen Strathmore drawing paper and Strathmore bristol drawing paper, because these surfaces handle applied tone quite differently. Both papers appear almost alike, with the bristol being slightly yellower than the drawing paper. The best way to make sure what kind of paper you have is to look at the embedded factory mark. Strathmore drawing paper will have a stamp stating "Strathmore Draw Either Side," while the other is marked as "Strathmore Bristol." The first will accept smeared graphite very easily, allowing for deep velvety tones, while bristol has to be used mostly for pencil work, remaining "warmer" through the application of tone.

Boards, whether illustration boards or some other type of compressed white material, must have a smooth or almost smooth surface and should be porous. Water should be easily accepted by the surface and should dry rather quickly. The surface itself cannot be washable, because dissolved and washed particles of fiber will interfere with rendering. Cold-press illustration board will handle wash technique quite well. Absorption of liquid will be in the medium range, allowing for control of wet areas. Whenever working in wash, remember that tonal differentiations should be applied slowly, always starting from the lightest and proceeding to the darkest.

The textured surface, especially coquille board, is best used for quickly prepared, imitation tone drawings to be reproduced in line. Coquille board is produced with differently textured surfaces ranging from uniformly organized textures imitating dots through smaller textures that are very irregular in their bump shapes. Carbon pencil is the best drawing medium for the desired effect. Because of the textured surface, coquille board cannot be used for very detailed work, as the raised portions of the board will interfere with precision. Taking advantage of bumps present on the surface, the density of tone is controlled by covering more or less of those raised bumps with carbon pencil without unnecessarily filling up the spaces between the separate raised bits of the surface. The final effect is a drawing similar in appearance to a photograph reproduced in a newspaper. Scratch board, as the name implies, is used for scratching the surface, thereby producing or correcting a drawn image. Two types of scratch board are available: white board and black board both with smooth surfaces. White scratch board is better suited to our purpose, because its white surface allows for transfer of a detailed drawing. Tone is produced by the density of black lines drawn in crow quill or other type of pen. Larger areas of dense tone are first covered with black ink, using a brush, and afterward scratched in order to produce a black image on white background. Using scratch board, it is possible to draw an illustration technically similar to the good-quality engraving of the past century.

PAPERS FOR PAINTING Among many types of papers used for painting, the best will be the ones that allow for implementing corrections during the process of painting as well as after illustration has been finished. It is hard to control pigment on a very absorbent surface.

The pigment will quickly penetrate the fibers of the paper, not allowing possibly necessary removal. In order to prejudge the quality of the paper, you will have to consider its porosity and smoothness of surface. Depending on the quantity of details, the size of the illustration, and pigments used for painting, appropriate stability of the surface of the paper must be the choice. Two-ply Strathmore drawing or bristol paper will be the best choice. In place of Strathmore paper, any other not-very-porous and water-absorbent paper will suffice. This leaves out all watercolor papers unless you do have very good control of painting techniques as well as are able to judge tonal differentiations without mistake. Naturally, the surface will have to be chosen according to the technique of painting. All tempera, watercolor, and acrylics will adhere to the unsized surface of the paper, while penetrating oil-base paints will require different surfaces or specially prepared sizing compounds before applications. Speed of production is a serious consideration; therefore, any slow-drying technique is out of the question. A lack of storage space will restrict the size and bulkiness of the surface chosen for painting. At all times, the surface must remain flat, during the process of painting as well as afterward. A color illustration is for reproduction; therefore, it must be prepared on material that will not interfere with platemaking procedures.

TRACING PAPERS These are semi-transparent tissues or papers which will be used primarily for sketching procedures. The most beneficial feature of tracing tissue is its transparency and ability to hold pencil lines. For our purpose, expensive vellum tracing tissues are not very good, although they are excellent for architectural or other drafting-type renderings. For the most part, cheap tracing tissue will suffice quite well. The medium-weight number 84 tracing paper is just about right in thickness as well as in its characteristic good acceptance of pencil and ink. It serves as an appropriate surface for pen-and-ink rendering, although it will not accept well continuous tones drawn with pencil. For pen-and-ink rendering, never use tracing tissues which do not have a slick, smooth surface. Ink will feather out during drawing and will produce irregular-quality lines. Try to avoid colored tracing tissues, as readability of a drawn image suffers when color is introduced. Also try to avoid good-quality tracing papers for pen-and-ink renderings, as many types of such tissues are produced for pencil line drawing and have surfaces treated in such a way that smudging of graphite is retarded. The treated surface of tracing paper will not accept ink as well as a less-expensive, not-so-specialized surface. Ink may bead up in an unpleasant manner, prohibiting proper rendering.

TOOLS FOR PREPARATION OF ILLUSTRATION FOR PRINTING

Preparation of illustrative material can range from so-called packaging and marking the reduction or enlargement factor for platemaking through retouching, changing, and fixing up the image itself. Depending on many unpredictable factors—such as a change in the direction of research, last-minute decisions pertaining to the presentation of coded information represented by type, or changes in quantity of illustrations per plate—the procedure, materials, and tools will vary and overlap. Specialized materials designed for only one purpose can and should be adapted to the needs of the situation. For example, instant transfer type can easily be used for retouching missing sections of pen-and-ink line drawing, although its primary function lies in the area of typography. During the preparation of illustrations for reproduction, such extreme measures are usually not necessary, but nevertheless some last-minute emergencies will occur. Materials such as instant tones or instant colors can be used during the preparation of illustrations, but because in scientific illustration their application is limited to finishing touches rather than primary areas of production of illustration, we may consider them equal to other preparatory materials. Final preparational procedures should be taken seriously. There is no need to present a finished drawing that is sloppy or contains misinformation. All instructions to the platemaker should be easily legible. The layout of the plate must be clearly marked, and some kind of protective material should be affixed to the surface of the illustration. This final stage is as important as the rendering of the illustration.

BLUE PENCIL Nonphotographic blue pencil is used for marking the parameters of a layout. It is used to draw a designated area on the paper or illustration board, assuring clear visibility to the artist but being invisible to camera. Any nonphotographic blue pencil is good; as a matter of fact, any light blue pencil can be used provided that drawn lines will not be embedded into the surface but will appear as a readable but gentle designation. Layout in scientific illustration does have a slightly different meaning, as it does not involve putting numerous pages together, but rather deals with the placement of separate images within one page or one designated area. Blue pencil can and should be used for designations of reduction or enlargement. In such a situation, the common method of drawing a straight line someplace away from the image is employed. The straight

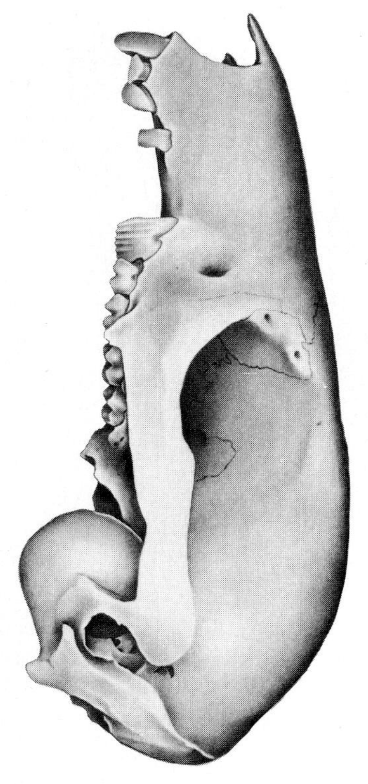

Figure 3-10. *Bettongia penicillata,* recent, FMNH 35330, left lateral view. Pencil on Strathmore drawing paper. Drawing by Zbigniew T. Jastrzębski, Field Museum of Natural History, Chicago. Published in William D. Turnbull, "The Mammalian Fauna of Madura Cave, Western Australia, Part VI," *Fieldiana,* Geology, New Series No. 14 (1984) by Field Museum of Natural History, Chicago. Drawn in 1983.

line will end with arrows pointing to vertical bows designating the beginning and end of the image. Handwritten notes stating R (reduction) or E (enlargement) in percentages, inches, or centimeters will be clearly marked along the line. The R or E should be enclosed in a circle, making it more noticeable to the platemaker or photographer.

BURNISHERS Burnishers come in various shapes but are used for one purpose only: to burnish things down. A burnisher is a piece of plastic used to apply pressure in a smooth fashion to the surface of an illustration. During dry-mounting procedures or during transferring instant letters or tones, such pressure will smooth out a given area as well as dissolve the backing by friction. At all times, make sure that some kind of protective sheet is placed on the surface of illustration while burnishing. It is very easy to destroy a rendering when using a burnisher directly on the surface of the illustration. Some burnishers are shaped similar to a pencil and are used for transferring instant transfer type. The same burnisher is the stylus used for holding paper down during stippling procedures. When a plastic burnisher is not available, use a strip of illustration board bent in half. Such a handmade burnisher is as good as a plastic one. When burnishing larger areas, make sure that the edges of the illustration will not be caught by the tool. Do your best to apply pressure using a motion directed from the center of the illustration toward the edges.

COVER-STOCK PAPER Cover-stock paper is any one-ply, usually black paper used as a protection for an illustration and as a packaging device. The black paper affixed to the surface of an illustration board and covering the whole area makes the final product more appealing to the receiver's eye. At the same time it offers very sturdy protection for the surface. When planning to use black cover-stock paper, try to obtain black photographic masking tape. Black tape and black paper will look better than when regular masking tape is used. Actually, any paper can be used for such a purpose as long as it is sturdy enough to withstand heavy use and is of a not-too-colorful nature. Packaging is important, as it projects your image. Associations produced by beautiful but unusual colors may not be the most beneficial for you.

DRY-MOUNTING PRESSES A variety of heat-producing presses are available on the market, with prices ranging according to the size of the press. A dry-mounting press is a contraption made basically of two parts—the stationary bed and the lid, which can be opened and closed. The upper section, the lid, produces the heat, while the lower section, the bed, will hold all materials used during dry-mounting procedures. Temperature is adjusted by a regulator placed on the upper section of the press. Some presses have lights for the purpose of telling you when to remove the illustration. If you can, do not pay attention to blinkers but count the seconds by yourself. It is safer to be personally in control of the time span rather than to rely on built-in times. Dry-mounting presses are to be used only for dry-mounting tissues that require heat. If a press is not available, a hand-held iron will do the job. Be sure that the iron contains no water, as steam will destroy the illustration. At all times, place a protective tissue between the surface of the iron or the press and the illustration. When applying heat and pressure by hand, start from the center of the illustration and proceed toward the edges in order to remove the air trapped under the dry-mounting tissue and under the illustration. Be sure that the temperature is set on the lowest designation, as too much heat will destroy the adhering capabilities of removable dry-mounting tissue. The best temperature for safe dry-mounting is approximately 180°. Naturally, if thicker paper or if permanent dry-mounting tissue is being used, the temperature and the time should increase.

DRY-MOUNTING ADHESIVES Dry-mounting tissues and other adhesives are controlled by temperature or pressure. Seal-removable and seal-permanent dry-mounting tissue both require heat and pressure. Twin-Tak or other similar adhesives do not. Besides dry-mounting materials, other adhesives are frequently used. Wax-

Figure 3-11. Strange-tailed Tyrant, *Alectrurus risora*, and Sharp-tailed Tyrant, *Culcivora caudacuta*. Painting by Dr. John W. Fitzpatrick, Field Museum of Natural History, Chicago. Published in Melvin A. Traylor, Jr., and John W. Fitzpatrick, *A Survey of the Tyrant Flycatchers, The Living Bird*, Nineteenth Annual, 1980–81, Cornell Laboratory of Ornithology, 1982. Painted in 1981.

ing is one of the newer methods. It is based on a procedure of application of hot wax to the backing of an illustration or a proof sheet with printed type. Hot wax can be applied with a hand roller or with a waxing machine. Once the back of the illustration is covered with wax, then through application of pressure combined with the heat produced by the friction of a burnisher, the illustration will stick to the desired surface. The great plus of this system is removability of material with ease of repositioning. Waxing is a very appropriate medium for positioning type but it is not feasible for illustrations. The main reason is that it is impossible to completely protect the front of the illustration from stray pieces of hot wax. If a bit of hot wax attaches itself to the surface of an illustration, it is impossible to remove without injuring the illustration itself. Bits of wax will attract dust and dirt, damaging the rendering. Twin-Tak adhesive, although very sticky and capable of destroying an illustration totally, can be removed with benzene. The safest way to attach an illustration to a board is with masking or other tape. Naturally, when pieces of tape are used during preparation of a plate containing halftone illustrations, all separate images must be masked, otherwise the pieces of tape will appear in the reproduction. Rubber cement will destroy an illustration after a few months. The illustration will yellow and will not look good. Yellowing may interfere with proper platemaking, again resulting in an unpleasant final printed image. All dry-mounting procedures are only temporary measures designed to hold a drawn image in position until the plate used for printing is made. In most cases, dry-mounting procedures destroy illustrations for posterity. If you are interested in preserving your work for the future, do not use dry-mounting techniques—although unfortunately, during production of illustrations, you will be forced to employ such techniques at least occasionally.

FIXATIVES Fixatives serve a two-fold function. The main purpose is to prevent a sketch or finished drawing from smearing; the secondary function is an enhancement of color or tone. For preservation of surface, use available spray, although you should test your fixative for possible staining of the papers used for final rendering. The Krylon products are better than other brands and are capable of drying in all circumstances without leaving a mark on the sketch or rendering. The acrylic spray coat produced by Krylon Crystal-Clear Fixative, when used on continuous-tone drawings, will change the density of tonal range for the better, making them deeper and stronger. The same applies to color illustration: The color will appear stronger and fresher after being sprayed with Krylon Crystal-Clear. Such changes will be noticeable by the camera, resulting in better reproduction.

ILLUSTRATION BOARDS These are boards made of compressed cardboard with surfaces covered by various coatings, used for drawing, painting, or as a backing for illustrations. Hot-press boards are used for pen-and-ink drawings because of their very smooth-plated finish, while cold-press illustration board is much more suitable for brushwork. All illustration boards can be used for a variety of purposes. The most universal is Crescent illustration board. Its sturdy surface and the very good quality paper mounted as the upper layer of the board make this illustration board best suitable for layout and positioning. Mistakes, fingerprints, and smears can be removed easily by erasing without causing serious injury to the board's surface. All boards come in various thicknesses designated just like papers, by number. One-ply board is the thinnest and most commonly used. Usually the surface of the board is presented in three finishes: smooth, medium, and coarse.

INSTANT COLORS Instant colors are transparent or semitransparent colors transferred from a preprinted sheet by cutting pieces and applying pressure, producing friction. Instant colors are printed on a thin film with waxy backing and come attached to an underlying protective sheet. In order to use instant colors, a section of printed thin film has to be cut out without injury to the backing sheet; after peeling, the cut-out piece should be placed in the desired position on the artwork. Gentle rubbing with your finger will produce enough friction to slightly melt the underlying wax, causing adherence to the surface. Then, using an X-Acto knife, a desired portion must be cut out and unwanted parts thrown away. A burnisher should be used for application of more pressure, thereby producing more heat by friction. After the procedure is completed, the instant tone will stay in its designated position. The scientific illustrator will use instant colors rarely, usually for preparation of artwork to be used as an explanatory aid during a scientist's lectures. It is possible to implement use of instant colors for scientific publication, providing that the subject will agree with the flatness of color as well as agreeing that the printing will be in color. Letraset Letrafilm Color System is one of the best on the market. It offers a wide variety of colors, halftones as well as solids, all in heat-resistant form. The heat-resistant, pressure-sensitive sheets of color will require, depending on the choice of color, four-color separation, making the Letraset Color System not very feasible for budget printing. The Pantone Color Selector may be a better choice for selection and designation of colors. Each color is designated by number, making for easier communication with the platemaker. All that is necessary is the delineation with blue line of the space to be covered by a chosen color. The blue line can be put directly on the artwork, but an overlap should be marked with the written number of the chosen color. Heat-resistant instant transfer color will not shrink or shift during dry-mounting procedures, and the heat from a dry-mounting press will not affect the size of cut-out areas. Be careful when using acetate as a surface for application of instant color. Removing unwanted sections from the surface of acetate may be difficult. Depending on the shelf age of instant color and the quality of the acetate, problems can range from minor inconvenience to a total waste of materials and effort. Besides instant colors in sheets, a selection of colored tapes of various widths is also available. Tapes are produced in opaque, matte, glossy, or transparent form and come in various patterns. The use of tapes is as rare as instant colors, confined to diagrammatical representation and restricted by problems of color separation.

INSTANT LETTERING Instant lettering is a pressure-sensitive, preprinted letter-transfer system operating on the principle of heat caused by friction. Sheets of various styles of letters, numbers, and symbols, printed in black or white, are used for quick dry transfer of printed material. A variety of brands as well as typefaces are available, some of them heat-resistant. Each letter is easy to position in the appropriate place and easy to transfer by application of pressure. Among the wide variety of products available, only some are of use in our profession. In order to judge the usefulness of a given typeface or symbol, readability must be considered. This leaves out a large quantity of very beautifully designed typefaces. Any typeface similar to Helvetica is simple enough to be readable, thereby making it of use for making numerical values of scales or applying designations to illustrations. Again, Letraset products are the most reliable. Letraset transfer type will not smear, will not dissolve under fixative, and will remain relatively stable under heat produced by a dry-mounting press or sunlight. The protective sheet supplied with instant type should be used to protect the backing of the printed sheet both during storage and work. Avoid placing the back of transfer type directly on the drawing table. Dirt will quickly accumulate and damage the undersurface, making the sheet of type unusable. Whenever a mistake in position or spelling has been made, use masking tape to remove the letter. Do not scrape the drawing surface unless it is absolutely necessary. Rubber-cement erasers also will remove unwanted particles from an illustration. During use of transfer type, be sure to check carefully the surface of an illustration or overlay for accidental transfer of portions of other letters. Such accidents occur frequently and must be corrected as otherwise they will appear in print. The size of letters or numbers should be chosen in proportion to the size of your finished illustration, so they do not appear too small or too large after reduction and printing. Avoid Geotype Instant Letters, because fixative will dissolve them, destroying your finished product. The blue backing sheet supplied with Letraset products can be used for Twin-Tak dry-mounting tissue, as it will not stick to adhesive. The sheet is a perfect protection for illustrations sprayed with fixative and to be dry-mounted using heat. Again, the sheet will not stick to the surface; thus it can protect a sprayed film of acrylic from melting under heat. While transferring large letters, it is proper to use the *prerelease technique*. The letter or symbol should be pressed by rubbing without removing the protective sheet from underneath. Afterward, when letter appears slightly gray, it is placed in position and rubbed again, accomplishing a smooth transfer. If this is not done, then there is a possibility of breaking the thin printed film during transferring.

INSTANT TONES This is a system similar to instant color, but available only in black and white. A variety of halftone screens as well as other patterns are available and will be used rather frequently by the scientific illustrator. Instant tone is a combination of thin printed film with an adhesive, waxy back side attached to an underlying protective sheet. It must be cut out and peeled off from the protective sheet before being placed in the appropriate position on the artwork. After cleaning-up procedures, consisting of finishing cuts and removal of unwanted parts, the instant tone must be burnished to the surface of the illustration.

MASKING TAPE Masking tape refers to rolls of paper or plastic produced in various widths with adhesive on one or both sides. The most commonly used tape is the all-purpose opaque masking tape. Another variety of all-purpose masking tape is the drafting tape, which differs from others by the smoothness of its surface and quality of its adhesiveness. It is irrelevant what type of masking tape you use as long as the tape will stick and will not damage the drawing surface. Photographers' masking tape is black and will be used mostly for holding down cover stock. Transparent cellophane tape is helpful when small drawings must be arranged into a large plate. Naturally fingerprints will not appear on cellophane, and if an arranged plate contains drawings executed in continuous tone, the separation of line and tone will occur. Without the separation, all pieces of tape will appear on the finished printed image unless the whole plate is to be reproduced in line only. Scotch double-coated tape number 666 can be used effectively whenever drafting film must be attached to the surface of the illustration board. In such a situation the protective white plastic liner should remain in its position and the tape is to be used as singularly coated tape. The whiteness of plastic lines adds to the quality of presentation. Otherwise this particular tape is of no great use to us. Whenever pulling a piece of tape from the edge of a paper, be sure the procedure starts from the surface of the paper and proceeds toward the edge. When the reverse is done, there is a possibility that paper may be torn in an unwanted manner.

PHOTOSTATS Photostatting is a photographic process that involves making a print of a subject using paper plates, resulting in instant developing of the positive. Photostats are of importance because it is possible to obtain better-quality camera-ready photoprint than when photographs are used. When continuous-tone paper is used for tone drawings, details and tonal range will be captured regardless of reduction. The quality of a Photostat will depend on the time used for exposure. Because of this and because of difficulties involved in leaving the process as well as access to the Photostat camera, it is better to ask a professional Photostat company for the service. A stippled line drawing will be nicely translated into a tone illustration when the continuous-tone process is employed. Again, such an image must be masked from the background for appropriate presentation of the drawing after it is printed. Even high-contrast Photostats produce better results then regular photography. In sciences it is customary to submit an 8½-by-11-inch glossy photograph of the illustration for publication. When properly executed and properly washed of developing liquid, a Photostat will contain more of the drawn image as well as lasting for a long time.

RUBYLITH OR AMBERLITH These are masking films used for tone and line separation. They consist of a sheet of acetate covered with a thin layer of red semitransparent plastic coating, used as an overlay on an illustration. They are an important tool for separating background from drawn image. The image has to be cut out exactly and the background peeled off, leaving a red coating on the surface of the drawn area. The red color is seen by the camera's eye as black. Therefore the first negative, using high-contrast film, will "remove" the background, leaving a white area where the tone drawing should have been. The second exposure with tone-sensitive film and Rubylith mask removed, shot through a screen, will capture the essence of the drawing, including the background. When both negatives are overlaid and properly registered, the resulting combined image appears as a negative of the drawing with a solidly black background. When exposed to the plate covered with photosensitive emulsion, a reverse will occur. Combined negatives will be transformed into positives. The background will be white with the image remaining intact, but translated into "tone" with the help of the screen. Masking should be executed carefully and with precision. Imperfections will affect the final outcome. Line, such as the scale and type, should be covered with a separate sheet of Rubylith and generally masked out. There is no need to cut all separate letters because the first high-contrast shot will differentiate between pure black and light tone. The line mask and tone mask must be separated by a very thin space; they cannot overlap. An overlap of line and tone mask will produce unpleasant results. Illustrations with a dark, strong-tonal unwanted background should have a separate overlay for all designations in line.

TACKING IRONS These are small, hand-held irons used for tacking an illustration to the illustration board before a dry-mounting procedure involving heat. Before an illustration is dry-mounted it must be attached to the surface of the board. When dry-mounting tissue is placed under the illustration, touch the center of the illustration through a protective sheet with a hot tacking iron. Attach the drawing only in one place, thereby avoiding the possible formation of air bubbles. At all times be sure the surface of the illustration is protected from dirt that has accumulated on the surface of the tacking iron.

GENERAL-PURPOSE TOOLS

Storage cabinets, flat-top drawers, lamps, and chairs are essential parts of any studio. During working hours you must be comfortable and should have relatively easy access to files and accessories. The general belief that drafting tables are the best does not have any ground in scientific illustration. Remember that drafting tables are designed especially for drafting, not for holding microscopes or jars containing liquid. This does not mean that you should not acquire a good-quality drafting table, as long as that table will remain in a horizontal position to properly support the necessary equipment as well as your arms during any drawing procedures. General-purpose tools are to be designated for a variety of uses. Storage space should have areas which can be locked, preventing accidental rummaging by others. Many specimens will have to be carefully protected from possible damage. Drawers used for storage of papers and finished illustrations should be as dust-free as possible, at the same time offering sturdy, flat support for materials. Lamps are best attached to the side of the drawing table, giving enough illumination to see clearly without interfering with movements of tools, specimens, and your own arms. All of this does not mean that in order to have a professionally set-up working area the most esoteric and specially designed components are required. Using logic, you will be able to acquire and organize furniture and other basic necessary objects. It is possible to obtain one of the best and sturdiest tables using two wooden horses and a sheet of thick plywood. Such a contraption is easy to move from place to place and costs only a fraction of what you would pay for a drafting table. If by any chance a drafting table is desired or needed for practical reasons, acquire an old-fashioned wooden model. Naturally when your budget will permit, feel free to spend money on things like a Bieffe Duolite drafting table, but in truth, there will not be much use for the drawing board and light box built into one unit. On the other hand, avoid all "stowaway" drawing tables simply because of their bad construction. Within a short period such a table will fall apart. In scientific illustration there is no need for designer-type furniture, although a good chair is a must. A wooden chair with a pillow placed for convenient seating is as good as any. Unfortunately, when working with a microscope attached to the drawing table, the distance from a regular chair to the oculars of the microscope is not adjustable. The peculiarities of that distance combined with the length of your own body, especially length of your neck, may prompt you to invest a sizable sum of money in an adjustable chair. Of all chairs, including drafting chairs, I have found that the Bieffe secretarial chair designated as BR5/P is the best. It offers sturdy support as well as the possibility of adjustment in height in crucial areas not attainable by other adjustable chairs. A place to draw and a place to sit should be complemented by a pair of Luxo LS-1 or Ledu C-10-D pair of lamps. The fluorescent and incandescent bulbs will provide a good-quality illumination of the drawing board. Both lamps should be used simultaneously, throwing light from two directions. Besides the light box, pencil sharpeners are a must during production of illustration. The electric pencil sharpener by Panasonic will last forever; besides sharpening pencils, it will also sharpen pencil-type erasers. A hand-operated technical pencil sharpener will be useful for a variety of technical pencils. An added benefit from owning a hand-operated Faber number 42 lead pointer or a Pierce Power Pointer is ground-up graphite. Remember to keep that graphite for reuse during pencil tone renderings. Naturally you can use an X-Acto knife for sharpening pencils, but it will be better if you use a supply of number 11 blades for cutting papers, Zipatone, or masking. Do your best to avoid things such as pencil and paper catchers designed to catch instruments and paper or drawing-board covers offering sponginess rather than a hard surface. Do not cut anything directly on the wooden surface of the drawing board, but use an old piece of illustration board to protect the new drawing table. Try to be simple and think about usefulness rather than designs of the drawing area. A simple, uncluttered area is better for production and therefore will help you reach success.

CHAPTER FOUR

Production of Illustration

DIVERSIFICATION of subjects does require differentiation of approaches during the production of illustrations. Methods of observation will vary as much as the techniques used for rendering a visual image. Within certain boundaries, scientists' and illustrators' personal preferences for continuity of undertaken steps may help or deter the quality of the final product. In all situations the requirements imposed by the need for presentation of certain information will influence the production procedures. Besides procedural steps imposed by the researcher upon scientific illustrator, the illustrator himself should have a preplanned and detailed organization of operation. The illustrator will have to appraise the situation and, acting accordingly with needs, should remove some procedures from the general plan of activity. Usually, during the production of illustration, only three basic steps are considered by involved parties. The more or less precisely described and explained initial introduction to the projects is followed by some kind of sketch, later rendered in the desired technique. Most in-between steps are not considered at all nor thought necessary to implement by the illustrator. Usually, the steps and procedures following the final rendering and needed for high-quality reproduction are deemed unnecessary by scientists. In both cases the cause is ignorance of preparational activities during the process of drawing and platemaking as well as printing techniques. From the illustrator's point of view, meticulous adherence to the steps involved in production of correctly presented visual material is a must. He or she is the one who controls most of the pictorial aspect of the information. He or she is the one who must organize the procedures of decoding the subject and drawing. He or she is the one who will have to inform the scientist about possibilities of control of printing quality, and he or she is the one who must implement necessary corrections. For those reasons the illustrator should do his or her best to adhere meticulously to the steps involved in the production of a correctly presented subject. The scientist and the illustrator must communicate during preparation and must work as a team. Their cooperative steps are as follows:

1. *Research and selection of a specimen by the scientist.* In the usual situation the selection of a specimen for research is conducted by the scientist, although in some cases such a procedure can be left in the hands of a competent illustrator. Naturally, such an illustrator must know the subject of research as well as the scientist. He or she must be completely trusted by the scientist and while personally involved in the selection of the specimen will be in constant communication with the researcher, needing his approval before drawing can start.

2. *Receiving the project by illustrator with an explanation from researcher.* All information regarding aspects of positioning as well as all informational inquiries will have to be cleared. Questions will have to be asked by both parties and the answers given. It is irrelevant who will talk the most, although in most cases it will be the scientist doing his best to introduce the subject to the illustrator. Anything that helps

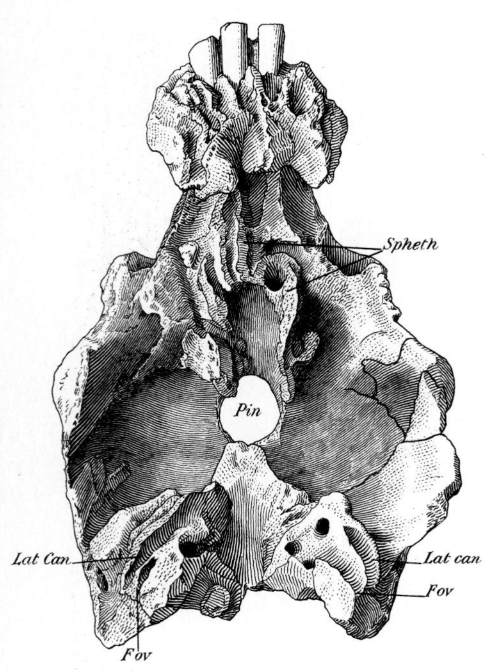

Figure 4-1. Ventral view of skull of *Diadectes* with basircranium and palate removed. Crow-quill pen on three-ply Strathmore paper. Drawing by John Conrad Hansen, Chicago Museum of Natural History (Field Museum of Natural History), Chicago. Published in Everett Claire Olson, "The Family Diadectidae and Its Bearing on the Classification of Reptiles," *Fieldiana*, Geology, 11, no. 1, 1947, by Chicago Natural History Museum (Field Museum of Natural History), Chicago. Drawn in 1942.

mutual understanding is desirable. Quick sketches and explanatory notes may be necessary. The illustrator should be alert and should know that now is the time for picking the scientist's brain.

3. *Observing a particular specimen and making a decision pertaining to positioning*. Before jumping into the project and quickly positioning the specimen, the illustrator must investigate the problems on his own. Observation is of extreme importance. Once the specimen is in its designated position it may be impossible to see all a specimen's structures. Lack of observation of structures can lead to very basic mistakes in drawing. The visible image of the specimen can be slightly distorted or obscured by used optics or a reflection of lights. The structure is important, but in order to know the structure the illustrator will have to spend a moment and just look.

4. *Technicalities of positioning*. Nothing should be damaged during handling and positioning. Taking time is beneficial for the project as well as for the specimen. The illustrator will have to handle the specimen with care, making sure that the subject is aligned horizontally and vertically as it was discussed with the scientist. In most cases, the specimen will remain in its designated position throughout the rest of the drawing procedures.

5. *Beginning of sketching procedures*. Techniques of transposing the image of the specimen will vary depending on the size of the specimen and the availability of tools used for observation and measuring. Large specimens can be sketched using the grid technique or their features can be traced with built devices or a variety of other devices available on the market. The quantity of initial sketches will depend on the complexity of the subject and the skill of the illustrator. It is better to have more information than less, therefore the illustrator should not be self-conscious and should try to obtain or squeeze all information onto one piece of paper. In case of any problems, the scientist should be consulted immediately. If meeting with the scientist is impossible, then questions should be clearly written on preparatory material.

6. *Clarification and organization of initially obtained visual data represented in the final sketch*. The final sketch is a clean version of the previous work. Clarity of the presented information will help in mutual understanding and will enhance communication between the illustrator and the scientist. The procedures of organization of previously obtained information will vary depending on the subject or the solved problems to be presented through rendering. Such clarification and organization can be a simple tracing or can be an involved and time-consuming procedure.

7. *Approval of final sketch by researcher followed by implementation of changes and, if necessary, redrawing of final sketch*. The final sketch will have to be approved by the scientist before the illustrator can proceed any further. The approval of the final sketch is an important step because it gives the scientist an opportunity to see the semifinished version of final rendering. All comments, written or otherwise, presented to the illustrator will help him to reach appropriate decisions during rendering of the illustration. At this point all seemingly final questions should be resolved.

8. *Decision pertaining to size of the finished drawing undertaken by researcher and illustrator*. The size of the printing area or other factors dictating the size of the rendering should be discussed. If necessary, the final sketch can be reduced by tracing on a reducing-enlarging viewing camera (Lazy Lucy) or through photostatting to the appropriate size. During the discussion the illustrator should obtain information pertaining to the width or length of the printing area. Such information may be difficult to obtain because the scientist himself may not know where and by whom his paper will be published.

9. *Transfer of the final sketch to the drawing surface*. Depending on the type of drawing surface used for the rendering, the transfer of the final sketch may or may not have to take place. When transfer is necessary, caution has to be exercised during the procedure. The illustrator must make sure that *all* pertinent data presented on the final sketch will reach the surface of

drawing paper in an unchanged state. The exactness will depend on the methods used and the illustrator's patience.

10. *Rendering of illustration*. During rendering, the illustrator cannot let his mind relax and cannot "improve" the subject's image. Constant referral to the specimen is a must in order to finish the illustration properly. The problems of quality must be overcome. Lines will have to be drawn clearly and the tonal changes kept under control. The illustrator cannot forget that the final product will be used for printing and not for display. Changes of tone occurring after the reduction of finished image will have to be anticipated, and the readability of the drawn image after the printing of the illustration must be considered.

11. *Preparation for printing*. All necessary steps leading to proper reproduction will have to be undertaken. Reduction factor, tone and line separation, as well as instructions how (line, halftone, finescreen) illustration should be printed are indicated at the lower portion of the illustration. If necessary, the drawing can be attached to the illustration board, allowing extra space for necessary instructions for the platemaker.

12. *Printing*. Unfortunately, printing procedures are difficult to control by involved parties. Nevertheless, a request for printed proof of illustration should be a standard procedure. The platemaker (photoengraving company) issues at all times three printed proofs from each plate. Proofs are for approval of the quality of the image; they are sent to the publisher. The scientist should receive one printed proof, should be able to write his comments about the quality of the image, and should not hesitate to reject an improper-appearing print. The best reproduction of a drawn image will be attained if the original illustration is used for making the printing plate. When use of the original drawing is not feasible, great care must be exerted in capturing the drawn image through photography or photostatting. The illustration should not be drastically reduced or enlarged. Such abrupt changes in size will destroy most of the subtleties of tone. Rephotographing such an image during platemaking procedures will create more problems. Throughout the whole process, cooperation between the scientist and the illustrator is a must. The project will not be finished properly if the illustrator resents seeking information and will not show incentive. It may be necessary to go to the library and measure the size of a printed page from a particular publication, and it may be necessary to discuss problems with the scientist in a convincing manner.

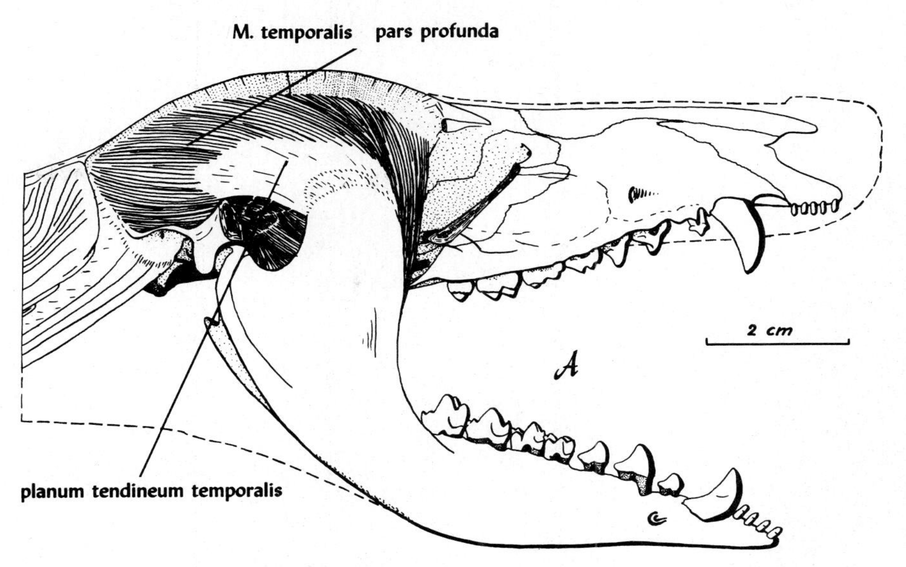

Figure 4-2. *Didelphis*, Virginia opossum. Pen and ink on Strathmore two-ply bristol paper. Drawing by Dr. William D. Turnbull, Field Museum of Natural History, Chicago. Published in William D. Turnbull, "Mammalian Masticatory Apparatus," *Fieldiana*, Geology, no. 18 (1970) by Field Museum of Natural History, Chicago. Drawn in 1963.

MEETINGS WITH THE SCIENTIST

Depending on your position—staff artist or free-lance illustrator—your initial meeting with the scientist will have a different beginning.

For a free-lance artist the communication will originate with the telephone call from the artist to the scientist. It is hoped that the conversation will lead to an initial meeting during which the portfolio will be presented. If the portfolio is satisfactory and if the scientist is in need of an illustrator, the rest of the relationship will be the same as for the staff artist. The staff artist will start with receipt of an official requisition. The requisition will be written by the scientist, signed by him and countersigned by the chairman of the department. It will state the date of issue as well as the deadline for the project. It will describe the project in a more or less detailed way. The description may be elaborate or may be very short. For example:

> Dr. ______________, Department of Geology,
>
> 10 pen and ink drawings:

or

> Dr. ______________, Department of Zoology
>
> 1 illustration, left lateral entire view of holotype of new species of *Vinciguerria* from Northern Red

Sea, left lateral view of head (enlarged) emphasizing dentition of upper and lower jaws. I plan to have finished manuscript by the end of November and need the drawing prior to submission;

or

Dr. ______________, Department Zoology/Mammals

date of submission, date of deadline, name of publication (*Mammalian Species*), instructions: please furnish occlusal and lateral view drawings of the skull and mandible of 60698 *Caluromysiops*, comments: see attached sample for finished products.

After receiving a requisition or free-lance project, a series of discussions will have to take place. The illustrator is the one who should contact the scientist. It is good manners and good business practice. You are working for him, not the other way around. Make sure that the time and place are convenient to the scientist. First, let the scientist introduce the material. Listen to the explanations and form questions in your mind. Do you understand what he is talking about? If not, do not feel bad; he is the expert in the field. Do not hesitate to ask questions. Remember that you will have to know what you are drawing. Ask about terminology. If this is too complicated, ask specific but simple questions: "Sir, is it this bump in the upper-left corner you are talking about? Do you mean this darker patch located on the center of the specimen? What do you mean by lateral position? Sir, if I understand correctly, you want me to draw this object from the top?" Make sure that you ask him about the rendering technique. If he has a favored one and this technique is suitable for the subject, you have no problems. If you do judge a technique as unsuitable, try suggesting another. If this does not work, you are stuck, as you must use his technique and come up with very good results.

Do not be afraid to say, "Excuse me, but for clarification I will draw a quick sketch and get your approval as to the position." Draw one yourself and ask the scientist to draw one. All scientists are able to draw a general sketch of materials they are familiar with. Be prepared to take notes. When everything is clear or seems to be explained, the meeting is over. Do your best to let the scientist end the meeting. You are not a guest; you are working for him. Your illustration is of importance to him.

After the initial sketch has been completed, make sure the scientist checks everything. Stay in touch with him. Do not make him look for you. Be available. The second important meeting will consist of his checking your progress. If during work you have a question, do not run madly after the scientist. He has his own work to do. Judge the situation and find out by yourself if you really cannot proceed further without raising issues. If you cannot; then do ask. When the final sketch is approved and all corrections implemented, your rendering will start. Make sure that the scientist will see your drawing in progress. It is important for him to know what is happening, and for you his approval will be reassuring. Mistakes or misunderstandings can be corrected instantly. Usually when a scientific illustrator has worked with a scientist for a long time constant meetings are not necessary, although it is good to have an extra eye watching over your shoulder. Naturally, when the project is finished, you will have to inform the scientist. Contact him first. Do not wait for him to call you. If you want to be a professional, behave as professionals do. The number of meetings necessary will vary depending on your experience, your employment status, and the location of your drawing desk within the compounds of the institution. In some situations you will work next to the scientist, shoulder to shoulder, so to speak. Such arrangements can be considered one long meeting, as the expert will be on the premises at all times and will make frequent comments. You will be able to ask questions as the work progresses. Through initial observation, sketching, and rendering, the scientist and you will become a stronger team. All in all, the following pattern should remain constant:

1. Initial meeting consisting of receiving the project, explanations, and questions
2. Working sketch or sketches presented for comments
3. Finished sketch or sketches presented for approval
4. Finished drawing and whole project delivered

It is important for you to remember that in some cases, when free-lancing, it will be impossible to work in your home studio. The scientist may be reluctant to give specimens to a third party. The specimens are fragile, can be easily damaged, or could be lost. Who can predict what will happen when specimens will leave the premises of the laboratory? It does not necessarily mean that the scientist does not trust you, although this also may be the case, but accidents may happen. You have to realize that for the scientist the objects you draw are basic to his research. No objects, no research, to put it bluntly. In such situations you will be asked to work in the laboratory next to him.

Receiving the project

Your initial introduction to the problem always calls for quick thinking. You should be courteous but at the same time assertive. Make sure you understand everything. Do not think that later, somehow, you will figure things out. If you attempt to solve the problem by yourself you will waste a lot of your own and the

Figure 4-3. Side view of head of *Ailuropoda*, showing pattern of vibrissae and hair slope. Crow quill on two-ply white cardboard. Drawing by John J. Janacek, Chicago Museum of Natural History (Field Museum of Natural History), Chicago. Published in D. Dwight Davis, "The Giant Panda, A Morphological Study of Evolutionary Mechanisms," *Fieldiana*, Zoology, Memoirs, Vol. 3, 1964, by Chicago Museum of Natural History (Field Museum of Natural History), Chicago. Drawn in 1941.

scientist's time, without producing a desirable illustration. Ask questions. In order to ask questions, you must be very observant. As the specimen and the context of illustration are explained to you, observe the subject and try to notice peculiarities which later may become problems. It all depends on the type of specimen and complexities of the drawing. It also depends on the scientist's ability to explain. Remember that what is obvious for him may not be easy for you to grasp. The fact of the matter is that for the scientist, some things may be so obvious that he may not even bother to explain or inform you about them. Do you really know which of those pencil or ballpoint pen lines on his sketch are valid? Can you read his handwriting? Can you spell Latin words? Do not hesitate to ask for an explanation of his vocabulary. You do have to know what he is talking about. Ask for permission to draw quick sketches during the meeting, and be prepared to hear a variety of comments. Have your own materials. Be sure to have a small pad of tracing paper and pencil. If a project is difficult, ask for additional information—somebody else's drawing, a photograph, more specimens. Above all, be a diplomat. Do not grab his specimens and look at them because you are interested. Ask for permission. Do not look through his books while waiting for him. The scientist's office may appear disorganized, but actually it is not. The specimens on his table are placed in a certain order, known to him. The books are for his reference, not for your pleasure. If you will be using his microscope, let him introduce you to the machine. You may be familiar or fascinated with it, but restrain yourself. Microscopes are expensive and in all probability, the scientist went through a lot of administrative pain before he received his. Do not break it. If you have wandering hands, keep them in your pockets and pinch yourself a lot. There is nothing worse than a stranger roaming through the laboratory.

Initial meeting, you say?

EXAMPLE 1. You are in the office of a mammalogist. Your assignment is to draw coloration patterns of South American monkeys. The scientist is very busy and brisk; he tells you that his previous illustrators could not draw. He hints that if you try hard there may be a possibility that something will become of you—naturally, that is, if you know how to draw. He states that you can pick any technique you want, but the scratch-board is the best. He shows you around. You like what you see, as you studied some biology. Should you impress him with your knowledge? If you really know biology then yes, but do not

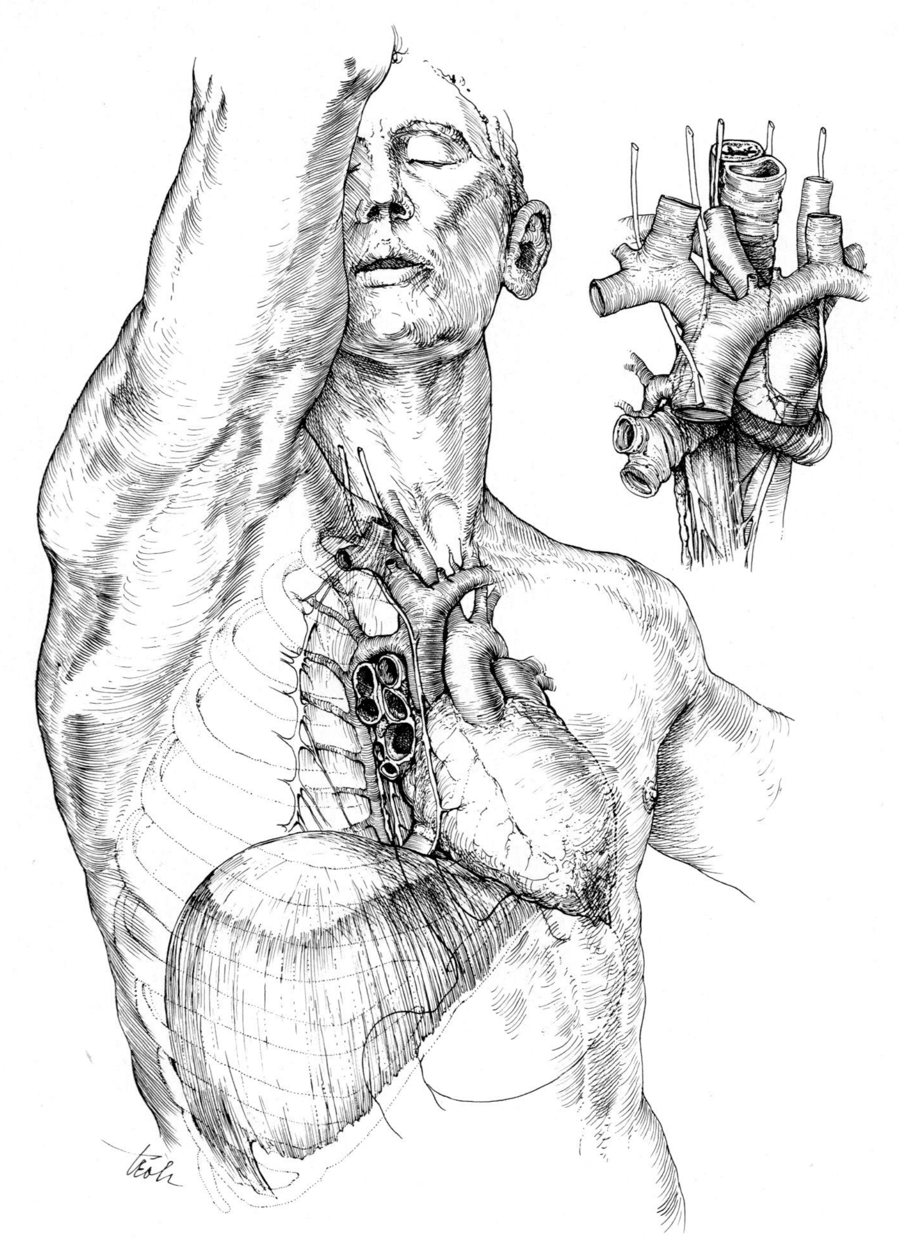

Figure 4-4. Orientation of the right lung: an anatomical orientation to a series of surgical procedures in the thoracic region. Pen (crow quill) and ink on Linekote board. Drawing by Alfred Teoli, Biocommunication Arts, University of Illinois at Chicago, Health Science Center, Chicago. To be published in Edward Beattie, M.D., and Steven Economou, M.D., *An Atlas of Advanced Surgical Techniques* by W. B. Saunders Co., Philadelphia. Drawn in 1983. In preparation.

get into scientific arguments. You are there to draw. The technique of drawing already has been stipulated. You do not have much choice but to use the scratch-board technique. Anything else will in all probability turn into failure. The scientist gives you a specimen and tells you to draw it, pointing to areas of interest. Try to ask specific questions. Ask for examples of drawings. Quickly compare the specimens with drawings of other subjects. Ask for a Xerox copy of the drawing he thinks is best. If the drawing is good, then spend some of your own time trying to learn that technique. It will be beneficial to you and to the scientist. Do not force new things immediately, even if you see a need for change. Truthfully speaking, you do not know why he prefers the other drawings. Is it his personal taste? Maybe the clarity of presentation of information? Perhaps to keep all illustrations for the publication within some boundaries of uniformity?

EXAMPLE 2. The computer is making noises and the tables are piled up with printouts. The scientist needs charts and diagrams. He has plotted out all the information and is giving his sketches to you for pen-and-ink rendering. Pay attention to his explanations, not to the computer terminal. Keep your hands off the buttons, even if you consider yourself a computer expert. Ask about the thickness of the black line. Assure the scientist that his sketches will not be destroyed by accident or by "purposeful" direct rendering on the sketches. Again ask about spelling. His handwriting may be legible, but the typographical symbols are probably new to you. Ask if he wants to see his sketches redrawn in pencil. Suggest appropriate materials such as drafting film to draw the finished version. Ask about feature reduction of the illustrations. Usually graphs and charts are reduced considerably for reproduction. Inquire about the possibility of typesetting all captions. Make sure that the text will be presented legibly.

EXAMPLE 3. The scientist's project consists of about 100 drawings. He gives you a list of all of the specimens. Naturally, all the specimens are numbered, therefore his list consists of columns of numbers, each with an appropriate description of position. The specimens are small bones, teeth, and parts of jaws, neatly placed in small containers. Instead of taking the boxes and the list and disappearing into your corner, make sure that each specimen is explained fully. Make sure that a quick sketch of each specimen will explain to you the exactness of the position. Draw quickly and efficiently, marking vertical and horizontal positions with arrows. Make sure notes of structures will help you to identify the top or bottom of each specimen. Use plain language. Make sure that you will know which groups of specimens will be placed together on each page of the publication. If something is not completely clear, ask questions. When the scientist is too busy to discuss all the specimens at one time, ask for a description of the first batch. Maybe some of the specimens will be drawn from two, three, or more sides. Make sure that this is clear. Ask about modification. Remember that finished illustrations of specimens drawn from more than one side will have to be aligned after drawings are made. Ask if preliminary, well-defined sketches have to be developed. If specimens are small, ask about the microscope. Ask where is your space for work. Ask about the supplies for drawing and necessary materials for positioning specimens.

Setting up the specimen

Setting up is an initial, very important stage in the production of the illustration. The specimen must be positioned correctly if it is to be drawn correctly. Such a statement applies to drawings of a specific view, especially when small subjects are in question. We can divide all subjects into four general categories. The first will consist of very large objects, mainly of an architectural nature. It is impossible to position an irrigation canal stretching for tens or hundreds of miles. Also, it is impossible to position the skeleton of a dinosaur. In such a situation the illustrator must position himself to the subject rather than having subject positioned in relation to him.

The second group will consist of much smaller objects, but still too large or just not feasible for positioning. The drawing of a weapon such as a harpoon may call for a specific view, but again the illustrator may not be able to position the objects with that specific view facing him because of lack of space or lack of equipment allowing him to trace the outline. Specimens such as baskets will be in this category. Deciphering weaving patterns calls for close scrutiny but not for precise positioning of the specimen. Drawings presenting muscular structures, coloration patterns, structural aspects, or theoretical conclusions will be based on actual specimens but will not call for specific positioning of the subjects.

The third category consists of reconstructions. Virtually all reconstructions are based on evidence. All evidence is based on actual objects, but the illustrations are a composite of information gathered together through various means. The illustrator will have an array of data varying from scanning electron microscope photography through scientists' sketches, verbal descriptions, and numerous specimens. The positioning of a specimen is out of the question.

We are greatly concerned with the fourth group. Here the illustrator will face one specific, usually small object ranging in size from approximately one meter down to microscopic. The skull of a rodent will have to be viewed from a specific side. Specifications will not change. A view from the left side really means a direct view from that side. No perspective, no foreshortening, no three-quarter views. In all cases, examine the object before positioning. If it is large enough to be viewed with the naked eye, spend some time just looking. Pay attention to changes in surface structure. Notice the fragility of the subject. See if it is possible to notice all the twists and curves. Does the specimen have some openings that go through the whole structure? You must notice this before positioning, before setting up the subjects. Later this important information may not be so easily available. Perhaps an opening in the rodent skull will be obstructed by the Plasticine supporting the subject. When looking from the prescribed view, such information may escape your attention. Note the shape of the side that will be supported by a foreign matter. This will help you to place the specimen in position without breaking it and without clogging it up with Plasticine. If the object is preserved in liquid, do not let the specimen dry, as dryness will destroy the tissues, the specimen will shrink and the surface will break. If you are allergic to ethyl alcohol or formaldehyde, use surgical gloves during removal of the specimen from the container and afterward make sure to keep the subject moist with water. Using distilled water assures greater safety for soft-tissued specimens. When the subject is submerged in glycerine, just transfer it from the original container into a small petri dish full of glycerine and observe it under a microscope. It may be advisable to use finely woven cotton gauze as a support. Specimens in glycerine tend to appear stable in their position but in actuality do rotate slowly. Subjects will be positioned through different means according to their structure. Let's start with specimens of a skeletal nature.

EXAMPLE 1. All small skeletal structures must be supported by Plasticine. In order to support larger specimens, the base must be relatively wide and firmly attached to the illustration board or petri dish. Most smaller skeletal objects, such as a tooth, will be glued to the metal pin by the preparator or a scientist before reaching the illustrator's hands. The tip of the pin should be inserted into the Plasticine. Do not attempt to place such a specimen on anything that would obstruct its surface. Bits and pieces of Plasticine attached to a small specimen will obstruct its surface, making observation and drawing difficult. Do not forget that after foreign matter has found its way to the surface of the specimen, somebody will have to clean that specimen. This is extra work for either the preparator or scientist. In order to set up such a specimen, shape the Plasticine into a small conelike formation. Be sure that you refer to initial sketches depicting the exact position of the specimen. If you find that the pin with the subject must be attached to the Plasticine at an angle, be sure that the lump of Plasticine is high enough for this particular position. The specimen itself should be above the surface of the base. It should have room for vertical or horizontal adjustments. Take the opposite end of the pin and insert it into the upper section of Plasticine. Make sure that the pin will not slip out by itself or during adjusting. Then, looking directly from above, using a microscope so you will not have any problems, delicately rotate and move the pin until the visible image of the specimen matches your descriptive sketch. Remember that if the specimen has to be drawn from more than one side, the next view should be positioned on the same vertical plane as the first one. Changes in height will affect magnification. Make sure the specimen is not damaged, scratched, crushed, or dirtied. If you are not sure about the position, ask for the scientist's approval.

EXAMPLE 2. Various liquids are used for preservation of the specimen: ethyl alcohol, glycerine, formaldehyde by itself or compounds of alcohol, formaldehyde with additives to cancel its unpleasant smell. Subjects such as fish are kept in an ethyl alcohol solution. Sections of soft tissue sometimes are preserved in glycerine. Parts of plants are stored in one-third part formaldehyde and two-thirds part water. Each of the liquids requires slightly different handling. Ethyl alcohol evaporates rather quickly and therefore should be added as needed to the container with the positioned specimen. The glycerine can stain your drawing permanently because it is greasy. Formaldehyde may be objectionable to some because of the strange, unpleasant smell and its damaging properties to live human tissue. When handling formaldehyde, use surgical gloves. Most specimens can be removed from preserving solutions and observed in their semidry state or in water. The only exceptions are those that have been preserved in glycerine or are too fragile to handle. You know that glycerine does not mix easily with water. Containers are needed to position a wet specimen. The most readily available container is a glass dish, the petri dish. Depending on the size of the specimen, its length, width, and depth, the size of the container will vary. When positioning bulky fish, you may need a large tray. Any trays can be used, but dissecting trays are best because of the thick coat of wax inside such trays. The specimen can be pinned down with thin stainless steel pins to the bottom of the tray and then covered with liquid. When using pins, be sure the specimen is not injured and that pins will not obstruct the view. You know that fish positioned for drawing are viewed from their left lateral side. Place the subject on the bottom of the tray and using a couple of pins, stabilize the main portion of the subject by placing the pins next to the body. Make sure that the pins will hold tight but will be at a slight angle facing away from the subject. Depending on the size of specimen, continue pinning the main sections of the subject by inserting pins in the junction of the fins and the body. Some of the pins can be bent to forming the letter *L;* when inserted into wax they will hold the bulk with their longer arms pressing the specimen down to the bottom of the tray. The fins will have to be judged separately. If in good condition (undamaged and strong), they may be stretched forward by inserting pins at the upper tip of each spine.

When the specimen is small, use an appropriately small dish and secure the subject between thin pins. The pins can be attached to the side of the glass dish with Plasticine or to the bottom if spongy transparent plastic is placed inside the dish. In some cases, depending on the specimen, the subject can be pressed against the bottom of the dish with horizontally located pins. The glass bottom will allow reflected light to pass through the specimen, helping you to distinguish certain features more easily. Soft-tissue specimens preserved in glycerine may require the very same setup. Sometimes a piece of gauze is placed under the subject for it to rest on. Constant slow movement of the subject is a problem. The only solution will be a securely pinned specimen unable to twist and move. When gauze is used the desired view may not be easily established, as the position of the specimen is still a bit accidental. In many cases you will be stuck with such a problem because the subject is too fragile or too small to be pinned. The gauze should cover the whole glass dish, sagging slightly at the middle. The specimen will be placed in the central part and a teasing needle will be used to turn it until a desired position is achieved. Naturally, glycerine will first be poured into the dish, and the gauze stretched later. The smallest specimens are prepared on pieces of glass called slides. Again, the preparation is performed by a skilled technician or the scientist himself. Here the illustrator does not have to position anything. The slide is inserted into the special holder or placed on the viewing base. The only problem will be finding small, sometimes invisible-to-the-naked-eye specimens, even when looking through a high-power stereomicroscope.

EXAMPLE 3. Insects are dry specimens, already prepared with a pin inserted through their bodies. Pins are necessary because insects are extremely fragile and are easy to damage. Such specimens are handled with the utmost care, holding the head of the pin instead of the subject itself. In actuality, positioning for drawing is a constant repositioning of the subject throughout the whole process of sketching. Starting from the top, there is the head of the pin, then an insect, and under the subject there are two or more very small pieces of paper called labels presenting available information. Those small pieces of paper are pierced by the pin, identifying that particular specimen. The labels cannot be damaged or removed by the illustrator. As the specimen is pinned to the bottom of the box, the illustrator has to exert great care when removing such subject from its storage. Delicately but firmly, take the head of the pin in your fingers and, holding the box with the other hand, pull the subject out. Be careful not to touch the walls of the box with the specimen, as this may result in damage. Transfer the subject to a previously formed cone of Plasticine or other material and stick the sharp tip of the pin into it. Assuming that the project calls for a drawing of a whole specimen, you will have to center the subject's parts separately. Start with the main portion of the body. Gently move the subject by touching the pin or by reinserting it into the supporting base. Check the position carefully. Insects are small and you will have to use a microscope in order to see and draw. A slight tilt of the specimen in any direction will result in an improper position and therefore an improper drawing. Ask the scientist for approval. After completing a preliminary sketch of that part of the subject, you will have to reposition the specimen in order to draw its head, abdomen, legs, and antennae. As a matter of fact, you will have to reposition the specimen more than a few times. During preparation of the specimen for storage, delicate parts such as legs and antennae will be twisted slightly. As an insect's legs and antennae are made of numerous sections, each section may be twisted differently from other parts. You will have to reposition the subject as many times as necessary in order to view its separate parts correctly. It is a delicate procedure requiring caution and care. The sections will be drawn separately and later will be spliced together to form a satisfactory presentation.

Figure 4-5. Fowl, comfort behavior and care of the plumage. Pen and ink. Drawing by Käthi Stutz, Zürich, Switzerland. Published in D. W. Folsch and K. Vestergaard, *The Behaviour of Fowl*, Birkhauserverlag, Basel, Switzerland. Drawn in 1980.

Importance of Observation

Without knowing the subject it is impossible to produce any illustration for science. Looking is just not enough. The process of selection is based on thorough investigation of the specimen. Our eyes can fool us. We can easily omit an important segment or structure. Never be afraid to examine the specimen completely. Look at the subject from all sides, including all areas that will not be presented in the drawing. Remember that all structures are connected and complete observation will lead to better understanding of the form. Knowing the specifications, the process of analytical thinking must take place. Questioning discovered areas will be a natural reaction. Is this dark spot of any importance? Is it caused by the dirt? Is it caused by the thickness of the bone? Is it an indentation? If it is an identation, how deep is it? In order to really know you must handle the specimen. You must rotate the subject around. What about the curvature of the walls? Two separate structures are easily noticed, but are they really separate? Maybe the appearance of separations is caused by foreshortening. When handling small specimens, investigation may be impaired by their size, but nevertheless it must take place. Do not take anything for granted. Carefully look through the microscope and do your best. Try to rotate the subject in order to observe all sides before positioning it for drawing. Such a procedure has to be performed carefully, as the specimen cannot be injured. Take your time, but do not indulge yourself. Remember that production of an illustration must take place. Examine the soft tissue with a teasing needle. Tissues may be transparent. What you see under magnification may be the internal structure, not the outside of the specimen. Naturally, if the project calls for depicting internal structures, then concentrate on its insides. Remember that once the specimen is positioned it will be difficult to observe "hidden" sides. Carefully touch the surface with the teasing needle and see if an apparent fold is really there. For observation, smaller magnification will be more desirable. A specimen that measures approximately one centimeter can be observed easily when magnified six times. When enlarged 25 or 50 times it will not be viewed as a three-dimensional entity. Such large magnification will allow you to see a very small section of the surface of the subject. The observation of more strucures will be impaired. Do not start drawing immediately upon receiving a project. Make sure you know what you're supposed to look for, then find it. Remember that lights used for illumination can deceive you. A reflection of lights can be taken for a structure. At other times, because of the angle of the light, some surface configurations may be completely obscured. Remember that lights are for you to use and move around. With large, obvious specimens not requiring the use of a microscope be as careful as with smaller ones. We may have a tendency to assume that if something is large, we therefore know how it is built. Do not assume anything. Examine the object, sometimes the surface, see how it is formed, and then begin the process of sketching. Never be afraid to ask questions. If there is something you are not sure about, call the scientist. Remember that the drawing you are about to produce is not being drawn just for visual pleasure, but for an explanatory purpose.

EXAMPLE 1. The specimen is a relatively large skull measuring approximately six centimeters. The project requires three drawings: the left lateral, ventral, and right lateral views. The specimen has been observed carefully and it appears to present a rather straightforward situation. No problems are anticipated. The specimen is positioned on its right side, using Plasticine, in order to draw its left lateral side. The sketch is completed. The scientist has asked you to render in pencil and you have complied. No problems. The next is a ventral view. The same happens. The scientist is satisfied and the project nears completion. The right lateral side is the last drawing. Naturally, for each new view the specimen must be repositioned. After placing the specimen on its left side you notice an opening in the skeletal structure. The opening was not noticed previously because of improper examination of the specimen and because during the first drawing it was covered up by the Plasticine. This time it is visible because Plasticine has not hidden the opening by accident. Actually, you are lucky. You have noticed an important structure before the project is finished. The bad part is the fact that the drawing of the left lateral side does not show this opening. In order to repair the first drawing, you have a limited number of choices, all of which depend on the technique of drawing and the materials used. You can paint the opening with white paint or cut out the paper. The white paint may dissolve the graphite, so be careful. If the drawing is not mounted to illustration board it is easier to cut. In any case, you have trouble on your hands.

EXAMPLE 2. The table is cluttered with specimens in glycerine. All are small structures. You have been asked to produce drawings uniformly magnifying structures 50 times. There are a lot of drawings to do and not much time in which to do them. You have observed a number of the specimens and have noticed some relatively visible structures on most. Rushing, you begin to prepare the sketches and trusting your experience, immediately transform the sketches into pen-and-ink renderings. After a while, when delivering the finished drawings, the scientist asks you about those prominent structures. "I draw what I see," you reply. "Not really," he says. "All of those structures are internal glands; they should not be drawn at all." He says, "You have rendered the drawings very well but the surface, the transparent surface, is not depicted properly." He states the fact that the project calls for drawing of the outside structures. Unfortunately, you did not notice the transparent surface when observing the specimen. You have used too high a magnification and were not methodical enough in the procedure. Now you have to start again from the beginning.

EXAMPLE 3. An anthropologist has presented to you various stone tools for drawing. He has stated repeatedly that the direction of cleavage must be presented clearly. He has said that close supervision will be necessary. Deep inside, you resent such a statement. You know that you can draw, and you know what he is talking about. As far as you are concerned, there is no need for close supervision. Feeling a bit edgy, promising yourself that you will show him a drawing, you begin. The specimens are small and are perfect for camera lucida. Feeling good, you proceed to trace the visible image. Trying to impress the scientist, you work fast, and are surprised when the scientist is not satisfied. The cleavage is not depicted correctly and numerous flakes have been entirely omitted. The reason? You have not investigated fully. The only way to observe the cleavage would be through constant manipulation of lights and constant change of the angle of light, illuminating almost every broken chip separately.

Choosing Tools for Measuring

Whenever possible, use good-quality measuring tools. Precision is a necessary ingredient of the drawing process. Unfortunately, you may not be able to own the whole range of measuring tools, and it may be necessary to use whatever you have. Just remember that simple measuring devices such as rulers and compasses will produce more discrepancy than a fine set of pointers or a proportional divider. Make sure that the correct measuring tools are used for a particular subject. It is more difficult to measure a large vessel with small pointers than with a measuring tape. Try to use rulers and tapes with metric designations. When such are not available, inches will be fine, provided that you are prepared to translate inches into centimeters whenever necessary. Remember that precision is a relative statement, but at the same time, be sure that discrepancies are relative to the size, enlargement or reduction of the specimen, and to the objective of the project. Naturally, a proportional divider will give you better and faster results than a beat-up compass, but at the same time, if you are careful you can use any measuring tool efficiently and with relative precision. Make sure that when measuring the width of the walls of a vessel you use some kind of calipers. It may be necessary to know how to operate a reducing-enlarging camera for tracing large objects. When hard-pressed for a measuring tool, make your own. It is possible to use a strip of paper with uniformly marked distances for measuring. In such a situation, it is not necessary to copy inches or centimeters, as long as the distances between designations are equal. Remember that in actuality you are using proportions. It will always be that "so much of this" will fit into "so much of that." The distances will be compared. Precision tools are very helpful but not that essential. Much more essential to the whole process is your patience. Without patience it is possible to make horrendous mistakes using very fine tools. Tracing devices such as camera lucida, and a variety of drawing tubes or projection devices will help to speed up the process of drawing. Remember that matching a measuring tool with a particular specimen can make a difference. Also remember that when using a crude measuring tool, mistakes you make will be compounded during enlargement. A very small initial mistake of one millimeter, when repeated six times during the enlarging procedure, will grow into more than half a centimeter. If this mistake is repeated when measuring a complicated structure, shapes will not match. You will be forced, because of the tool, to calculate an average of one initial and repeated measurement and to spend more time during preliminary stages. Regardless of the tools used, you must remember that when measuring specimens the conventional way and without any tracing devices, four basic rules have to be applied.

1. The specimen must be positioned correctly.
2. The position of the head in relation to the specimen must be constantly correct during the whole process.
3. The measuring tool must be held parallel to the base supporting the specimen.
4. The tool should be kept at approximately three quarters of the height of the specimen.

You must remember that any tilt of the measuring tool will produce distortion. The tool must be kept at approximately the same height in order to minimize the discrepancies produced by the depth of field.

Two-pronged measuring tools such as a compass or pointers with thin, sharp-ended tips will give you a better reading and therefore more precision. Do your best not to use small pointers when measuring larger specimens. Because of the small span of such a tool, it will be impossible to measure the whole length of the specimen at once. This will lead to measuring smaller sections of the desired distance. The smaller the sections, the greater the chance for major mistakes. Whenever possible or feasible, use tracing devices, but do not blindly trust prisms and optics. How do you know what tool to use?

EXAMPLE 1. The project calls for drawing a large skull, about one meter in length. Do you think that using pointers will be feasible? A measuring tape or a ruler will be sufficient. When drawing such a large specimen, all measurements will be reduced, therefore all your mistakes will be harder to notice.

EXAMPLE 2. The specimen is fossilized matter approximately ten centimeters in length. Tape or a ruler will be too large a measuring tool, so pointers or a proportional divider will be the proper tool to use. With fine pointers it is possible to measure all necessary details within a tolerance of one millimeter.

Exercise After drawing coordinates, vertical and horizontal, and positioning the specimen correctly, take a crude measuring tool. Use a cheap compass for an exaggerated effect. Do your best to measure a horizontal distance, magnifying all measurements six times. Mark the distances carefully. When finished, take a good-quality tool such as a fine set of pointers or a proportional divider. Place the new sheet of tracing tissue over the first one and secure it with tape. Repeat the same process of measuring. Make sure that the distances on the specimen measured with the better tool are the same as those measured with the crude compass. Compare the results.

Importance of Sketching

There is a similarity between the purpose of sketching and the finished product, the rendered illustration. Sketches are essential in establishing meaningful communication between the illustrator and his customer, the scientist. The finished rendering is essential in establishing the very same between scientists. A sketch is usually the only method of coherent presentation of the visually depicted data. How else would it be possible to build the image? Do you think it is possible to look at the subject and draw it correctly? I would like to meet an artist who could do this. Do not misunderstand: It is possible to draw an image resembling the subject. It may even be possible to present the basic characteristics in such "eyeballing" drawing. Nevertheless, a drawing produced just by looking at an object will not be precise. True, that definition of precision will vary, but the closer you get to the rightful proportions of the subject the more meaningful your drawing will be. It is certainly easier to correct discrepancies on a sketch than on a finished rendering. Remember that the product, the drawing, is for the service of science. There is not much room for feelings or impressions in sciences of today. Measuring is a must. Precision is a must. Sketching is a slow but correct way to build the visualization of the subject. It always leaves room for changes, and some changes will always be necessary. It is rare to find a subject that can be sketched on one piece of paper. Usually it will take two or more overlays. Guessing will be eliminated and the facts presented when the subject is properly viewed and sketched. Usually the word *sketching* is applied to the freehand, quick drawing, and usually such a drawing is to be considered as the ultimate end. In scientific illustration sketch is something that should properly be called a *study*. It is a conglomeration of information carefully noted and, through the selective processes of the mind, arranged in a comprehensive fashion fitting the designated specifications. Such a study presents the true proportions for the very sake of truthfulness. It is of utmost importance to struggle with your mind to overcome inclinations leading to "improvement" of the drawn image. The sketch, or study, is used as a base for communication between you, the illustrator, and the scientist; it must be readable. Do not expect extra powers to come from the scientist. Forget the "quality of the line" and gentleness of the tones. The sketch is comprehensive information, not meant to be an end in itself. It is a step in the whole process of illustration. This does not mean that your sketch cannot be visually appealing. A lot will depend on the subject you draw, and a lot will depend on your know-how. As a rule, remember that a functional, clean sketch will look good. Naturally, any writing will enhance the looks of the sketch, and you will have plenty of opportunity to write. If you want to have a nice-looking sketch, start with light lines. Do not draw heavily at the start. Do not use very soft pencils; they produce uneven, thick lines. Later, when the presentation of the subject is more coherent, you can use a softer pencil. Just remember that soft pencils smear easily. When the sketch is all smeared up it does not look good. What about your handwriting? It should be legible. Write nicely. Look at Da Vinci sketches—they are functional all the way to the writing. Did he draw his sketches for study or did he produce them for the sake of drawing? You know the answer. This does not mean that you should not practice drawing for the sake of drawing. The more you draw, the better you will be. Keep a sketchbook for this purpose. Force yourself to draw three meaningful sketches a day. Sketches of ordinary subjects like this will be one of your ways of practicing drawing. You must develop a skill and patience. You must conquer your urge to do things fast. Do not try to get exercises "over with"; it does not work. If you want to draw, practice drawing. Devote yourself to it.

Exercise 1. Keep a sketchbook with you at all times. Draw subjects such as fruits, pans, chairs, or animals. If you can procure a three-dimensional representation of a human face to be used on a subject, the better for you. Produce three sketches a day. Use lines; do not use tones. The object is to practice drawing. Do not draw from imagination. Subjects created by imagination are not realistic. In such situations you really do not know what you are drawing, and your mind does not have to work as much as when you draw from real objects. Do not throw your sketchbooks away. Keep them so you can compare early sketches with later ones.

Exercise 2. Patience is a must for a good illustrator. Analytical thinking has to be acquired. Take one large sheet of paper and attempt to draw slowly, thinking about the sub-

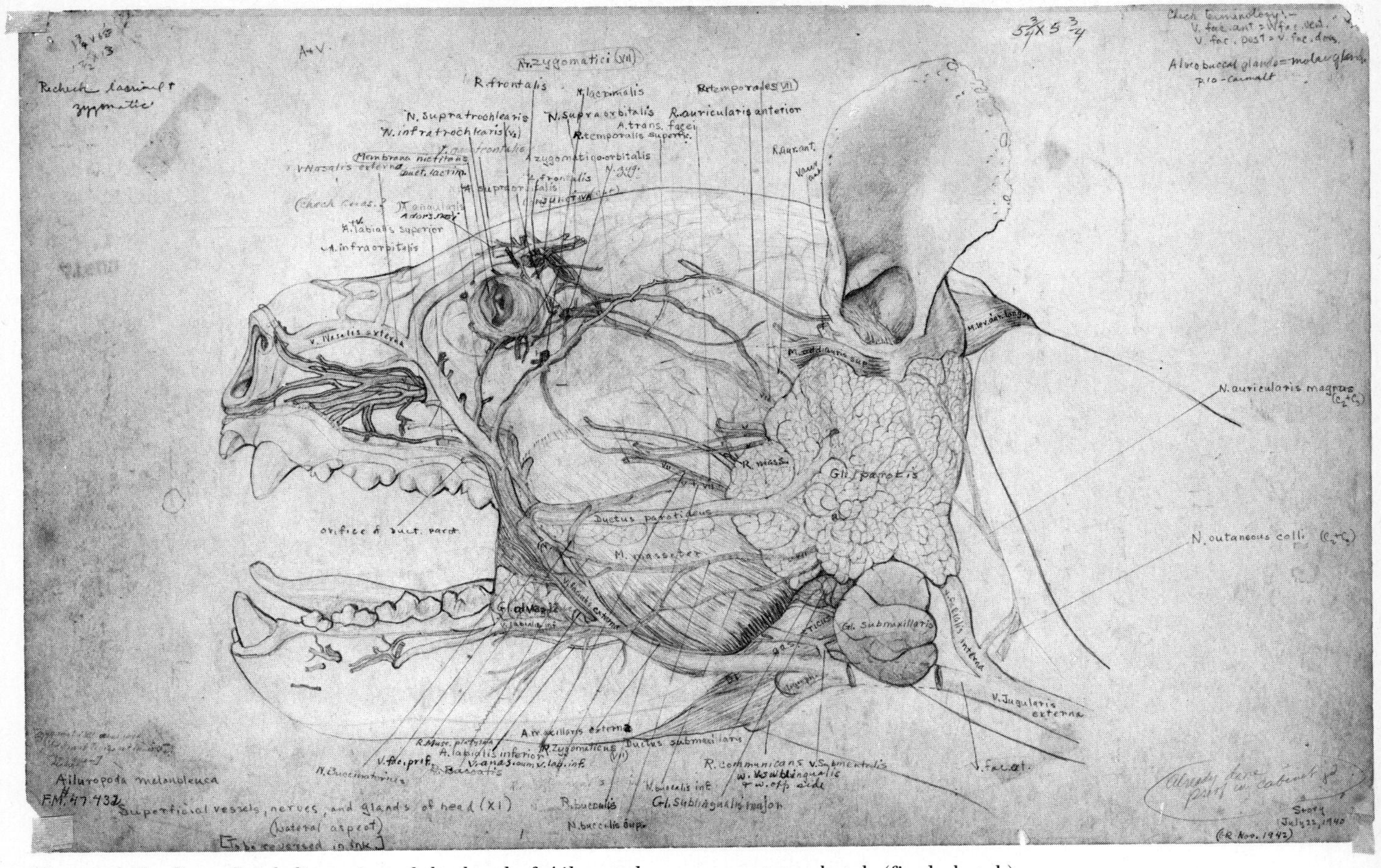

Figure 4-6. Superficial dissection of the head of *Ailuropoda*, a preparatory sketch (final sketch). Pencil and watercolor on off-white paper. Drawing by John C. Hansen, Chicago Museum of Natural History (Field Museum of Natural History), Chicago. Published in D. Dwight Davis, "The Giant Panda, A Morphological Study of Evolutionary Mechanisms," *Fieldiana*, Zoology Memoirs, v. 3, 1964, by Chicago Museum of Natural History (Field Museum of Natural History), Chicago. Drawn in 1940.

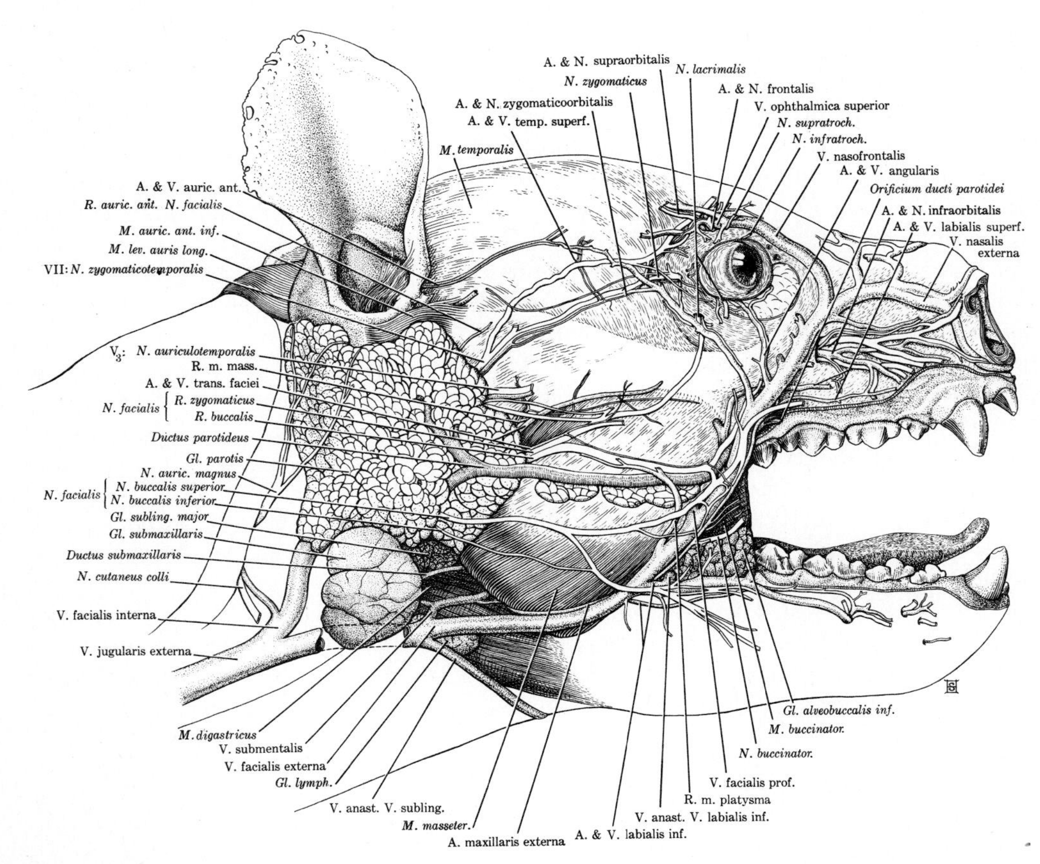

Figure 4-7. Superficial dissection of the head of *Ailuropoda*. Pen and ink on two-ply white paper. Drawing by John C. Hansen, Chicago Museum of Natural History (Field Museum of Natural History), Chicago. Published in D. Dwight Davis, "The Giant Panda, A Morphological Study of Evolutionary Mechanisms," *Fieldiana*, Zoology Memoirs, v. 3, 1964, by Chicago Museum of Natural History (Field Museum of Natural History), Chicago. Drawn in 1940.

ject and the drawing. Remember that this is a sketch, not a final drawing. Do not measure. Look at the subject and draw. The key portion of this exercise is to use *only* one piece of paper and to draw on it for a long period of time. The minimum time will be 35 hours of work. The object is to draw during that 35 hours using one sheet of paper and one tangible subject, without overdraining either the subject or the paper. You must not erase anything. Once you place a line on the paper it will stay there. The more lines, the less white areas. You will be forced to think selectively. In all probability, you will not make it the first time. You will discover how hard it is to think. You will discover how important your decisions are and how conscious you must be during drawing.

Exercise 3. Draw a complicated but not-too-small subject, preparing the sketch for a scientific study. Choose your subject carefully. A broken piece of pottery—or, if you can obtain it, a tooth or a skull—could be the best. Make sure that the subject is positioned correctly and follow the basic rules for this type of sketching.

First Sketches

Initial drawing procedures should be informative and functional. That means presenting the basic information on the paper. I strongly suggest tracing tissue and one of the harder pencils as a medium. The functional visualization should give you the base into which later refinement can be implemented. What are the reasons for using tracing tissue? The medium is transparent, allowing for easy tracing of the image onto another surface. An H or HB pencil is hard enough to produce a visible thin line without much pressure and without breaking the tip of the pencil. Make sure that the pencil is sharpened correctly. If you like to use a technical pencil, do so. When using regular pencils it may advisable to sharpen a whole bunch of pencils at one time; then, during drawing, when the top loses its sharpness, switch to another.

Depending on the project, the size of the specimen, and the availability of materials, it is best to have a sheet of smooth-surface illustration board to place tracing tissue on. Do not erase. I know it is easier to say that than to do it, but remember that the series of first sketches is produced more for your information than the scientist's. They are working sketches from which you will be able to "construct" the image of the subject. Whenever you notice a wrong line, ask yourself if it really is a wrong line. Question yourself at all times. Make sure that you do not produce thick lines, as they can add to the mistakes.

When measuring or tracing basic proportions, drawn lines are in actuality coded information. The less precision in drawing, the more miscalculations. Mark unwanted or questioned lines with the typographical "delete" symbol (₰). By using this symbol you will be able to clearly mark unwanted information without obstructing everything. Remember that even if there is a mistake in the initial sketch, that mistake is valuable information for you. You will know what part is wrong. When erasing mistakes, you will be left with a twofold problem. First, you will have to erase sections of correct parts of the sketch. Second, the reference to the mistake will disappear. Because of this and because of the stubbornness of the human mind, there is a distinct possibility of repeating the very same mistake. You may think that it is correctly drawn but in reality it may not be. Remember that functionality is fundamental to the first series of sketches. When overcrowding of lines begins to obstruct clarity, take another sheet of tracing tissue and place it on top of the first one. Tape it to the illustration board. Make sure that registration marks, a cross within a circle, are drawn on the first sketch. Transfer those registration marks by tracing them onto the new sheet of tracing tissue. Remember to write necessary information: the number copied from the specimen, job number, and enlarging factor. This way you will not lose track of sketches for various projects. Trace all of the good parts of the previous sketch onto the new sheet of tracing tissue and continue your work. Depending on the project, the original size of the specimen, the enlargement, and the space used for work, your first sketches may be drawn on one sheet of tracing tissue or on many. When using many smaller sheets, make sure the image will overlap. This will be of great help to you when small sheets are put together to form the complete visualization of the subject. Do not be afraid that the first sketches are too messy. The beauty of sketching is of secondary importance, although cleanliness and readability are vital. This does not mean that you should not try to produce visually pleasant drawings. You will quickly find that a working sketch drawn for the purpose of gathering and organizing information will look good. It will look purposeful, because that's what it is. Generally speaking, there are four steps in the preparation of the first series of sketches:

1. Careful observation of the specimen
2. Measuring or tracing the general dimensions of the subject
3. Working within the designated parameters of established dimensions
4. Checking the truthfulness of observations and measurements

Remember that the first series of sketches will lead to the final sketch, which is a clean tracing of all pertinent information used for presentation to the scientist and, when approved, is used as the base for the rendering.

Naturally, there are exceptions. You may be asked to produce a rendered illustration without preliminary sketches. This is possible to accomplish when the subject is relatively small and when you are using camera lucida. You may be asked to draw a sketch in its final form on the surface of the paper used for the rendering. Again, the easiest way to accomplish this task is to use a tracing device. In such a situation, remember that when an illustration is to be rendered in ink, the underlying pencil lines will have to be erased. Draw such a sketch using light lines. When you erase the pencil after ink rendering is finished, the pressure of the eraser will remove some of the blackness of the ink. If you can write your observations or questions on the sketches, this will help you to understand the forms you are drawing and will help you communicate with the scientist.

EXAMPLE 1. The project calls for the reconstruction of a skull from various fossilized remains. Available materials include four or five fossils partially cleaned but embedded in rock, the rubber molds of those parts, the scanning electron microscope photographs of details found on the surface, and drawn and written comments by the scientist. Following the instructions, the illustrator observes and measures the subjects, writing basic proportions: the length of fossilized subjects and the width, measured in at least three or four places. By adding the dimensions and dividing the result by the number of measurements, the average length and width are obtained. The first step in sketching consists of drawing parameters for the subject. This will be a straight vertical line representing its length, and the horizontal lines representing the widths, to be placed at right angles to the vertical line and at their respectively appropriate levels. The enlarging or reducing factor is noted and all results of measuring are accordingly enlarged or reduced. When the initial dimensions are depicted, the illustrator knows that the rest of the subject must fit into the designated area. The registration marks are drawn, and the number or numbers found on the specimens are noted on the side of the initial sketch. The enlarging or reducing factor is written. Using the overlay, more details are implemented. The separate bones are presented; all unknown, questioned areas are marked. Because the illustrator is part of a research team, his work is closely scrutinized after a relatively coherent image is drawn. The scientist decides that the posterior width does not fit his calculations, and the sketch has to be changed. Make sure that all information pertaining to changes is written down on the sketch. Ask the scientist to write and draw. Using successive overlays, the necessary changes are implemented. A new meeting takes place and the work is reviewed. More changes follow. More overlaying sketches are produced. By now the first of the series of sketches appears to be of no importance but do not throw anything away. Later it may turn out that some of the initial informations presented on those sketches is important. Having by now numerous layers of tracing tissue, remove some from the bottom, as they may interfere with upper layers. Place these sketches in some kind of folder, with necessary information written on the folder. Because of the constant removal and addition of sheets of tissue, the usefulness of registration marks becomes apparent. The work proceeds with constant additions and changes as well as redefinitions of the subject until the scientist is satisfied. Through retracing and through cutting and splicing various sections of sketches, the final version of the reconstruction is reached. The very last step is the preparation of the final sketch.

EXAMPLE 2. The specimen is a well-preserved and well-prepared skull of a marsupial. The illustrator has a microscope with a tracing device attached to the ocular. The lateral dimension of the specimen equals 75 millimeters. The problem lies in the fact that if a microscope's minimum enlargement is 6×, then obviously the specimen cannot be visible through the microscope as a whole unit. The most that can be viewed will be approximately 20 millimeters of the surface of the specimen. The viewed section will be slightly distorted, the degree of distortion depending on the type of microscope and the precision of optics. The distortion will be there, and you, the illustrator, will have to eliminate it. First, measure by hand, using pointers or a proportional divider, the length and width of the subject. Make sure you enlarge all measurements according to the magnification of the microscope. Transfer the measurements to a sheet of tracing tissue. Draw a base sketch of the subject. This sketch will be used for checking the distortion of optics.

Now set the first sketch aside and look at the specimen through the oculars of the microscope. Do not forget to adjust the focus for camera lucida. Make sure that parts of the microscope such as the main body containing the lens and upper section with oculars are aligned. Look for parallax displacement. If there is any, eliminate the apparent displacement of the image. After all adjustments are checked, start drawing. Looking through the oculars, you will see the object as well as your hand holding the pencil, because camera lucida will superimpose the image of your hand on the specimen. Start at the front or back of the specimen. Draw—or rather, trace—the outlines of visible shapes, but avoid all images near the edge of the lens, as the greatest distortion is produced there. The 20 millimeters of visible distance is not much, especially when actually you will be able to trace much less. Remember in your mind what shapes you are tracing. Make notes if necessary. I strongly advise you to use rather small sheets of tracing tissue, as a large sheet will interfere with drawing. It will have to be rolled up or folded in order not to obstruct the area of drawing. With large, rolled-up sheets there is a danger of accidentally pushing the specimen out of its set position by the paper. After finishing that small section, move the specimen under the microscope in order to see the new section. Remember that part of the previous section should be visible, as you will have to maintain continuity of the image. Using a new, clean, small sheet of tracing tissue, continue drawing, making sure that part of the previous area is included into the new sketch. Repeat the process until the whole image is transferred on separate sheets of tissue. Do not forget about the depth of field. You will quickly notice that when your microscope is focused on the structures closer to you, the others below will be out of focus. In order to draw them, you must change the focus. You will notice that shapes and lines do not match. Here you must figure out an average for the specific dimension, and you must mark this clearly and cleanly on the sketch. Preferably during the work, take separate sketches and place them on top of each other, matching the overlapping sections. Tape the sketches together without obstructing the drawn image. While in the process of splicing and taping, take the first measure sketch and place it on top of the traced sketches. Compare the results. You will quickly be able to see how much you are off. If you are correct, congratulations! If there is a discrepancy, correction is necessary. Such correction may mean resketching that particular section of the specimen; if the distortion is great, you must recheck the microscope for its optical accuracy. Do not hesitate to write your comments on the sketch. At the same time, make sure that writing will not obstruct the drawing.

Continue until the whole subject is sketched. At this point, the first sketch will consist of a "mosaic" made up of overlapping sheets of tracing tissue. Place that sketch on illustration board and tape it to the surface, making sure that the tissue is not bent and that it remains flat. Place a new, large-enough sheet of tracing tissue on the top. Secure it with tape and begin retracing the image. The new traced image should be executed cleanly, as this is the image that will be presented to the scientist for approval. When all pertinent information is transferred onto the new sheet of tracing tissue, you have the final sketch. Naturally, all pencil lines are clean and sharp, and all questionable areas are marked with appropriate comments. Of course you will ask the scientist to implement any corrections he feels are necessary. Make sure to point out to him or her that the final drawing will be smaller than the sketch. A rendering of such a size is a time-consuming work. Remember that 75 millimeters enlarged by six will equal 450 millimeters. This means that the sketch is 45 centimeters long, approximately 17¾ inches. When the sketch is approved and everything is explained, reduce the image to a more manageable size by photostatting the sketch or by retracing it, using a reducing-enlarging apparatus. When this is finished, spray the sketch with fixative. Remember, there must be no dirt on the paper used for rendering. Using a light box and without applying much pressure, trace the image onto the surface of the paper used for final rendering.

EXAMPLE 3. A pair of leggings is lying on your table. The project asks for a pattern depicting all pieces of leather, their respective shapes and sizes, as well as a clear presentation of the arrangement of pieces forming that particular garment. You cannot place the object under a microscope and trace it, as it is too large. You cannot place such an object on the bed of a reducing-enlarging camera because all pieces forming the garment bend around, making tracing impossible. Besides that, the heat from the lamps would damage the material. First, you must realize that the subject must be reduced. Second, you must be aware that your sketch will produce a flat, two-dimensional representation of a clothing pattern. The only tool for measuring will be a measuring tape with metric designations. After careful examination of the specimen, establish two basic points of reference. You will quickly find out how many separate pieces have been sewn together, and will realize that reference points must be fairly far apart. The most stable and most visible will be the inseam. Measure it and, after appropriate reduction, transfer the results onto the tracing tissue. Having two basic points will allow you to find a third one. To start, concentrate on one of the segments of the specimen. Having two points, find a third pertinent point on the garment. Measure the distance from first to third and from second to third. Remember to write down the measurements. If you do not, there is a possibility that you will forget the measured distance. Remember that you may have to measure around the specimen. It is not likely, but the third point may be located on the underside of the garment. Once you have the measurements, remember to reduce the results by the same factor as you first used. Take the end of the tape or ruler and place that end at the first point clearly marked on the sketch. Using a measuring instrument like a compass, draw a circle. You may have to mark the edge of the circle with dots and then connect the dots by hand. Repeat the same with the measurement existing between the second and third points. Where the two circles cross, you will have the so-far-unknown third point. Make sure to check all measurements a few times to compensate for the stretching of material. If you are not sure, then measure a distance a few times, add all the measurements, and divide the result by the number of measurements, obtaining the average. Repeat this process until the shape is found and drawn. Clearly mark the most important points of contact using numbers. Proceed to the next unknown shape, repeating the same process until you have documentation completed. Use as many sheets of tracing tissue as you need. Do not try to squeeze everything onto one sheet. At this initial stage this would be impossible. Make sure that the shapes of the pieces making the garment are drawn separately, but that their points of contact are clearly and visibly marked. When everything is finished, you have numerous sheets of tracing tissue with two-dimensional images. Make sure that before tracing all information into a semifinal or final version, the correct pieces of garments are placed next to each other. You will have to arrange and move around most of the initial sketches. When organization of the depicted information is completed, take one large sheet of tracing tissue and, using rulers and French curves, trace the separate images onto one sheet. Make sure that everything is clearly marked and that corresponding numbers are placed in proper place. When finished, you will have a final sketch, ready for checking and approval of the scientist.

EXAMPLE 4. The project calls for drawing a very small object, using the microscope. The microscope does not have a camera lucida, which means it is impossible to trace the image. Because the specimen is very small, it cannot be measured with any measuring tools. As always, examine the specimen carefully. Make mental and visual notes of the structure. After you are familiar with its complexities, pick one of the segments of the structure for a measuring unit. The length of this segment will be used by you to measure the rest of the specimen. Draw a straight line on the side. The line will be arbitrary. Keeping your eyes fixed on the specimen, figure out how many lengths of the chosen segment will fit into the width and length of the specimen. Using the previously drawn line, construct the sketch by mechanical process. Use the first straight line as a known distance. If four segments will fit into the width of the specimen, the four lines placed consecutively will do the same on your sketch. Continue until the image is delineated. Retrace it, then have a talk with the scientist. He will tell you the correct size of the specimen. Once you have this information, measure the drawn image and divide the result by the truthful size of the specimen. The result will be a magnification factor. After approval of the final sketch, proceed with the rendering.

Exercise. Take an ordinary but relatively complicated object such as a small piece of tree branch. The piece should not exceed ten centimeters. Observe the specimen carefully and position it correctly, using the usual methods. Make sure that:

1. The specimen *is not* too far away from the small sheet of illustration board.
2. Plasticine *does not* obstruct any details and is used for support only.
3. The horizontal and vertical lines *are not* drawn too far from the specimen.
4. A right angle *will* be maintained between the horizontal and vertical lines.
5. The lines *will* be drawn on the same illustration board on which the specimen is positioned.
6. The specimen *is not* tilted in any direction.
7. The lines *will be* drawn on the sheet of tracing tissue before transfer of measurements takes place.

Place the specimen on the table, near the edge, in such a position that when you bend your neck automatically you will be looking directly at the specimen with both eyes. Directly means looking at the center of the specimen. Measure the subject using pointers or a compass. Be sure to mark all measurements as the sketch progresses. During the process of measuring, remember that:

1. First you must measure the distance from the vertical line to the end of the specimen.
2. You must measure the distance between both ends, left and right, of the specimen.
3. You must measure the distance between the vertical line and the closer end of the specimen to that line.
4. You must measure the distance between the horizontal line and all points making the outline of the specimen.
5. You must mark all of the measurements as they are taken.
6. You must hold the measuring tool with both hands.
7. The measuring tool must be held in a position parallel to the base.

8. The specimen must not be accidentally touched or dropped.
9. You must pay attention to details, but first you must obtain general proportions.

The most common problem is double vision, which you may experience to some degree. This means that the tips of the measuring tool, when aligned with prescribed points, will appear as a double image. If the problem persists or if the doubling occurs with regularity, it means that one of the muscles controlling the movement of your eyes is not working properly. In such a situation, in order to produce a correct drawing, use one eye. Position your head correctly and close one eye. Depending on which eye you close, left or right, draw the respective half of the specimen. In order to draw the other half of the specimen correctly, you will have to turn the specimen and your sketch around, positioning both upside down. The object is to draw with a minimum of distortion. When using only one eye, again depending on which eye it is, the opposite end of the specimen will not be viewed directly. It will be viewed from the side. To be familiar with the problem, look directly at the center of the specimen with both eyes open. Close one eye, then open it, at the same time closing the opposite. Observe how the specimen appears to jump from left to right. Observe the details at both ends while closing your eyes. You will notice that when the left eye is open the left end is more visible; when the right eye is open the right end of the specimen comes into full view. In order to maintain stereoscopic presentation of the specimen, you must keep both eyes open or, if this is impossible, you must draw only half of the specimen at a time. For practice, repeat the sketching procedure with various objects.

Taking Notes While Sketching

Why should we write on sketches? What should we write about? Writing is an aid to help us remember encountered problems. It is a collection of good and bad information. It is an important step in communication between you, the scientific illustrator, and the scientist. In many cases, during preparation of the sketch it is impossible to present everything without written information. Just imagine a complex structure that is unknown to you, a reconstruction of creatures compiled from various sources. Do you think you will not have any questions? Well, the scientist, himself an expert in such matters, will have continual questions. You, the illustrator, will have some doubts about shapes, lines, dents, and formations of the structure. As it is impossible to remember a passing thought, it is better to write it down. When coherently expressed, it may help you when drawing and it may also help the scientist in his research. Do not be afraid that your thoughts are worthless. Do not be afraid that the scientist will make fun of your writing. He will not, as he greatly values such observations. It may happen that your notes are of no help to him, but at the same time all notes made during preparation of the sketch will be of great help to you. After approval of the final sketch, you will know what should be drawn and how it should be presented. This does not mean that writing on the sketches should become more important than the sketch itself. Try not to write long essays. The object of writing is the preservation of clarity and mutual understanding of the subject, primarily for you, but also for the scientist. Again, writing does not have to appear on all sketches. Be informative, especially when you have a question. Write some information for yourself. Do not try to lecture the scientist about the subject you are drawing. Above all, write legibly. Naturally, if you are interested in the visual appearance of the sketch and have some knowledge of calligraphy, then use it. Remember that written information will fall into seven basic categories.

1. The shape of the subject, meaning removal or addition of lines depicting the subject
2. Presentation of the proper count of sections of the subject
3. Notes pertaining to details, which can be supplemented with additional drawings of those details
4. Surface of the structure, explaining the ups and downs on the subject as discussed with the scientist
5. Coloration, to be presented in color or in black-and-white tonal values
6. Comments for yourself only, such as "see the structure from the other side for reference," "this is OK," or simply "wrong conclusions"
7. All pertinent data, such as the number found on the specimen, enlargement or reduction factor, job number, number of the sketches, and anything else needed for identification of your sketches

The true value of taking notes will become apparent when working on numerous projects at the same time. In such a situation some of the projects will receive priority over others for a while. This means that you will find yourself placing some unfinished projects aside to be worked on at a later date. After a few months it may be very hard to recall a lot of information. We are forgetful. Instead of making a pest of yourself and constantly calling scientists for the same basic information, it will all be there in writing.

This does not mean that you will not discuss the project again. In all probability you will do so, but at the same time the basic information will be preserved for reference.

EXAMPLE 1. The project calls for a color illustration of a plant. The illustration will appear in a guidebook, therefore color is important. Material received by you consists of 35-millimeter slides as well as a live specimen. Recording the color is a must; describing the stiffness of stems is absolutely necessary. Discussion with the scientist is essential. The slides, in color, will vary a lot, depending on the exposure, the type of film, and developing procedures. The plant itself will die during preparation of the drawing. It can be placed in preserving liquid, where it will virtually instantly lose all its coloration, or it will be injured by you during handling. In any case, the color of the specimen will change. Take notes in writing or paint actual patches of color with appropriate designations to help you remember and reach an understanding with the scientist. The scientist will describe the colors and you should write down all he says. If you do not preserve all information, the correct presentation of the specimen will be impossible.

EXAMPLE 2. Drawing of a fish. The specimen is well preserved and not damaged by dissection or handling. In order to compare your observations with those of the scientist, you must present in writing an accounting of spines and rays per fin. Any questions as to the location of breathing pores should be noted. The problematical placement of soft tissue between spines and rays should be noted. Write respective numbers for each spine and ray. It will help you to check the correctness of the sketch and will help the scientist to recheck the proper count when the sketch is compared with his notes.

EXAMPLE 3. A reconstruction of a crustacean for paleontology. The project has been in your drawer for a year. You have no notes, no information. During sketching, make sure that "This is correct view" has been indicated clearly. Otherwise you will not remember what you should draw. Details are marked "See sketch from camera lucida." Small things such as the distance between internal structures should be marked with an arrow pointing to the detailed section and with explanatory writing: "Important, this should be *V*-shaped." Parts that are not to be drawn are marked "Out." A ridge is noted "Rounded ridge with shallow lateral grooves becoming more prominent posteriorly." The questionable sections are marked "Omit?" Naturally, the view, ventral or dorsal, is clearly written on the sketch. How can you remember all details after a year? You cannot. Therefore, write information down.

Make sure that lines with arrows point the information into appropriate places. Remember that you will have to read what you write, so write legibly. Do your best to place all written information away from the drawing so as not to obstruct the sketched image.

Preparation of Final Sketch

The final sketch is, in most cases, the last of all sketches. You may be surprised how many changes can be implemented in to the final version. The final sketch should be drawn on one piece of tracing tissue. Try to avoid presenting to the scientist a composite paste-up made of small sheets of tissue held together with tape. It is not that such presentation is insufficient; it is definitely a workable presentation. But as a professional, you should care about projecting that image. A composite does not present clarity and readability like the final version will. It will not have the cleanliness of a retraced image. The preparation of the final version means extra work for you, but it is worth all your effort. The main reasons for preparation and presentation of final sketches are:

1. Readability of the image
2. Clarification of the information for you and the scientist
3. Allowance of space for necessary corrections and changes
4. Presentation of yourself as a professional illustrator

It is much easier and better to present the arrangement of a full-page illustration, a *plate*, when separate sections are neatly traced with a sharp line within designated parameters. Naturally, the technical preparation of the final sketch will vary from project to project. It depends on the complexities of the subject and it will depend on your experience with that subject as well as your experience in the field of scientific illustration. With so many variables, you may wonder what to do. Should the final sketch be prepared, how should it be prepared, and how long should preparation take? Remember that when starting in any business, your diligence will be a great plus for you. This means that in virtually all situations except those when the scientist states a definite "no," you should prepare a final sketch. Projects involving reconstruction or visual description of a new species will always require a final sketch.

The length of time spent on the final sketch will vary. Complicated subjects will take more time than simple ones, but do not judge too quickly the complexities of a subject. It does not mean that a simple-looking specimen will produce a simple-looking drawing, or that when working on a structurally complicated subject you will end up with a complicated result. In many cases, the final sketch is necessary to clean up excessive and unwanted information. Let the scientist be the judge of that. Refinement of the information may be necessary. Remember that your drawing is produced as an illustration depicting a specific set of problems. With the first batch of sketches, get ready to prepare the final sketch by arranging all pertinent visual information in accordance with specifications. Be selective as to what will be presented in the final version. At this time you should have a good idea of what is important. Tape the separate sketches together, making sure that their surface is smooth. Tracing can be accomplished with the use of a light box or without, depending on the transparency of the tracing tissue you are using. You know that you must see what you are drawing. This is essential. If the tracing tissue is relatively opaque, use of a light box is necessary. When thin and very transparent tissue is used, make sure that smooth-surface white cardboard or illustration board is placed under the first sketches. A rough surface will produce slightly ragged lines. The final version should be clean and the lines straight—no bumps and no wiggles. Place the new sheet of tracing tissue on top and tape the tissue to the illustration board. Remember to wash your hands and keep the surface of the final sketch clean. Do not tape the whole thing to the table. In order to produce even lines, you will have to turn around the illustration board while you draw. When using a light box, be sure that the new sheet of tracing tissue is taped to the sketch or sketches to be traced and refined. Again, do not tape the tissue to the light box. It is definitely easier to turn the paper around rather than yourself. Depending on the subject, use an aid when tracing curved lines. A French curve is a must when drawing spines and rays on a fish. Be selective when writing any information on the final sketch. Write what is pertinent: identification of the specimen and all important questions you may have. If necessary, number the segments of the structure. Never forget to write the magnification factor and draw the scale. Preparation of the final sketch should not take much time. After all, it is a clean tracing of previously compiled information. Naturally, when you are asked to draw the sketch of the subject directly onto the paper used for rendering, there will be no need for a final sketch. In such a situation, your first sketch must be prepared very cleanly and very precisely. There will be no room for indecision. Remember to use the appropriate hardness of pencil for the paper you use. Try to match the right tools for better results. The pencil can be neither too hard nor too soft—a very hard pencil will injure the surface of the paper; while a very soft one will not give the required precision of line, and will also smear easily. Make sure that the weight of the line is consistent.

EXAMPLE 1. The subject is a pictorial reconstruction of pottery. Various preliminary sketches have been prepared. The shape of the vessel has been established; the sketch has been drawn in appropriate positions. The design found on the surface of the remains has been deciphered and presented separately. The final sketch will have to combine all of the information. The size of the final sketch will be dictated by the initial enlargement or reduction of the subject. Proper arrangements for retracing of the basic data have been made. The pencils are ready and clean illustration board has been placed on the drawing table. The initial sketches are arranged in order designated by the process of deduction and are sprayed with fixative. Remember about cleanliness: If the initial sketches are not sprayed with fixative, the image you trace will appear on the back of the final sketch. The graphite will be transferred by the pressure of the pencil. When fixative is not available, take another sheet of thin and transparent tracing tissue and place it between the first sketches and the sheet used for the final sketch. It is hoped that the image can be seen through the layers of tissue. Trace the sherds by hand, but do not shade. Do not apply any values or tones onto the final sketch. Remember, the final sketch must be easily readable. Use only line. If you want to practice application of tone, use a separate sheet of tracing tissue. At this stage the tone will obstruct essential information. Use a French curve to draw the walls of the vessel. The missing sections will be drawn in short lines and parts aligned with sherds in continuous line. All designs on the sherds will be drawn in line. It is true that later the design may be rendered in tone acquired by stippling or other technique. The flat reconstruction of the design running through the whole length of the vessel will be depicted on the right side and also drawn in line. The lines must be clear. Erasing is neither advisable nor needed. After drawing the scale at the bottom of the final sketch, you are ready to call on the scientist for final approval. Be prepared for changes. Ask him to draw on your final sketch. This is the safest way to communicate. When the sketch is approved, prepare yourself for the production of rendering.

EXAMPLE 2. Project for botany calling for one drawing of a habit of a fern, details of construction of the small segments of the specimen, and presentation of construction of one of the leaves. The sketches are prepared. The information pertaining to the size of the printed page should be obtained from the scientist. This is very important because virtually all botanical illustrations of this type are full-page illustrations. You must know the size of the page before arranging any images, as it is necessary to produce an illustration that will be proportional to the space available for the printed image. Knowing the size of the printed page, measure the length of the longest first sketch. This will give you a rough base for establishing the width and length of the illustration. Take the proportional wheel. The scale on the larger wheel represents the size of the reproduction. The smaller has a scale for the size of the original. Find the length of your measurement on the smaller wheel and align this measurement with the known *length* of the printed page. Hold the wheels together with your fingers and find the known *width* of the printed page on the "size for reproduction" scale. Read the corresponding number from the "size of original" scale. This will be the width of your illustration. You will not be permitted to place any images outside those boundaries. If you need more space for your first sketches, you will have to adjust the length and width in proportion to the size of the printed page. Write the measurements and draw, using a ruler, a rectangle corresponding to the sizes read from the "size of original" scale. Tape the sheet of tracing tissues with the rectangle to the illustration board, placing the tape at the top of the sheet. You will have to lift this sheet in order to place the first sketches under it. Take the first sketches and place under the rectangle, making sure that the arrangement of images corresponds with the scientist's requirements and that none of the images will be visible outside the drawn borders. Leave some space between the border and the images. You may be asked to draw the border on the finished rendering. If images will touch or be too close to the border, the readability of the illustration may be affected after the finished drawing is reduced for printing. When satisfied with the arrangement of first sketches, tape the sketches to the illustration board, making sure that their positions have not been changed in this process. Trace all of the first sketches onto the sheet with the border drawn. When finished, the final sketch will represent the *plate*. The final version of the illustration has been reached. Here, as in all situations, shortcuts can be taken. When the project calls for a large quantity of full-page illustrations, do not try to figure out separate proportions for each illustration. Try to establish some uniformity; you will save time by doing this.

EXAMPLE 3. Drawing soft-tissue specimen using the camera lucida. As in all cases, the preliminaries are finished. You have nothing to arrange, having one specimen drawn from one view, but you may have an image made with shaky lines. There may be too many lines that are unnecessary, although they may be helpful during the initial stage of sketching. Clean the sheet by retracing the image. Drawing through camera lucida will produce unevenness of the line. I would strongly suggest that you present the final version in a clean, professional fashion.

Exercise 1. Take your first sketches of any subject and trace the image. Work slowly but consistently. Make sure that the density of line is consistent. Pay attention to the shapes on the first sketches and make sure that the new image will *not* differ from the previous. Use two methods of transferring:

1. Use a light box.
2. Trace directly without any light coming from below, seeing the image through the tracing tissue.

Exercise 2. Obtain a proportional wheel. A large art-supply store should have one. Choose the size of the printed page from a printed publication, be it a scientific journal or any book. Make sure that you will not confuse the size of the page with printed area. Measure the width and length of the printed area. Draw the rectangle, using a ruler and the triangle. Make sure that the lines are at right angles to each other. Choose a length at random, making sure that it is longer than the length of the rectangle. Do the following:

1. Find the length of the illustration on the proportion wheel. The smaller wheel usually is designated "size of original." The original is your illustration.
2. Match the length of the illustration with the length of the printed space. The larger wheel is designated "size of reproduction." Reproduction is the final printed image.
3. Find the width of the printed space on the larger wheel.
4. Match the respective width of the printed space with the number found on the smaller wheel.
5. Look at the cut-out window placed below both scales. Two designations will be found: the "percentage of original size" and "number of times of reduction." When the small wheel is rotated, notice that "number of times of reduction" stops at 1. This means that all of the numbers on both wheels will be aligned. The image will not be reduced or enlarged. It will be "same size" or "actual size."

Repeat the process until you are familiar with it.

Exercise 3. Take a sheet of tracing tissue. Using pencil and a continuous line, draw a curved, swirling line all over the tissue. Tape another sheet of tracing tissue over the initial image. Take a French curve and a pencil, and retrace the image, using the French curve. When drawing, pay attention to the ends of the curve. The ends have a tendency to slip under the tissue and rip the paper when the curve is moved from one position to another. After drawing a line, pick the curve up rather than sliding it. Make this a habit. When drawing, turn the pencil around in your fingers. This will produce a more uniform line. Try your best to continue tracing that swirling line, starting from one end rather than jumping around. It is hard to join lines when drawing from two opposite directions. Do not align the edge of the curve exactly with the part of the image to be traced. Leave a bit of space. Keep the pencil at a slight angle except for very tight turns. Make sure that the new traced image will not be smeared by the French curve and your hand. Do not attach any coins or tapes to the undersurface of the French curve, as coins will raise the curve above the tracing tissue but will also dent and damage the surface of the paper. Tape will attract a lot of dust and graphite and will smear the pencil. A raised curve will make it difficult to align the edge of the curve with the drawn line. The result will be more difficulties and less precision.

Approval of Final Sketch by the Scientist

Having arrived in the scientist's office with final sketches in your hand, do not immediately place them on his table. Let him direct you where to find a space. Be polite, but behave functionally. How would you like to have somebody come to your drawing table and without asking for permission, place some kind of material on top of your drawings? If you would not have any bad feelings you are an unusual person. When meeting with the scientist, remember that he is your customer. Your position and salary depend on his satisfaction. When delivering final sketches, take care in presenting them. Do not crumble papers; do not rip edges or simply dump the whole thing on the table. Lots of work has gone into the drawing. Remember that you will be judged by your presentation of the sketches as well as by your ability to draw. Let the scientist take time to look at the sketches and compare the drawn images with the specimens and his notes. Make sure that he feels free to write or draw on your sketches. When he participates in corrections or changes, life will be easier for you. The scientist, by drawing or writing directly on the final version, will be able to communicate his information more easily. Talking may lead to misunderstanding, while visually presented information is more coherent and easily remembered. If the final sketch does not require any changes, feel satisfied and proceed with the production of the finished illustration. On the other hand, whenever any changes are necessary, make sure that all are implemented. Make sure that the final version of the sketch is redrawn as many times as necessary. In some situations it will be your fault and in others the scientist will simple change his mind. Remember that visual presentation of the subject is very important, and that in the scientist's mind, the concept for the presentation of a particular problem may take place after the final clean version is produced.

Why should the scientist change his mind? Your drawing, the presentation of the subject, will influence the scientist's thinking process. In many cases he may have a certain idea of how to explain his research before the final sketch is produced, but afterward, seeing the result, he may discover that it does not match the context of his writing. Producing a scientific illustration is a cooperative venture. It is the work of a team made up of yourself and the scientist. You are helping him as much as he helps you. Naturally, discussion about the drawing will take place. If you have any questions, now is the time to ask. Now is also the time to explain the intricacies of rendering. At this point the illustrator knows what technique will be used. If you think that the scientist may expect too much from the required technique, be sure that both of you reach an understanding. The production of the sketch has been based on pretty tangible evidence, but the rendering in tone remains in your mind. The scientist, unless he is familiar with your work, does not know how it will come out. It may be wise to suggest that you will produce a trial version of the finished product. Then the scientist will be assured as to the final look of the illustration. Are you too shy and cannot talk? Overcome such a tendency. In order to maintain your position, talking is necessary. If this is a real problem, then before going to the meeting write, point by point, all necessary questions, suggestions, and comments. Remember that most scientists do this because they realize it is possible to forget some things. Do not be afraid to ask the scientist to write an OK, with his initials, on the sketch. It is a good prevention against the future. When the final sketch is approved it must be rendered properly.

EXAMPLE 1. Project calling for one drawing, a lateral view of a fish. The final sketch has been drawn clearly and everything in the illustrator's mind is ready to go. During the meeting, the scientist checks the sketch by comparing his notes and the specimen with the sketch. The scientist is taking his time counting spines and rays in each fin. He has discovered that the last ray on the dorsal fin is presented as a singular structure when in actuality it is divided into two separate structures. When counting scales it is discovered that there is one too many. He is making notes on the sketch and is asking you to look through the microscope to see this. You, the illustrator, are a little bit tense and, in actuality, do not see what he is pointing at. Do not agree blindly. Tell him that you do not see the required structures, but you will carefully investigate the problem. Make sure that you listen carefully to all comments and make sure that the areas in question are marked on the sketch. After the meeting, investigate the matter and implement all corrections. If corrections are small, such as the ray on a dorsal fin, do not redraw the subject. When major areas are to be changed, the final sketch will have to be redrawn completely.

EXAMPLE 2. The scientist is very busy and cannot spend much time discussing your final sketch. He simply agrees that it is just fine, very fine. The meeting is rather short, and as you are preparing to leave, he points to a section of the drawing, stating that the depicted bump cannot be there. Ask him to look at the specimen and observe. If he refuses, stating that he knows what he is talking about, you do not have much chance of convincing him that you are right. Again, tell him that the problem will be investigated carefully and you will remove the bump. Take your sketch to the drawing desk and carefully compare it with the specimen. If the bump is there, draw another sketch without the bump and ask the scientist for another meeting. Present both sketches; his choice will be final. After all, the bump may be present on your specimen but may not be a characteristic feature of the whole structure. Render the drawing according to the scientist's specifications. Do not try to force your ideas. He knows what he wants.

EXAMPLE 3. The project had a specified number of drawings of soft-tissue structures. The scientist stressed that the sketches must be produced on the same paper, which will be used for rendering. The rendering will be in black and white, using line and stipple. Naturally, you realize that because of such technical limitations it is impossible to use a lot of lines. The more pencil lines on the surface of the paper, the greater difficulty in erasing all pencil from under the final pen-and-ink rendering. All in all, the final sketches are ready and the scientist is about to make changes. Make sure that a sheet of tracing tissue is placed on every final sketch before the scientist's corrections are indicated. If the surface of the paper will be damaged, you, the illustrator, will have to struggle during the final rendering. Ask the scientist to draw on an overlay. Ask him to draw lightly, without much pressure. In such a situation you must be assertive and prevent accidental interference with future rendering. Prepare tools for drawing before the scientist reaches for his ballpoint pen. Have sharpened pencils ready. When he picks up the overlay and is about to draw on your sketch, stop him. Tell him not to touch. Say, "The surface of the paper is very sensitive, let me do this." Draw yourself; erase yourself. Naturally, if he is very insistent and is about to put up an argument, do not engage in verbal violence. In such a situation let him use his ballpoint pen, and let him draw or erase on the sketch. You are simply stuck in an unpleasant situation. Remember that at all times it is possible to repair a damaged drawing and it is possible, although difficult, to draw on a damaged surface.

Corrections on Final Sketches

The final sketch has been prepared. All necessary designations are noted and the meeting between you, the scientific illustrator, and the scientist is nearing an end. Be prepared for corrections and changes of the final sketch. Do not regard the final sketch as an end in itself. It is not. The final sketch is one of the steps in the production of a high-quality, professionally prepared illustration. A lot of work goes into such a sketch, so take care and restrain yourself from placing a sandwich on top of it. As a matter of fact, try to make a blueprint copy of the sketch if semitransparent paper has been used. It is good for records as well as for reassurance. If a copier is not available, do not waste time redrawing the sketch; just be a bit careful. Remember that the scientist is your customer, therefore you should implement all changes indicated by him. Now, depending on quantity of changes, make up your mind and get to work. How changes will be achieved depends on the materials used for preparation of the final sketch as well as on the technique of rendering. If tracing tissue has been used, corrections will be easy regardless of the choice of technique used for final rendering. When using opaque papers for the final sketch, corrections will require altering that surface by erasing or painting over unwanted lines. Both situations will damage the surface, thus interfering with future rendering. Why? Remember that opaque papers may be used for the final sketch because of the scientist's specifications. He or she may ask you to produce the sketch and the rendering on the same piece of paper. It is impossible to predict what type of opaque paper will be specified, as this will depend on availability of papers as well as his or her personal experience. It may be Strathmore with kid or glossy finish, some kind of white cardboard, or even slightly textured paper. The difficulties of correcting the drawn image will multiply if the surface is too absorbent or too hard, and especially when a drawn line will be embedded into the surface. For that reason you should remember the following:

1. Never press a pencil into the surface during drawing or correcting the sketch.
2. Ask a lot of questions, thereby escaping major changes.

If you make sure that the final sketch is drawn lightly but legibly, you will not have to live through a lot of stress. Remember at all times that whiting agents such as acrylic, Liquid Paper, or Pro White will interfere with final rendering. In situations when small corrections have to be implemented, use whiting agents, but with great caution. When the final sketch is beyond repair, a new one will have to be produced. In some cases it may be possible to project the image through an opaque projector and carefully trace the good section. Whenever the retracing technique is used, remember the enlargement factor. A projector will enlarge or reduce the image, seriously interfering with implementation of necessary corrections. In such a situation it will be just like drawing the whole thing from the beginning. Remember that all shapes and lines must match correctly. You will not be allowed to stretch reality for the sake of personal convenience. When using tracing paper, you will have a lot of flexibility. The final sketch can be retraced very easily without losing its original scale. It can be cut and later spliced. In order to change a small section, take a new sheet of tracing tissue and place it over the unwanted area. Secure the sheet in two places to the underlying final sketch with tape. Make sure that the tape does not obstruct the image and will be placed along one edge of the new sheet of tissue, making the tissue a sort of flap. Draw necessary corrections on the new sheet. Do not erase anything from the old sketch. Let the scientist see both versions. Do not remove such corrections for presentations. If a major section has to be redrawn, then, depending on the subject, the size of the sketch, and the size of the specimen, you have the following choice:

1. Redrawing the incorrect section and splicing a new part with the previously drawn sketch using transparent Scotch tape
2. Drawing new corrected lines and marking the old incorrect ones with the typographical "delete" symbol
3. Rendering the whole sketch over from the beginning

Naturally some subjects may be very difficult to correct. The difficulty will not be in drawing or taping papers together, but in finding the mistake. The scientist will indicate that certain structures are not in proper order or quantity. The problem will be to find the correct alignment or finding that missing part. In such a situation you will have to spend additional hours observing the specimen and correlating the observations with the drawn image. It may happen that when a mistake is discovered, one small section will ruin the previously recorded organization of structures.

EXAMPLE 1. A final sketch has been drawn on opaque material. The paper is a three-ply Strathmore, making tracing impossible. The subject is small, only visible under a microscope, therefore the drawn image is in actuality a "tracing" of the subject with the use of a drawing tube. The rendering is to be produced in ink. The scientist has specified that in order to speed up the production and cut costs, initial sketches should not be produced, and the only paper available will be selected from leftovers from previous projects. This specification results in production of one sketch, the final one, on the same surface used for the rendering, to be accomplished in ink. After rechecking the final sketch, the scientist has indicated that certain anatomical structures are not drawn correctly. He has clearly stated that imperfections found in the drawing are not produced by incorrect observation during the process of drawing, but by changes in his research. The corrections are important, and happen to be on virtually the center of the sketch. You are lucky that during the preparation of the sketch pencil lines were not deeply embedded into the surface, and that the finished illustration will be rendered in ink. Because of this, erasing unwanted structures is possible. You have the following choices:

1. Erase unwanted structures with a hand-held eraser.
2. Erase unwanted structures with an electric eraser.
3. Lightly overpaint unwanted structures with white paint.
4. Draw corrections over the previous drawing without any erasing.

Remember that after rendering, all pencil lines will have to be erased. If too much white paint is used, the paint may come off during that process or ink placed over whited areas may smear. The ink smearing will depend on the type as well as the age of the white paint. In some cases, aged, premixed whiting compounds will not dry or will not hold ink. Try the paint on the side of the paper and see how it works. Do not use untested materials directly on the drawing. If initial pencil lines are drawn lightly, you may draw new structures over the old ones with a slightly darker line. In this case, make sure that you explain to the scientist what the new changes are. Later the whole pencil sketch will be erased with art gum. When using an eraser, be sure that a bit of the good structures is erased. Implementing corrections will be easier because you will not have to align the sketch so precisely with the

Plate 1. *Bembidion alaskense* ♂ (Coleoptera: Carabidae). Mixed media (mainly colored pencils and Pelican inks with some watercolors and Pantone markers). Drawing by David R. Maddison, The University of Alberta, Edmonton, Canada. Published as a cover illustration of *Entomologica Scandinavica*, Supplement No. 15, and to be published in David R. Maddison, "Systematics of subgenus *Chrysobracteon* of *Bembidion*." Painted in 1981.

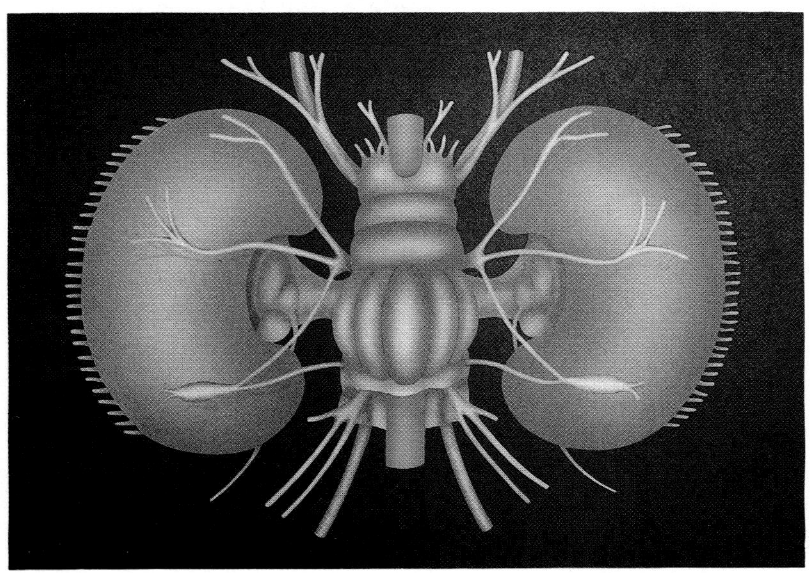

Plate 2. Central nervous system of a *Sepia* (Mollusca, Cephalopoda). Airbrush with acrylic enamel paint on wood panel. Painting by Willy Lauwens. Produced for display at the Koninklijk Belgisch Instituut voor Natuurwetenschappen, Brussels, Belgium, New Natural History Museum, Invertebrates. Drawn in 1983.

Plate 4. *Gastritis*, possible result of excessive use of the common over-the-counter drug aspirin. Airbrush and hand painting on colored drawing board using acrylic paint. Painted by William R. Schwarz, University of Illinois Health Science Center, Chicago. Produced for a TV health program titled *Over-the-Counter Drugs*, with guest Dr. Robert G. Mrtek, College of Pharmacy, and F. Keith Fearon, producer of *Consultation*, The University of Illinois Health Television Series, Chicago. Painted in 1977.

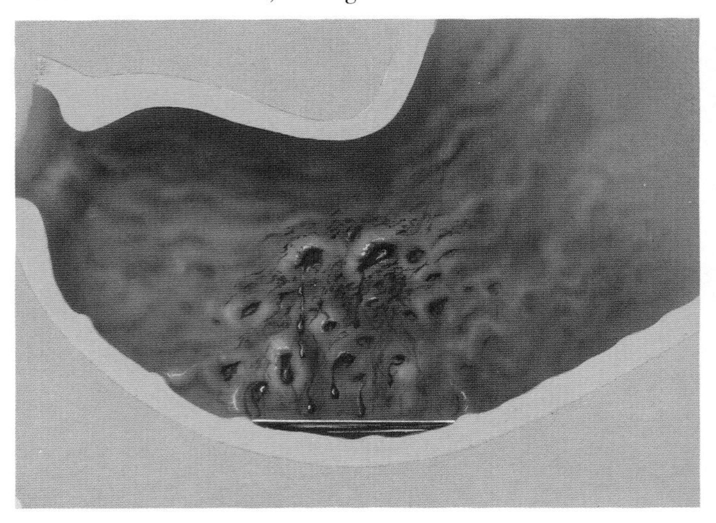

Plate 3. *Lyrurus tetrix* (Black Grouse), *Tetrao urogallus* (Capercaillie), *Mergus merganser* (Goosander). Watercolor and watercolor with pencil. Paintings by Matthias Haab, Zürich. Published in *Vogelschutz in der Schweiz* (*Schweizerisches Landeskomitee für Vogelschutz*), Switzerland. Painted in 1982.

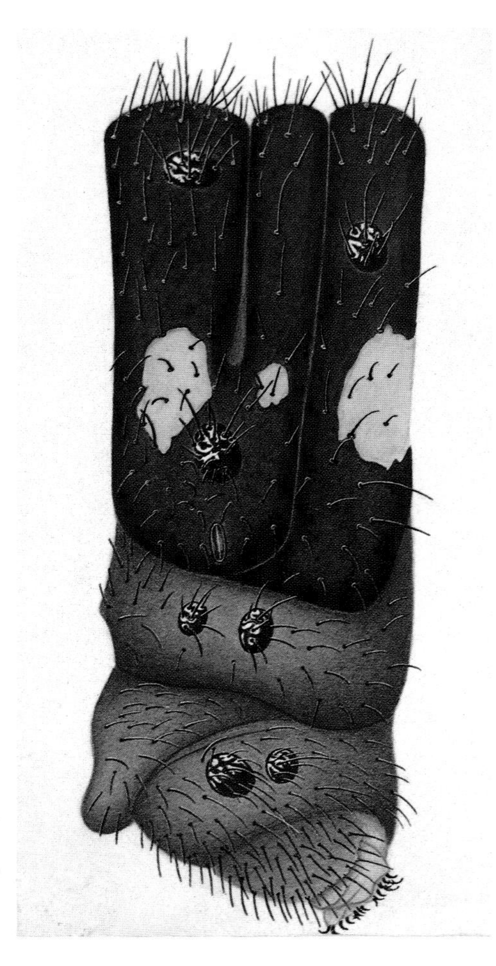

Plate 5. *Parnassius apollo pardoi*. Watercolor and gouache on matte-finished cardboard. Painting by Ricardo Abad Rodriguez, Spanish Institute of Entomology, Consejo Superior de Investigaciones Científicas, Madrid, Spain. Published in various publications.

Plate 6. *Fugu exascurus*. Watercolor. Painted by M. Shirao, Japan, World Life Research Institute, Colton, California. Unpublished. Painted in 1968.

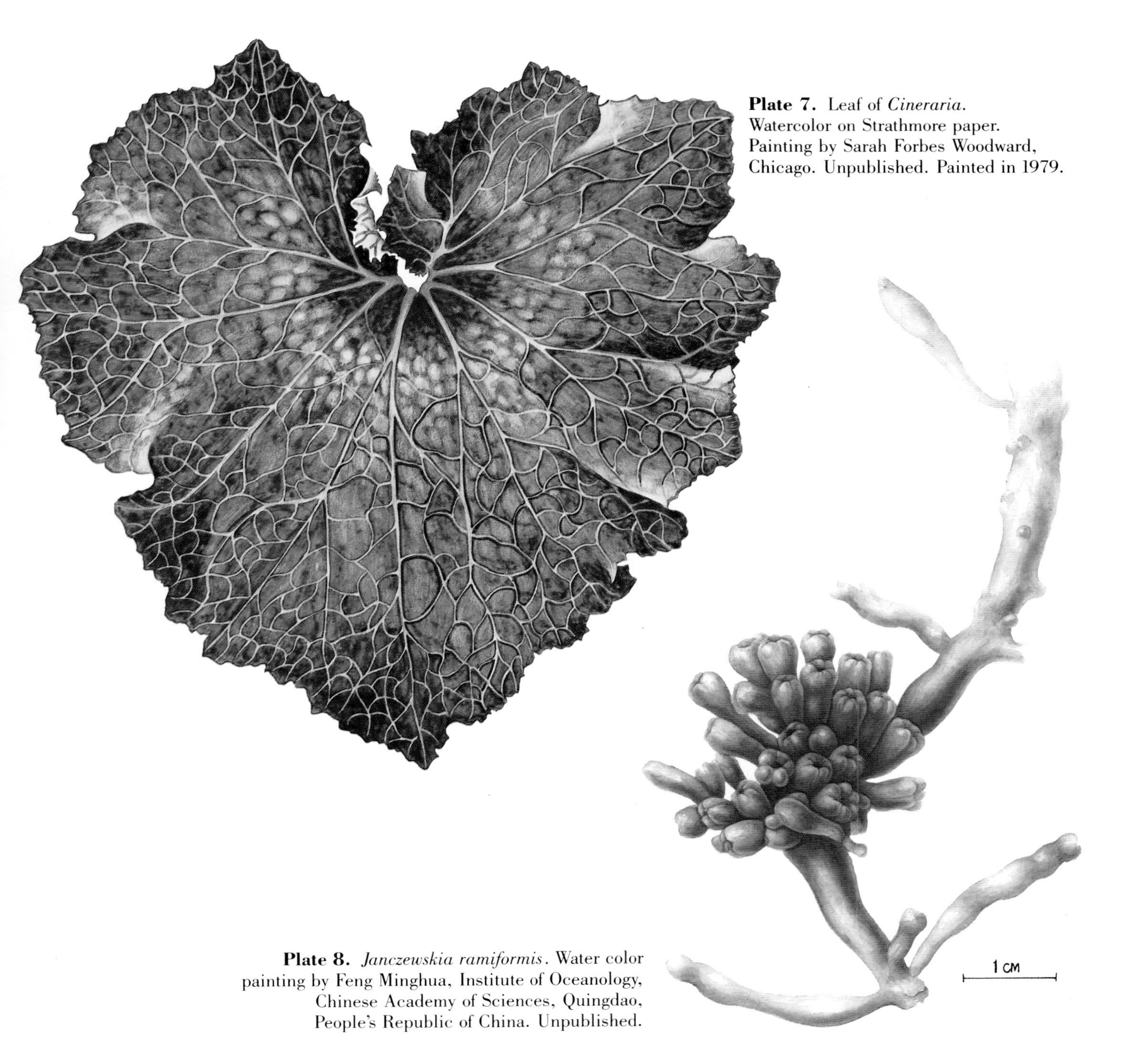

Plate 7. Leaf of *Cineraria*. Watercolor on Strathmore paper. Painting by Sarah Forbes Woodward, Chicago. Unpublished. Painted in 1979.

Plate 8. *Janczewskia ramiformis*. Water color painting by Feng Minghua, Institute of Oceanology, Chinese Academy of Sciences, Quingdao, People's Republic of China. Unpublished.

Plate 9. Woman's dress from Laesø, Denmark, 1850–1900. Tempera paint. Painting by Beth Beyerholm, Institute of Cell Biology and Anatomy, University of Copenhagen, and Danish National Museum (Antiquities), Copenhagen, Denmark. Published in *Folkedragter (Old Danish Costumes)* by Lademann, Copenhagen. Painted in 1978.

Plate 10. Illustration of Solar Maximum Mission (SMM) satellite. The SMM was the heaviest satellite to be lifted into orbit by a Delta rocket. Delta 3910, the SMM's launch vehicle, utilized nine solid propellants and strap-on boosters, in addition to first- and second-stage motors to place the 2315 kilogram (5105 pounds) spacecraft into circular orbit at an altitude of 574 kilometers (357 statute miles). Tempera paint. National Aeronautic and Space Administration, Goddard Space Flight Center, Greenbelt, Maryland. Painted in 1980.

specimen when the sketch is viewed through a drawing tube. If an electric eraser is to be used, make sure that the final sketch is well secured and that you are familiar with the power and motion of the eraser. It is very easy to remove too much of the drawing or too much of the surface. When using hand-held erasers, make sure that whatever you use will not smear good sections of the sketch. For precise erasing, use pencil-type erasers such as Eberhard Faber Stenotrace eraser number 1507. Try to avoid highly abrasive erasers because of the possibility of excessive damage to the surface of the paper. When the surface is very damaged the fibers constituting that surface will make the ink run during the rendering. Remember that overpainting or erasing may not produce a visually pleasant rendering, but at the same time will *not* interfere with the quality of reproduction. Why? The finished product will be a black-and-white ink line rendering, and as such will be photographed on high-contrast film during platemaking. Such film will remove all visually unpleasant blemishes from the original artwork.

EXAMPLE 2. Proudly presenting the finished and well-designed sketch of a fish to the scientist, you are not anticipating any problems. Everything is as it is supposed to be. The scientist, while comparing his notes with the specimen and the sketch, has noticed that the number of scales presented on the sketch does not coincide with his notes and with the specimen. With great patience he has counted all scales on the specimen again and again, coming out with the same conclusion: One is missing. The problem is which one and where it is located on the fish. You have to find that scale before progressing further. The final sketch is drawn on tracing tissue. Making a correction seems easy. The drawing part may be easy, but finding where and what the error is may be difficult. Two possibilities are facing you:

1. Pretend that you have found the missing scale.
2. Redo the study from the very beginning.

Do not try to add an extra scale, as this will eventually be discovered and your reputation will disintegrate. Truthfully, the only way to repair the final sketch is by redrawing the initial sketches. You have to go back to the microscope armed with a teasing needle and a fresh sheet of tracing tissue. Draw the scales one by one. Make sure that you pick up each one with the teasing needle, making you aware where each particular scale ends. Number all scales as the drawing progresses. When finished, place the new sketch over the previously finished final sketch and register both drawings properly. Carefully examine the placement of all scales. Upon finding the troublemaker, look for the misplaced arrangement of that particular row of scales. If it is a simple addition, then draw the newly found scale in the previously drawn final sketch with a softer pencil, making sure that it is visible. If the whole row of scales has to be redrawn, take a new sheet of tracing tissue and place it over both sketches. Tape the tissue securely. Draw the new scales in their correct position. Draw parts of the general outline of the fish on the same tissue. When finished, remove all tapes and separate the tissues. Take the corrected version and place it over the final sketch. Make sure the images register by aligning drawn parts of the outline. Now you can tape the second version to the final sketch, making an overlay, or you can splice the redrawn section into the final sketch. When you are finished, ask the scientist for approval before starting the rendering.

Transferring Sketches to Drawing paper

Whenever transparent or semitransparent surfaces are used for the finished drawing, the final sketch does not have to be transferred to the drawing surface. As a matter of fact, it should not be. The sketch will be visible through nonopaque materials. In order to begin rendering, simply place the sketch under a semitransparent or transparent drawing surface and proceed with work. Numerous very-suitable materials are available and can be grouped in three categories:

1. Tracing tissues
2. Drafting films
3. Specially treated or coated acetates

An opaque surface such as regular paper or board will require an additional step in the production of an illustration. Transferring the final sketch to the surface of such materials is a necessity. This process can be frustrating and requires patience. It is too easy to accidentally change the image or to damage the surface of the drawing paper. Never apply muscle power. Watch for excessive pressure of the pencil. Once the surface is damaged by the pressure of the pencil, there may be difficulties during rendering. Do not use soft pencils. Soft pencils such as 2B or softer will produce an uneven line. A line that is too wide obstructs information. Remember that the tip of a soft pencil loses its sharpness very quickly, thereby changing the width of the drawn line. Remember that after the image is transferred, soft pencil will smear more easily, producing unpleasant effects. If you want a clean background for the finished drawing, there are three basic ways to transfer the image:

1. Using light coming through the sketch and the drawing paper,

transferring the sketch by direct tracing

2. Using an opaque projector, again transferring the sketch by tracing
3. Using carbon or graphite methods, transferring the sketch by applying pressure while tracing

The choice depends on the type and thickness of the drawing paper as well as available equipment. In order to achieve proper results, a light box will be required, although a window will also suffice. When using a light box, be sure that the sketch and the drawing paper are not taped to the surface of the glass. Depending on the size of the drawing paper, whether it is longer than the sketch or smaller, be sure that the sketch is attached to the paper and taped, using masking tape, in only two places. Be sure that masking tape is attached away from the drawing area. Naturally, if tracing of a part of the sketch is necessary, do not waste a large sheet of drawing paper to trace a small section of the sketch. Judge the situation according to needs, but make sure that you will not use too much tape. It is not necessary. Excessive use will trap air between the two sheets of paper and produce slight bubbles, making it difficult to trace the image with precision. Do not use Scotch tape or any other highly adhesive tape. When removing the traced image from the taped sketch, a highly adhesive tape may tear the back surface of the drawing paper.

Once you have the sketch and drawing paper secured, place both on the surface of the light box. Restrain yourself from taping both papers to the surface of the light box. It does not serve any purpose. When papers are securely attached to each other, they will not fall apart. During tracing, turn the papers around, aligning the line to be traced with the natural movement of your arm. The result will be a better and more stable line as well as less stiffness in your neck. When papers are taped to the surface of the light box, you will have to twist and contort. Such unnecessary exertion will produce general tiredness of the muscles, resulting in a shaky hand and imprecision. Do not press hard while tracing. The only object is to transfer the image. Rendering will follow later. Naturally, you do know that the thicker the paper, the harder it will be to see through. For that reason, use one-ply paper. Depending on the strength of the light, the weight of the paper can be increased, but it is pretty hard to see everything through three-ply paper. In a case when a light box is not available, use a window. Because it is a vertical surface and you will not be able to turn the sketch around, you must tape the sketch to the surface of the glass separately from the drawing paper, and live with the difficult situation.

Having an opaque projector and a darkened room will help with thick papers. Be sure the sketch is well secured to a sheet of illustration board before being placed in the projector. A sheet of glass, when placed on top of the sketch, will help to stabilize the surface of the sketch: Tracing paper will be flattened, allowing for less motion caused by the projector's fan and for better focusing. Make sure not to tape the drawing paper directly to the wall, as all walls have texture, and when drawing on a textured surface you will transfer all of it to the surface of the drawing paper. Make sure that a sheet of illustration board is placed under the drawing paper and secured properly. Watch for unnecessary enlargement or reduction of the traced image. Naturally, with a large sketch it may be difficult to place the whole sketch in the projector. Do the tracing in small sections. In such situations, depending on the equipment used, it may be possible to reduce the large sketched image to the desired final size. The direct-tracing method using graphite as the transferring medium can produce a mess if caution and patience are not exerted. The sketch must be turned image side down, and on the back all lines will have to be covered with graphite. The easiest way to achieve this is to use a soft pencil such as a 4B. Place the graphite with short strokes, holding the 4B pencil more or less parallel to the surface of the sketch. Cover only pertinent lines, and remember that soft graphite can smear easily on the drawing surface. When finished, carefully place the sketch on the drawing surface and secure it with masking tape. Do your best to restrain yourself from applying pressure to the sketch with your hand. Remember that any strongly applied motion of your wrist, palms, or fingers will produce smears. Trace the image gently, using an H pencil. Do not press too much, as excessive pressure will cause a crease on the surface of the drawing paper and can rip your sketch. At intervals, lift the sketch and see the transfer. Did you forget a part of the sketch? Make sure that everything is transferred as it should be. Do not remove the sketch completely during transfer, as repositioning will be very difficult and in most cases unsuccessful. In order to avoid excessive application of graphite to the drawing surface during the transferring procedure, prepare a separate sketch traced from the final sketch. In order to do this you must turn the first original final sketch with the image facing the illustration board. Tape the sketch and place a new sheet of tracing tissue on top. Retrace the whole image using a B pencil, making sure that nothing is lost and nothing has been exaggerated. When finished, take the new "reversed" tracing and use it as a base for the transfer. The "reversed" image will be placed image side down on the drawing surface. Because tracing tissue is semi-

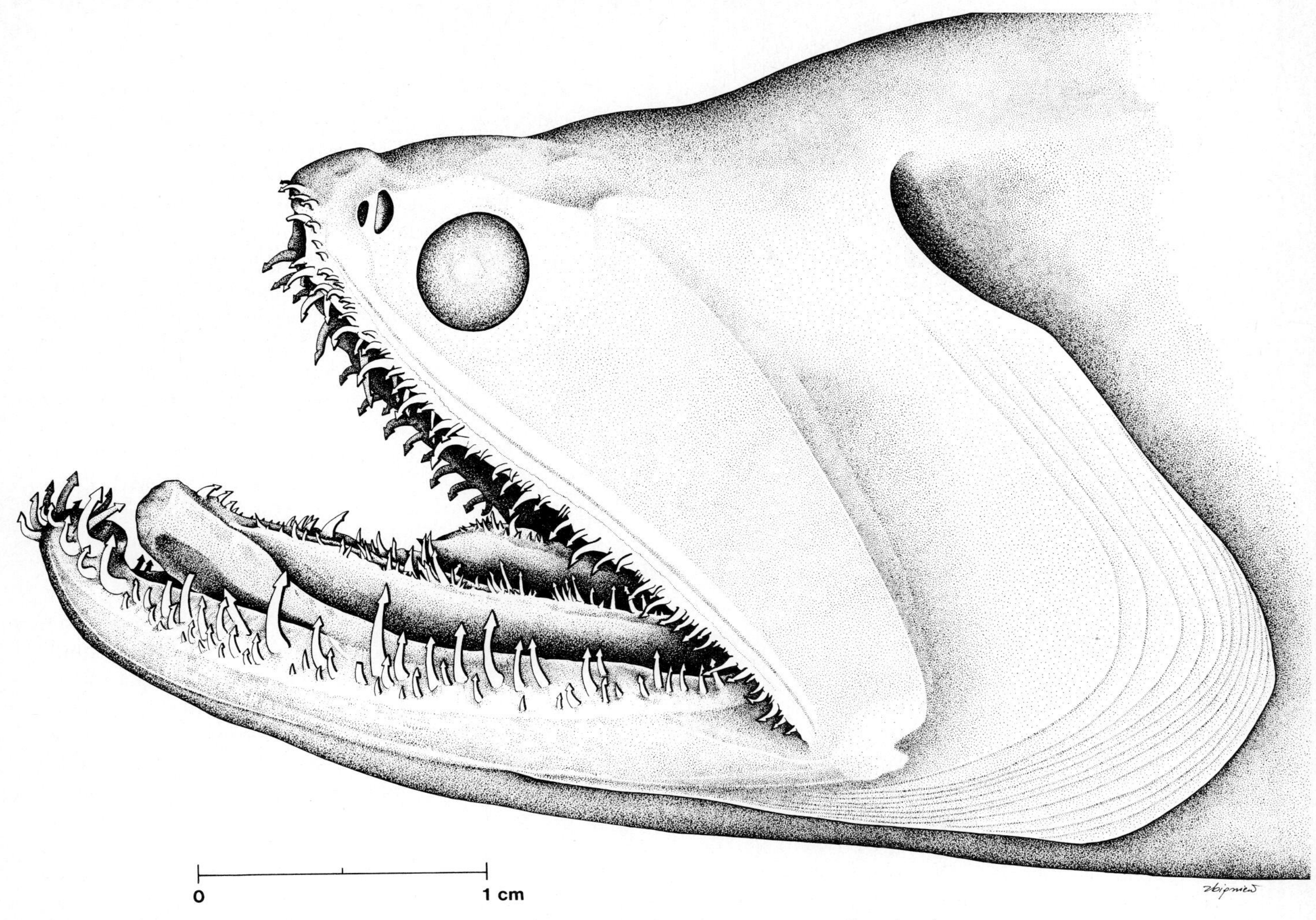

Figure 4-8. *Harpodon translucent* (Bombay duck). Pen and ink on tracing tissue. Drawing by Zbigniew T. Jastrzębski, Field Museum of Natural History, Chicago. To be published in Johnson and Schmitz, "Review of the Fish Genus *Harpodon*," *Fieldiana*, Zoology, 1985, by Field Museum of Natural History, Chicago. Drawn in 1980.

transparent, you will be able to see all the lines. After securing the reverse-image sketch to the paper used for final rendering, draw the image on the so-far-untouched surface of the tracing tissue. The previous top side is facing the drawing paper, allowing the transfer of the graphite when pressure is applied. Use of carbon paper will ensure a proper transfer. Choose from a variety of papers manufactured for this purpose and use it by placing such paper between the sketch and the drawing paper. Again, by applying pressure to the image, a transfer will be achieved. Whatever the tracing method, remember at all times to:

1. Keep the surface of the drawing paper clean.
2. Make sure that the produced transfer is an exact copy of the approved final sketch.
3. Avoid damage to the surface of the drawing paper by excessive pressure of the pencil.

EXAMPLE 1. The sketch is approved and the scientist strongly suggests that "some kind of board" be used for rendering Unfortunately, an opaque projector is not available and the sketch is too large for a direct transfer. The only way out is to reduce the image through photostatting or photography to the required final size. A new tracing is necessary because it is impossible to use papers as the base for transferring process. A light box is out. The direct-transfer method must be used. It is unfortunate that the illustrator did not ask the scientist about his preference of materials. Such a question would have eliminated extra steps in production. Remember that some scientists and some institutions may not have budgets for photostatting. If they do not, then you will have to pay.

EXAMPLE 2. Knowing all preferences and knowing all of the available materials, you are proceeding with a transfer using a light box.

Cheerfully drawing, you are using a 3H pencil, making sure that the transferred image will be visible. It must be, otherwise you may make a mistake. The subject is a complicated structure of some fossilized matter. After the transferred image is completed, the rendering begins. During rendering, when applying soft tones to an area delineated with line transferred from the sketch, you notice that this damned line picks up more tone than the surrounding area. You have not planned for a line to be visible. The line was drawn on the sketch to indicate a change of shape to be rendered by a change in tone. You are picking that darker-than-planned line with a kneaded eraser without positive results. Every time a new tone is carefully applied, that line reappears. Grabbing a hard eraser, you are solving the problem by muscle power. The line is gone forever, and again you try to apply a light tone. This time, to your surprise, a larger area of the paper becomes too dark. Struggling, you realize that nothing will help and the desired light tone will have to be changed to a darker one in order to camouflage the damaged surface of the paper. You also realize that next time the pressure you exert on the pencil during the transferring process will not be as great.

Exercise 1. Take a small sheet of Strathmore one-ply, smooth-surface drawing paper and transfer a small section of one of the more complicated final sketches, applying reasonable pressure to the pencil. Do the same a second time, but without much pressure, making sure that the transferred lines are visible but very light. Render both in continuous tone. Observe the possible interference of deeply pressed lines on the first transfer during rendering. Compare both renderings and reach your own conclusions.

Exercise 2. If you have access to a light box, transfer the final sketch to the drawing paper by taping the sketch and the drawing paper to the surface of the light box. Again, transfer the same or another image by taping only the sketch to drawing paper. Compare the ease of procedures and the quality of transfers.

Exercise 3. Transfer the finished sketch to the surface of the thick white board by the direct method. Be as precise as possible. After the transfer, compare the final sketch with the result. Look for all imperfections and unwanted smears.

Preservation of Specimen During Drawing

Specimens should be handled with great care at all times. None can be injured, broken, scratched, or dropped. From the moment the specimen is given to the illustrator, he is responsible for its safety. Why such a fuss about a specimen? Remember that scientists cannot conduct their inquiries into the mysteries of life without specimens. Be aware that the "only one" may mean exactly that. The subject you are about to draw is the "only one," even if it is selected from many. The specimen in your hands has been carefully chosen, and many hours of hard work were invested before you received that subject for drawing. You must realize that one fish does not necessarily indicate the structure of all other fishes; this may be the only example available. The fragility of specimens will force you to adopt a number of precautions.

1. Do not handle the specimen with one hand unless you absolutely have to.
2. Do not empty a container containing the specimen by turning it upside down.
3. Keep specimens away from the edges of the table.
4. If you can, lock up the specimen for the night.
5. Be aware of strangers and housekeeping personnel.

Remember that for somebody unfamiliar with research procedures, the specimens may appear to be unwanted garbage. A lot of broken pottery sherds in a box full of dust may be thrown away by well-meaning cleaning personnel. During positioning procedures, be extra careful. Do not press or try to force anything. If something breaks, it is your fault. Instantly report any damage to the scientist, and do not expect any sympathy from him. If the damaged specimen can be repaired it will be pieced together, but in all probability such a process will take a number of hours. If the specimen is lost or damaged beyond repair the research may be affected. Do not carelessly leave soft-tissue specimens without protective liquid. Liquid keeps tissues moist, while air and light, by drying the surface, will inflict permanent damage. Once a specimen is positioned for drawing, do your best not to change its position. It is an impossibility to exactly reposition a moved subject. Do not think that large objects are easy to handle. As a matter of fact, large objects create more problems. Where are you going to store ten American Indian garments in your cramped space? Where are you going to place large mace heads or harpoons? Do you have enough space to store thirty boxes containing fossilized remains? What about extremely fragile vessels? Small specimens can be locked away safely, while large ones will have to stay outside the drawers. Be extremely careful when examining specimens. Watch for any signs of glue or wax. Many may have been reconstructed by the scientist or his preparator. Specimens glued with wax will disintegrate under the warmth of light. If you are lucky, your microscope will be equipped with a cold light source. The light transmitted through fibers of glass will not be as hot as regular light, therefore parts of glued specimens will stay together. When working with regular lights, make sure any suspected specimen is examined before it is placed under the microscope. If it is glued with wax, you will not be able to work for long periods of time. The lights will have to be turned off about every five minutes to allow the specimen to cool. Naturally, no lights, no work. Make sure that trays, petri dishes, and other containers with positioned specimens have airtight covers. Do not leave a specimen submerged in ethyl alcohol uncovered for any length of time. When working, make sure that liquid is added to the container continually. Ethyl alcohol will evaporate quickly. If the container is left uncovered overnight, the specimen will be damaged. Be sure to add enough water to the botanical specimen when leaving your office for any length of time. Naturally, you will not draw any pencil or ink lines on the specimens. Besides injuring the subject, you may be the cause of the scientist's heart attack. In numerous situations you will have to use a teasing needle to probe various openings. Be sure that you do not scratch the surface or rip pieces of tissue. If any thin wires are attached or inserted into the specimen, do not remove them. Wires designate continuation of canals. By removing such important marks, you destroy the scientist's research. Do not attempt to clean any specimens. Leave all "dirt particles" in their respective places. It may be very tempting to see what is underneath, but realize that dirt particles have been left on purpose. In most cases such foreign matter helps to keep parts of the specimen together. Leave all cleaning and preparation of specimens to the experts. Remember, you are there to draw. If certain areas are obscured by the foreign matter, ask the scientist about it. He will be more than happy to explain. He will tell you what to omit and what to draw. If by any chance the specimen cracks and pieces drop, remain calm. Remember that the piece or pieces are somewhere around, and you might step on them. Be extremely careful in your motions. Try to recall the direction in which the cracked pieces fell. You will have to find all of them, regardless of how small those pieces are. In some cases it may take you a day of looking around before all missing parts are back in the petri dish. Make sure that you scrutinize every inch of the table and floor. Look at the clothes you are wearing; you may find missing pieces there. Do not attempt to glue the broken specimen yourself.

All specimens are numbered and all containers bear respective designations. Never misplace the specimen. Never place a specimen into the wrong container. Everybody is used to the fact that numbers on the containers correspond to the numbers on the specimens. If a specimen is misplaced it may take days or months to find it. Your carelessness may cause a great waste of time.

EXAMPLE 1. You are a busy illustrator. The project calls for reconstruction of a vessel, and four or five dirty boxes full of pottery sherds, dust, and bits of sand are on your desk. During the process of reconstruction you have arranged sherds which you think are the most exemplary and useful. The rest, the remaining bits, are in the boxes and look like typical garbage. You come to work the next day to discover that your area has been neatly cleaned. The "garbage" has disappeared, the floor is swept, and the arrangement of the sherds does not look the same. A well-meaning soul from maintenance has done his or her best. The missing pieces will never be found. They are gone forever. The rearrangement of the remaining sherds will take some time. Call the scientist and report the incident. The project has been severely disrupted. After all, you as well as he may have needed the remains for accurate reconstruction. Next time, be sure to inform all cleaning personnel not to touch anything. Tell them not to clean your area. Write a sign, "Do not touch," and place it on the desk every time you leave your space.

EXAMPLE 2. A paleontologist has given you a small specimen to draw. He has stated repeatedly that it is a fragile specimen. "Handle with care," he said. When looking at the subject, you grow in confidence. You touch the specimen; you turn it in your fingers. Satisfied with your initial investigation, you are attempting to establish a proper position using a bit of Plasticine attached to the bottom of the petri dish. Preoccupied with the procedure and looking through the microscope, you forget about the distance between the specimen and the glass. You forget that the glass is hard. When looking through the microscope the glass is invisible to you because the microscope is focused on the specimen, not on the bottom of the petri dish. You are being very careful, but the specimen does not want to stay in its designated position, as you press it into the Plasticine. You hear and feel a crack. You see a speck of something flying out from under the lens of the microscope. The specimen is broken; one piece is someplace outside the petri dish. Where? It is really impossible to tell because you were looking straight through the oculars of the microscope at that moment. The only way to trace the resting position of the missing part is to try to recall the direction of its movement and the momentum. Carefully look around in that direction. Look for any objects off which the missing part could have bounced. You do feel badly, but brush all feelings aside and find the missing piece. Hopefully, this will be possible. Remember that it is hard to judge the distance of the specimen from the underlying base when looking through the microscope.

EXAMPLE 3. The botanist brought some small fern leaves and has asked you to place the specimen in a few drops of water so the leaves can absorb the moisture and uncurl. You do this; then looking through the microscope, you are adjusting the lights when somebody calls you. Leaving the whole thing, you run out; when you come back 20 minutes later, you face disaster. The drops of water have evaporated and the small leaves are completely dried out and cracked. Remember to place a tight cover when leaving a specimen submerged in liquid, and remember to turn off the lights of the microscope.

Size of Finished Drawing

Do you know how to choose the size of the finished illustration? If the scientist tells you how large or small the final size has to be, you have no problems. In most cases the scientist will tell you the magnification or reduction factor, but he or she will not venture into the size of the rendering itself. It sounds contradictory, but actually it is not. You may be told to magnify a particular specimen fifty times. The reason for such an enlargement lies in the presentation and clarification of details found on the specimen. Naturally, you will do this, and everything will be presented on the final sketch. The sketch itself will be of considerable size. Usually the scientist will give you the information about the printed size, but the size of the rendering will be left hanging in the air. You may hear, "Perhaps the rendering should be smaller." But smaller by how much? The final size of the finished drawing will be left to your judgment, especially in the following situations:

1. The scientist does not know the size of the printed image.
2. The final sketch is very large but the size of the printed image will be very small.
3. The deadline for the finished product is quickly approaching.

Remember at all times that the finished, rendered illustration must be larger than the printed image. Avoid producing a drawing that is equal in size to the printed area. Why? Nobody is perfect; all artists make small mistakes during rendering, mistakes such as placing too many dots in one area or not smearing pencil tone to the utmost perfection. Remember that imperfections in the rendering will remain visible when the illustration is produced to a one-to-one ratio. The one-to-one ratio is referred to as the *actual size*. At all costs, avoid producing an illustration *smaller* in its final drawn size than the printed area. In such a situation, all the imperfections will be enlarged. Such an illustration does not look good when printed. This does not mean that such a drawing will not contain its intended scientific value. If it is rendered relatively well, it will. The biggest disappointment will be its printed presentation. Remember, all mistakes and imperfections will be enlarged. On the other hand, when an illustration is reduced, all blemishes are also reduced and are harder to detect on the printed image.

Illustrations larger than the printed area also present some dangers. Do not take for granted that all mistakes will disappear. It is obvious that improper rendering will not be improved by photographic manipulations. Tone, regardless of technique used, and line must be handled well in order to reproduce well. The biggest problem facing you will be accidental disruptions of tone appearing after the reduction. For this reason, the illustration should be prepared for reproduction purposes and not for display. You have to remember that tonal range on a large illustration, when reduced considerably, may change drastically for the worse. Usually dark areas become darker and light areas become lighter. In such situations there is a distinct possibility of losing readability. This will be especially noticeable when drastic reduction of line illustrations takes place. Continuous tones will behave markedly better because of the screening process involved in transferring the image of the plate.

When considering the size of the rendering, you must take time into account. It does take more time to render a large drawing. Depending on technique, your time, and the designated or assumed printed area, the size of the final rendering should be established. Naturally, not all illustrations will be printed as a separate unit. A great many will be designated to appear together within the assigned printed area. This itself will shed some light on your decision. Most plates or full-page illustrations will contain anywhere from three to twenty separate drawings. Again, you should inquire and find out the size of the printed area. The scientist should be able to supply the information. If not, then find out the name of the publication. Go to the library and measure the area. Knowing the final printed size will be necessary for proper production of illustrations and preparation of the plate for production. Because you will be dealing with many variables, it is impossible to predict the size of the rendering. Keep in mind the quality of the printed image as well as possibilities of miscalculations and sudden changes. After all, the scientist's work may be designated to be published in one journal, but for various reasons will appear in another. Be prepared for

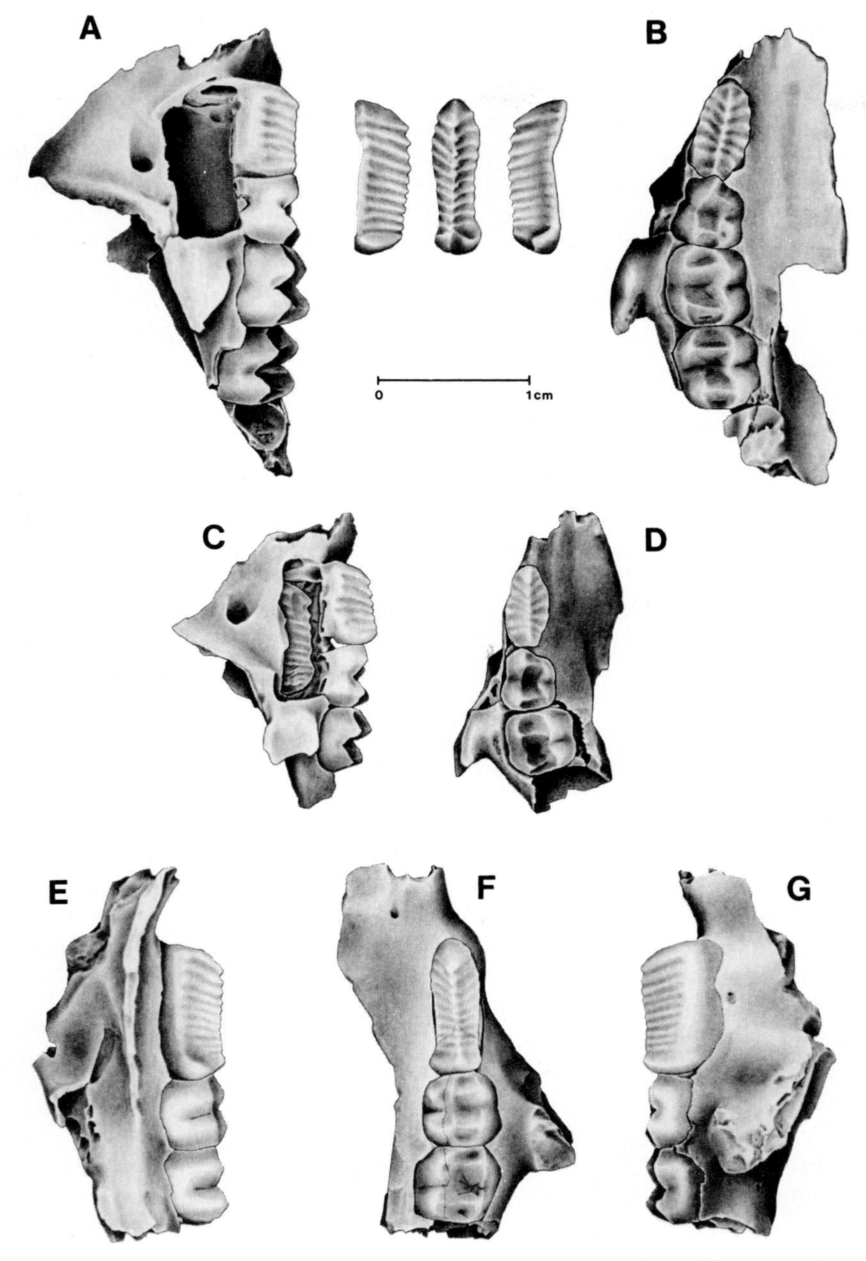

Figure 4-9. *Bettongia lesueur* from Madura Cave. Units 1 (post Pleistocene) and 2 (late Pleistocene) A,B, PM 4786, right maxillary with P^3, dP^4, $M^{1\ 2}$ in place removed 1 inch, and crypt for M^1, shown in labial and ventral views. The P^4 is shown in lingual view. C, D, PM 4787, right maxillary fragment with P^3, dP^4, exposed P^4, M^1, and crypt for M^2, shown in labial and ventral views E–G, TMM 41106-117, left maxillary fragment with P^4–M^2, shown in lingual, ventral, and labial views. Pencil on Strathmore paper. Drawing by Zbigniew T. Jastrzębski, Field Museum of Natural History, Chicago. Published in William Turnbull, "The Mammalian Fauna of Madura Cave, Western Australia, Part VI," *Fieldiana*, Geology, New Series No. 14 (1984) by Field Museum of Natural History, Chicago. Drawn in 1983.

very drastic reduction of the drawn image. For this reason, remember:

1. Do not draw too large unless asked.
2. Do not draw too small.
3. Do not overrender the illustration.
4. Make sure that all pen-and-ink lines are of equal strength.
5. Do your best to find out the size of the printed area.
6. Think about deadlines.

In some situations relatively large drawings, perhaps 30 centimeters in length, will have to appear together with very small ones measuring approximately 2 or 3 centimeters. In such a situation the size of the renderings will be dictated by the enlargement of the main specimen. The main specimen is an object from which the other, small ones have originated. The large and the small drawings will have to be aligned in such a way as to produce a clear visual correlation between all structures on one plate.

EXAMPLE 1. A botanist needs a series of illustrations presenting ferns. Each illustration, he says, will consist of one or two drawings of a habit plus a number of details approximately enlarged. He is very specific as to the size of illustrations. He says that all illustrations must be the same size as herbarium sheets. This means you have to find out what size the herbarium sheet is in his collection. The sheets are of more or less uniform size throughout all the scientific world, varying only slightly, from approximately 11½ by 16½ to 11¼ by 18¼ inches or so. The specifications call for "illustrations," but in actuality the "illustrations" are full-page composites consisting of numerous separate drawings. By knowing the final size and knowing the size of specimens as well as the enlargement of details, you will be able to estimate the size of the finished product.

EXAMPLE 2. The mink skull lying under the microscope has to be drawn in four separate views: left lateral, ventral, right lateral, and dorsal. The scientist has found an important abnormality consisting of a baby tooth retained within the structures of the jaw. The tooth has been removed and at present is considered a separate specimen. This small separate subject will have to be drawn from three sides and the drawings will have to be properly aligned with the larger specimen when mounted on the plate. This means that all drawings—the four views of the skull and three views of the tooth—will have to be enlarged by the same factor. The different size of the specimens will produce respectively different-size drawings. If we assume that the lateral measurement of the mink skull is 7 centimeters and the tooth 3 millimeters, the size of the drawing of the tooth should not exceed 2 centimeters. Remember, both specimens will have to be magnified by the same factor. Two centimeters is not a large area to draw within, but the drawing of the larger specimen, the skull, will occupy a lot of space. You have to keep in mind that the final rendering of the skull cannot be too large because of prolonged drawing time. At the same time it cannot be too small because the smaller subject, the tooth, will not have enough space for meaningful rendering. In such situations the size of the drawing of the larger specimen should be established by finding out how much space will be needed for the drawing of the smaller subject. Let's assume that our 7-centimeter-long mink skull will be enlarged six times. The size of the drawing will equal 42 centimeters, or approximately 16½ inches. Magnification by six will produce a pretty large drawing. At the same time the tooth, measuring 3 millimeters, when enlarged six times will produce 18 millimeters of horizontal distance, a rather small area of about ¾ inch. If it happens that your project has such a variety of sizes, make sure that both large and small drawings are drawn in such a way that all will reproduce properly.

EXAMPLE 3. The subject is a small, complicated structure, dictating large magnification and therefore a large sketch. The scientist has stated that the illustration will be reproduced in a journal that has a text printed in two columns. Each column is 2⅝ inches wide. The width of the printed page equals 5½ inches. The scientist has said that it is impossible to predict the final reproduction size. The printed illustration may have to fit into the width of a column or may be allowed more space. The decision as to the size of the printed illustration is in the hands of the editor of the publication. You, the illustrator, know the maximum, 5½ inches, and the minimum, 2⅝ inches, of allowed space. Obviously, the reduction of the finished illustration for the purpose of printing cannot be too great. Such a situation leaves you with no choice. The sketch must be readable and will be as large as necessary, but its final version will have to be reduced considerably, making the size of the rendering more fit for reproduction. Naturally, a lot depends on the technique used for rendering. If continuous tone is used, then the size of the rendering may be larger than when dots are used. Stipples, or dots, do not reproduce as well when reduction of the finished product is too great. Generally, reduction of the rendering should not exceed one fifth of the original size of the drawing on the plate. If great reduction is necessary, the rendering will have to be prepared with greater care and the illustrator will have to judge its tonal value according to the size of the printed page.

Exercise 1. Obtain a subject of a skeletal nature such as a chicken bone. Prepare it by cleaning and position it for drawing. Draw the subject in actual size. Note how much time this drawing takes. You will prepare the sketch, transfer the image to a smooth-surface, good-quality paper, and consciously begin to draw. Render the subject in two separate techniques, first in pencil and then in pen and ink using line and stipple. When you have finished both illustrations, draw the same subject, presenting the same view, again. This time enlarge the chicken bone six times. Make sure that you start from the the beginning. Prepare a new set of sketches, measuring your specimen carefully. When the final sketch is finished, transfer it to the drawing paper. Render the subject in two techniques, continuous tone using pencil and pen and ink using line and stipple. Keep track of passing time. When the project is finished, compare the time spent on the small drawings with the time spent on the large ones. Compare the precision of measuring as well as the precision of rendering.

Exercise 2. Take four prepared drawings to a photostat company and ask for the following:

1. Reproduce all drawings in actual size using continuous-tone paper and high-contrast paper. This will give you eight separate prints.
2. Reproduce all drawings, reducing the image by half, again using continuous-tone and high-contrast paper. You will receive eight separate prints.
3. Reproduce all drawings, reducing the image by three quarters, using continuous-tone and high-contrast paper. Again, you will receive eight separate prints.

Spend some time comparing the results. Make sure that you compare the prints with the originals as well as with the other prints. Look for:

1. Loss of details;
2. Loss of tone;
3. Gain of darkness in unwanted places;
4. General change of the image.

Presentation of Finished Illustration

Your finished product should be packaged properly. It cannot be given to your customer crumpled up, or with ragged edges, fingerprints, or tape attached to the edges. It is bad policy to deliver any product that is not your best work. Do not hand a piece of paper with a finished rendering to the scientist unless he specifically asks for such a procedure. The illustration should have a clean background and a stronger backing. Remember that the drawing will be handled by various people. It will not stay in the office or a laboratory indefinitely. The illustration will travel to the photographer, the publication department, the platemaker, and again back to the office. The surface can be injured in many ways. The most common damage will be inflicted by bending, rather than by smearing something on the surface. Other potential problems are dust and sunlight. Do your best to protect the surface of the illustration and to enhance your image as a professional. Affix a finished illustration onto harder board, illustration board being an obvious choice. Use hot-press, smooth-surface board produced for key-line and paste-up procedures. It is sturdy and pretty hard to damage. Make sure that the illustration is attached securely to the surface, best done with a dry-mounting technique. When for some reason dry-mounting is not feasible, use tape. Make sure that when tape is used it is attached only to the top of the illustration. Remember that the illustration must be flat for good-quality platemaking. Do not tape all sides of the illustration. This traps air under the paper, producing bubbles and irregularities of the surface, which in turn create a variety of problems for the cameraman. Make sure that a clean sheet of tracing tissue is placed on the surface, again attached only at the top with masking tape. Place a sheet of cover-stock paper over the illustration, attaching it again with masking tape at the top of the board. For your and the scientist's convenience, affix a white label on top of the cover stock in the lower-right corner. The label will be used for writing the number of the specimen and for consecutive numbering of the plates. That way it will be clear what is inside. When presenting properly packaged illustrations, make sure that the scientist knows why you went to all the trouble. The tissue can and in all probability will be used by the scientist for writing comments as well as for marking any further changes. A properly packaged product exerts a psychological impact on the receiver. It becomes more precious; it shows that you care. How would you like to buy something and have it handed to you, unwrapped? Would you like to go to a gallery, buy an artwork, and be given the product without an envelope or without some kind of packaging? The scientist is your customer. Treat him with proper respect and treat your product the same. Naturally, if the scientist asks you for changes, touch-ups, or redrawing, do not create obstacles. Agree to do the necessary corrections; don't tell him that he approved the final sketch and now it is his tough luck. After all, he is the expert, and you like to earn a living. At all times remember that:

1. You must present your product the best way possible.
2. You must be polite and keep all emotions under control.
3. You must be able to detect the likes and dislikes of your customer and act accordingly.
4. You are making money by drawing and therefore being paid for something you like to do.
5. If you do not like the situation you can get out.

EXAMPLE 1. The illustrations are finished and neatly packaged. The scientist looks with great amusement at all of the covers and asks you, "What is that?" Be patient; explain; tell him the advantages of proper packaging. Very likely this is the first time he has ever seen a professionally presented illustration. Tell him what the labels are for. Tell him why cover-stock paper has been placed on top of each illustration. Ask him if he has seen old illustrations dirtied by dust. In all probability he has. With some patience on your side, he will eventually understand. Remember that at such a moment the roles have been reversed. When the scientist patiently explained the intricacies of a structure he was the expert. Now you, the professional illustrator, are in the position to explain. The scientist may not know the procedures of printing. He may think that "the stuff" is sent to the publishing division of the institution and "something happens there." He may not realize that at least four or five people will handle the illustration. He may not think about accidents at the photoengraving facility. Tell him about the cameraman's potentially dirty fingers. Tell him about the stacks of artwork dumped on tables in the platemaking shop. Make him aware of proper storage and therefore proper preservation of the illustration.

EXAMPLE 2. The scientist complains about his storage facility. There is no room, he states, for all the covers. Explanations do not

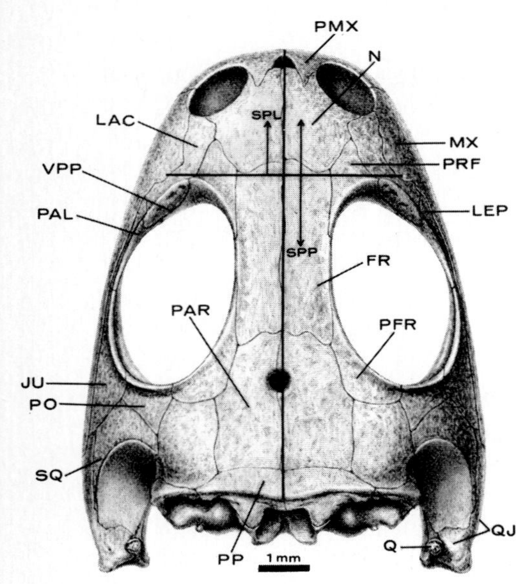

Figure 4-10. Composite drawing of skull of *Doleserpeton annectes* in ventral view. Scale equals approximately 1 mm. Pencil on two-ply paper. Drawing by Robert F. Parshall, University of Illinois at Chicago, Department of Biocommunication Arts, Chicago. Published in John R. Bolt, "Lissamphibian Origins: Possible Protolissamphibian from the Lower Permian of Oklahoma," *Science* v. 166 (1969) by American Association for the Advancement of Science, Smithsonian Institution, Washington, D.C. Drawn before 1966.

help, but rather appear to make things worse. Take the covers off. Do not try to force anything. If he does not want covers and tracing tissues over the illustrations, do not give them to him. Remove all the covers and tracing tissues right there on the premises. Ask him if he would prefer unmounted illustrations next time. Obviously there was a gap in communication. It is a little bit too late for you. Make sure that the scientist receives what he wants.

EXAMPLE 3. While receiving a finished, packaged illustration, the scientist commends you on the quality of the product. Have a good conversation with him, sharing your knowledge about printing, preparation for production, and various techniques of drawing. It will be important. Obviously, he is interested in technicality. He may even draw himself. If you do not know this, ask him. Lots of scientists do draw and are sincerely interested in professional exchange of information. Just remember that you should know what you are talking about. If you do not know something, do not pretend. Be honest and you will sound better.

Exercise. Prepare two well-rendered drawings. Package one and keep the other without backing and without covers. Prepare or choose one drawing of not-so-great quality. Make sure that it is presented correctly and cleanly. Do your best to show all three illustrations to somebody who is not an expert in the field. Observe their reaction. You will quickly learn that if a not-so-good drawing is presented properly, it will be received better than a good drawing without proper packaging.

Codes Used in Scientific Illustration

A set of agreements pertaining to the presentation of various subjects has been established among scientists. The agreements or codes have one common objective: clarification of the subject. Universally agreed-upon presentation does enhance mutual understanding. The codes can be grouped in three general categories:

1. Position of specimen
2. Technique of rendering
3. Direction of illumination

The illustrator will have to comply with generally accepted standards; because of this, he may encounter some technical difficulties. In some cases the codes of presentation are very rigid, while in others the parameters of visual depiction will be more relaxed. Skeletal structures will have to be presented precisely in their respective designated viewing positions. Therefore, when drawing a skull in a dorsal position, the subject will have to be positioned properly, viewed properly during observational procedures as well as during all sketches, and rendered in such a way that there will be no question as to the exactness of presentation. This means that dramatization of the subject will not be allowed. This means that a three-quarter view, regardless how minimal it is, will alter the generally accepted standards of readability. You know that when looking at an object drawn from different directions, that object will appear in different proportions. How is it possible to represent the truthfulness of the nature if standards in communication are not agreed on beforehand?

Some knowledge of anatomy will help you to understand pertinent terminology when working with skeletal structures, but at the same time it may be of no help during production of drawings for an anthropologist. Remember that all skeletal structures will be presented in top, bottom, side, front, or back positions. Make sure that you know how you should present a particular structure; ask questions. There are plenty of exceptions, as a lot depends on the structure itself. For example, a tooth drawn from the side will have to be presented slightly different from a three-quarter view. Why? Because the tooth has cusps and all of the cusps have to be visible. All fishes are drawn depicting their left side. Looking through scientific publications, you will notice that heads of fish are at all times on the left side. The fins from the right side are not drawn at all, because generally, there is no need to present them. Fins on the right look exactly like those on the left side. If the scientist decides to describe an anomaly in the structure or coloration of a particular fish, then addition of some structures from the right side will be permitted, with appropriate annotation in the text. Insects are drawn from the top, which is actually the back of the insect. Such a position is ideal for clear presentation of legs as well as basic segments of the insect. Drawing of the insect is slightly stylized, as in nature insect legs and antennae are not symmetrical. Drawings for an acarologist will be executed in outline only. Rarely will a three-dimensional depiction be attempted. The subjects, a variety of mites and ticks, being extremely small, are impossible to see as a whole structure when magnified suitably for research. All one can see while focusing the microscope is layer upon layer of structures ever changing in their configuration. In order to convey the correct message, outline is the most suitable medium. A three-dimensional presentation of plants is rarely found in botanical publications. The whole plant is especially prepared and preserved on an herbarium sheet. In order to present the specimen truthfully, the plant is drawn as it is preserved. If the plant is bent and crushed, it will also be drawn in that manner, unless the specimen is too badly damaged by previous handling.

The most common anthropological drawings are pottery sherds and stone tools. Pottery sherds are always drawn with the outside wall of the sherd facing the viewer. The cross section of the sherd is placed to the right of the direct view of the sherd in America and on the left side in Europe. Stone tools are drawn in line only. The lines are the conveyors of important information. An outline tells about the shape. The solid lines within the outline present the shapes of separate chips and the direction of breakage of matter from the chips. The curved lines do not represent actual bulges or dents, but a direction of cleavage. In a good drawing of a stone tool, the illustrator will attempt to represent the three-dimensionality of the subject by controlling density as well as thickness of those lines.

Techniques of drawing such as line or tone renderings are almost universally adopted by scientists to represent certain subjects. Subjects like stone tools or mites and ticks are drawn in line at all times. The techniques applied to other subjects may vary, but certainly color will not be used. Color is very hard to represent. The biggest problem is truthful reproduction of a color illustration. Very good color separation costs a lot of money, therefore it is used only in the most extreme situations. The direction of illumination of the subjects has nothing to do with illuminating that subject for observation. Most drawn subjects are illuminated from the upper left side. Such illumination is not mandatory, with the exception of drawings of stone tools, where light must appear to shine on the object from upper left side. In other cases, such as drawings of fish or drawings of skulls, the direction of light is undetectable. The light appears to come from the viewer's position, illuminating the subject directly. In most cases shadows are not included in a drawing. Shadows can be mistaken for part of the depicted structure and therefore can be misread as a coloration pattern. Presentation of shadows is helpful to the artist but not to the viewer, unless the drawn structure is so well known by all scientists that additional dark tones will not be misunderstood.

Line Drawing

The most common form of visual presentation is a line drawing. Line drawing is executed in black ink on some kind of white, semitransparent or transparent background. You know that the term *line drawing* does not necessarily mean a drawing executed only with lines. It means any black-and-white illustration that can be reproduced without use of the screen. Line drawing is produced in four basic techniques:

1. Using line drawn in black ink to convey information
2. Using separate dots, also drawn in black ink with changes in their density, to imitate the effect of continuous tone
3. Using black washes and scratching white lines
4. Using carbon pencil on a rough surface such as coquille board to imitate the effect of separately drawn dots

The procedures used in rendering line drawings vary depending on the available drawing tools as well as on the surface used for application of the ink. Technical pens such as Koh-I-Noor, Castell, or Mars are capable of producing a precisely designated thickness of line or a dot; they have built-in containers for ink and do not have to be dipped into a bottle. On the other hand, crow-quill pens will have to be submerged into an ink container quite often. Crow-quill pens produce lines of various thicknesses. The thickness of the line will depend on the size of the pen, the type of metal alloy used in production of the pen, and the pressure exerted during drawing. Both types of pens can produce a line as well as a dot. Control of the technical pen will be achieved easier and faster. Crow quills will require more practice as well as patience. Remember to take proper care of all drawing tools. Pens which are well taken care of will last for years.

When using a technical pen, remember to:

1. Check the level of ink inside the cartridge;
2. Never shake the pen over the drawing;
3. During shaking or filling up the pen use Kleenex wrapped

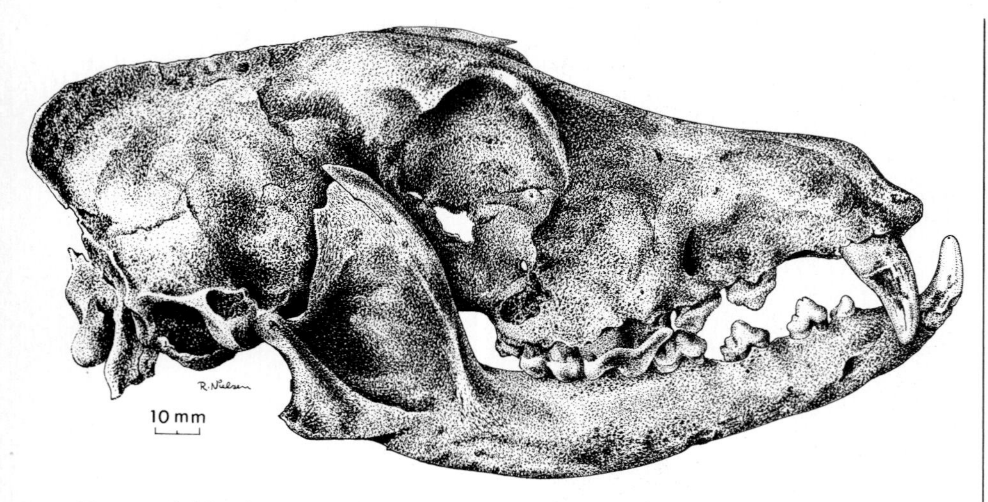

Figure 4-11. *Canis familiaris* (domestic dog), lateral view. Pen and ink (Rotring technical pen). Drawing by Robert Nielsen, Zoological Museum, University of Copenhagen, Copenhagen, Denmark. Published in Dr. Tove Hatting *Antikvariske Studier 2* by Fredningsstyrelsen, Denmark, 1978. Drawn in 1978.

around the tip of the pen to stop accidental spillage of ink;

4. Check the flow of the ink by drawing a short line on scrap paper.

During the process of drawing, all problems encountered with the pen can be placed into two groups:

1. The ink will not flow.
2. The ink will flow too much.

The first is the result of a lack of ink in the container or a clogged point. The solution is obvious: Add ink or clean the pen. The cause of the second problem is more complex. The flow of the ink is regulated by pressure inside the pen. A sudden change of barometric pressure will affect the flow of ink. If the pressure inside the pen is greater than outside, the ink will flow uncontrollably, sometimes through the point and other times around the edge of the holder. Be sure to notice overflow of ink, as your drawing is at stake. In order to alleviate the problem, open the back part of the pen, remove the container, and replace it. Be careful when doing this. Never take the pen apart over the drawing—the ink will drip on the surface and destroy the drawing. Make sure that you wrap a paper towel around the tip of the pen before attempting to place the ink container back in place. At other times the ink will flow through the tip because the ink container is almost empty. Follow the same procedure and refill the container. Be sure that you never shake a technical pen over a drawing.

Preparation of new crow-quill pens is simple. Before drawing, wash all pens well, using soap and water. The surface of the pens is slightly greasy and the coat of grease will have to be removed. If it is not removed, the ink will bead up and will not stay on the underside of the pen, or the ink will move upward on the metal surface but will not flow when the pen is pressed. In either case, the result is unpleasant. No ink on the pen, no lines on the paper. When the ink attaches itself to the upper section of the undersurface of the unwashed pen, the pen will not produce any lines. Usually you will be inclined to shake the pen in order to make the ink flow. Do not do this. The result will be tragic. Pay attention to the holder into which a crow-quill pen must be inserted. The shape of the holder will affect your drawing. Remember that all of us have a different muscular structure and will place fingers differently on the holder. If you do not feel comfortable it will be hard to achieve full control of the tool. Match the holder for crow-quill pens with your fingers.

Drawings in which the image is conveyed without tone are called outline illustrations. There are two possibilities facing you when producing such illustrations.

1. Using a free hand
2. Using accessories aiding in production of the outline

Do not be afraid to use French curves and rulers. Make sure that you have very good-quality accessories. When necessary it will be easier to produce a desired effect using a French curve rather than just a hand. Make sure that a Rapidograph pen is used; hold the pen slightly at an angle in relation to the ruler or French curve. Do not hold the pen vertically, because ink flowing from the pen may bleed under the surface of the ruler. When the tip of the point is slightly away from the edge you will not produce a smear. Be sure that the ruler is held well and will not move under the pen's pressure. Be sure that you place the pen on the desk before releasing pressure from the ruler. Using a free hand, pick up one end of the ruler and, bending it slightly away from the surface of the paper, remove the ruler without sliding it across the surface. The same applies to French curves. Avoid placing tape on the bottom side because the tape will collect dirt and after a short period will smudge the drawing. It will be difficult for you to unpeel the tape after a few weeks, and the bottom of the ruler will have to be cleaned very well before use. If you happened to apply tape, use benzene or alcohol to dissolve any particles of adhesive stuck to the ruler. Make sure that the edge of the ruler is smooth. Measuring designations are etched into the surface of the metal, sometimes producing a ragged edge. For best results, use metal rulers without measuring designations. Before starting to draw, make sure the sketch is sprayed with fixative and attached with masking tape to a smooth-surface illustration board. Make sure the drawing surface is also attached with masking tape and has no creases or air bubbles underneath. The surface must be smooth. Remember that the paper will move downward slightly when the pen makes initial contact with the surface. This will be of primary importance when drawing lines or making dots without the help of a ruler. In order to alleviate this major problem, be sure that you have a stylus. Before touching the drawing surface with the pen, press the paper with the stylus near the place where the pen is about to touch. This will help you make a good start when drawing a line or will help you to produce a round dot. When you place a pen on the paper without a stylus the paper will "give" under the pressure of the pen and will bounce when the pen is removed. Naturally, such a problem will not exist if a heavy-weight paper is used for drawing.

Make sure that the illustration board with papers taped to its surface is *not* secured to the drawing table. It is easier to move the drawing around, adjusting the area about to be rendered to natural movement of your hand, rather than straining your arm and back muscles. Avoid unnecessary difficulties, thereby minimizing the possibility for mistakes. Keep your hand on the surface of the drawing rather than above. Sliding the edge of the hand on the surface will give you more stability and therefore more precision. Naturally, there will be many situations where keeping your hand on the surface will not be feasible. Watch for wet ink. Do not take for granted that "by this time" ink must be dried out. It is very painful to smear well-drawn lines. Look at the drawing from a low angle and, if you can, move the desk lamp down and toward you in order to detect any wetness. Remember that different drawing surfaces will respond differently to the pen. The width of the drawn line will expand on drafting film. Whenever using drafting film, make sure that it has been produced for ink or pencil. Read the specifications printed by the manufacturer before investing in large quantities of this material. Remember that any drafting film will be scratched slightly by a Rapidograph pen. Small particles of the film will interfere with cleanliness and the width of the drawn line. Any scratched particle attached to the tip of the pen will cause a small smear. In order to avoid any significant buildup, restrain yourself from applying too much pressure to the surface. Be sure that the tip of the pen is cleaned periodically. Wipe the tip with Kleenex, with a downward motion, assuring yourself a clean pen. After each cleaning, shake the pen and wipe the tip again.

Tracing tissue will vary considerably. A good-quality vellum may not assure you a good drawing surface. Some highly-priced tracing tissues have too much greasy repellent in their coating, making them unsuitable for ink application. Ink may tend to bead up on some sections of the tissue, while on others it may adhere quite well. As a rule of thumb, use cheap tracing tissue for your work. Remember that you are using materials that were not really manufactured for the type of drawing you are producing. Cheap tracing tissue will give better results, assuring more uniform adherence of ink to the surface. Opaque papers such as Strathmore, one- or two-ply kid or gloss finish will accept ink very well. The only problem encountered with papers may be their texture. It is difficult to draw a straight line on a bumpy surface; therefore do not use textured papers for ink drawings. For best results, smoothness of surface is required. When using pens, the imitation of tones will be achieved by:

1. Making dots;
2. Drawing short lines;
3. Drawing long lines;
4. Crosshatching;
5. Scratching.

When producing the effect of tone, always remember that all tonal values will be affected by reduction. The processes of photography, platemaking, and printing will change the original drawing into something else. Your object is to control the effect of tone and anticipate all future possible changes. Make sure that you are aware of this during the drawing process. In order to notice an anticipated change of tone, you have to be very critical and observant. There are two standard ways to appraise the situation:

1. By looking at the drawing from a great distance
2. By using a reducing lens

In order to look at the drawing from a distance, you need space—depending on the size of the drawing, sometimes a lot of space. The drawing will have to be placed rather far away in order for you to discern its anticipated reduction. In most situations the necessary space will be impossible to arrange. Drawing studios, or actually drawing places, are small. Obtaining such space is a luxury. Such space must be lighted properly. Assuming that space is no problem and this is the only way to obtain the information, place the drawing far away and look for areas

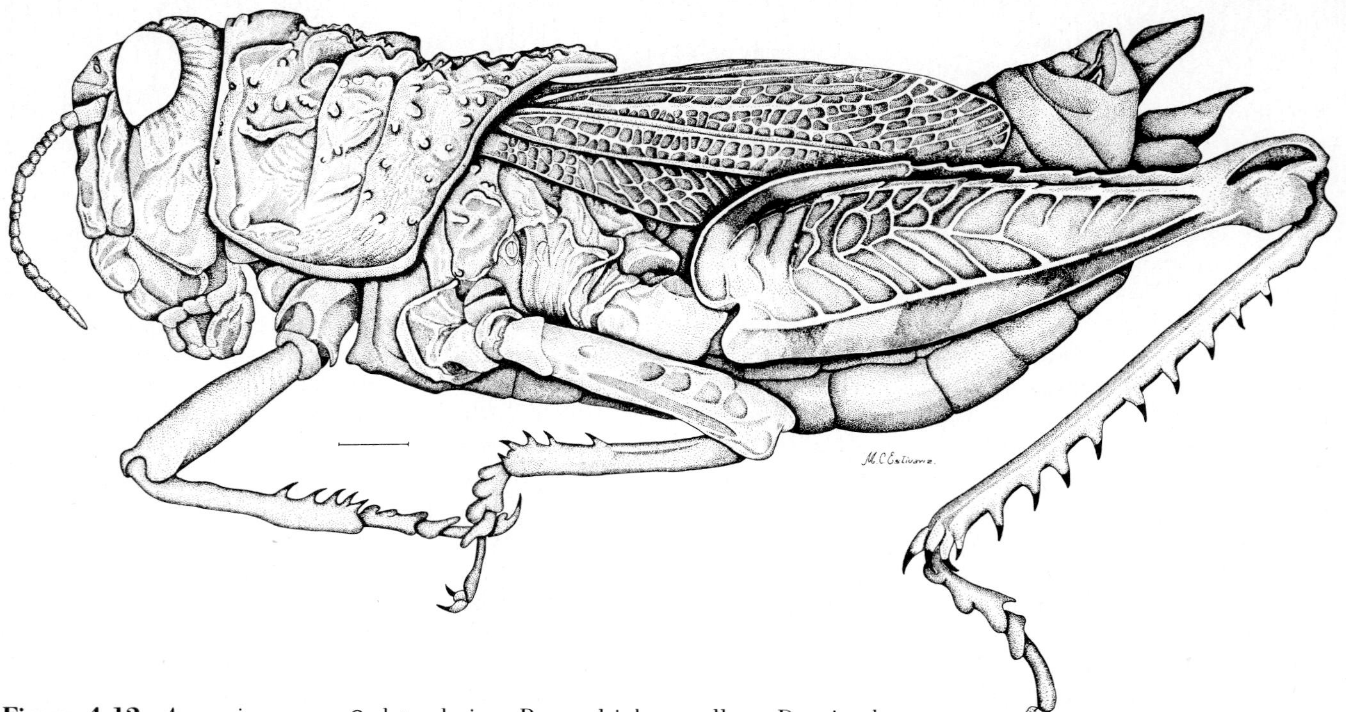

Figure 4-12. *Aucacris eumera* ♀ lateral view. Pen and ink on vellum. Drawing by Maria Cristina Estivariz, Museo de la Plata, Buenos Aires, Argentina. Published in R. A. Ronderos, *La Familia Ommexechidae, Arcrida* by Association d'Acridologie, France. Drawn in 1979.

that tend to disappear or that appear darker. The disappearance of dots or lines is the natural process when reduction takes place. Narrow lines or small dots will diminish in size. At the same time, dots or lines placed close to each other will flow together because the distance between them will narrow. The flow of dots or lines will be greater in dark areas because of congestion. Appraisal of the drawing must take place during rendering, not after the illustration is finished. This means a lot of handling of the illustration. Do your best to be careful with drawings, as accidents happen. The other way to appraise the density of black markings can be achieved by looking at the produced drawing through a reducing glass. This will eliminate the need for space and will also leave the drawing on the desk. A reducing glass allows for instant great reduction. The drawing, regardless of its size, can be "reduced" to the desired size by establishing the correct proportion between your eye, the lens, and the drawing. The flexibility is greater and the result is better.

When preparing line illustration, remember that it is impossible to achieve the same three-dimensionality as when using pencil. All pen-and-ink tone drawings will appear flatter than pencil renderings. Do not force too much contrast because of the problems caused by reduction. Remember that such illustration will be shot on high-contrast film, which does not allow for too much tonal flexibility.

When stippling, making dots, the density of dots will designate the desired tone. Make sure that each dot is neat and round. The elegance of the final result will depend on this. Use a stylus to hold the paper before placing a dot. Make sure that when using semitransparent material for rendering you place a thin but opaque white sheet of paper between the surface used for drawing and the final sketch. The sheet of white paper will help you to see the rendered area better, separating it from otherwise visible underlying pencil lines. Naturally the white paper will be removed when you continue rendering. Remember not to produce accidental visible patterns. When stippling, you may have a tendency to place the dots in a uniform fashion. Do your best to avoid this unless the project calls for presentation of some sort of uniform pattern. Stipple at random, making sure that the produced pattern is uniform in desired value without accidental black lines. Watch carefully for the appearance of white semicircles or streaks produced by the spaces between black dots. Do your best to restrain yourself from overrendering. Remember at all times that you can add dots, but it may be difficult if not impossible to remove them. Make sure that the technical pen is held at a slight angle to the surface of the paper. Hold the pen with one hand and the stylus with the other. Place both hands firmly on the table. While pressing the paper with the stylus, make a few dots; then change the position of the stylus, continuing to make dots. Do not hold the pen vertically. It is impossible to see what you are drawing when the pen obstructs your view. Do not change the position of the stylus for every dot, because it will take you too long to finish the drawing. Try to avoid making short lines unless this has been specified by the scientist. Producing lots of short, squiggly lines is proper if a given area has to appear textured or unbroken. In most cases short lines will make the drawing look like it has been produced by an inexperienced illustrator. Such a drawing will not have the desired sharpness and cleanliness. It will be too fuzzy. It is harder to control the density of tone after the reduction and printing when you use short lines. When using long lines, the only way to control the tone will be by their length and their density. While technical pens can be used for such a purpose, they do have limited capacity because of their built-in inability to produce variable thickness of line. Crow-quill pens are much more suitable for such purpose. With crow-quill pens the control of the thickness of the line can be achieved, adding an important ingredient. The tone, and this means value, will depend on variable thickness as well as density of the lines. You are aware that crow-quill pens respond to pressure; by pressure, the differentiation of thickness of the drawn line is achieved. Choose the right-size pen for the size of the drawing. It is hard to produce a continuous-tone line with a small pen. Crow-quill pens will have to be dipped into the container in order to replenish the ink on the pen's metal tip. Be sure the ink container is not in your way. Be sure that you have a folded rag or some absorbent paper to wipe the tip of an excess of ink or to place the pen on. Do not allow the pen to roll on the drawing table. During drawing, hold the pen at an angle to the paper; do not hold it vertically. Rest the pen on the third finger, controlling the rotation of the instrument with thumb and index finger. Do not attempt to draw very long lines, as there will not be enough ink on the metal pen. Do not press too hard—the black line will become very wide and there is a possibility that all the ink will suddenly flow onto the surface of the paper. Watch for wet ness. Crow-quill pens leave more ink on the paper than Rapidographs. Be careful, and work with premeditation. The direction of line will suggest three-dimensionality to the viewer. Take advantage of this situation whenever necessary; remember to watch for the ends of the lines. The ends of lines should stop uniformly on each pertinent area of drawing. Keep in mind nineteenth-century engraving as a model for excellence when placing lines next to each other. Be very careful when density of lines becomes great. Drawing a new line right next to a previously drawn, still-wet line may affect the ink flow if by accident you touch the previously drawn line. Remember that ink does not lie flat in its designated place. Wet ink bulges slightly and is held in its place by the paper. It is easy to touch that bulging, overflowing, previously drawn line and drag the ink by the edge of the pen across the surface of the paper, producing one very thick line. When using paper as the drawing surface, keep in mind that the only way to correct mistakes will be by overpainting with white paint or by cutting out parts of the paper. Cutting out will be feasible only when the paper is thin and the illustration is to be dry-mounted on white background. Otherwise you will have visible holes in the drawing.

Figure 4-13. *Homalodotherium,* Forefoot, FMNH P13092. Crow-quill pen on three-ply Strathmore bristol board. Drawing by Carl F. Gronemann, Chicago Museum of Natural History (Field Museum of Natural History), Chicago, *Fieldiana,* Geology (1937) by Chicago Natural History Museum (Field Museum of Natural History), Chicago. Drawn before 1937.

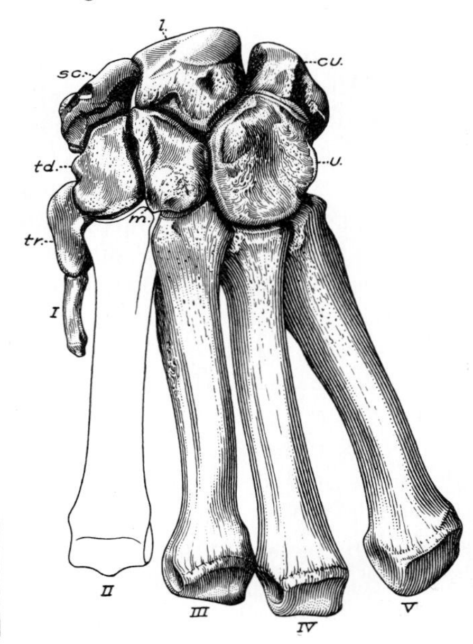

When choosing a pen, be sure to try the metal nib on the paper used for rendering. The pen should not scratch when moved across the paper's surface. It should move smoothly, following the natural movements of the arm. Crow-quill pens are not to be fought with. The pen should feel like a natural extension of the illustrator's hand. When drawing, hold pen slightly sideways. Such a position will allow for better transition between the volume of the flowing ink, producing a better line. Crosshatching can be produced by both pens, the Rapidograph as well as the crow quill. Again, the elegance of the lines will be a natural outcome when using crow quill.

It is impossible to firmly suggest a proper size of pen for a particular drawing, as the choice of size and type of the pen will depend on the size of the drawing, type of subject, and on particularities of your muscular structure as well as habitual movements of the arm. Generally speaking, reserve small pens for small or very detailed drawings and large pens for large drawings. When crosshatching, use deliberately placed short strokes rather than randomly grouped lines. When depicting a three-dimensional form, be sure to use curvature as well as density of line. Tracing tissue is not very good for crow-quill pens, as it will wrinkle easily under a quantity of ink. Strathmore papers and drafting films will give you a better choice of drawing surfaces. Good-quality drafting film allows for a quantity of repairs during rendering. The surface can be scratched easily with an X-Acto blade, so be cautious and do not overdo such repairs. When the frosting is removed from the surface of drafting film, the adherence of regular India ink will be affected. Patent Black ink produced by Grumbacher will help to solve any emergency problem. This particular water-soluble ink will adhere to most acetate surfaces, producing nonreflective but high-gloss lines. When large areas will be covered by this type of ink, you may have to place a thin coat of Vaseline over the ink on the surface of acetate. Otherwise this particular ink may stick to anything, injuring the drawing. Vaseline should be rubbed into the ink with a soft cloth after it has dried. Afterward, any excess of the greasy substance must be removed by further wiping using a new, clean, soft cloth. Scratching can be turned into an advantage. The scratch-board technique is based on removal of the ink from the surface, thereby producing white lines. Two materials are suitable for this effect: scratch boards and drafting films. Both offer sturdy surfaces and both are easily scratched. To obtain the proper effect, you must plan the final sketch with this drawing technique in mind. The object is to produce tone by scratching into a flatly painted black surface. Again, the result will be similar to a hand-engraved plate. The tone will have to be in the illustrator's mind or very clearly designated on the final sketch. Density of the scratches, actually white lines produced by scratching, will modify the change of tone. In addition to the tone produced by scratching, an outline of some shapes can and should be used. Therefore the whole subject will first be drawn in outline using a crow-quill or technical pen, then areas of tone will be painted black. Make sure that the black is opaque, as changes of ink density will affect the final result. White lines will be scratched by an X-Acto blade or any other sharp tool. An engraving tool can be used on a scratch-board but not on the drafting film, because the board's coating will be softer and the board itself thicker, allowing for adjustment of pressure applied to the tool. The pressure will produce a line similar to crow-quill pen, but it will be in negative, therefore a white or black background. It is possible to buy special scratch-board gravers producing simultaneously multiple lines. Did you ever wonder how drawings can have such exquisitely parallel lines? A sturdy hand combined with the proper tool will produce five or eight exactly parallel lines. When using gravers to scratch drafting film, be careful not to cut the surface too deep. A rough surface is very suitable for limitation of the stippling technique. Coquille board is produced with a variety of surfaces. In order to render a tone which later can be photographed on high-contrast film, special pencil has to be used. Carbon pencil will make a dark enough stain similar to black ink in its density. Two manufacturers produce reasonable carbon pencils, Wolff and Koh-I-Noor, with Koh-I-Noor carbon pencils being capable of producing darker tones. Carbon from such a pencil can be applied to the textured surface of coquille board with pencil itself or with stomps. The object is to smear carbon on the bumps but not much into the indentations between the bumps. When using pencil as the primary tool, this can be achieved easier but not necessarily faster. The tip of the carbon pencil should be sharp when drawing lines or retouching tonal imperfections. For general tone, use the side of the sharpened pencil. Remember that this technique should produce a tonal drawing in a short time. General subjects such as presentation of a tumor or the brain will be most suitable for coquille board. The changes of value will be achieved, similar to stippling technique, by density of marks on the surface of paper. Remember that a textured surface will catch all marks only on its upper sections, producing a pattern of darks and lights. The stomp may smear carbon residue into the spaces between embossed textures, making the drawing not very suitable for line reproduction. For that reason, in order to avoid unnecessary smearing, erasing cannot take place. The only way to remove unwanted tone is by gently stabbing that area with a clean kneaded eraser. Actually the best way to remove anything is by overpainting mistakes with white paint. Remember that too much smearing will destroy the feasibility of reproduction of an illustration such as a line drawing. At all times, be sure that only the upper surface of the textured paper is the area to which carbon pencil will be applied. Remember that gentleness of application will produce the light tones; pressure will produce dark ones. Areas of light tones have less carbon on the textured surface, while dark tones are made by congestion of carbon particles. In actuality, this technique is very good for producing fast, general tonal drawings with a minimum of outline. Naturally, a textured surface does inhibit the clarity and cleanliness of line so needed for an outline drawing. At all times, when working on line drawing, do not use Magic Markers, ballpoint pens, or lettering quill pens. At the same time, feel free to use a nonreproducing blue pencil as a tool for sketching on opaque surfaces. Blue nonreproducing pencil will not be noticed by the camera, but will help you to delineate information. Lettering pens can be easily mistaken for crow-quill pens. Actually, the same construction principle is applied to lettering quill pens as to crow-quill drawing pens. Both are basically built the same way, with lettering pens having a much more stable tip, prohibiting the nib from too much flexibility. A lettering pen will produce a line relatively stable in its width, while a drawing crow quill is capable of various thicknesses.

EXAMPLE 1. After discussing the specimen as well as the size of the designated space for printing, the scientist asks your opinion about possible technique of rendering. Do you think stipples will do for the project, or is there perhaps a better way to present the content without losing too much of the details? He adds, "You are the artist, so you must know; I await your opinion." Judge the situation according to received information. Do not try to impose too much of your personal artistic ego upon the scientist and in the project. If he has stated that the illustration will be reduced to the width of one column during printing, forget superbly and precisely stippled renderings. In most cases a drawing that is extremely well stippled will lose most of its precision, information, and charm after being reduced to the width of one column. Look at a science magazine and see for yourself how much space the finished drawing will occupy. In such a situation you must be sure that the scientist is aware of the fact that the drawing will not present all details he would like to include. Do tell him that some stipple can be applied, but that the illustration in its original size cannot be overrendered. Inquire into the essence of presentation and suggest an outline, with appropriate verbal designations; scratch board without much detailed rendering; or a crow quill. The choice of technique will be difficult, as in reality the quantity of details will matter much more than the technique itself. Be inquisitive during discussion, but at the same time be firm in your decisions. When facing such large reduction, keep in mind factors that will influence the outcome: the type of specimen; details to be presented; enlargement of the specimen for the purpose of the drawing; stubbornness of the scientist, which you have to live with; and your own technical inabilities. In order to present visual information in a small space the drawing, regardless of its original size, must be relatively simple.

EXAMPLE 2. Upon receiving a job requisition you find that all drawings must be produced in a specific technique, be it stipple, scratch board, crow quill, or other. It is obvious that the scientist has made a decision, although you may not know the reason behind it. Do not try to force your opinion and your demands. As always, ask pertinent questions, but at the same time do not lose your employer. If you do not know a specific technique, tell this to the scientist. Obviously you cannot draw if you do not know how. Nevertheless, facing a serious problem, you should produce required drawings in that specified technique. Try that technique before rejecting the project, before admitting defeat. There is no better way to learn than under real circumstances and real pressures. You, a professional illustrator, cannot use a crow quill? This is not true—of course you can. Yes, it will add more burden to your busy schedule, but the result will be very beneficial. Find examples of somebody else's drawings in scientific publications and scrutinize, analyze the results. Look at the general presentation of the subject. Notice the type of lines, dots, or scratches being used to present the image. Try to figure out what tools have been used in production, and how those tools have been held by other illustrators. Think about various papers, boards, or other drawing surfaces. Practice on your own time and begin the project, being more critical of yourself than usual.

EXERCISES For all exercises choose subjects that are simple in general form but rather complicated in detail. Easy-to-find items like a pineapple, cabbage, or plain comb will present an abundance of challenge. For those who can obtain a specimen of a skeletal nature, I suggest an animal skull of small to moderate size. Before proceeding with a rendering, you must prepare the final sketch. When using different subjects for each technique, you will provide a variety of illustrations for portfolio.

Before starting the main exercises, do simple tests of materials in order to familiarize yourself with technique, tools, and various drawing surfaces. While drawing, remember that the illusion of three-dimensionality can be achieved only by manipulation of values. Do not be afraid of strong contrasts. Remember that you must continually change the tone regardless of the placement of the general dark or light areas. Assume that the direction of the light will be from the viewer's point of observation. Make sure that desk lamps have no bearing on the presentation of direction of illumination. This is a mental, imaginary process not related to the actual illumination of the subject, drawing space, or paper. Remember that the easiest way to solve this problem is to assume that the closest parts of the specimen to you will be the lightest and all receding sections will be darker in proportion to the established light tone. Do your best to maintain light-dark changes of tone. Do not subdue sharp edges, but make sure that in such situations values are exaggerated. Be sure to use a reducing lens for observation of the rendering during the process of drawing. Above all, be sure that you are conscious during drawing and that you avoid distractions. Think about encountered problems and be very, very critical about your work.

Exercise 1. Outline

A. Place your final sketch under the tracing tissue and tape the sketch and tracing tissue to the surface of a smooth-surface illustration board. Draw the outline of the subject with an 00 technical pen without using French curves or other similar accessories. Make sure that all lines are joined together properly and that the drawing is clean. Move the illustration board with the sketch and tracing tissue taped to its surface around while you are drawing. Make sure that the beginning and end of the produced line are of the same thickness as the middle. Do not try to draw long stretches of line; three to four inches should be the maximum you draw at one time. Do your best to draw a continuation of the line rather than attempting to join two lines originated from opposite directions. If you make a mistake during rendering, leave it alone. Do not try to retouch or erase during rendering.

When finished, dry-mount the illustration on hot-press, smooth-surface illustration board, using seal-removal dry-mounting tissue. After the dry-mounting procedure take a sharp number 11 X-Acto knife and, gently cutting through dry-mounted tissue, remove unwanted bits of black line by lifting the cut tissue with the tip of the blade. If you prefer, using Pro White or acrylic white, paint over mistakes, blocking unwanted darkness from the camera's eye. It is hoped that your drawing will not require any corrections.

B. Take drafting film and repeat the same process of drawing, using a technical pen. Do not dry-mount the finished rendering. Using masking tape, attach the sheet of drafting film onto the surface of the illustration board, but only at one edge of the finished rendering. The most appropriate edge is the top of the drawing. Make sure that the drafting film will remain as an easily bendable flap along the attached edge. Now remove all unwanted mistakes by scratching blotches of unwanted ink out with an X-Acto knife or a razor blade. You cannot use white paint as it will crack or peel after a while due to drafting film flexibility.

C. Use an opaque paper such as two- or three-ply Strathmore. In order to see the image, you will have to transfer the final sketch to the surface of the paper. Repeat the same process of drawing. Heavier paper can be dry-mounted to the surface of illustration board or can be attached the very same way as drafting film. Mistakes will have to be painted over with white paint; the underlying drawing must be erased without gum. When erasing, do not apply too much pressure because ink may be affected by this. It is better to leave some of the pencil marks on the surface of the paper rather than to remove the blackness of ink.

D. Repeat the very same process of drawing, using a crow-quill pen. Pay attention to the nib, which should be chosen in accordance with the size of the drawing as well as with tests you have previously made. While drawing, make sure that relatively equal pressure is applied to the drawing tool, producing a line of almost equal thickness. Remember that a crow-quill pen is a difficult tool to master; do not be disappointed with results.

Exercise 2. Stipple

A. Prepare yourself and the drawing tools for rendering as usual. Make sure that your stylus is nearby. Using tracing tissue, make sure that the tissue is pressed down by the stylus during drawing. Using a technical pen with number 00 point, draw the outline of the subject. Do not draw any solid lines inside the outlined area. Everything enclosed by the basic outline will be rendered in tone by making dots. Place dots carefully, paying attention to their roundness and their placement. Make sure that you start with the lightest area. This will establish the base for necessary addition of dots. While drawing, check the contrast of values with a reducing lens. Remember that you can add dots easily, but removing them will be very hard. Safely place dots next to the previously drawn outline, touching that solid line. Do not leave a thin white secondary outline on the drawing. A thin white line next to the solid black line will give an appearance of fuzziness all around the subject. You will then be actually drawing three separate outlines of the subjects: the solid line, the thin white one, and the edge of the tone. When necessary, the dots can be overlapped and placed on top of each other. Areas with a great density of dots will appear to be very black after reduction. Remember that darkness on the drawing does not have to be in direct relation to the darkness found on the specimen. At all times, pay attention to the accidental formation of various unwanted patterns produced by improper placement of dots. When you are finished, dry-mount the illustration properly and then correct all mistakes.

B. Repeat the same process of drawing with the same pen on opaque paper. Remember that the size of the dots may be different because of the different drawing surface. Make sure that the ink is not diluted because blackness of dots is needed when you erase the underlying pencil drawing. The process of drawing is the same except that the drawing surface is different. When you are finished, proceed to package the illustration in the usual fashion.

C. Using drafting film, again repeat the same process of drawing. Make sure that the tip of the pen is regularly cleaned while drawing. Drafting film is easy to scratch and residue may block the flow of ink, affecting the size of the dots. Do not be surprised if dots appear to be larger than on previous drawing surfaces. Drafting film has a tendency to enlarge the width of a line or dot. You may produce fewer dots than on tracking tissue when using drafting film. The length of time of contact between the pen and the surface will also affect the size of the dot. You will have to find a golden middle: A too-short or too-quick touch will produce a smaller dot or thinner line, while a too-long or slow motion will produce a larger dot or thicker line. Again, when you are finished, be sure that the illustration is properly prepared for production.

Exercise 3. Crow Quill

A. In order to familiarize yourself with the tonal range offered by crow quill, prepare a second final sketch.The first will be drawn as usual, but the second will be a tracing of the first using H pencil for outline, with tone drawn in line using 2B pencil. The pencil lines simulating the effect of crow quill will be applied with changes of pressure exerted on the drawing tool in more or less the same manner as with crow quill. Each line will start wider and end in a narrow tip. Starting to control the pressure while using a pencil will be beneficial for you, as first, you will be accustomed to the idea of pressing the tool, and second, you will not be overwhelmingly inhibited by the possibility of ink spillage. Work on the second final sketch slowly and place the lines in a deliberate fashion. It will help you to build the image by drawing sections of the subject and placing them over each other to see how well you are doing. Do your best to draw the lines following the curvature of the surfaces of the portions of the subject. Remember that surface changes will be depicted by the change of thickness of the lines. Try to maintain parallel configuration of the lines. Remember that parallel lines with space between them produce a secondary affect—a white line, a natural by-product of such a process. Use that white line to your advantage. The less space between the line, the narrower the white line will be, therefore the darker the area. Try this a few times with soft pencil and proceed to finish your second final sketch. When it is completed, be sure to spray the surface with fixative so it will not smear. During the process of drawing, remember that the crow-quill technique will produce more abrupt contrasts. The harshness of contrasts will be slightly unnerving because while drawing, you view the illustration from a very close distance. Do not forget to use a reducing lens for more objective appraisal of the illustration. After that sketch is finished and secured to illustration board, place drafting film over the sketch and trace all pencil lines using a real crow-quill metal pen. Do your best to maintain the same width and spacing of the lines during ink tracing. Make sure that nothing is smeared. Do not have too much ink on the pen because when pressing the tip of the pen it is easy to produce an unwanted thickness of line.

B. Use opaque paper such as Strathmore two-ply kid finish. Draw the same subject while looking at the first illustration. The first final sketch should be transferred properly to the surface of the Strathmore paper. This time pencil tonal line rendering will not be present, therefore all lines will have to be drawn using your mind, the second final sketch, the previous rendering, and the specimen as a reference.

Exercise 4. Crosshatching

A. With a technical pen on Strathmore or drafting film, render all tonal values using short parallel lines. For darker tones, overlap previously drawn lines with new, short lines, producing greater density. Feel free to draw a number of layers of crisscrossing lines until the pertinent area is saturated with darkness. Be sure to restrain from making dots, but at the same time, in order to lighten the harshness of crosshatching edges, do break some of the short continuous lines into shorter dashes. Remember that fewer lines or broken lines will produce a lighter area. While drawing, make sure that lines are short, neat, and straight.

B. Repeat the same process using crow-quill pen. Choose a smaller-size pen, as the lines should be short. When crosshatching with crow quill, take advantage of the variable thickness of the produced line. This will result in fewer overlappings of lines, but at the same time will create the desired darkness.

Exercise 5. Scratching

A. Having a choice of two surfaces, start with white scratchboard. Transfer the final sketch to the surface and use technical pen or crow quill to draw the outline. Next, using a good-quality brush, paint all tonal areas flat with black ink. Scratch black areas using an X-Acto knife or graver. The quality of the scratched white lines will be somewhat similar to the quality of crow-quill lines, especially if the graver is used for scratching. Take your time and scratch carefully, as all lines should be parallel and straight. Remember that if a mistake is made you can add more black ink to the surface of the scratch board with the brush and rework the area. Using a pen, extend the newly produced black lines whenever necessary. Remember that you do have control over the thickness of the white, scratched lines, thereby having control over the density of tone. The resulting black lines will vary in thickness, producing an effect similar to engraving.

B. For further enhancement of your knowledge, use a black-surface scratch board and experiment with the results. Remember that the resulting white image will be harder to control than on a white-surfaced scratch board because of lack of major background contrast. In order to produce an effect similar to white-surfaced scratch board, use more scratches or paint the surface with Pro White. Remember that the resulting image is being produced in negative. In order to reverse the image, photograph or photostat the drawing and use a negative for black-on-white illustration.

C. Repeat the same process with drafting film. The technique is the same but the material is more limited. After lots of scratching, drafting film will not accept regular ink. In such a situation use inks that are produced for use on acetates.

Exercise 6. Coquille Board

Transfer your final sketch to the surface of textured coquille board without applying much pressure. Using carbon pencils, draw the outline. Make sure that you do not break the tip of the carbon pencil too many times. Using the side of the pencil or a stomp, smear the textured surface with carbon, producing a tone. Remember that the density of resulting dots or patterns will produce quality of tone. Carbon pencil will reproduce black, regardless of its lightness. The apparent lightness is a result of few carbon particles sticking to the surface of the board. Do your best to avoid smooth smearing. Leave white all indentations in the board. Use textures as the base for carbon to stick onto. Use a kneaded eraser to remove mistakes and white paint for retouching.

Exercise 7. Photostatting and Photographing

Assuming that you are wealthy enough to spend a few dollars, choose a selection from among your produced drawings and have them photographed or photostatted on high-contrast film or paper. Make sure that all are reduced by at least half. Compare the final results with your renderings and draw your own conclusions.

Tone Illustration

What is tone drawing? It is an illustration that can be properly reproduced *only* with the help of a screen. Do not be puzzled at this point, because you know that the printing process has only two technical possibilities available for transforming the drawn image into a print. One is line and the other is tone. Truthfully speaking, a tone is called a halftone because the printed image consists of areas solidly covered with printed ink dots. Various densities of dots produce the appearance of a continuous tone and simulate the changing tonal values present on the drawing. The size of dots is also important because small dots are more suitable for reproducing delicate tones, while large dots will give a general idea of values. For that reason, the quantity of dots per square inch denotes the quality of reproduction. In order to obtain the proper "dotted" image, the tone illustration must be photographed through an appropriate screen. The resulting negative is a translation of unbroken areas of tone from the drawing into dot pattern on the plate. During such translation the tone drawing will change into a halftone with white spaces between the dots, regardless of how small the dots are, remaining white. All illustrations produced with continuing and changing values are obvious subjects for halftone reproduction. It does not mean that only tone illustrations can be reproduced through this technique. The line drawing also can be printed using halftone. The harshness of a black solid line will be diffused by an array of dots, resulting in somewhat decreasing sharpness and loss of clarity. Unfortunately, the tone illustration cannot be as easily printed using high-contrast film as a transformation medium during the platemaking process. On the other

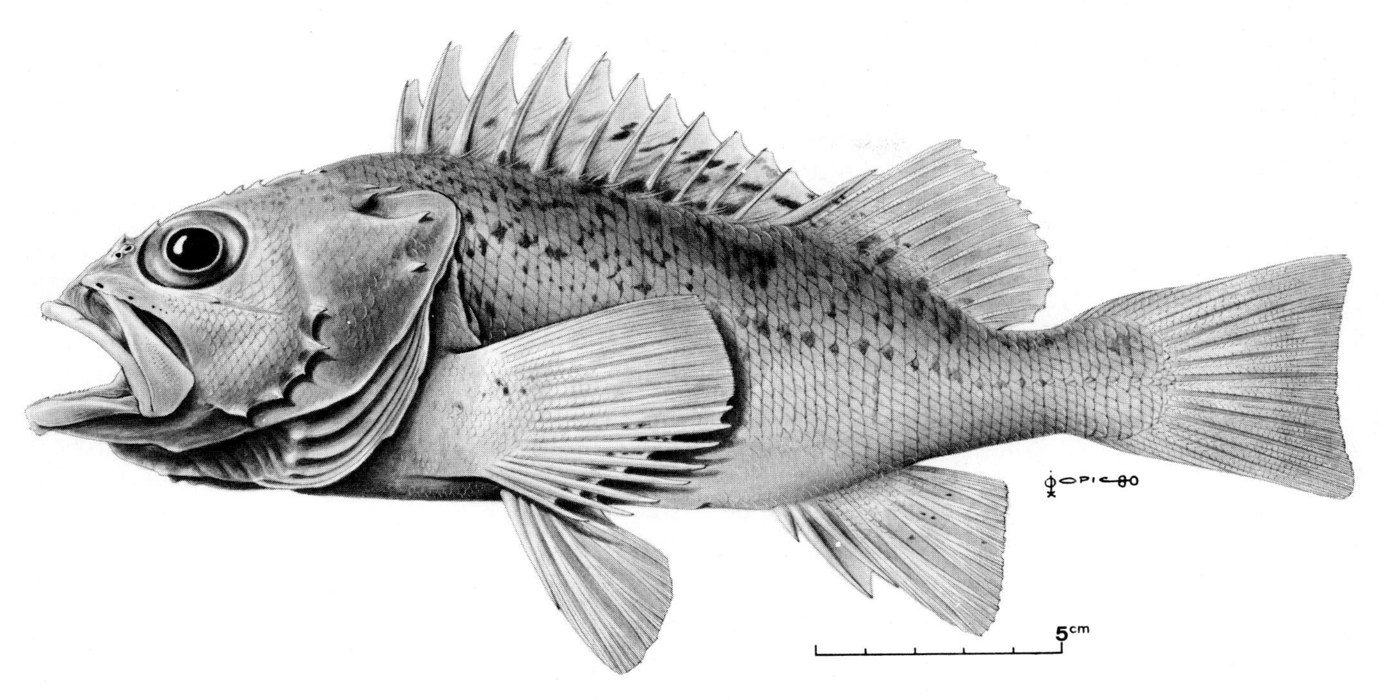

Figure 4-14. *Helicolenus dactylopterus*. Wash on bristol board. Drawing by Pierre Opic, Muséum National D'Histoire Naturelle and Office de la Récherche Scientifique et Technique Outre-Mer, Paris, France. Published in *Poissons de Mer de l'Ouest African Tropical (West African Tropical Fishes)* by ORSTOM, Paris, 1981. Drawn in 1981.

hand, line drawing or line illustration can be printed with the halftone technique. Only the tone, black-and-white or color, illustration *cannot* be reproduced in line. The delicacy of tonal values will be lost without the translation of those values into dot pattern. From our point of view the production of tone illustrations can be grouped into three basic categories.

1. Images executed, rendered, or drawn with tools used for production of any kind of drawing such as pencils or powders.
2. Illustrations produced with brush as the drawing or painting tool.
3. Illustrations in which mixing drawing and painting techniques is the source of visual effect. From a technical point of view the mixed-media illustrations have only practical boundaries, as you can use everything available to you so long as the desired effect is achieved.

The most readily available drawing tool is a pencil. You are aware of various hardnesses but nevertheless may not realize that different brands of pencils do vary considerably. The composition of graphite, which is a mixture of graphite, clay, and other binding substances, is in the hands of the manufacturer. The mixture, easily distinguished by brand name such as Venus or Koh-I-Noor, precludes the use of pencil in a given technique of rendering a drawing. Remember that if pencil and paper are not matched properly with each other and with the technique of drawing you are facing difficulty and frustration.

Did you ever wonder why friends of yours can achieve such nice pencil lines or such smooth and delicate tones while you, trying diligently, could not achieve the same? In most cases the answer lies in the types of tools you have used. The knowledge of the tricks of the trade and experience are both secondary in such a situation. Smoothing of the tone should be achieved regardless of the length of personal experience. In scientific illustration there is a need for line as well as tone pencil drawings. Be aware that the line produced by pencil can have and in many cases should have as good clarity as line produced by pen. As a matter of fact, the tonal variations of such line can be very beneficial in presentation of some subjects, just like the use of graphite or carbon dust for other subjects. When facing a given project and having a choice of techniques, be sure to experiment with drawing tools in order to obtain the maximum quality from materials. Pencil drawing techniques can be grouped in accordance with use and type of pencil and the type of surface used for drawing. We will concentrate on technical aspects and problems encountered in most common techniques.

1. Line drawing, executed with pencil on Strathmore kid-finish paper
2. Continuous-tone drawing, executed with pencil on Strathmore kid- and glossy-finish paper
3. Continuous-tone drawing, executed in graphite obtained from pencil on kid smooth-surface Strathmore paper
4. Continuous-tone drawing with pencil lines designating some areas of the subject, executed on mylar drafting film
5. Continuous-tone drawing executed with carbon pencil and carbon dust on mylar drafting film

When using pencils or powders, graphite or carbon, exert caution and avoid accidental smears. Fingerprints may not be easily noticed and slight unwanted changes in tone can appear on a white background. Make sure to keep a can of fixative handy and use it to protect the surface of the sketches from smearing. Be sure that your hands are clean and that powders will not spill or transfer onto the table or the surface of the material used for drawing. When using powders, you must be careful to avoid draft. A gust of air will spread a layer of fine powder all over.

Before rendering starts, prepare the necessary drawing tools. When using pencils, be sure to sharpen all properly. It is advisable to have more than one pencil of a particular hardness ready to be used. By having more drawing tools you will avoid a mess during rendering. If you can, buy an electric pencil sharpener. It is a good investment. Besides sharpening pencils, such a tool can also be used to sharpen pencil erasers. Obviously the most important tools are the pencils, therefore for rendering techniques described here use Venus brand pencils. At all costs, avoid pencils that will not smear easily, such as so-called office pencils—Eagle Filmograph, Mars Duralar, Ebony, Microtomic, and others. The delicacy of tone and the quality of line will depend on your drawing tool. When using technical pencils, use Pentel products. The graphite sticks are produced in two sizes, 0.5 mm and 0.3 mm. Range of hardness extends from 2H to HB for 0.3-mm leads and from H to 2B for 0.5-mm leads. It may not appear to be a wide range, but in actuality it is more than sufficient. Such technical pencils are good for fine lines, retouching areas of continuous tone, and actual rendering. Bear in mind that because of the very small diameter of lead, the process of rendering will be prolonged when using such technical pencils as the only drawing tool. You may feel it is necessary to use a larger-diameter lead, for which an appropriate lead holder must be found. For example, Koh-I-Noor number 5617 or 5616 lead holders will give you an opportunity to use leads up to 3 mm thick. In order to sharpen such lead you will have to use an X-Acto knife, sandpaper, or a technical pencil lead pointer. The lead pointer is a manually operated sharpening device capable of preparing the tip of lead to the desired point. In the container there is a cutting wheel that shaves the tip of the lead. The shavings are so fine that they can be used for drawing processes. For that reason, do not empty the lead sharpener into the garbage, but gently transfer all lead shavings into a small container and preserve for future use. The shavings will be a mixture of various pencil hardnesses, because once the composition of graphite and clay is broken, the resulting powder will behave about the same when being smeared on the surface of the paper.

A tone drawing executed with line needs an appropriate hardness of lead to match the subject. In simple words, when drawing soft tissue do not use very hard lead. Do not try to obtain reasonable changes of tone within the thickness of line while using 2H or harder lead. Do not try to fight with drawing tools, but use them for the benefit of the presentation of the subject. A soft tissue such as the palm of a monkey's hand will require some tonal delicacy. With the delicacy the crispness of rendering must be combined. Do not use very soft lead such as 3B or softer because the thickness of the line will be adversely affected. As you control the situation, you are responsible for the width of the drawn line. When the tip of the pencil becomes too rounded, the resulting line loses its definition. Do not forget that the illustration must be readable besides being beautiful. Do not forget that the portrait of the subject must be truthful to that subject. Remember that when rendering using line you are not sketching. The lines must be definnitely finished in appearance. This means that a tracing of the final sketch must be made.

When preparing the transfer of the final information to the surface of Strathmore kid-finished paper, you must make sure that all the transferred lines are drawn very lightly. The surface of the paper cannot be damaged by pressure. Remember that you will not be able to erase transferred lines from under the finished rendering because both are in pencil. While removing the information you will also erase the finish. Instead of tracing the final sketch onto the surface of drawing paper, it is possible to use a light box to illuminate the final sketch from below, making it visible through one- or two-ply paper.

When using a light box, be sure that the final sketch is drawn with dark, solid lines; also be sure that the surface of the final sketch is sprayed with fixative so it will not smear. Affix the final sketch to the drawing paper with two or three small pieces of masking tape and place both sheets on the surface with the drawing paper facing you. Do not tape papers to the light box. Remember that you will have to turn the drawing around during the process of rendering. In order to really see how your drawing progresses you will have to switch the light off. In order to continue drawing, the light will have to be switched on. When the light is on the sketch, visible through the drawing paper, it will interfere with proper appraisal of the quality of rendering.

During the process of rendering you will have to apply and release the pressure on the drawing tool in order to actuate the changes of value of the produced line. Naturally, the thickness of the line will also be of some importance. A perfect outline, without changes of tone, produces a rather dull visual result although it will depict precisely all necessary information. Such clarity can be accomplished better using pen-and-ink technique than uniform pencil outline. Using pencil to produce the lines invites the artist to introduce a certain delicacy and finesse into the drawing. Make sure that the quality of the line will match the subject correctly. Strathmore kid-surface paper is smooth enough to assure a nice continuity of the line but at the same time has reasonable "bite" built into its surface, allowing graphite to congest when a pencil is pressed. During the process of drawing it will not be possible to escape from problems encountered in the presentation of a three-dimensional image. The direction of illumination will have to be established and followed consistently. Because a continuation of areas of tone is nonexistent, the only way to present an illusion of the direction of illumination is by the thickness of the line. This means that if the light in the illustration is forthcoming from the upper left side, the right side and especially the right lower side of the image must have a darker and wider line than the left side. It also means that in most cases any lines representing the form or forms will have to be darker and wider on the right side and narrower and lighter on the left. When drawing from right to left, be sure that the pencil is pressed harder and then slowly released as the tip moves toward the left side. When drawing a line from left to right, do the opposite. Start with little pressure and slowly apply more until the desired thickness and darkness are achieved. The diameter of the lead used for drawing should be in proportion to the size of the original as well as the anticipated size of reproduction.

Do your best not to erase any lines during the process of rendering. If you have to remove some parts, be sure that the smudges resulting from erasing will also be removed. You must remember that pencil lines will be reproduced in halftone, which means that any not-completely-erased smudges will appear in print. Lines which have been applied with considerable pressure will be very difficult to erase. In such situations a remnant of that line will remain embedded in the paper. Again because the halftone process will pick up everything within the drawn area, such marks will appear on the printed page. The pencil-type erasers are the best because precision of erasing is assured. Any large chunk of gum will begin to smear pencil lines within the vicinity of the unwanted feature. For even greater precision, sharpen the eraser to a point and use it slowly, with premeditation. Try to avoid using your hand to remove eraser and graphite residue, as it will smear the drawing. Use a soft brush and brush all particles off the drawing. Do not forget to brush off your drawing table periodically to remove the eraser's particles. Keep your drawing area clean, therefore assuring yourself of a properly finished, clean illustration. Remember that the quality of pencil line drawing will depend on the final sketch. Preplanning positions of all lines before beginning the final rendering clarifies and answers all questions pertaining to the form of the subject as well as all tonal

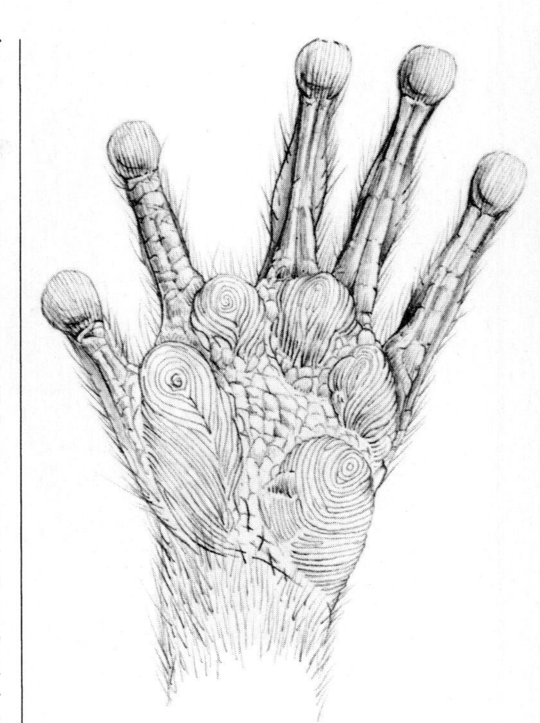

Figure 4-15. Volar surface of right hand of *Marmosa murina* (Didelphidae, Marsupialia). Pencil on Strathmore drawing paper. Drawing by Samuel Grove, Field Museum of Natural History, Chicago. Published in Phillip Hershkovitz, *Living New World Monkeys*, Vol. 1, by University of Chicago Press, 1977, Chicago. Drawn before 1977. ©1977 by the University of Chicago. Copyright reserved.

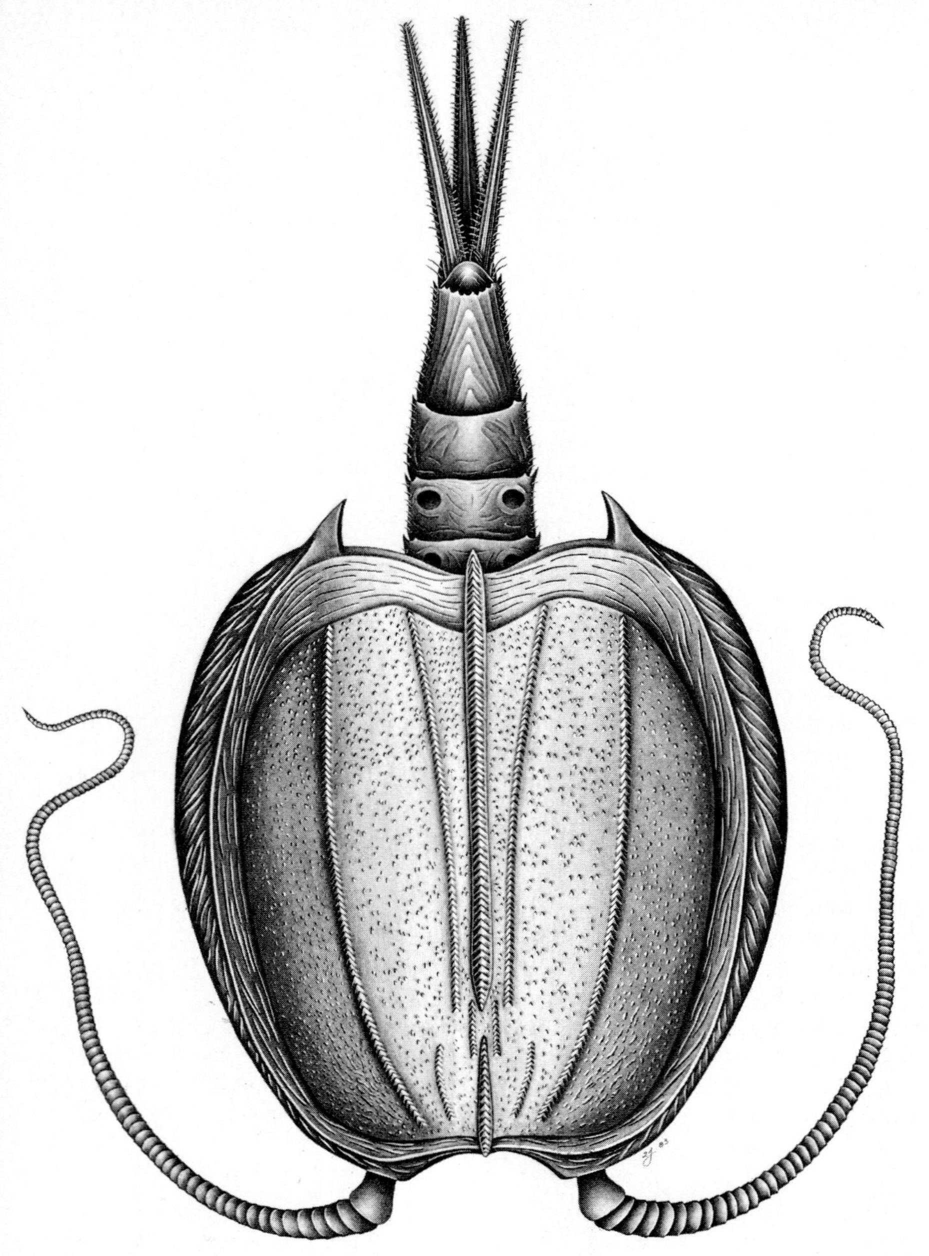

Figure 4-16. Reconstruction of fossil crustacean, *Dithyrocaris schrami*. Pencil and Pro-White retouching paint on Strathmore drawing paper. Drawing by Zbigniew T. Jastrzębski, Field Museum of Natural History, Chicago. Unpublished. Drawn in 1983.

values. A good final sketch leaves you only to perform the technicalities of the rendering, as all other problems have already been solved and clearly depicted on the final sketch.

After the rendering is finished, depending on the complexity of the subject, you have two choices in preparing the drawing for reproduction. The first is the simplest: The drawing will be produced with tone on the background. This means that the hand of the monkey will be enclosed in a square or rectangular gray shape. The hair that protrudes outside the main outline of the subject will not have a white background after the illustration is printed. In order to prepare the drawing for reproduction, take a sheet of Rubylith and affix it to the illustration board. By now you have attached the drawing to the illustration board with some sort of adhesive or with a strip of tape, making sure that drawing remains flat on the illustration board. The sheet of Rubylith is placed over the drawing and is also attached to the illustration board with masking tape. When doing this, be sure that a strip of the thin red film that forms the coating of the Rubylith is removed from one edge of the sheet, exposing the acetate base. The tape is affixed to the acetate base of the Rubylith and to the illustration board. Remember that when masking tape is attached to the red plastic coating, it will not hold the sheet of Rubylith in place because the plastic coating is removable and will easily peel off. After this, cut a square or a rectangle, making sure *not* to cut through the sheet of acetate but only through the red plastic coating. Make sure that the edges of the square or rectangle are pretty close to the drawn image. When this is finished, peel the background off. The masked area is the area that will be printed. The other way to prepare your finished drawing is to make sure that the background around the main drawn area is white. This takes more time but is more rewarding. Use a Rubylith sheet just like the one previously described, but when cutting through the red coating make sure that you follow the outside edge of the main outline of the subject. Here you must exercise caution, because if your cutting is not performed correctly it will affect the final outcome. When finished, take Grumbacher Patent Black ink, designed to stick to the acetate faithfully, and paint exactly over all lines found outside the main outline. You must be sure that the ink is opaque, therefore do not add too much water to it. After the ink dries, take a bit of Vaseline and apply it to the ink, wiping the excess with a soft rag. Instead of cutting through Rubylith, especially when the subject is complicated and you and the scientist agree that some areas within the major outline such as fingernails should be pure white, use Patent Black masking ink to mask the entire drawing. The procedure is the same as with Rubylith except that here a brush will be used instead of an X-Acto knife. Regular acetate can be used for such ink application. Why first draw in pencil and afterward paint the very same lines in black ink? The reason for such a procedure lies in the charm and quality of halftone print. The drawn lines, when masked from the background, will retain their pencil qualities after the printing plate is prepared because the mask serves the purpose of separating the previously drawn image from the background, thereby allowing the softness of the pencil lines to appear on the white printed page rather than on the grayness of the rectangle.

Continuous-tone drawing executed only with the tip of a pencil is much more time-consuming that a pencil-line illustration. The importance of the correct association of values produced by areas of continuous tone does make the drawn image more difficult to appraise during the process of rendering. Quality depends on the illustrator's technical ability of application of the graphite to the surface of the paper, the choice of papers and the pencils, as well as his patience. In order to produce such a drawing, pencils of various hardnesses are needed, ranging from 2H to 4B. The surface of paper will not readily accept harder pencils than 2H because excessive pressure is needed to produce a visible line without damage to the surface of the paper. It would be very good if we could appraise correctly all tonal values beforehand and be able to produce the desired tone in the appropriate place, but unfortunately we cannot do this. During the process of drawing, certain changes will occur, therefore the tone application will have to be performed cautiously and carefully. First of all, the paper must have a smooth surface. Any paper with texture will interfere with the smoothness of the tone, and the illustrator will have to fight all ups and downs present on the paper. Naturally, the thickness of the paper is of importance. You know that if the paper is too thick, the transfer of the final sketch will be more difficult. There are various methods of transferring a final sketch to drawing paper, but usually transfer achieved by anything else besides direct tracing will not be as precise as desired. The availability of materials such as a light box and the definition of precision established by the scientist will dictate this or the other method. The most important factor during such rendering is your own patience. You must make sure that all the pencils have propery sharpened tips, as the tips of those pencils will be used in rendering. During the application of the tone the Strathmore kid-finish paper will behave differently from Strathmore glossy drawing paper. The kid finish will take the tone more readily than glossy drawing paper, but at the same time the glossy drawing paper is more susceptible to the delicate changes of the tone. The Strathmore bristol unfortunately does not lend itself to the application of very dense, velvety dark tones. Values placed on bristol will remain much lighter than on the other two types of paper, removal of unwanted tone is also much more difficult. The tone itself, at all times, will be applied with the tip of the pencil. Smearing with fingers or stomps will speed up the process of drawing, but at the same time will produce a different quality of tone. During tone application you will have to pay close attention to the tips of the pencils. Pencils should be sharpened and should be moved in circular motion, producing small, slightly furry areas of tone.

To start, minimal pressure must be applied to the drawing tool. The density of tone is achieved by layers of graphite applied with various hardnesses of pencil. I do not advise the use of very soft pencils because they produce too thick and too coarse a line. Soft pencils such as 3B or 4B can be used only in the areas where you are positive that tone will be very dark. There pressure exerted on the pencil can be considerable. After all, you know that you will not remove the tone from such areas. It will be placed to stay. Remember that once the surface of the paper is damaged by pressure or by scrapping, future corrections will be difficult if not impossible. You may wonder why you should not use short, horizontal, parallel lines during tone application. Why the need for a light, circular motion of the pencil tip? Try it and find out. The sketchy parallel lines will vary in final density because of uneven pressure. It is very hard, if not impossible, to draw a number of short lines that are equal to each other in darkness of the tone. Regardless of how much you try, there will be a few that start too dark or end too dark. When such lines overlap, accidental dark patches will appear on the drawing. Such patches are not welcome in our type of rendering. First of all, they are accidental, therefore not controlled by you; second, they appear to represent a surface discoloration of the specimen. If discoloration is the main subject of the drawing, then do make use of the accidents, but this will rarely be the case. The scientist will be interested in the portrayal of the structures rather than anything else. The circular motion assures you of a much more even application of tone and better control of the marks left by the pencil. Whenever presentation of the surface coloration is needed, you will be in a better position to control the tone as well as its parameters. While drawing, be sure that the outline, the line designating the separation of the subject from the background, is drawn completely before the rendering takes place. If you are afraid that smearing of the drawn line may occur during rendering, then draw the outline of the area you will render. This will assure and help you in defining the utmost parameters of the drawing. You will be able to stop application of the tone right on the line and escape future corrections and erasing. The kneaded eraser is the best, being the only one that will remove tone without smearing pencil marks. Make sure the eraser is soft by stretching the rubber in your hands until it is possible to mold the eraser into a desired shape. Depending on the area to be removed, the eraser should be molded in your fingers into a more or less sharp cone. Then with the tip of that cone touching unwanted parts of the rendering, apply some pressure, embedding particles of graphite into the eraser. Ideally, after two or three removals of graphite, the eraser should be reworked between your fingers, assuring you that a fresh section of the eraser will be used next time.

While drawing—that is, adding or removing tone—you must remember the whole image. The usual problems involving properly placed values and underlined contrasts will have to be solved in addition to the technicalities of tone application. Pay attention to the direction of illumination pertinent for the drawing. Naturally, the scientist will tell you how his subject should be presented. As always, ask questions, as your questions will guide you through the rendering. Should light illuminate the subject from the left or right side? What shapes are of importance? What should be included and what should be excluded from the drawing? All of this should be and will be cleared up during preparation of the final sketch, but nevertheless, do not be afraid to ask about anything during rendering. Remember to keep the reducing lens near your side. Use it during the rendering—not after the drawing is finished, but during the production of the image. Use it right from the start so you will be able to appraise the situation immediately. Remember not to overrender the areas that should be light. As in all areas, it is easier to add tone than to remove it. When using just the tip of the pencil, make sure that excessive pressure is avoided and the surface of the paper will not be damaged.

Figure 4-17. Golden Lion—tamarin, *Leonthopithecus rosalia, rosalia*. Pencil on scratch board. Drawing by John E. Pfifner, Field Museum of Natural History, Chicago. Published in Phillip Hershkovitz, *Living New World Monkeys*, Vol. 1, by University of Chicago Press, Chicago. Drawn before 1977. ©1977 by the University of Chicago. Copyright reserved.

Make sure the paper is stored properly. Creases will produce problems. Dust will add unwanted grayness. When taking a break or leaving the working area, place a clean sheet of tracing tissue over all, covering the surface of the drawing as well as parts of the illustration board onto which the drawing paper is attached. Tape it, but avoid excessive hand pressure in the drawing area, as this may cause smearing. Do not use the same protective tissue next time. Always use a new clean sheet.

While drawing, you may encounter a few problems. The most important will be your own patience. You will have to consciously fight with yourself to keep working slowly on the project. You will have to remind yourself that slow progress is for the benefit of the drawing as well as the customer. The other biggest problem will be a technical one. The tone application will be highly concentrated to one small area. You will not be able to cover more than half a square inch at a time, and usually it will be less than that. At the same time the overall tonal quality of the image will have to be kept in your mind. After a reasonable area is covered with tone, you will have to go back to the rendered part and inspect that area, smoothing out some of the uneven tone, adding more contrast or perhaps removing some. Remember that any removal of tone will call for reworking that particular area. You will have to keep uniform pressure on the pencil so all marks on the paper will blend together. When using an eraser, be careful not to smear existing tone too much. Smearing will produce a tone of different quality. Naturally you will not forget about the sharpness of those features that should remain sharp. Do not make sharp edges of a bone fuzzy. Be very careful and remember that the crispness of the finished rendering depends on the definition of the features existing within the subject.

The other type of continuous-tone pencil drawing is rendered through use of smears. In order to obtain proper results, prepare graphite from a 4B pencil by smearing the tip of the pencil on a piece of illustration board. Make sure that the graphite covers about a 2-by-4 or 2-by-5-inch area and is of reasonable thickness. Use cold-press illustration board for this purpose or any board which has a smooth but hard surface. A board with a soft surface on paper will absorb too much graphite, not allowing enough small particles of graphite to remain on the surface for future use. The resulting powder will be the prime source of drawing material. Make sure that graphite from the 4B pencil does not break into splinters. The object is to obtain uniform residue; irregular splinters will interfere with drawing. During application of the graphite to the surface of the paper with drawing tools such as stomp or pencil-type erasers, the irregular splinters will unexpectedly injure the surface. If you do have a hand-operated technical pencil sharpener, use the graphite powder collected from the pencil sharpener. Because in all probability you have broken some tips of technical pencils during sharpening, make sure that such ready-made powder is ground in a pestle. As a matter of fact, feel free to break a graphite stick and, using some kind of mortar, grind the particles into a powder. Look in hardware stores, as sometimes it is possible to obtain ready-made graphite powder. Be cautious with ready-made powders, as they are not produced for drawing but for other reasons, and therefore they may be too finely ground. The proper powder cannot be extremely fine because then it will not adhere in sufficient quantity to the surface of smooth papers, which means that dark tones will be impossible to obtain. Besides the major component, the powder, you will need a good quality of stomps and pencil-type erasers. The stomp should not produce anything besides a smooth tone. Avoid those that leave multiple lines on the paper. Depending on the size of the drawing and complexity of the subject, choose stomps of the right size. Can you render a small, detailed drawing with a large, flat stomp? You know better than that. The pencil-type erasers are of importance. Large, thick erasers will bend too easily in your hand, not allowing proper control of the tool. Will you remove tone with such an eraser? Not all of the time. The tone will be applied to the paper with soft erasers. Only hard ones should be used for removal of unwanted graphite. Because control of the eraser is needed, try to obtain rather thin erasers set in wood. Wood will not bend during application or removal. The rubber tip will bend slightly, allowing you to feel the pressure and control the quality of tone. Do not forget to sharpen erasers in the pencil sharpener. Electric or hand-operated sharp-

Figure 4-18. *Acheloma pricei*, MCZ 1485. Right orbital region seen from above and slightly laterally. Crow quill on white scratch board. Drawing by Robert F. Parshall, University of Illinois at Chicago, Department of Biocommunication Arts, Chicago. Published in John R. Bolt, "Evolution and Functional Interpretation of some Suture Patterns in Paleozoic Labyrinthodont Amphibians and other Lower Tetrapods," *Journal of Paleontology*, by Society of Economic Paleontologists and Mineralogists and the Paleontological Society, Tulsa, Oklahoma. Drawn before 1974.

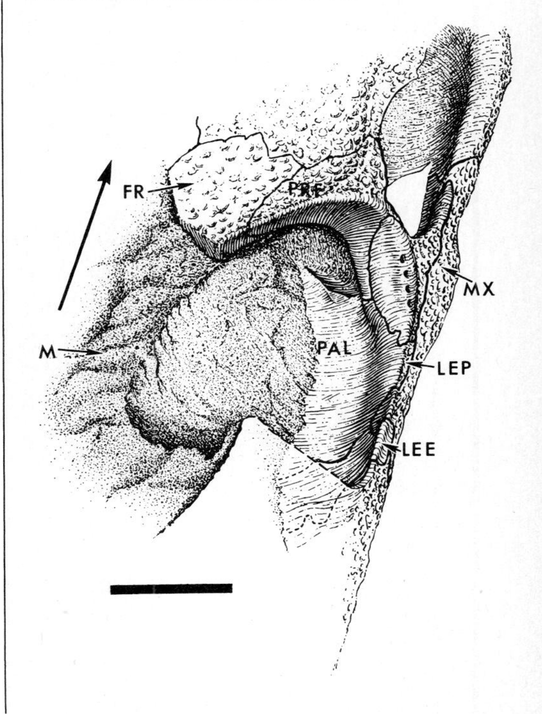

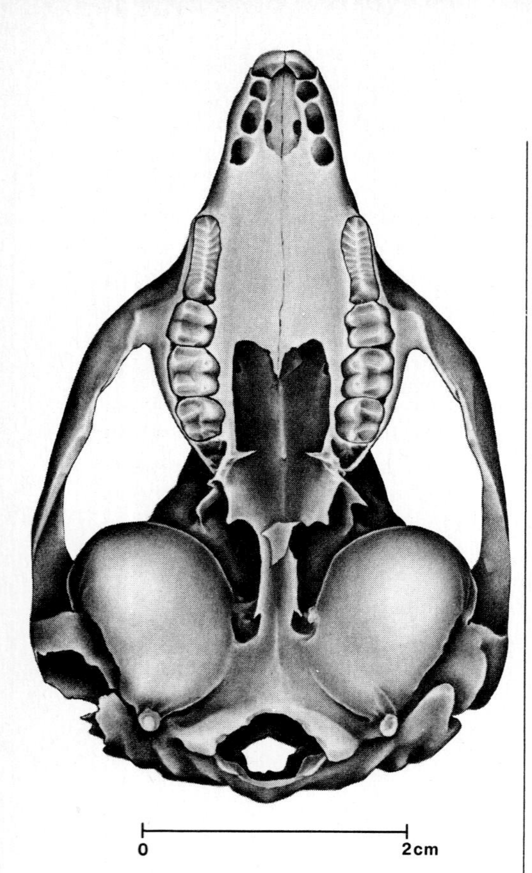

Figure 4-19. *Bettongia lesueur* from Madura Cave, surface (probably recent), TMM 41106-20, skull, ventral view. Pencil on Strathmore drawing paper. Drawing by Zbigniew T. Jastrzębski, Field Museum of Natural History, Chicago. Published in William D. Turnbull, "The Mammalian Fauna of Madura Cave, Western Australia, Part VI," *Fieldiana*, Geology, New Series No. 14 (1984) by Field Museum of Natural History, Chicago. Drawn in 1983.

eners will do the job. Naturally, when the eraser is thick and set in a paper covering, sharpening is impossible. In any case, you must cut the tip of soft as well as hard erasers at an angle, using a sharp X-Acto blade. The cuts must leave a smooth surface, especially on the soft eraser, as the majority of tone will be applied with this surface. Make sure that the edge is without burrs, because irregularities on the edge will interfere with precision of tone application.

When your tools are ready and the final sketch gently traced on the surface of smooth paper, start rendering by drawing the outermost outline in pencil. The pencil should be well pointed so you will not produce a thick line while drawing. As you know, the outline is the outer edge of whatever will be drawn, and therefore must be drawn with caution. You can apply some pressure while drawing the outline, knowing that there will be no further corrections necessary to that line (assuming that during tone application the outline will not be smudged or erased). Feel free to draw the whole subject or just a section, but you should not draw some parts here and some parts there. Remember that you will have to present yourself to the scientist as a professional. Therefore make sure that while rendering, your drawing looks like it is being controlled by you and not by accident. When using stomps, be sure that the whole side of the pointed tip is rolled or worked into powdered graphite. You know that in order to apply tone to a larger area the very tip of the stomp is not enough. How much powdered graphite has to be on the stomp? You are the judge. The more graphite, the darker the tone. For that reason, have some of the stomps covered with a thin or light layer of graphite powder, while others will have to be well worked into the blackness. Do not apply too much pressure on the stomps. It is easy to break or bend such a fragile tool. At the same time, do not throw away so-called "bad" stomps. If the tip becomes too fuzzy, as will happen with some because of fiber separation, then use it for larger light tonal areas. For extremely light tones, use Q-tips. The cotton consistency is just right for gentle values. It is possible to sharpen stomps after the original has disintegrated? Yes, but with caution. The best way to sharpen them is to gently rub the stomp against medium-grained sandpaper, rotating the stomp during the procedure. The movement as well as the rotation should be *with* the direction of the fibers of the paper from which the stomp is made, otherwise the side of the tip of the stomp will become more of a mess.

While applying tone with the soft, pencil-type eraser, make sure that the diagonally cut tip of the eraser is well covered with powdered graphite. In order to accomplish this, take the eraser and, while applying pressure, work the flat, cut side of the tip of the eraser into the powdered graphite. Try not to damage the eraser while doing this. Your object is to cover the flat side with graphite without damaging the sharp oval ridge surrounding the flat cut side. During application of tone, that sharp edge will give you the possibility of producing precise areas of tone. It will allow you to get to the outermost, or any other, outline with precision. While applying tone, place the eraser with its flat side down on the drawing paper, making sure that the sharp oval edge is touching the outline. Apply enough pressure while moving the eraser across the paper to produce a tone. The more graphite on the eraser and the greater the pressure exerted on the tool, the darker the produced smudge. Problems? To answer bluntly, yes. Do not expect to achieve smoothness of tone while using a pencil-type eraser. Do not be surprised if you are unable to reach exactly to the outline with its tip. The smoothness of the tone will be achieved after the first layer of tone is already applied. You will have to use a kneaded eraser for removal of some small darker patches and a combination of stomps and pencil to even out as well as to add tone to the lighter sections. One thing you must remember: Do not work on large areas at the beginning of the rendering. Confine yourself to relatively small areas until the desired smoothness is achieved. Why use the pencil-type eraser for tone application at all? Because that particular tool will produce very deep, velvety, dark tones not available to you through any other drawing tool, and because you will be able to get to the small and structurally dense areas and still apply a gradation of tone, maintaining sharpness of the edge of necessary structures without spending a lot of time. This does not mean that retouching will be abandoned. In such a precisely defined area as scientific illustration, definition of the subject's shapes is of utmost importance. Therefore you will have to go back to the underdrawing and finalize it with other available tools. Do not expect instant perfection; you will have to work at it. You will have to remove some tone and add a little in order to achieve clarity as well as complete smoothness. Such a process is a continuing fight with the technique, tools, materials, and above all with yourself. Use a reducing glass during rendering for proper appraisal of values. Remember that in order to draw a three-dimensional image that looks as three-dimesional as it is in nature, you will have to exaggerate contrasts and underline or backlight numerous structures which otherwise would not be visible. Make sure that for this type of drawing you use proper materials—paper with a smooth surface and graphite that smears easily. Any textured paper, even if slightly textured, will prevent you from achieving airbrushlike smoothness of tone. During application the graphite will catch the ups of the texture, leaving the downs as white specks. Do your best not to fight with materials, but use them for your benefit.

For drawing, instead of paper you may choose to use a drafting film. Your biggest problem will be to obtain good-quality drafting film. Remember that not all drafting films will accept graphite well enough to produce a good-quality drawing. Write to manufacturers for specifications and then choose. On the average, a better-quality drafting film will be frosted on both sides. Why use drafting film for pencil drawing? Drafting film is semitransparent and, as such, will help speed up the process of rendering by allowing you to see the final sketch placed underneath. It is a good medium for rather quick but fine-looking drawings, but at the same time you will not be able to achieve the precision of pencil-on-paper techniques. Any detailed tonal changes are not to be considered, therefore graphite on drafting film should be used only when generalizations are asked for. The tools for drawing are essentially the same as for any graphite tonal techniques: regular pencils, stomps, a kneaded eraser, and cotton Q-tips. The pencil-type erasers can be used only for removal of tone, not for the application. During rendering, make sure that the outline of the subject is drawn sharply with a well-pointed pencil, and the tone applied as needed with softer tools. When using drafting film in conjunction with graphite, you will encounter one major problem: The darkness of tone will be in a much narrower range than in other rendering techniques. You will have to get used to this. It will be impossible to produce the same dark values achieved so easily on paper. The whole tonal range will shift toward light. Keeping this in mind, you will adjust the light values accordingly. When applying tone, periodically place a white opaque piece of paper between the final sketch and the drafting film. The whiteness of the background will help you appraise produced values properly and the sketch, usually visible through the drafting film, will not interfere with your decisions. Do not worry about the seemingly light, almost translucent, tonal qualities, because this translucence will have no bearing on reproduction. You may have to use pencil to add graphite to the darker areas because through other tools such as stomps or Q-tips you will not be able to achieve these darker areas relative to the light tones. Leave the lightest sections of the subject's structures completely without any tone. During translation of the tone drawing into a halftone plate those areas will be covered with dots anyway, therefore appearing as if drawn in tone. Drafting film has a sturdy surface and can be scratched with an X-Acto blade. Make use of this when trying to obtain sharp white lines. At first such an area will be rendered in designated tone and only after the tone has been applied, then you can scratch the surface of the drafting film. Once the surface is injured it cannot be repaired. As a matter of fact, it will pick up more tone when smudged across the finger or a drawing tool, adding to the problem.

The changes of surface colorations of the specimen can be achieved with a kneaded eraser. The eraser should be shaped between your fingers to the desired thickness, then designated areas can be touched with the tip of the eraser. Such a process will remove the tone. As long as the surface of drafting film is not injured, the process of addition and removal of tone can be repeated indefinitely. Naturally the precision will suffer, because kneaded eraser is not as precise a tool as the tip of a pencil. Make use of this after general tone is already in place. Whenever darkness has to be added, the only resort is the pencil. For best results use a very finely sharpened, soft (2B, 3B, or 4B) regular pencil or a technical 0.5-mm pencil with 2B lead. Move the pencil in a circular motion when larger areas of dark are to be achieved. Try to avoid visible pencil lines unless that is what you and the scientist have planned from the beginning. After the rendering is finished, you must spray the drawing with fixative, as otherwise everything will smear. While spraying, be careful not to spray on too thick a coat of fixative. Because drafting film is more absorbent, the fixative will stick and stay only on the surface. Too light a coat and the tone will not be preserved; too heavy a coat and the fixative will stick to anything you happen to place on top of the drawing. Even though you are not planning to place anything on top of the drawing, consider others who will handle your illustration. Under pressure the fixative will lift from the surface, destroying parts of the rendering. Repairs are hard to achieve because the fixative will not accept pencil, and its thickness will prevent you from matching the rest of the tone. You know that mylar drafting film should not be dry-mounted to the surface of the illustration board. Attach the sheet with masking tape along the longer side of the drawing. Then trim the ends of the masking tape to the size of the mylar sheet. This will allow for flexibility in handling as it will be easy to pick the sheet away from the illustration board. Naturally, during rendering, you may substitute the pencil outline for pen and ink. The process of tonal application will be the same when the outline of the subject is to be more outstanding. If the scientist wishes to have various structures within the subject underlined more prominently than the whole subject, pen-and-ink line may be the answer. Such line will be very prominent, as it will be much darker than the general tone of the whole rendering.

The carbon-dust or carbon-pencil technique can be used on a variety of surfaces. Be careful, as carbon residue is highly smearable.

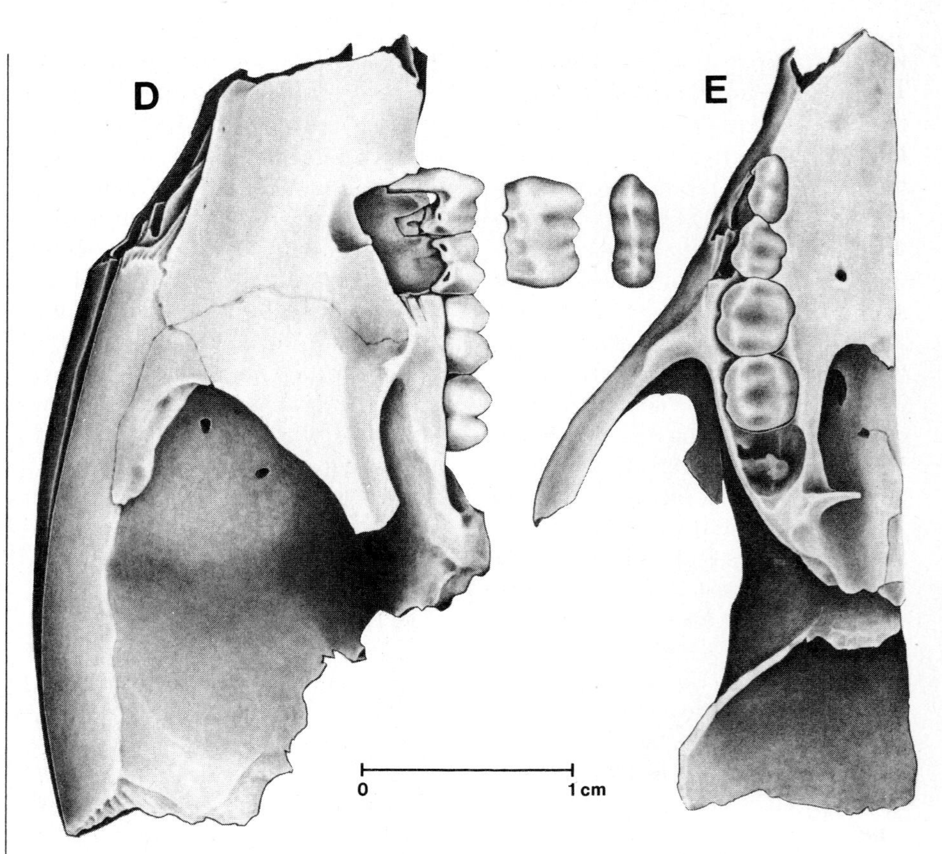

Figure 4-20. *Potorous platyops*, PM 6313, subrecent partial skull from Hastings' Cave, Western Australia; (left), right lateral view with P^4 removed from its crypt; (right), ventral view of right side with P^4 removed from its crypt and shown in occlusal view. Pencil on one-ply Strathmore drawing paper. Drawing by Zbigniew T. Jastrzębski, Field Museum of Natural History, Chicago. Published in William D. Turnbull, "The Mammalian Fauna of Madura Cave, Western Australia, Part VI," *Fieldiana*, Geology, New Series No. 14 (1984) by Field Museum of Natural History, Chicago. Drawn in 1982.

Some papers may be more absorbent than others, thereby reducing the danger of accidental smears. When using mylar drafting film as a base, you can choose from two variants of this technique. The first is quite similar to pencil on drafting film. That means that an outline designating particular information must first be drawn in ink and, after the ink is dry, carbon dust is applied with the same tools—stomps, cotton Q-tips, chamois cloth, or anything soft, such as a brush. By exerting certain caution you will not exceed the boundaries drawn in ink. The tonal qualities will have a greater range than when using graphite, as with carbon pencils it is possible to achieve very dense blacks. Highlights can be produced by removal of tone using kneaded eraser. For best results use Koh-I-Noor carbon pencils. Wolff's carbon pencils will not produce such a great variety of tones and will tend to scratch the surface of drafting film. In order to obtain carbon dust, you will have to grind the pencils into the desired consistency. This technique is very good and very effective for subjects with a lot of outline designations such as fish or snakes. Because of the importance of build in structures, scales, rays, and spines, the general tonal presentation of the whole subject is rather secondary. The other, much more time-consuming technique consists of building tonal differentiations with carbon dust and pencils alone. It is more time-consuming and much more messy, but the final result can be quite overwhelming. Because of the gentleness of tones and possibilities for very impressive contrasts, this technique lends itself toward very three-dimensional and complicated subjects. Unfortunately, as with all tone application on mylar, the precision of definition of small parts of the subject is hard to achieve. Again, be careful and do not sneeze into the carbon dust, as it will fly all over. The primary tools used for application of the tone are stomps and brushes. Pencils themselves should be used for retouching imperfections, for production of very dense blacks, and when necessary for drawing a line separating light tone from white background. The general tone should be achieved first; afterward, do the detailed work. A kneaded eraser will be used for the removal of tone. The biggest problem will not be rendering itself, although it will require some patience and delicate handling of drawing tools, but your own working habits. If you are sloppy and disorganized while working, problems will abound. Do your best to be careful and do not add unnecessary burdens to such delicate technique. The tonal range will depend on the subject itself. You know this. You should also be aware of the fact that not all subjects will lend themselves to a particular drawing technique. After rendering is finished the background will be cleaned up and the whole surface of the drawing will be sprayed with fixative.

Drawing in tone, using pencils or powders, depends entirely on your patience and proper understanding of the range of values that can or should be produced. Remember never to place two tones of the same value next to each other. By doing this you will cancel the illusion of three-dimensionality on your drawing.

Illustrations produced with a brush used as a drawing or painting tool can be classified into innumerable techniques depending on the type of inks, paints, and painting or drawing surfaces that can be used. You know that it is possible to paint on the wall as well as on a piece of paper. You know that each set of materials will have its own technical requirements as well as its own drawbacks. As long as you are aware of this, everything is fine. Being a thinking human being, you will test given materials and quickly learn the possibilities. Nevertheless, some painting or drawing surfaces do not lend themselves to such precise work as is needed in scientific illustration. Watercolor paper will be of no use, because such paper is very hard to handle and does not allow for corrections of the drawn or painted image. Changes or corrections are part of the production of scientific illustration. It is not just your decision to do this or to do that; it is a team effort. In most cases what you do and how you do it depends on the decision of a third party. For this reason, try to choose proper surfaces for working with brush. An image painted on watercolor paper is virtually impossible to change, but the surface of illustration board or Strathmore two- or three-ply drawing paper can take more beating without noticeable damage. For this reason the most commonly used surface, besides special and unforeseen projects, is Strathmore drawing paper or illustration board. Sometimes drafting film will be used for some wash drawings, usually because it will be on premises, not requiring the acquisition of new materials by the research institution. Strathmore drawing paper, illustration board, or drafting film are not the easiest or best surfaces for application of watercolor or acrylics, but at the same time are among the best for quick changes, removal of unwanted pigment, and addition of images. As always, before a wash drawing or color illustration reaches the final stages of production, a series of sketches and a final sketch will have to be approved by the scientist. In order to prepare a color illustration for printing, the research institution must have money. Any color printed image is composed of four overlying separate primary colors. Red, blue, yellow, and black are the source of all the hues and changes visible after the printing. This means that the photoengraver, when given tonal artwork, will have to photograph each color illustration four separate times through four separate filters. The result is called *color separation*. Besides added color-separation expenses, four separate plates have to be prepared so the printer, whenever he receives the project, can print the four separate basic colors, registering each separately on top of the others. The result of such printing is color illustration on the page of the publication. As you can see, expenses can multiply rather quickly depending on the quantity of illustrations as well as the desired quality of the end result. Besides the budgetary problems, the accuracy of representation of color illustration is of importance. The accuracy depends on three factors: color separation, registration of separate plates, and the materials used for printing, such as inks and especially paper. If a color separation is not prepared correctly, the printed illustration will appear too yellow, red, or blue, and won't be a truthful representation of the painted illustration. When registration of the four plates is imprecise, the printed image is fuzzy. Check some color

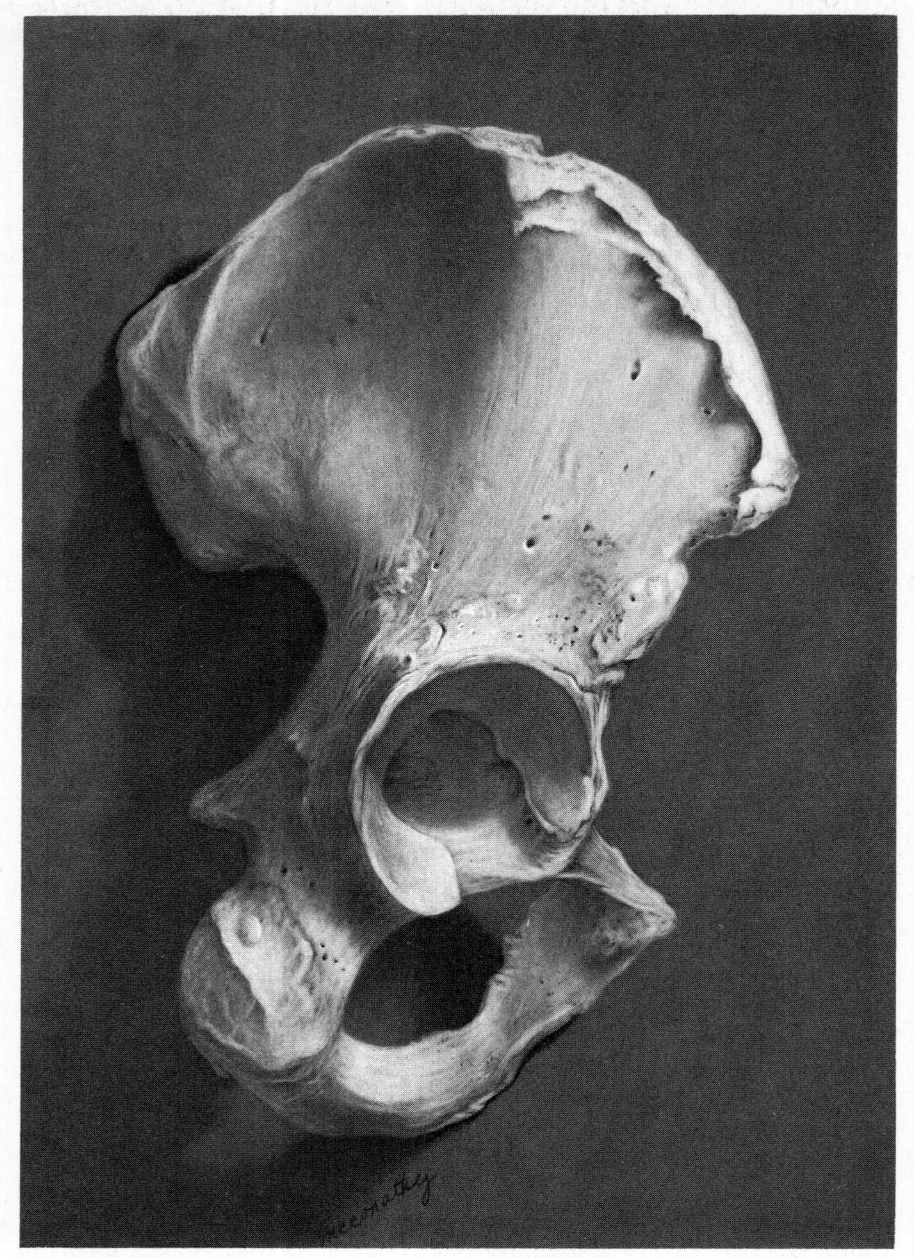

Figure 4-21. Os Coxae (Innominate bone). Carbon dust on color match paper. Drawing by Deirdre Alla McConathy, University of Illinois at Chicago, Department of Biocommunication Arts, Chicago. Unpublished. Drawn in 1979.

reproductions and look at the edge of the printed image. Most times you will find one or more plates printed slightly to the side. Naturally, the quality of the paper used for printing is important. All of this means that color is rarely used in the preparation of illustrations for research journals. First of all, it is very expensive when compared with black and white; second, when printed it will not represent the true colors of nature. However, subjects in the natural sciences have to be printed in color, and therefore have to be painted. Some research institutions have fewer budgetary problems and can afford color reproduction. Besides NASA, some government and private research institutions will also use color. In natural science there are areas of subjects—namely birds, plants, and insects, especially butterflies—that are represented as color illustrations in publications. During preparation of a color illustration you will have to struggle with two important problems, the first conceptual and the second technical. You will have to logically and consciously separate color from value. This is of extreme importance. If your illustration paper appears to be flat, that means that value—darks and lights—has not been separated from color. One of my old painting teachers once told me that if I could draw, I could paint. I did not believe him. How is it possible to mix painting and drawing together in such a manner? But he was right—if you can render a black-and-white image of a three-dimensional object in continuous tone, you do know how to paint. The concept is the same. The application of the values and contrasts is the same. The confusing factor is the color. When using color, you may be afraid of contrasts. You may assume that because a particular leaf is uniformly green, it must be presented that way on your illustration. But the subject cannot be painted as it appears in front of you. Keep in mind that you are used to thinking and reacting in three-dimensional space. You know that this uniformly green plant is three-dimensional. The viewer will react differently because he will be facing a flat piece of paper, knowing in his mind that it is flat, that it is two-dimensional. You, the illustrator, will have to manipulate the viewer's mind through exaggeration of values accomplished only with manipulation of color. In other words, if you have two overlapping leaves, then regardless of their original, natural color, you will have to paint portions of one darker than it appears in nature. In order to find a suitable darkness, you will have to add red, brown, or maybe even a touch of purple to the green mixture you are preparing. Remember that ready-made, factory-produced colors have a limited tonal range. All you can do is dissolve densely packed pigment with liquid, making it lighter, but this will not solve all of the problems. In painting, the principles of tonal changes remain the same as in black-and-white illustration. If there is a shape with a dark tone, the nearest other shape must be light or at least lighter than the neighboring dark tone. There must be a continual change of light, dark, light, dark tones followed with your mind, not with the subject's accidental presentation as illuminated by the table lamp. The subject is only for reference. It is there to remind you about its colors and to keep you on the right track.

When working on Strathmore smooth-surface drawing paper, the greatest technical difficulty will appear to be making the painted surface smooth; it may seem an impossible task to preserve already-painted areas. Remember that you are working with layers of pigment. The first layer, whether thin or thick in actuality, is an underpainting to be worked on later. In order to accomplish this task, use a bit of acrylic white while mixing colors and apply this as a very thin layer over accidental scratches with a virtually dry brush. Wait until the area dries and repeat the same process until the surface is smooth. The acrylic white, being a plastic base paint, will hold the watercolor in its place, allowing you to place a quick wash with almost pure color over its slightly whitish surface. Do your best to avoid a lot of water. Water will make the paper buckle and will quickly destroy its surface. Keeping the surface of the paper as intact as possible will allow you to remove unwanted paint by washing it out

Figure 4-22. One of several different concepts showing a possible configuration of a space station that could be used in servicing future missions and satellites. Tempera paint. National Aeronautic and Space Administration, Goddard Space Flight Center, Greenbelt, Maryland. Painted in 1983.

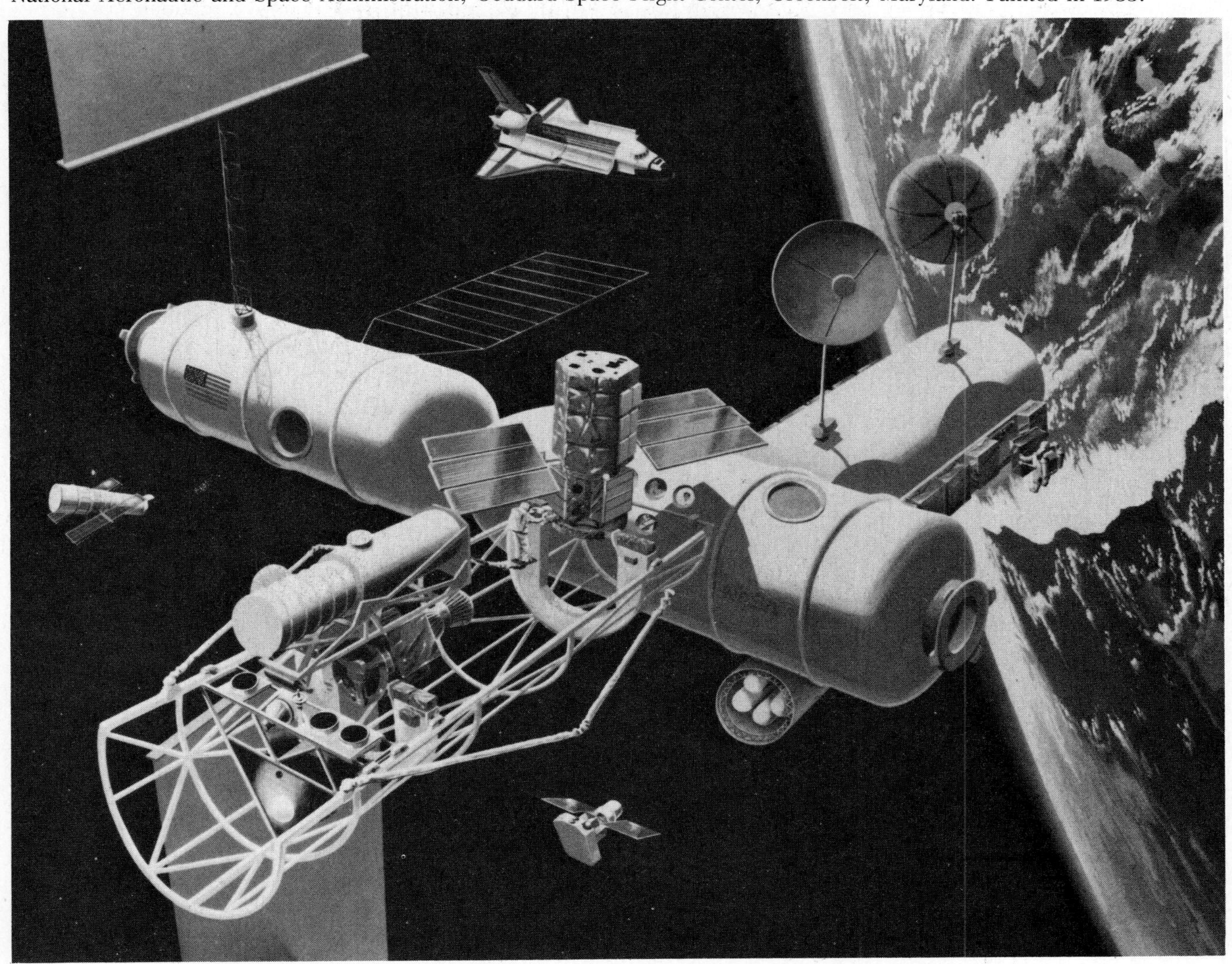

and will not produce any large surface bubbles. Use the best brushes possible. A sable brush is the best choice providing its tip is very well pointed. Before buying the brush, check its tip. If it is rounded, do not buy it. The finely pointed tip of a number 6 brush will allow you to draw fine lines, and the body of the brush can be used to cover larger areas of the subject. When using watercolors on Strathmore, try to avoid very finely ground pigments. Such pigments, usually associated with good-quality watercolors, have been produced for instant penetration of the paper's surface. This we should avoid. Once the surface is penetrated well by the pigment, a change of color or form is very difficult to accomplish. It is safer to use a cheap watercolor paint, as those pigments will stay more on the surface of the paper. In order to alleviate a problem created by a layer of pigment placed on the surface of the paper, have ready a tube of white acrylic. The white acrylic should be used carefully and sparingly, usually as an additive to watercolor pigment or as a separate layer of thin wash placed on top of a previously painted surface. In any case, do not use a lot of water and do not paint with the paste produced by mixing acrylic with watercolors. Why not use acrylics only? Acrylics are plastic-base paints, that produce, after drying, an impenetrable surface. All you can do to make corrections or changes is to overpaint with another layer of pigment. With watercolor, washing out of some pigment is possible, producing effects that are impossible to achieve any other way. This does not mean that acrylics do not have virtues of their own, especially when used on thicker surfaces or when a larger-than-average illustration is to be produced. The thin, almost transparent coat of white acrylic will stabilize the watercolor pigment, making it easier for you to apply a second coat over that area. When using colors on Strathmore, do not expect instant perfection. When painting you will encounter the following problems:

1. The paint will appear to dry in irregular patches.
2. During application of the second or third layer of pigment, the first underpainting will wash out in some places.
3. The paper may buckle and crease.
4. You may have difficulty drawing thin lines with the brush.
5. You will have problems distinguishing tone from color.
6. The painted image will be too faithful in overall color when compared with the subject.
7. When the subject is finished, the painting will appear dull in color.

When a mixture of pigments is applied to the surface of the paper, it will produce darker and lighter streaks or patches. Leave those patches or streaks alone. Let the paint dry out completely before going back to that area. Make sure that the mixture of pigment is controlled by you during the mixing process. Do your best to avoid plastic or glass as a base for mixing the paint, because when water beads up on a water-repellent surface, you do not know the percentage of pigment in a particular puddle of water. Regardless of the quantity of pigment, the mixture will appear the same. You will discover the truthful proportion of pigment and water—the value—only after application to the surface of the paper. It may be too late to change. It may be too much to correct. Not knowing what you have on the brush may cause trouble. Mix paints on an absorbent white surface such as illustration board. Such a surface will give more information about the color, value, and density of a particular mixture. Yes, it is a wasteful procedure, but you do want to paint correctly, ending with a product that satisfies both you and your customer.

The patches left from the first application can be removed or smoothed out in two ways. The more obvious and more difficult is to wash those areas out using only the very tip of the brush. While doing this, be sure that the brush contains very little water. It should be moist enough to dissolve the pigment from unwanted areas without injuring surrounding areas. The quicker and easier way is to use white acrylic. Mix some on the side with water in such a way that it will produce a very thin wash. Make sure that the brush is "dried out" on the board by running it until the very tip is flattened and will produce faint separate lines. Apply the thin layer of acrylic wash to the surface of the desired area. Be cautious. Do not press and do not try to cover a large area. Repeat the process, making sure that each layer has completed dried out, until you will have a thin whitish layer of pigment overlapping the first one. You can and in most instances should mix white acrylic with other watercolor pigments, matching the tone, not the color, of the area you are concerned with. This process will smooth out unwanted irregularities and will form a second layer of underpainting. If you manage to use thin coats of such mixture, the first layer of underpainting will still be visible but with a presently stabilized surface, appearing dull and whitish. The area will have to dry completely.

The next step will consist of an application of a thin wash of relatively pure colors. Remember, the tone should be considered first in order to create an illusion of three-dimensionality. Mix a desired color with water. A green leaf will call for virtually pure yellow. Apply a thin wash of that color over the area, restoring brilliancy. If not much water is used, the paper will not

Figure 4-23. *Tetrastes bonasia* (Hazel Hen). Watercolor and pencil. Painted by Matthias Haab, Zürich. Published in *Vogelschutz in der Schweiz (Schweizerisches Landeskomitee für Vogelschutz),* Switzerland. Painted in 1982.

buckle. Naturally, the thicker the paper, the less worry during painting. The thickness of the paper may be very desirable for painting but not for transferring the final sketch to its surface. Depending on materials available, such as transfer paper or a light box, you can be the judge and choose the thickness. When a mistake is impossible to correct during painting, the only way to solve the situation is to cut out the painting and splice a new addition, pasting the whole thing on a new background. You know that the white background will be necessary for proper reproduction. Cutting through three-ply or four-ply Strathmore can be very difficult. If the brush has a good, well-pointed tip, drawing a fine line will be no problem. When the tip is even slightly rounded, the thinnest line you can draw will equal the diameter of the tip of your brush. How about number 2 or number 1 brushes? Small-number brushes are fine for retouching but difficult for general use. In order to draw a relatively long line—for example, a line of two or three inches—such brushes have to be dipped into paint numerous times. If you can match and join short strokes without visible effect, then you have no problem. I strongly suggest a number 6 or 8 brush because those sizes can hold a reasonable quantity of moisture, allowing you to continue one brush stroke for some distance.

The biggest problem you will face is not technical; it is neither the application of paint nor the washing out of pigment. It is a conscious recognition of values. The tone cannot be mistaken for color. The tone, or value, is a primary source of an illusion of three-dimensionality of the image. This, bluntly speaking, means that you will not be able to represent all colors as they appear on the subject. You will have to exaggerate in contrasts between overlapping parts of the subject as well as between distant and closer ones. Which should be light and which should be dark? Truthfully speaking, it does not make any difference, although it is helpful for the viewer as well as for the illustrator to present closer sections as lighter and farther ones as dark. Watch out for unexpected situations produced by an ever-changing dark, light, dark, light chain reaction, because when you proceed without a general plan you may run into a major tonal controversy someplace in the painting. Being aware that the color of the subject will have to be altered in order to present a feeling of three-dimensionality does not mean that subconsciously you will not try to represent everything as you see it. You must make an effort to paint using logic in addition to eyesight. If your painting appears flat it means that tone, not color, differentiation is insufficient. Remember that the general tonal as well as color range will change when your finished illustration is sprayed with Krylon Crystal-Clear fixative. Do not overspray at first. Show restraint. Slowly build four or five layers of fixative coats, making sure that each coat has dried before the next is applied. The values and colors will change, appearing stronger and brighter by about 30 to 40 percent over the previous state. Remember that after a fixative coating is applied, hot dry-mounting is not advisable. A hot press will melt the fixative, making it stick to whatever material is used for protection from bare metal, removing pigment and destroying the illustration. For dry-mounting, use Twin-Tak or with extreme caution substitute tracing tissue with backup sheet from Letraset instant type. As a medium, acrylics should be used in about the same way as watercolors. Painting with a thick paste is not advisable and is hard to control. Scientific illustration calls for a precise representation of the form, and a thick mixture of pigment will not allow you to control the form of the subject with precision. Mixed techniques can involve a conglomeration of materials such as airbrush, color papers, photographs, various paints, pencils, and crayons. What is to be used is dictated by the availability of materials and the subject. Color graphs and charts are easily produced by using color overlays. A number of manufacturers produce instant color transfer sheets sold under various brand names. Pantone Color/Tint Overlay by Letraset or the Presfilm Color Sheets produced by Prestype are typical examples. A pen-and-ink outline with airbrushed tonal effects can become a superb representation of the subject. In natural sciences, airbrushing is not widely used because of the great diversity of subjects, the complexities of structures on one subject, and the average time allotted for a finished illustration. Cutting stencils or using frisket can be very time-consuming, especially when the precision of airbrushing leaves something to be desired. In most cases mixed techniques will result from the instant needs of the scientist rather than from the desire of the illustrator. Mixing pigments, Magic Markers, and other image-producing tools call for a bit of experimentation. If such has to be done, take a bit of time and try whatever mixed technique you are using on the side before starting the project. It may be a good idea to keep some kind of notes pertaining to behavior of materials as a future reference.

EXAMPLE 1. A requisition from the department of geology has just arrived. It states that one of the scientists needs a drawing of a map of a certain area with necessary differentiations of various stratigraphic features represented by tone. Do not jump blindly into the project. You have all necessary information. The base map has been supplied, the scientist's sketch with stratigraphic lines is included, and even the percentage of tone to be used on each stratigraphic level is indicated. Do inquire into the scientist's definition of tone. Did specified tone mean a continuous value or does it refer to the densities of darkness? Is the scientist thinking of line illustration or real tone illustration? Maybe the truthful meaning of tone is enclosed in some dotted or crosshatched pattern. It is very likely that in the scientist's mind the tone is something which in actuality is an imitation of the tone. Perhaps preprinted so-called instant tone is the answer. In such a case the scientist's tone illustration in actuality is a line illustration. Ask him how this illustration will be reproduced, how it will be printed. Make sure that you inquire in detail, because most likely that scientist will not know the illustrator's professional vocabulary, and he does not need pencil or airbrushed tone on his map. When everything is explained and clarified, proceed with the project. You know that fake tone, the tone represented by a formation of black dots or black lines, can be found among Letraset or Prestype products. Such "tone" will be applied to the illustration with neatly drawn pen-and-ink lines in respective areas, using an X-Acto knife and burnisher. The actual, real continuous tone should be produced with the pencil, brush, or airbrush. Depending on the complexities of base illustration as well as on the type of surface used for drawing outlined areas, the procedure will follow a routine of frisket or stencil masking. As you know, the areas that are not to be sprayed with airbrush will have to be protected by something. As the map can be elaborate or simple, depending on the scientist's needs, two basic procedures should be considered. The necessary stratigraphic lines can be drawn at first with acetate overlays used for tone placed over the basics, or the tonal range will be prepared before the lines representing stratigraphy. With the second possibility facing you, make sure that the final sketch of the structures does represent configurations approved by the scientist. Invite him to your drawing table and ask about that linear final touch. Also, you will have to make sure that the overlays with the tones or overlay with the lines will have proper registration marks, allowing platemakers and printers to reproduce such an illustration correctly.

EXAMPLE 2. Finally it happens. Color is to be introduced into your portfolio. An ornithologist asks you for a series of illustrations. As in most cases you will be presented with the question, "Can you do this?" The scientist will usually show you somebody else's work and will ask if you can handle his complicated project. There are many illustrations to be prepared and after seeing your pencil and pen-and-ink drawings, you have been approached as a possible illustrator. "Oh," you say, "positively I can paint." is your answer. To your surprise, he looks at you and states, "My illustrations are birds." Do you feel uneasy? Do not feel bad that your abilities have been questioned. After all, he had not seen anything produced by you in color. Besides this, his problems and his doubt may not be of a technical nature. The position of each subject may be what is on his mind. Remember that it is one thing to paint for yourself and quite another to do the same for a picky customer. You will be expected to produce color illustrations representing an image that is in his mind. Among various technical problems encountered by you will be the form and the color of the subject. Most likely the illustrations will have to be grouped in plates, full-page illustrations representing various interrelated subjects. Most likely you will be expected to paint all subjects included on one plate as one composition. Through sketches and through your conversation with the scientist, any questions about the placement of subjects, their position relative to each other, and the size of the plate will be solved. That is fine. But the biggest problem will be that of color. What is real to you may be darkish, brownish real to him. Communication and mutual clarification of meanings will be essential. In addition to making notes, I strongly suggest quick sketches with explanations, making sure that a sample of colors will be presented for approval. Do not forget the proportions of pigment and be prepared for lots of changes. Why? Usually good bird illustrators know their subjects in the natural state. They have observed birds in flight and have noted colors as they appear in nature. In work, you will be presented with prepared specimens. Such specimens differ considerably from their natural state. The position is completely altered and the colors are off. The truth is that you may have slides or photographs to work from, but you know that the nature of film as well as the exposure will alter the coloration of the subject. You must ask a lot of questions to clarify such matters. It would really be best to know the subject as it appears in its original state. If you do not, then in all probability you will have problems.

EXAMPLE 3. Invertebrates present a challenge. Numerous shells await your attention, each better-looking and more complicated than the last in structure. Surface color patterns are delicate yet well-pronounced. The scientist has specified tone illustration. Tone appears to be a good conveyor of message. The question is, what kind of tone is achieved through what kind of technique? You must consult with the scientist. If he is open to a variety of techniques, then suggest to him to talk to the printer. The information you both receive will be valuable. You will learn the possibilities. Does this particular publication use color? Can it reproduce a black-and-white halftone? If the publication is a small scientific journal, then in all probability black-and-white, line-and-stipple illustrations will have to be produced. If it is not, then pencil or color can be used. With luck, it turns out that halftones can be used. Now a technique of drawing should be established. Will it be a continuous-tone pencil, carbon dust, or wash? Again, discuss the details of the project with the scientist. During the meeting, make sure that the scientist explains his objectives and that you make quick notes and suggest various possibilities. If the basic features of the specimen—that is, the curvature of the structures and the shape—turn out to be the primary concern, then suggest carbon dust for delineation of the three-dimensionality of structures and pen and ink for outlines. Such a technique will allow you to draw faster, and thereby produce more illustrations. A coloration pattern may not be represented with great precision, but an overall representation of the subject will be easily achieved. On the other hand, if coloration patterns are of equal importance with representation of the structures, then pencil will be the best medium. Pencil is always the medium for greatly detailed and very three-dimensional-appearing illustrations. Once the technique is approved, the images of the subjects will have to be drawn. Because of complexity, regular measuring and grid techniques will have to be abandoned. Tracing forms as well as coloration patterns will be necessary. In order to do this, you will have to use a camera lucida attached to the microscope when drawing small subjects and some kind of viewer-projector type device for larger ones. The type of camera lucida will depend on the kind of microscope you will use during the project, while the viewer-projector will remain basically the same as for any reduction-enlarging procedure. When using viewers, the initial result will be a sketch on tracing tissue, later to be translated onto the drawing paper.

With camera lucida you will have two choices. The first will result in a similar sketch produced with the help of the viewer, although such a sketch can be drawn on the surface of drawing paper. Again, the rendering will follow, with the subject remaining under the microscope for continual consultations.

The other choice will be an immediate rendering without the initial sketch. This means that the tonal rendering will start without a preliminary sketch. Naturally, the choice of drawing procedure will depend on circumstances as well as on the scientist's decision. When using wash as the medium used for representation of the specimen, you will not be able to begin rendering in the same way as when using pencil. Wash will be more difficult to control, and the materials, especially the surface used for the rendering, must be chosen carefully. An initial sketch is a must, and after it is approved it will have to be transferred to the drawing surface before actual rendering will take place.

EXERCISES The most important part of a well-rendered and well-produced tone illustration is the conscious differentiation of light and dark tones. The play of values must be ever-changing. The illustrator's shyness must be overcome. Every artist preparing a three-dimensional illustration has to overcome the same problem. The personal struggle with decision-making processes can be very scary when great differentiation of contrasts is called for. Dark and light must be represented as shapes, although in scientific illustration outlines are very hard to escape from. The line presented in the area of continuous tone will have flattening tendencies. In other words, the more lines, the flatter the image of the object will appear. It is a challenge to exaggerate the contrast on a drawing. You must remember that such a process is based strictly on thinking and does not depend on the superficial range of values visible on the subject. By thinking of processes, you have to realize that the object under your scrutiny is illuminated by the light, which is not under your control. Such lighting or illumination of the subject is strictly accidental. You must realize that a table lamp, microscope light, a light fixture attached to the ceiling, or the sunlight does influence what you see. Even when trying to manipulate such illumination by hand, taking the table lamp and shining the light on the specimen from a certain direction, for example, will not do. The light will bring out some features, while others disappear in produced shadows. Assuming that under "controlled" conditions you draw what you see, you still may obtain unpleasant results. Why? Because lots of shapes and forms will not be noticed by you and will not be represented on the illustration by the necessary tonal variations. Am I trying to tell you to draw something fake? In a sense, yes. The form will be unchanged, presented on the illustration as it is in nature, but the illumination of that form will have no relation to the actual situation. The illumination will be conceived in your mind, and your mind will make the necessary decisions rather than the lamp. In order to present a three-dimensionality you will be forced to back-light numerous structures. Such back-lighting will be contradictory to the main direction of illumination. It will be contradictory but necessary. How else can you show hidden structural form to the viewer? Use a reducing lens during any type of tonal rendering. The simple device will tell you what is prominent and what has disappeared on the drawing. During any rendering, do not forget that the illustration will be used for printing. This means that the size of the illustration should be seriously considered. What about perfection of techniques? Techniques are secondary. The more you draw, the better you will draw, assuming that the process of drawing or painting is a conscious one. Do not feel frustrated and do not assume that perfection is easily attainable. Do the exercises a number of times. If I can continually exercise my hand, so can you.

Exercise 1. Pencil Line

A. Take a sampler of various papers and H, HB1, B, 2B, and 4B pencils. Place one of the papers on the illustration board in the proper manner. Sharpen all the pencils. Draw straight lines with each pencil, trying to induce changes in the thickness and density of the graphite on the paper. The pressure applied to the pencil in use will have to vary in order to produce a thinner or thicker line. Discover how different pencils behave on different papers. Note that harder pencils such as H and HB will not produce a great range of tonal changes within one drawn line. Notice that softer pencils such as 4B will quickly lose their sharpness. The object is to learn about the behavior of the most essential tools when used under the conscious manipulation of pressure. The quality of the line will be apparent. The line produced by the pencil should be clear and sharp as well as varying in tonal density.

B. Choose the best pencil, such as B or 2B, and make sure the tip is sharpened. Draw an outline of an imaginary apple. Apply the tone, using only pencil line. The light should appear as if forthcoming from a definite direction such as the upper left side. Draw tonal lines, following the surface of the three-dimensional object. Do not cross-hatch or smear. Start with light, thin lines and, continually drawing, change that line into one that is darker, and slightly thicker. Place tonal lines close to each other, covering the bulging surface of the apple. Remember that the surface of the subject you draw curves in two directions, horizontally and vertically. Make sure that tonal changes of drawn lines are adjusted for the curvature of the subject.

C. Draw your own or a friend's

hand with palm upward. Make sure that it is presented in frontal view. Prepare a good-quality sketch just as you would have done for the scientist. After outlining, implement papillary lines into the sketch, paying attention to the curvature and direction, and giving an idealized version of what is present on the subject. Note all structural changes and all bulges as well as major wrinkles and creases. Prepare the final sketch and transfer it to the surface of the Strathmore kid-finish drawing paper. Make sure that the sketch is transferred lightly. Using a pencil line, draw the papillary lines, changing the lightness and thickness of the drawn line according to surface configuration. Assume that darker parts of the lines will represent depth and the lighter and thinner parts of the lines will represent forward structures. Again, do not place the lines too close to each other. When finished, prepare the presentation of the drawing in the usual fashion.

D. Take one of the drawings, preferably the representation of the hand, and separate the image from the background. Use an overlay of Rubylith or Amberlith. The Amberlith is more transparent, therefore it may be more desirable. Cut the outline of the hand with an X-Acto knife, making sure that only the very outside edge of the line is used for guidance. Peel off unwanted portions of Amberlith, leaving the hand covered with red. If you have some money to spend, take it to a quick-print office and have your illustration printed in tone. To assure proper handling, write specifications on the edge of the tracing tissue covering the illustration proper. Write "Reproduce in halftone, reduce by 50%" or "Reproduce in halftone, reduce to 5 inches." In the printing office you will have two types of plates to choose from, a paper plate or a metal plate. The paper plate is cheaper but will not give you a good resolution in light areas. The metal plate is more expensive but the printed image will be better.

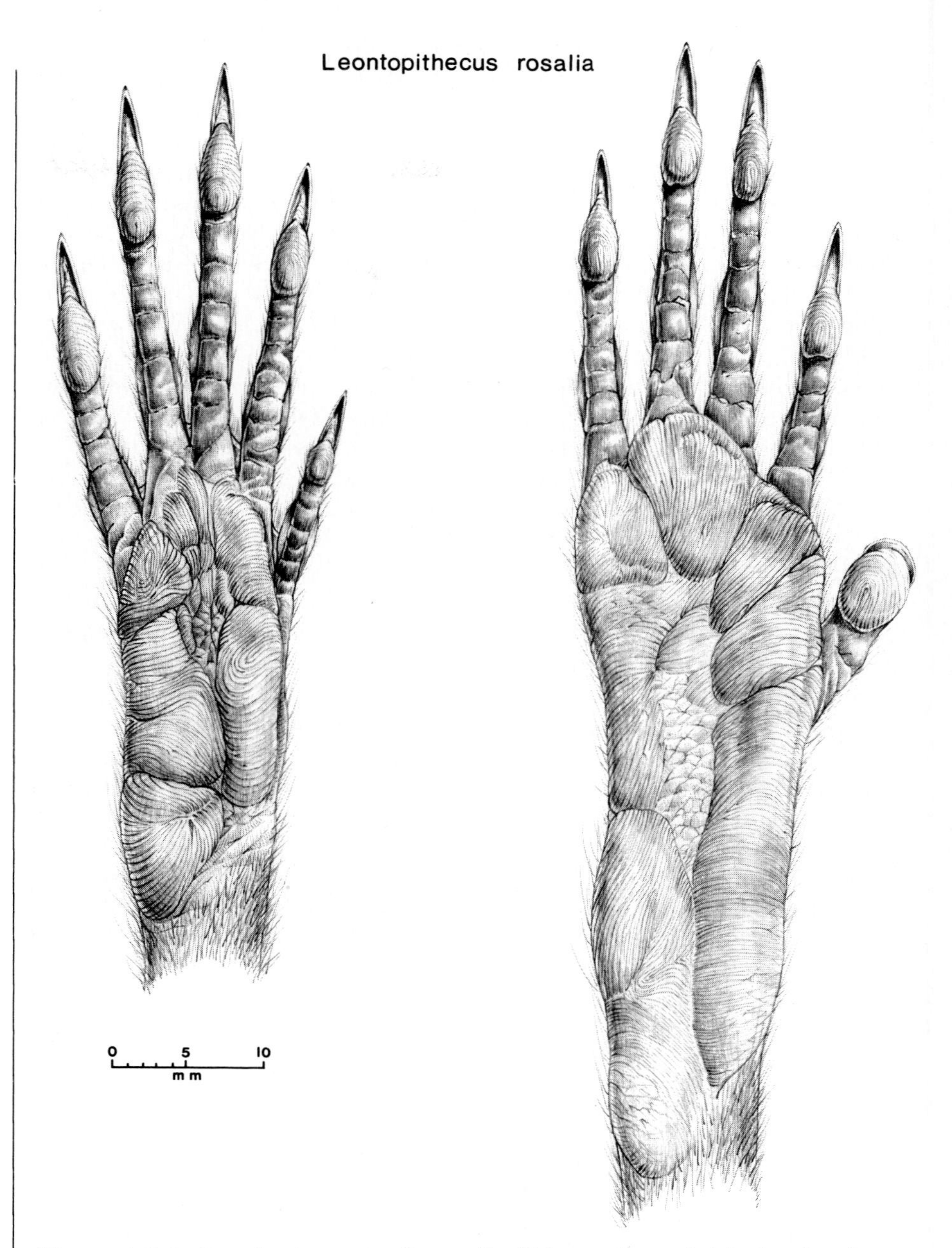

Figure 4-24. *Leonthopitecus rosalia rosalia:* Volar surfaces of right hand and foot. Pencil on Strathmore drawing paper. Drawing by Samuel Grove, Field Museum of Natural History, Chicago. Published in Phillip Hershkovitz, *Living New World Monkeys*, Vol. 1, by University of Chicago Press, Chicago, 1977. Drawn before 1977. ©1977 by the University of Chicago. Copyright reserved.

Exercise 2. Continuous Tone, Pencil on Strathmore Paper

A. Take Strathmore kid surface and Strathmore bristol papers. Draw a two-inch square on each of the surfaces with pencil, making sure that the line designating the squares is very visible. Using a well-sharpened B pencil, cover the designated area with tone. Do not smear the graphite. Use small, circular motions and do not rush. Make sure that the tone within the designated boundaries is perfectly smooth and will match in its value the value established by the outline. Have a kneaded eraser handy and when necessary, use it to remove the unwanted specks of darkness. Compare the behavior of two different drawing surfaces. Discover that Strathmore kid paper can produce darker values than bristol. When applying tone, do not press the pencil into the surface. Take your time; build the tone with layers of graphite. When finished, each should look like a patch of airbrushed tone. The outlines should not be visible at all and the surface should appear completely smooth.

B. Find a suitable subject such as a turkey or chicken bone. Prepare the sketch and transfer the sketch onto the Strathmore kid drawing paper. Looking at the subject, render the surface tonal changes using a tip of the pencil. Make sure that existing contrasts between upper and lower surfaces of the subject are exaggerated. Do not leave any outlines visible. Even in the uppermost parts of the subject, render the changing surfaces using areas of tone produced by the tip of the pencil. Be prepared to be forced to use back-lighting in some small areas. Depending on the subject, you may encounter small openings within the basic structure. Be sure that you will treat those as separate entities. Part of the edge of openings will have to be lighter than the surrounding background of tone and the openings themselves will have to be darker.

Exercise 3. Graphite on Strathmore

A. Prepare a sketch of two simple overlapping structures such as two rectangles positioned over each other, or a form similar to a spiraling belt. Transfer the sketch to the drawing paper. Prepare the graphite on the separate piece of illustration board. Prepare the necessary pencil-type erasers. Make sure that the tips of erasers are cleanly cut at an appropriate angle. Have two or three stomps in the vicinity and make sure that the kneaded eraser is worked up in your hands until it can be shaped easily. Start your work by rubbing the angled surface of the stomps into the graphite powder. Be sure that all of the angled surface is saturated with graphite. Take the soft pencil-type eraser and rub its flat cut surface into the graphite until when pressed into a white section of illustration board the eraser will produce a reasonably uniform tone. Reapply the graphite to the eraser during the drawing procedure. Start by applying tone with the pencil-type eraser. Hold the eraser in such a way that the flatly cut section will face the paper. The roundish tip should touch the outline and, by pressing the eraser into the paper, make a smudge. By making sure that the direction of application is away from the outline, you will not cross into the white background. Continue making smudges for about half an inch along the outline. Take the kneaded eraser and roll part of it into a cone. With this cone, pick up the darkest spots from the tonal area. Take small number 2 stomp and gently even up the tone. You will have some irregularities within the tonal area. Continue working until that part is smooth. If necessary, use a well-sharpened H or HB pencil to cover very small white spots as well as to bring the tone exactly to the outline. In the dark area, the outline should not be visible. It should become a part of the shape of the tone. During rendering, make sure that the bottom drawn rectangle is much darker in the vicinity of the top rectangle. Overlapping rectangles or shapes cannot be separated by an outline. The tone must become an integral part of the outline. The more contrast between rectangles, the more feeling of distance you will produce. During drawing you must remember that changes in value between those simple shapes and within enclosed areas of those shapes will depend on how you have drawn the outlines as well as on how you want to present the shapes. This means that if there is a three-dimensionality to be presented; as in the looping belt, the changes of value within the shape must exist. If the changes of value are not present, you will have a flatly rendered area.

B. To improve your skills, prepare a good-quality final sketch of a structurally complicated subject. The skull would be one of most obvious and more decipherable specimens. Render the subject using the very same tools, such as the graphite powder, pencil-type erasers, stomps, kneaded eraser, and on occasion, a pencil. During rendering take your time evening the tone as necessary. Be very picky about the presentation of the surface of the specimen. Make sure that enough contrast exists between dark lower areas and light or white upper parts. You may have to leave some outline as a necessary means of separating sections of the shape from the white background. Remember to use the reducing glass *during* the rendering. During the process of drawing do not smear graphite all over the place, and do not evade the possibility of making decisions. Compare the subject with the final sketch and do some thinking. The person who is in charge of placement of appropriate tone is you. Remember that it is easier to add tone than to remove it, therefore do not overrender the drawing. Assume that the closest sections of the structure will be lightest and the farther ones proportionally darkest.

Exercise 4. Pencil or Ink Lines with Graphite or Carbon Dust on Drafting Film

A. Use the same old subject—a chicken bone is just fine. If you have access to more complicated and more intricate specimens, then choose one that will not require a great deal of detail work. As always, prepare the final version of the sketch. Spray the sketch with fixative and when the sketch has dried, tape it to the illustration board. Take a sheet of drafting film of appropriate size and place it on top of the sketch; also tape it to the illustration board. The sketch will be visible through the semitransparent drafting film. Using a pencil, draw an outline of the subject on the surface of the film. When finished, apply tone using stomps or Q-tips. The tone will be very light, much lighter than on the paper. During application of the tone place an opaque sheet of white paper between the drafting film and the sketch in order to see tonal rendering better. For the darkest sections use the pencil. Remember that the light or lightest areas should not receive any tone. Use a kneaded eraser to remove tone from the darker areas. Do not forget to spray the finished drawing with fixative.

B. Ink outline works very well with dust techniques, especially with carbon dust. Prepare a sketch of a suitable subject such as a fish or snake. Place the sketch on the illustration board with drafting film over the sketch. Take the technical pen with appropriately chosen point size relative to the size of the drawing and draw *all* outlines. Draw all scales and other solid structures in unform line. The coloration pattern as well as the volume of the subject will remain to be rendered with dust. Apply the dust, graphite, or carbon to the surface of the drawing after ink lines have dried out. Spread the dust with a good-quality—preferably sable—brush or with chamois cloth. Again, use a sheet of opaque white paper placed between the drafting film and the sketch in order to check all tonal values. The scales will appear to overlap when the forward edges of each scale have a very light tone on the inside of the scale placed next to the black ink outline. Naturally you will follow the usual rules of values and contrasts. This means that when the subject is very three-dimensional, the darkest tone will be placed toward or at faraway areas. When drawing a snake, keep a simple rule in mind. The forward (that is, closer to you) area will be in the middle, while the farther, darker tone must be placed near the edges. Again, tone will be removed with a kneaded eraser. It is possible to scratch the drafting film with an X-Acto knife, producing a very fine white line, but be cautious, because once the surface of the film is scratched it will pick up graphite or carbon dust readily, and it is impossible to work over a scratched area.

Exercise 5. Carbon Dust Without Lines on Drafting Film

Obtain a specimen such as a starfish and prepare the final sketch. Prepare necessary tools for drawing such as carbon dust obtained from Koh-I-Noor carbon pencil number 3, the drafting film, kneaded eraser, stomps, and Q-tips. Place the sheet of drafting film on top of the final sketch that was previously taped to the surface of the illustration board. Secure the drafting film with masking tape. Draw the outermost outline of the subject with a well-sharpened carbon pencil. Being very careful, apply an overall, general tone to the whole area of drawing. Do not render details at this point. Make sure that three-dimensionality of the subject is captured. The closest parts of the subject to you will be very white or extremely light, while farther sections of the subject will be dark. Make sure that tone is applied smoothly and that any accidental streaks are removed. Be patient; do your best to keep the background of the image free of dust. At first this may appear to be an impossible task, but in actuality it is not. Do not blow any air from your mouth on the drawing during tone application. Little bits of saliva will damage the surface by producing dark, impossible-to-remove spots. Do not delineate any details in outline, and wait with rendering of small surface configurations until the whole subject is covered with gently changing tone. Make sure that an opaque piece of paper is placed between the drafting film and the sketch in order to properly appraise the range of tones. Depending on the type of carbon and the type of drafting film you are using, it may be necessary to add darkness with the carbon pencil. When application of general tone is finished, then using a kneaded eraser remove the tone from surface irregularities such as bumps. Form the eraser into a cone and by pressing the tip of the eraser's cone into the desired area, pick up previously applied carbon dust. The bumps should be much lighter and in most areas white. Then, using a stomp rubbed into the carbon dust, underline each surface irregularity with a darker tone, making sure that applied tone is darker than the background. Make sure that each surface bump will *not* be outlined with this darker tone. The tone should be applied only to the receding wall of each irregularity. Make sure that sharp, accidental outlines are gently fused into the previously applied general tone as well as into the white removed area.

While drawing, the biggest problem will be the carbon dust itself. It is a messy technique, so be careful not to smear unwanted areas of the drawing. After illustration is finished, spray the surface with fixative. Apply the first coat very gently. Let it dry and afterward apply a second coat. Do not overspray because a thick coat of dried fixative will stick to anything placed on the surface of the drawing.

Exercise 6. Ink Wash

A. Use illustration board as the surface for wash application. Trying is the best learning procedure, therefore make sure to obtain different types of smooth-surface illustration board and make tests before proceeding with the project. I suggest the cold-press, one-ply smooth-surface illustration board, but this does not mean that this type of board is good for presentation of all subjects. Depending on the texture of the subject on details to be rendered, as well as on the type of the ink, another board may be more suitable. Naturally, you should avoid textured surfaces because such a surface will interfere with control of the drawing tool. Prepare the sketch and transfer it to the surface of the board, making sure that transferred lines are not very dark. Choose the best brush. The size of the brush will depend on the size of the illustration. In any case, do not use smaller than a number 6 sable, well-pointed brush. Prepare a fair amount of water and a small, flat container for the ink. A small dish will be the best. Prepare a large quantity of soft paper towels or Kleenex and stack the paper up. The paper will be used to wipe the brush and to remove excess water from its body and tip. Pour some of the ink into the mixing dish, dip the brush into the water, and then into the ink. Wipe the excess on the towels and try the tone on the piece of the same board you are using for the illustration. This is important because various boards have different absorbencies. Before applying tone to the illustration you should know what to expect. You should know the density of the mixture as well as its behavior on the surface of the board. Start rendering the drawing with light washes. Do the lightest areas first, carefully applying darker tones later. Make sure that the brush does not contain too much water and that applied washes do not flow uncontrollably into each other. Wait until the first wash is dry, then place the second layer of mixture over it. The hard edges of layers of washes will be removed after the ink dries out if you use just a bit of water on the brush. Do not press the brush into the surface too much. The tool is used for application of the mixture to the surface and not for anything else. Remember that it is easier to add darkness than to remove it. During tone application do not damage the surface of the board. Such damage will be caused by repetitious movements of the wet brush over a wet area. Place the wash and wait until a layer of ink dries out, then correct it by adding or removing. Waiting for the wash to dry is important, because once the surface of the board is damaged it will produce particles of fiber that are then spread by the brush into other areas.

B. Use drafting film to achieve the same objective. Apply wash to the frosted surface of the drafting film more carefully than to the surface of the board. Remember that the uppermost coating of the drafting film is washable. It will dissolve slightly when wet. This process does solidify applied wash, therefore be prepared for more streaks than the board would have produced. Work slowly, waiting until a layer of wash has dried out before implementing corrections. It is possible to wash out unwanted streaks from the surface of drafting film using very little water and lots of patience. As usual, place the lighter tones first, slowly building up to the desired darkness. Remember that when looking at the tonal range placed on drafting film, that range will appear much lighter than any other surfaces. The reason lies in the absorbency of the surface. There is very little space for moisture to penetrate, therefore most of the tones will lie only on the surface.

Exercise 7. Watercolor on Strathmore Drawing Paper

A. After transferring the final sketch to the surface of the smooth Strathmore drawing paper, begin the painting process by observing the transferred drawing, thinking about the light and dark tones. Do not mix up color with tone. Do your best to preplan placement of tone. Think about the color illustration the same way as you do when drawing a continuous-tone pencil illustration. Remember that once you start with some kind of tone you may set a chain reaction of ever-altering light and dark areas, leading to a situation that will be very hard if not impossible to solve. For this reason the great painters of previous centuries have prepared studies of the subject. The tonal changes have been ironed out in such studies, leaving the artist with full knowledge of what to do. Do not expect the first try to be easy going, and do not expect that following paintings or color illustrations will be much easier. Painting is more strenuous than drawing because it is easy to confuse color with tone. Be prepared to lie about colors. For example, you have a light green, very three-dimensional shape to paint. That light green color is the actual color of the shape as it appears in nature. While painting, you are dealing with a flat surface which is read by the human mind as a flat surface; the painted image is only an illusion of the light green three-dimensional shape found in nature. Observing nature, our mind compensates, knowing that surrounding areas as well as the object itself are in actuality three-dimensional. Having a flat surface to work with, we must convince the viewer about that fact and will have to cancel his information input about the flatness of the surface. The light green cannot be represented in a uniformly painted light green color. If this is done, the shape will lose its three-dimensionality. In order to maintain a shape's three-dimensional feeling we have to change its natural color into one that is much darker in some areas. The receding parts or the forward sections of the shape will have to be darker than the general overall natural coloration. The problem lies in the fact that colors do not have a great range of values. A light green by itself will not produce strong enough contrasts even when pure pigment is used. In order to solve the problem you will have to mix some other color, usually darker, than the overall natural color of the shape to produce a desired strong contrast. This means that red, blue, or brown will have to be mixed with light green, producing under some circumstances a color completely different from the natural one. Now if more of the basic light green color is visible than the other "fake" one, then the shape will be read by the viewer as a three-dimensional light green form. For that reason, do think when mixing pigments. Use all colors with the exception of black. Black will flatten the desired range of tones. You will have a tendency to overemphasize details found on the subject. Keep yourself in line with the help of a reducing lens and the knowledge that the viewer is not prepared to see as much as you are. The paper you are using will behave differently from watercolor paper. Its surface will not absorb as much water and pigment, allowing you to change tones by washing out as well as by adding. Make sure that you are patient. Do not wet too much of an area you are working on. Be aware that during the process of application of the pigment most of the rendering is actually underpainting. Mix pigment on the white surface of a sturdy board, not in the plastic containers. Use a little white acrylic to stabilize the surface of a layer of watercolor. In order to do this, mix white acrylic with water or with the desired combination of pigment and by brushing the excess amount of moisture, apply the remaining mixture on the brush very lightly. It should produce a multitude of small, thin lines. *Let it dry*, then repeat the same process. An area will become whitish after such treatment, but afterward, when it is dry, apply a wash of desired color over such underpainting. The later mixture should be rather transparent, making it possible to see the tonal range produced previously. Work slowly and deliberately, keeping in mind the most essential dark and light tonal information.

B. Obtain reproductions of seventeenth or eighteenth century Dutch still life painters. Analyze and scrutinize the paintings; pay attention to dark and light tonal changes. Ask yourself why such-and-such section is lighter than the surrounding area. Look for the general direction of illumination. Find contradictions in the paintings. Pay attention, strictly from a professional view, to the fact that objects or parts of painted subjects hidden from the main source of light are still visible and tangible. It is the lighting or is it the change of values that is the cause of such effects?

Exercise 8. Instant Color Overlays

A. Prepare a pen-and-ink outline drawing of a skeletal structure such as a portion of human anatomy. A drawing of a spine or a hand will suffice. Depending on the type of paper used for drawing, permanently attach the drawing to the surface of smooth, sturdy illustration board. Use removable dry-mounting tissue for tracing paper and Twin-Tak for opaque heavier paper. Drafting film should be secured with a strip of tape only. Obtain three different colors of Pantone instant color, prepare an X-Acto knife with a number 11 blade, and a burnisher. Take one sheet of instant color and before cutting through, peel the corner off. See that the color sheets are made of two layers: the backing and the thin sheet printed with actual color. The upper sheet has a waxy substance on the side facing the backing. Knowing this, you will be able to judge the pressure on the X-Acto knife and you will restrain from cutting through both layers. Place a sheet of instant color over the previously rendered and dry-mounted illustration. Through the color it will be possible to vaguely see the pen-and-ink outline of the drawing. All colors will be reasonably transparent with the exception of dark reds and black. Holding the sheet in position, use the X-Acto knife to cut around the visible section of the drawing. Do not cut through the backing. Do not cut the drawn shape exactly. Remove the sheet of instant color from the drawing and, using the tip of the knife, peel the cut-out section of color. Gently pick it up and place it over the desired part of the drawing: Lightly press a portion of the color into the middle of the matching section of the drawing. Smooth out the cut-out piece of instant tone by hand, using minimal pressure. During the process, be sure that the desired area is completely covered by the color and there is an excess of instant tone. Using the X-Acto knife, cut through the instant tone right along the black line. Do your best to cut right on the middle of the line. In any case, do not cut too close to the inside edge of the black outline. Any cuts on the inside will be visible as white when the excess of instant tone is peeled off from the surface of the drawing. Remove all unwanted parts of the instant tone and take a sheet of tracing tissue. Place the tissue over the area and burnish the newly placed tone with reasonable pressure. The pressure, combined with the motion of the burnisher, will produce heat. The heat will melt the waxy underlying layer, making the thin semi-transparent or completely transparent tone stick to the drawing. The black line will be visible through the tone. When finished with one color, proceed to do the same with the others. Be careful—sometimes the instant tone may stick all by itself to the surface of the previously placed piece. This happens especially when old sheets are used, although it may also happen with freshly bought materials. For that reason, do your best not to overlap the second differently colored piece over the first too much. Continue until the designated areas are colored, then consider the illustration finished.

B. Using the same or a similar black-and-white outline drawing of a skeletal structure, apply the instant tone to an acetate overlay placed over the black-and-white drawing. Make sure that the overlays are attached securely to the illustration board along *one* edge of the acetate. Also make sure that when closed, the overlays will not interfere with each other. The three separate overlays, each for a different color, will have to be attached on three different sides of the illustration. It would not make any sense to attach large overlays to a small drawing or small overlays to a large drawing. In any case, the overlays should completely cover the illustration. The edge of acetate should not run across the black-and-white image. Apply instant tones the same way as in the previous exercise, making sure that you do not cut completely through the acetate during cutting, nor scratch it, nor make too many visible small cuts. When you are finished, prepare the illustration as usual by placing the protective sheet of tracing tissue over the overlays and by attaching the black cover-stock paper as a final touch.

Exercise 9. Cutting Stencils and Using Frisket

A. Prepare a color illustration representing a plant, using watercolors or acrylics. Plan to include a color background, to be added later. Secure the illustration to illustration board using dry-mounting techniques or masking tape. When using tape, remember to tape only one edge of the illustration to the board. Take a sheet of Rubylith and place it over the illustration, making an overlay. Using an X-Acto knife, cut the outline of the plant and remove the background, leaving the red coating covering the image. Protect the surrounding area with newspapers secured with masking tape. Take an airbrush and spray the surface of the overlying acetate with the desired color. Do not forget to mix the paint appropriately. Applying thin coats of paint, build the density to the desired level. When everything has dried, remove the newspapers and peel the remaining coating of the Rubylith. Depending on the type of paint used, the substance will be more or less stable. If you desire, coat the acetate with a spray of fixative. The result will be a color illustration on a white background or a color illustration on a colored background. When the overlay is in position, attach a registration mark to at least three corners of the illustration board and do the same with the overlay. Registration marks will assure you and others handling such an illustration that the overlay can be positioned properly.

B. While painting or drawing, use frisket to protect white areas of the illustration. Take liquid frisket and using a good-quality brush, spread the frisket just like paint over areas that will be painted later or that should remain white. Depending on the type of paper used for painting or drawing, the frisket can be removed with an eraser. Be cautious with very smooth, not-very-absorbent papers, as a coat of frisket applied to a painted area may be difficult to remove. Naturally pencil, carbon dust, and similar drawings will be coated with fixative before removing the frisket.

REMOVING MISTAKES

What will you do if there is something to be changed on the finished illustration? Let's examine the possibilities. Basically, there are two reasons for changing small or large sections of finished illustration. Namely:

1. Your own fault
2. Change of scientist's mind

The first—that is, your skill in observing the specimen, following directions, drawing, painting, and preparing the illustration for printing—will be challenged by a variety of unpredictables. The difficulty of

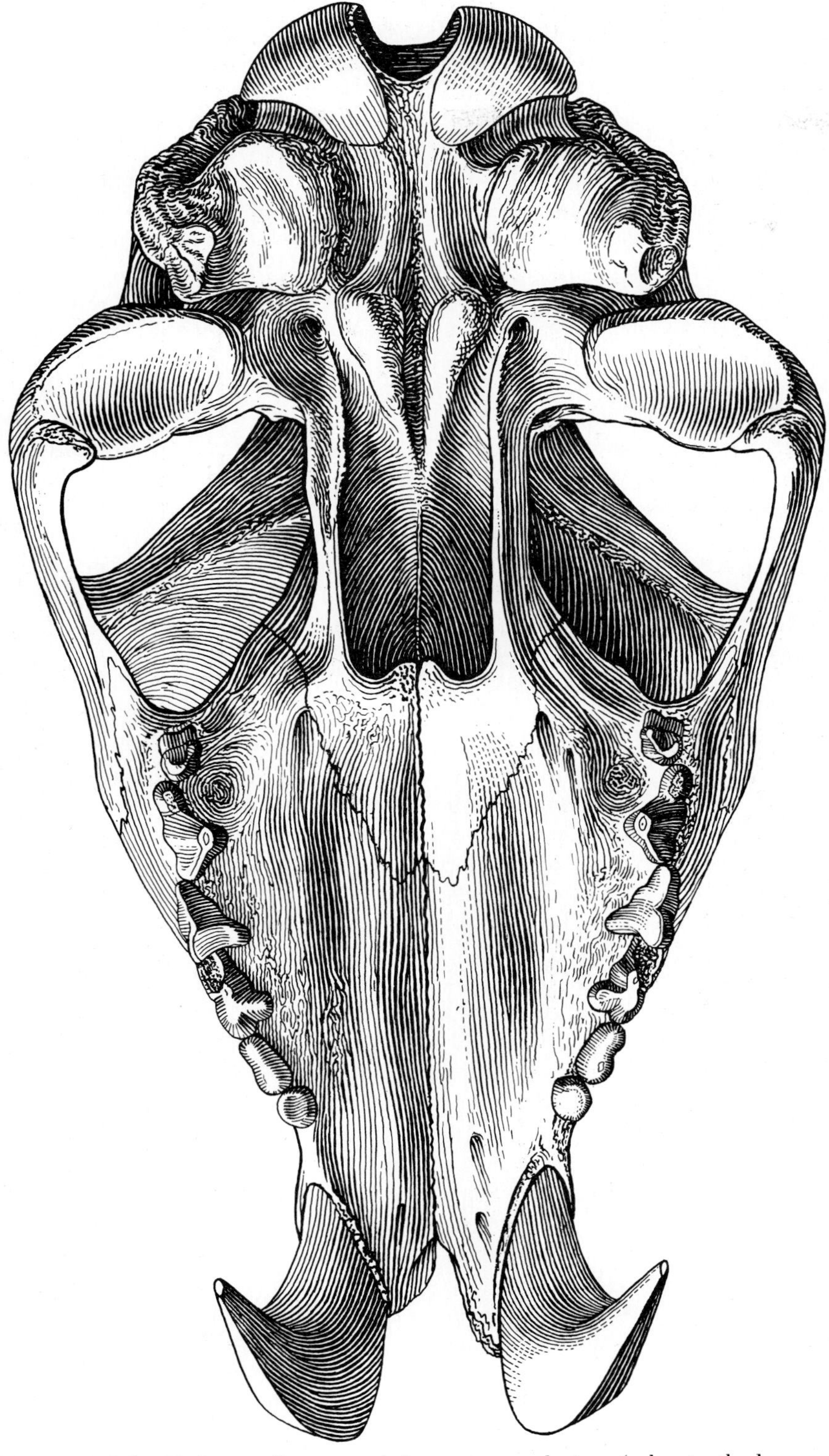

Figure 4-25. *Thylacosmilus atrox*, holotype, ventral view, (sabre-toothed marsupial). Crow-quill pen on two-ply Strathmore paper. Drawing by Sydney Prentice, Chicago Museum of Natural History (Field Museum of Natural History), Chicago. Published in Elmer S. Riggs, *Transactions of the American Philosophical Society*, New Series, Vol. XXIV, 1934, by the American Philosophical Society, Philadelphia. Drawn in 1931.

the subject and lack of communication with scientist will lead to an improperly although perhaps very well-rendered drawing. Dirty hands or a greasy sandwich are simply impossible to excuse. Dry-mounting, depending on the technique, materials, humidity, and amount of time spent, can damage the whole illustration. The drawing or painting is completed, but being only human, we all make mistakes which, when discovered, must be corrected. Most of the time you will have to make corrections on your own or somebody else's illustration because the scientist changed his or her mind, dictated by changes in research. Think about the processes involved in the publication of a scientific paper. Think about the waiting time before such a paper will be printed. You know that it may take anywhere from a year to ten years before the scientist's work will be on the printing press. Lots of things can and will change during this time. Think about the research conducted by teams of scientists, sometimes working in adjacent rooms and sometimes divided by continents. All illustrations may have to be corrected, changed at any moment. The degree of difficulty will depend on the materials used for the original drawing and the quality of rendering. The changes or corrections can be conducted by the following methods.

1. Overpainting
2. Scratching
3. Cutting things out
4. Pasting
5. Overlaying
6. Splicing

Whatever the method and regardless of how much care is taken during such procedures, it will be visible on the illustration that something has been changed. Visibility of marks, cuts, and overpainting on the illustration are of no importance unless they destroy the drawn image or get to the printed page. The freshness of rendering will be lost. As we are not really interested in untouchable artwork but the final result, the printed image, as long as we can cover up the edges of the paper or white spots from the camera eye we are successful. The decision of what type of method to be used must be left up to you because there are too many variables involved in appraising the situation. Overpainting is most popular because it is the quickest method. Pro White retouching paint, white acrylic, or in situations where there is nothing else to use, Liquid Paper correction fluid will do the job of covering unwanted features. It is simple to apply, but do not forget that Liquid Paper will crack easily; when this happens, the features underneath the overpainting will be exposed. Do not forget that Pro White will not dry well if it is old. It will remain sticky, attracing dirt as well as papers to its surface. Try it before you use it. When a given area is overpainted the desired feature should be rendered in such a manner that it will match the technique used for the drawing. The corrected portion should not be distinguishable. This brings us to an important point: If, in your mind, the illustration you have received for changes or corrections is of inferior quality, suggest to the scientist that you redraw it. He may not agree because of deadlines or because he does not want to repeat the whole explanatory process. When the scientist says no, be sneaky and redraw anyway, gambling that you can do a better job than the previous illustrator. If you succeed in such a project, you will bring back a happy customer. Cutting through a finished illustration, especially your own, can be a traumatic experience. Preplan, by sketching, what is to be removed and what will remain. Make sure that the scientist approves all corrections before you implement them. If the illustration is rendered on thin, one-ply, two-ply, or three-ply paper, you will need strong backing in order to join the good old section of the illustration with the newly rendered portion. Use illustration board or three-ply paper for such a purpose. First adhere the section of old illustration to the new surface, using Twin-Tak as a dry-mounting tissue. Then, positioning properly the new sketch of the missing portion, do the same. After dry-mounting, render whatever is necessary.

Whenever overpainting or cutting and dry-mounting are out of the question because of whatever circumstances, do not be afraid to paste a piece of opaque paper directly over the area to be corrected. The edges of the paper will not be visible after printing if the illustration is to be reproduced in line. For halftones, splicing will make more sense. Tone negatives are much more sensitive to delicate changes of values and may pick up the edges of paste on paper.

In some situations scratching is actually the only means to remove wrongly rendered areas. Drafting film can be scratched easily with an X-Acto blade or straight-edge razor blade. During such a process do not apply too much pressure—remember that under a coat of frosting which holds the ink there is smooth acetate. India ink will not adhere to a very smooth acetate surface. If necessary, use a different ink especially produced for acetate during rendering.

Overlays, which open as a kind of flap, are another possibility. With the illustration firmly attached to a sturdy surface such as illustration board, affix a white paper over the section of the rendering. Use masking tape to hold the flap tightly to the background surface. Naturally, you would not think of attaching an overlay to the surface of the illustration. Make sure that the overlay is taped only along one edge. When a correction is implemented, the illustration will have two versions. One is the original, completely unattached and the other, with the overlay down, is changed. Remember that illustration board is made from thin layers of pressed paper and can be separated easily. The uppermost layer will peel off when an X-Acto blade is inserted under the first layer. By slowing pulling the upper part of the board, it is easy to remove rendering drawn on the board's surface. Then through splicing and dry-mounting old and new sections a correct version will be produced.

EXAMPLE 1. There you are! You thought everything was fine and the project definitely finished. "Would you mind correcting this pen-and-ink drawing?" the scientist asks. "All I need is removal of the stipples from the upper sections of all the teeth and a strong outline designating the damage to dentine caused by wear during eating." The illustration is precise and well stippled. As a matter of fact, you had really put yourself into this one while drawing. It is a jaw with numerous teeth drawn from a lateral view. When you look at the carefully applied dots and think about damaging your own drawing, something snaps inside of you. Hold it! The change—or, as the scientist calls it, the correction—may be your own fault. It may not be caused by anything else but your own lack of communication abilities. Did you ask enough questions at the beginning? Did you pay attention to the answers? Did you make notes? Whatever the reasons, none of them are of any importance now. Appraise the situation as a professional. You have a job to do, so find out what has to be done. Ask for the scientist's sketches. Request a meeting. Discuss the problem thoroughly. Be sure that you know what he wants; be sure that he knows you are aware of the situation. Control your temper and be nice. If the drawing is executed on paper, use white paint to overpaint or retouch all unnecessary areas. When everything that has to be removed is gone, place a tracing tissue overlay and draw the changes in pencil on the tracing paper. Ask for approval. If everything is OK, then transfer the corrections to the actual drawing. Use a transfer sheet or graphite. When using graphite, first draw the very same sketch on the other side of the tracing tissue without removing the tissue, using softer but not too soft pencil. Then draw the very same from up front with little pressure. Try not to damage the painted areas, while at the same time transferring necessary information. Do not remove the tracing tissue overlay. Bend it back and tape it to the back side of the illustration board with one piece of masking tape. Draw the necessary corrections using the proper tool. If the original drawing was rendered with crow quill, then use the same. If a technical pen was used, then Rapidograph must be used. Be prepared for difficulties. The lines may be too thick and the paint may cause the ink to clog the pen. Leave imperfections whenever they happen to occur. Later, when the ink dries out, use white paint and a fine brush to retouch the area. When finished, again arrange a meeting with the scientist and *ask* his opinion. If he is not satisfied, rework the illustration. If the illustration was drawn on mylar drafting film, place the new version of the final sketch underneath the drafting film, making sure that it is properly aligned with the rendering. Using a straight-edge razor blade and a ruler, remove the upper sections of all the teeth from the illustration. Place a metal ruler's straight edge slightly underneath the wrong area and press the razor blade into the drafting film, scratching its surface with the blade. During the process, hold the ruler well so it cannot move. Make sure that one edge of the razor blade will slide against the ruler. The result will be a clean, uniform removal of unwanted features. For drawing, use ink produced for acetate. When you have finished, do not forget to wash your drawing tools, especially the technical pen. The ink used in a technical pen contains a stronger adhesive than regular drawing ink and may cause damage to Rapidograph pens by permanently clogging them.

EXAMPLE 2. Dr. X has come to your drawing table with an old pen-and-ink map. He wants some of the features changed. The illustration was drawn by somebody else so many years ago that nobody remembers the artist's name. It is thick with overpaintings. Various illustrators must have worked and reworked that thing. For some of the overpaintings Liquid Paper has been used. By now it is cracked and crumbled. Ask the scientist if he wants to redo the whole map. As it is, this illustration has been worked on to its limits. If he agrees, you have an extra project on your hands. Inquire as to the size he wants. In such cases the scientist will want to have the new version equal in size to the old one. Arrange a good-quality Photostat. Inquire if it would be possible to have a negative made and who would pay for such expense. Obviously this map is precious and of importance, as otherwise it would not have been kept that long. When retouching the original is the only way out, then get to work with care. Do not repair broken and damaged areas on your own. By doing this you can inflict great damage to the specimen. Before anything is done, ask him for very specific instructions. Perhaps some of the whiteout liquid was scratched on purpose? If you do not know what to do, restrain yourself. Be in touch with the scientist during the whole repairing and changing process.

EXAMPLE 3. An anthropologist has brought a second version of the pictorial reconstruction of the vessel. Says he, "The form and shape are fine; truthfully, this is all we could have done. The presentation of the theoretical information does support the evidence, but... the design, the configuration of marks found on the surface of the vessel does leave something to be desired. I have told you there is something peculiar about those more or less uniformly embedded markings. You know that their origin lies in the tool used for forming the walls of the vessel. Yes, all of the markings are pronounced, but somehow," he says, "I do not see the similarity between the drawing and the object." You, the illustrator, know that a lot of time has been spent trying to figure things out. Naturally, pertinent questions have been asked and the scientist has described the tool used for forming the vessel. Sketches were made of the tool and of possible design combinations which that tool could have left. The truth is that the scientist's descriptions did not quite fit into this peculiar pattern. At that time you had some misgivings about his explanations, but you took his word. During the observation, again you noticed certain peculiarities that did not coincide with possibilities offered by the scientist. Drawing is a joint effort: yours and the scientist's. Together, you form a team. Perhaps you did not ask enough questions at the crucial moment. Whatever the reason, at present both of you know there is something wrong. It is obvious that you have both reached the same conclusion. Now it is time to reexamine the whole drawing from the beginning. Ask the scientist if it is possible to obtain a list of materials used in the area where that particular vessel comes from. "Naturally," he states, "this is easy. Wood, wood, I have been telling you for some time. The wooden paddle with the carved design has been used to "spank" the walls of the vessel. All researchers agree on this," he says. There is nothing else. Think now. Is there a possibility the scientist is doing the same as you have done? Perhaps he is taking other people's words for granted. Inquire more. Is there anything else that grows in that area? "Well, naturally," the scientist responds, "reed grows in abundance, but other researchers have concluded that wood was used for pottery building." He states suddenly, "A reed? Well, this would be a first; this would be a major discovery. Maybe the reed has been used as you are suggesting; I did not think about such a possibility. Prove it," he says, "otherwise I will not believe you." Now the illustrator has ventured beyond stipples and techniques of drawing and has actually become an active investigator. After making a vessel from Plasticine and imitating the procedure used by the original potters, the newly produced pattern resembles the old one. By matching design configurations, both of you are happy because a little bit more knowledge has been added to studies of old civilizations. Now that you know the real cause of oddities in the pottery surface design, redraw the whole drawing from the beginning.

Exercise 1. Using one of your black-and-white line practice drawings, try to correct all the mistakes. The drawing should be dry-mounted and should appear as a finished product. Using Pro White retouching paint and a fine, number 6 sable brush, paint over all mistakes such as improperly drawn features or irregular lines. Wait until the paint is dry. Take the technical pen and redraw overpainted sections. Be careful and do not apply too much pressure on the pen because the paint will be scratched and the tip of the pen will clog up. If ink should spread by some accident, then leave it alone. When finished repairing the drawing, retouch new imperfections with retouching paint.

Exercise 2. Prepare a drawing with carbon pencils on coquille board. Dry-mount the illustration and spray its surface with fixative. Take a strip of coquille board and tape it to the illustration board above the finished drawing in such a way that it will form an overlay covering part of the image. Choose any section of the drawn image. Redraw the covered section on the newly attached overlay. Make sure that the flap can be picked up with ease. When finished, you will have two images—one underneath the overlay and one with the partially visible original and the overlay. While drawing, make sure that the newly applied carbon matches the background of the original in density and that the tone will extend exactly to the edges of the overlay. Do your best to obtain a Photostat of the drawing with the overlay in proper position. Examine the Photostat for signs of repair.

Exercise 3. Using a pencil, render a square in continuous tone. Dry-mount the square onto illustration board, then cut the square in half. Remove one half of the square from the illustration board by peeling off the upper layer of the board if you have used permanent dry-mounting tissue or Twin-Tak. If removable dry-mounting tissue has been used, reheat the whole drawing under the dry-mounting press or with a hand-held iron. When it is very hot, peel one half off from the surface of the board. After half of the drawing has been removed, carefully dry-mount that part again onto new illustration board. Take a new, clean sheet of drawing paper and dry-mount it onto illustration board *if* the removed portion of the drawing has been peeled off with a layer of illustration board. The thickness of papers must match or else a splicing line will be visible after reproduction. Make sure that the new portion of the paper is larger than the area designated for drawing, and make sure that the "missing" image is properly traced onto drawing paper. Make the splicing line with care and, using a sharp X-Acto knife, cut through the drawing paper *after* some kind of dry-mounting tissue has been attached or tacked on the back. Cutting dry-mounting tissue separately will do no good. It is almost impossible to align the precut dry-mounting tissue with precut paper. When cutting, exercise great caution, making sure that the knife does not cut the edge of the paper at an angle, as this will interfere with further retouching. After aligning and splicing the new, clean sheet of paper, try applying pencil tone to the spliced area. If the line is too visible, then use white paint to fill up the distance between both papers. When doing this make sure that paint will slightly overlap both papers. If you discover that paint is not needed, then proceed with rendering the new section. When paint is used, be careful in tone application, as that surface may pick up more graphite than drawing paper.

Preparation for Printing

Finished illustrations are submitted to the scientist, who in turn sends them along with his typed text to the editor of the publication. This process is known as submitting his research for publication. Before it is possible to reach this point, all the illustrations must be prepared for printing. Improperly prepared drawings may cause the scientist's work to be rejected and will cause you much grief. All illustrations should be prepared according to the specifications given by the publisher. The scientist is the person who knows the details; he will tell you the publisher's requirements. Usually illustrations are photographed and then submitted, although in some situations the originals will be sent for platemaking. You are well aware that for best printing results, the original drawings should be used for photoengraving. Before your drawings are submitted for printing, there are a number of steps in production of illustrations as well as various procedures which you may have to undertake. Depending on the circumstances you may have to be involved in the following:

1. Organization of drawn subjects into coherent visual presentation
2. Preparation of the scale
3. Keyline and paste-up
4. Application of explanatory designations in some kind of type to the illustrations or the board
5. Placing cropping or registration marks
6. Masking
7. Marking reduction or enlargement of the illustration
8. Arranging protection for the surface of the illustration

At this point connotation of the word *illustration* may mean the same as the word *plate*. The plate is an area of space into which a drawing or drawings will have to fit. If many drawings will be on one plate, all will have to be positioned and properly aligned in relation to each other. The *plate* also refers to the space designated for printing one or more illustrations. It is usually applied to the space needed for more than one drawing, especially when one specimen has been drawn from various sides. In such a situation, separate drawings of that one specimen will have to be properly aligned in relation to each other while the plate is being prepared. The whole unified visual description of that specimen must be presented in a logical, coherent way. A plate is prepared for two reasons.

1. To present a clearer visual explanation of one subject
2. To save space in the publication for more text

There are many ways to cover a given space with drawings. The scientist will be the guide, as he knows what drawings should be placed together and how they should be organized. He will know how close or how far apart each image should be placed, although the illustrator will have to take an active part in helping with the scientist's decisions. Each separate printed image, be it one drawing or many, will be designated in text as a figure. You know that illustrations in textbooks are marked "Fig. 4," or "Fig. 10," the same way as in scientific publications. You, the illustrator, will have to know the size of the printing area for the plate. Without this information, it will be impossible to prepare the plate in proper proportion to the printing area. In scientific illustration, most plates will be full-page illustrations, meaning that the whole page will be taken up by drawn images. *Full-page illustration, illustration, plate,* or *figure* all mean the same. Different terminology is used to represent the same concept. There are two types of full-page illustrations: The first will cover the area that is called the *printed page*. The printed page is different from the regular page because it refers only to the area covered by text on the regular page. This means that the white border around the printed text is not taken into consideration during preparation of the plate for a full-page

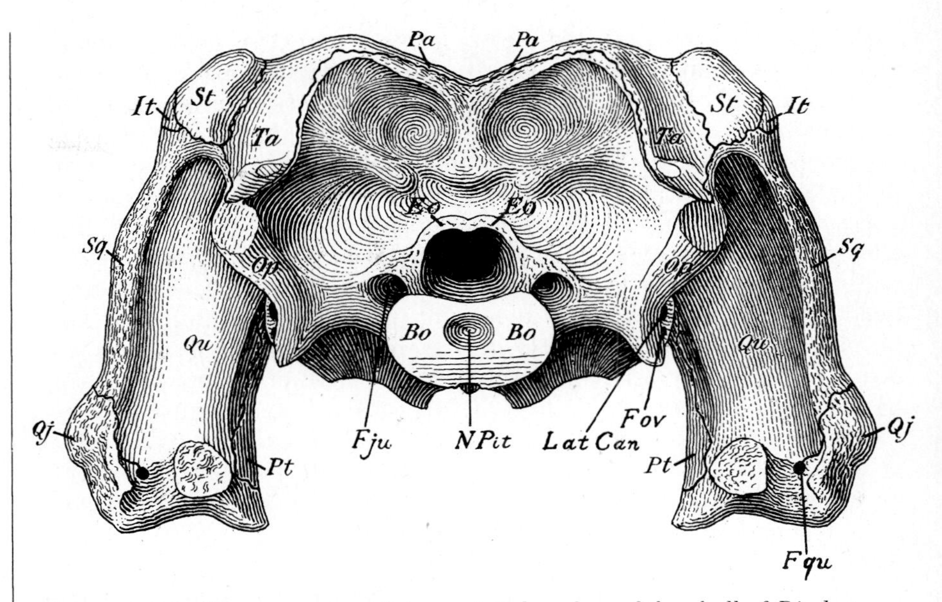

Figure 4-26. Reconstruction of the occipital surface of the skull of *Diadectes*. Crow-quill pen on three-ply Strathmore paper. Drawing by John Conrad Hansen, Chicago Museum of Natural History (Field Museum of Natural History), Chicago. Published in Everett Claire Olson, "The Family Didectidae and its Bearing on the Classification of Reptiles," *Fieldiana*, Geology, 11, no. 1, 1947, by Chicago Museum of Natural History (Field Museum of Natural History), Chicago. Drawn before 1947.

illustration. The other type will extend into the white border around the text. Usually in such situations the illustration or plate will extend beyond the size of the regular printed page. Such a plate is called a *bleeding illustration*, because it is allowed to bleed outside of the parameters of the page of the publication. In scientific illustration bleeding is rarely done because it does not serve any other purpose besides beautification of the page.

Scientific publications are printed as clearly as possible. Knowing the size of the printed area, a proportionally larger space will have to be deduced in order to prepare the plate. To make the necessary calculations, two systems of deduction are available, the proportional wheel or the trial method. The proportional wheel is a single device based on two independently rotating plastic or paper wheels. The larger gives the size of reproduction and the smaller the size of the original. In our case the original refers to the area designated for printing or organized and aligned images. The trial method will give the same results but will take a little more time. Knowing the size of a printed area, draw that area in its original size on some kind of paper, preferably tracing tissue. The result is a square or a rectangle. Extend the left vertical line and bottom horizontal line considerably. After this, connect the lower left and the upper right corners with a diagonal line; again extend that line considerably to produce the base for enlarging the original printed area. All lines parallel to vertical and horizontal original designations and connected to the diagonal line will produce the enlargement or reduction of our original space. Once we draw a variety of proportionally larger spaces on one sheet of tracing tissue, by placing that tracing tissue over renderings of specimens we will know how large a space we need in order to paste separate drawings together. The process of designating the plate, choosing the right size, and pasting separate illustrations in the proper order is the same as in any preparation for production. It is known as keyline and paste-up. The name explains the technicalities of the process. In scientific illustration there will be little or no type to paste up. In most situations the drawings themselves, the scale for drawings, and the basic designations will be pasted up in a keylined area. Once the enlarged size of the plate is known, organization of the subjects belonging to this plate can begin. The scientist will play a prominent role, as he knows exactly what should be grouped together. You can help him by preparing Xerox copies of separate drawings, thereby allowing him more flexibility in handling the images. The scientist will cut excessive paper from the copies and will paste them together in groups with some kind of type. Having Xerox copies will allow him to write comments and designate separate images without worrying about injuring original drawings. Usually the designations are simple. Capital letters will be used to "number" separate images or groups of images. He may indicate his choice of placement for the scale and may draw arrows to desired features.

Usually in the corner of taped copies he will indicate what figure number that particular plate or, as he may call it, illustration will be. Therefore he may write Fig. 3 or Fig. 4b. When this material is back in your hands, you will have to follow all designations to the letter. Knowing the enlarged size of the plate, you will have to prepare a keyline by taking illustration board and drawing an appropriate rectangle, using nonreproducing blue pencil. Make sure that all blue lines extend beyond their intersections for about two inches. Then take a technical pen and using a ruler, draw extended lines in ink, making sure that the ink line stops short of touching the main body of the rectangle. Those ink lines will be visible on the rectangle after the cameraman in the photoengraving company finishes his job, allowing another person in the composing room to position the plate in its proper place more easily.

Finishing with preliminaries, you will have to apply Twin-Tak to the backs of each separate drawing and arrange all on the plate by placing them down and securing them with small pieces of masking tape. Do not remove the backing from Twin-Tak. During that process you must make sure that none of the images extend outside previously drawn blue lines. The paper on which the images are drawn can extend, but the images themselves will have to be within the designated area. You will have to make sure that all pertinent parts of images are aligned properly. In order to do this, you will need at least a ruler and the right-angle triangle. A T square will be of great help. Do not paste anything down permanently as yet. Organize the plate thoughtfully, taking into consideration the placement of all necessary designations such as scales, letters, or arrows. All pieces of white paper that overlap, especially overlapping drawn images, should be cut away. Having made sure that everything is as it is supposed to be and nothing is obstructing drawn images, see that each separate piece of paper is secured with two pieces of tape to the board.

Because every separate drawing will have to be untaped and removed from the illustration board, you will have to make sure that the position of each piece of paper is secured. It is possible to do this in two basic ways. The first consists of drawing

Figure 4-27. *Eretmochelys imbricata*. Pen and ink on vellum tracing tissue. Drawing by Nancy Ruth Halliday, Florida State Museum, Gainesville, Florida. Published in Walter Auffenberg, "The Herpetofauna of Komoto, with Notes on Adjacent Areas," *Bulletin of the Florida State Museum*, by Florida State Museum. Drawn in 1980.

outlines of cut papers with non-reproducing blue-pencil. The blue pencil will not be seen by the eye of the camera, therefore it is safe to use; nevertheless, the keylined plate will not look very good in the eyes of the scientist. If you are desperate to upgrade your image, do the following: Tape the illustration board to the drawing table. Take a ruler and align its edge with one of the edges of the drawing, making sure that one of the corners of the edge of the paper is aligned with a stable mark on the ruler. Use the designations for inches, because those are easily remembered. Tape the ruler to the table at both ends. Remembering which corner of the paper with the drawing on it was aligned to the designation on the ruler, and how it was done, remove the drawing from the surface of the illustration board. Remove all pieces of tape holding that drawing. When finished, use an X-Acto blade to loosen the backing of a Twin-Tak and then, with your fingers, pull off completely. Carefully align the drawing with its sticky back against the illustration board with the ruler's straight edge, making sure that the proper corner of the paper matches the proper designation on the ruler. Do not drop that drawing! You know the powers of Twin-Tak. Once you drop it, the sticky backing will adhere. Slowly lower part of the drawing to the illustration board, applying a bit of pressure to the surface of the drawing with clean fingers. Continue that process until the illustration is glued in its proper position; when finished, turn your attention to another. Make sure that all illustrations are burnished down. Naturally, the finished plate should be protected with a sheet of tracing tissue and a sheet of cover-stock paper. Before returning the finished plate or plates to the scientist, do not forget about all necessary designations. Drawings on the plate should be marked with letters or numbers and, if necessary, arrows leading to previously marked features on his pasted-up composite should be affixed to the plate. You can do this in two ways. Such information can be directly applied to the illustrations and the surface of the board or it can be placed on the acetate overlay. Such overlay will have to be securely taped to the surface of the illlustration board with masking tape, but only along one of the acetate's edges. Choose the largest dimension because it will provide more security. Also, make sure that the acetate overlay extends over the whole area to be printed. Edges left within that area may appear on the finished, printed image in the publication. For small quantities of type, the instant-transfer type will do. For lots of designations, especially when many plates are being prepared and designations repeat themselves, having type set by a typographer will cut down on your production time. Make sure that such type is only affixed to the acetate overlay, because otherwise damage to the illustrations will be a strong possibility. Naturally such type will have to be cut out from the proof sheet. In order to glue a small section to the acetate, you will have to apply some kind of adhesive to the back portion of the proof sheet. The most commonly used adhesives are wax and rubber cement. Do not use Twin-Tak because the adhesive is too strong and, once placed, will be impossible to remove from the surface of the overlay. The wax will have to be coated with a hand-held waxer or a waxing machine. A hand-held waxer is a portable roller which has a container for a paraffin stick and a built-in heater. The heater will melt the wax. After placing the proof sheet back side up on an old piece of illustration board, wax is applied by rolling the roller over the sheet. Before cutting the proof sheet, let the coat of paraffin cool off. Then take the proof sheet and place it, waxed side down, on the surface of the new illustration board, very gently burnishing its surface. Using an X-Acto knife and a ruler, cut the necessary sections and transfer them to the acetate overlay. Make sure that all type is aligned with the bottom line of the plate, or you could end up with crooked type on the printed page. When finished, again burnish down all the pieces of paper with type printed on. Use a piece of tracing tissue as a protection, as otherwise the type will smear and some of the freshly pasted lines will move from their positions.

How do you specify the type and how do you order necessary proof sheets? Any typographer will be more than glad to help you learn these skills. He will provide the type specification books from which you will choose the typeface and will help you to specify the size as well as spacing of letters or numbers. The procedure is not very complicated; virtually every type specification book will contain a short description of how and what to do on its first few pages.

Why not use the waxing method for illustrations? The wax will melt easily when exposed to warmth. It will not hold for a long period of time. In scientific illustrations there is rarely a need for a large quantity of pasted type, therefore any designations which will peel off can be easily and efficiently replaced. The illustrations cannot. If one falls off and gets lost during what may be long periods of storage, it will take much more time and effort to redraw and repair the plate. Now the plate is completely ready to be sent to the photoengraver for platemaking. Do not worry about the reduction factor, because this will be designated by the publisher. In case you have to do this, draw a single line, using a ruler and the nonreproducing blue pencil, along one of the edges of the plate, ending that line with the arrows indicating the point where that particular side of the plate ends. To reinforce this information extend the designations of the size of the plate until they cross the newly drawn line. Write a large R and enclose it in a circle. A capital R stands for reduction; capital E designates enlargement. Next to the R, write the size of that edge of the printed page. If the printed page is 7 by 4¾ inches, then assuming that you have chosen the larger side, write R—to 7″. Such information is quite sufficient for the platemaker. Complicated plates will be encountered when working on projects for zoologists, paleontologists, anthropologists, ornithologists, and botanists. In many cases botanists or ornithologists may like to have you draw all of the separate illustrations on one piece of material, be it a paper, board, or drafting film. In such a situation the sketches, prepared ahead of rendering, will be arranged in a similar fashion and then transferred to the appropriate surface, rendered on drafting film, or painted as one composite illustration. If all the illustrations are in tone, masking will be necessary; that will be the final step. Using Rubylith, all of the images will be cut out and separated by the stark contrast between red overlay and white background. It is best for you to mask your own drawings. Somebody has to do the job, and you really know what the image is all about. A newcomer might overlook certain details because he or she is are not familiar with the structures. Such a mask may become an additional overlay but will not interfere with platemaking. It will help to clarify the image by removing all those edges of cut paper. If letters have been placed on the overlay, you do not have to prepare a separate mask for the line portion of the plate. As a matter of fact, you should worry about tone portions of the plate rather than the type. In a photoengraving company there will be a masker and a tone and line separator, which if necessary will separate the line for proper platemaking. When a single illustration is prepared for printing, the very same printed page will not be of importance nor will such illustration be marked for reduction. This will be done by the publisher, who will know the exact placement of the drawing within the publication.

EXAMPLE 1. The project is almost finished: 100 drawings of pottery sherds are ready for some kind of organization. The scientist states that from the planned 60 figures he will have to cut down to less than that because he needs more space for the text. The text is not finished and he would like to know how much space can be saved. You both sit down and, going over Xerox copies of drawings, reach a decision that the only way to save the space is to overlap most of the drawn images when preparing plates. The problem lies in deciding what drawings can overlap, which portions of the drawings will be chosen, and how to do this. The first two decisions should be left to the scientist. The technicalities are the illustrator's responsibility. Give the scientist all of the copies to play around with. He will have to decide what is most important. In the meantime, you can have a nervous breakdown yourself if you have included too much space between a pottery sherd and its cross section. This will mean a lot of cutting and splicing for you. If you have used semitransparent material for drawing such a drafting film or tracing tissue an additional problem will arise. How can images overlap if the bottom one will be visible? How will it be possible to secure a lot of small pieces of drafting film? Upon receiving the rearranged smaller number of plates, follow the scientist's directions and secure images in appropriate places. For tracing tissue and especially for drafting film, transparent Scotch tape will have to be used. Assuming that your fingers are clean the pieces of Scotch tape will also remain clean and without fingerprints. Some of the overlapping images will have to be cut out with an X-Acto knife, and the spaces between separate drawings will diminish. To the eye the plates will not look superbly well prepared because of the number of pieces of Scotch tape, but the camera will not notice and will not record anything besides the black-and-white pen-and-ink renderings.

EXAMPLE 2. All plates are arranged and the separate illustrations are securely pasted in their appropriate positions. You can proudly say that all of this work has been done well. The scientist is satisfied and now you are waiting for the printed result. It will be a nice addition to your portfolio. The tone drawings were rendered well and most of the subjects happened to produce visually interesting images. To your dissatisfaction, a minimum of five drawn images have been placed on every plate. It would be nice if every drawing had received more exposure, but nevertheless, you know that this will be a publication to show to others. In time, the scientist brings a copy to you and anxiously opening the pages, you look at the plates. Oh no, you say to yourself, how can all of the edges of the papers be visible? This really has destroyed a lot of the visual impact. The scientist is satisfied, but you are not. Why did you not separate tone from background? Why did you forget about this? A roll of Rubylith does not cost that much. The extra time needed for cutting masks would have paid off tremendously. Yes, the images are visible, but all those edges of paper will bother you for a long time. The next project will not end the same way—you know this, even if you have to buy that Rubylith yourself.

EXAMPLE 3. The end of the project is near and you are fully aware that illustrations should be properly prepared for reproduction. The drawings you have made are not to be arranged into plates. All of the drawings will be printed separately somewhere within the text. The technique of rendering has been dictated by the scientist; the drawings contain lots of details. There is a possibility that the reduction of images will be great. The black-and-white crow-quill illustrations may not hold all of the details after large reduction. Knowing this, do your best to convince the scientist about the need for halftone plates. Do your best, but diplomatically try to acquire his agreement for tone separation. Yes, the drawings are in black and white, and therefore theoretically are suitable for line reproduction, but because of the large quantity of details included in the image and the unforeseen large reduction factor, the line reproduction may result in a disaster. Details will close, delicate black-and-white lines will run together or will disappear. If you can ask the photoengraver for professional assistance, inquire about the density of the screen. If one company will not answer, call another. It is a crucial matter which should not be left hanging in the air. The scientist may know nothing about platemaking and printing procedures, but you should have some knowledge. Very detailed black-and-white drawings designated for large reduction should be reproduced in halftone. Halftone will help to maintain the clarity of the image, while line reproduction will destroy it. With luck, you can convince the scientist, as the final decision is up to him. If this fails, be prepared for bad final results.

Exercise 1. Call a photoengraving facility and inquire if it would be possible to see how plates are made. Introduce yourself properly, tell the manager that as an illustrator, you are seriously interested in learning something about platemaking. Remember that any company will have a schedule to maintain, therefore be very flexible with your time. Suggest that you are willing to visit the facility at night. Photoengraving companies are open all night. Work goes nonstop, all the time. Once inside, be respectful but inquisitive. Remember that this is your chance to talk to experts in this complicated field. Do your best to see all the steps of production of the printing plate, and ask questions.

Exercise 2. Practice tone and line separation. Choose some kind of image with uneven edges, such as a skull or a photograph of yourself. Properly affix the image to illustration board and place a sheet of Rubylith over it, forming an overlay. Cut the Rubylith carefully, following the outside edge of the image. Separate the drawing of the skull from the white background. Do the same with the photograph of yourself. Cut only on the outside of the edges of the body, then peel the background, leaving the red film covering the image but exposing everything on the background. Make sure to check for cutting imperfections, as it is easy to alter the important image by improper cutting.

Exercise 3. Pay a visit to a typesetting company. Arrange an appointment ahead of time—do not suddenly, at the spur of the moment, run into the shop. Again, ask questions. There may be a chance for you to obtain a copy of instructions on how to order type. Make a good impression, as you will do business with such a company in the future.

Working with Optical Instruments

As a scientific illustrator, eventually you will have to use some kind of optical instrument. Knowing how to operate it, what to expect, and how to apply gained know-how to the production of illustration will be crucial and important. In many areas of research use of optical instruments is extensive—just as extensive as the variety of such devices. Microscopes, viewing and photostat cameras, camera lucidas, and projectors are not difficult to operate. For you the need will arise in two areas of work, namely observing and drawing.

Your most important tool will be the microscope. It will become a close friend of yours while observing specimens and during some processes of drawing. Naturally, it must be treated with great care. All microscopes are delicate instruments, containing within their structures various combinations of lenses and prisms. Basically there are two types of microscopes, monocular and stereoscopic. A monocular microscope differs from a stereoscopic by the number of oculars. Obviously, a monocular has one tube to look through, while the stereoscopic has two. The number of oculars influences three dimensionality of the specimen and therefore clarity of viewing. A similar analogy exists between a record player with one speaker or with two. You are well aware of the fact that a stereo record player will give you much better sound and therefore much better clarity. Just like speakers, oculars on stereoscopic microscopes can be "tuned up" or should we say focused separately. Monocular microscopes will not allow for this, but nevertheless can be very impressive and important tools during observation of the specimen as well as during drawing. A good-quality monocular microscope will magnify an object in excess of 500 times. The quality and complexity of the optical components enclosed within the housing of the microscope determine the usefulness of the tool.

What are the major sections or parts of the microscope? Every microscope has a lens attached to the lower section of its main body, and the ocular or oculars on the top. The lens, called the *objective*, can be a grouping of lenses or a singular entity. The objective can be attached in various ways: It can rotate around or it can slide. The rotation of the objective containing a grouping of lenses of various powers can be obtained by turning the objective around with the fingers, by rotating a knob mounted away from the objective, or by sliding back and forth or sometimes sideways the housings containing the lenses. It all depends on the construction of the microscope. When the microscope has more than one lens in the objective it is more versatile. The primary function of such lenses is to magnify an object placed underneath. The microscope will produce magnification in steps, but it will be noticed by the viewer as one unit. First the lens within the objective and secondly the eyepieces will magnify the subject. The eyepieces are a separate configuration of lenses enclosed in their own housings. All eyepieces, just like the lens within the objective, have a preset magnification factor, usually enlarging the first magnification accomplished by the objective by ten or fifteen times. The combination of an eyepiece magnification with the lens in the objective will give the total magnification of the subject. In order to see clearly, a focusing mechanism is provided. The complexity of the focusing mechanism depends on the complexity of the internal structures of the microscope. In order to see through a microscope, you will have to look through oculars. When using a monocular microscope, you will use only one eye, but when a stereoscopic microscope is on hand, both eyes must be open and used to view the subject. In anticipation of possible problems, manufacturers have provided a mechanism to control the horizontal distance between oculars. You know that interpupillary distance, the horizontal distance between the centers of the eyes, does vary from person to person. Oculars on a stereoscopic microscope can be moved horizontally until the viewer can match his interpupillary distance with the centers of the eyepieces. If the interpupillary distance on the microscope is not adjusted correctly, the viewer will not be able to see the object with both eyes. It is a simple thing but usually overlooked by beginners, resulting in an unclear image. If at any time you feel the need to move your head horizontally, turn the head slightly sideways, or close one eye while using a stereoscopic microscope, the problem has arisen from improperly adjusted interpupillary distance on the microscope.

Light travels within the microscope housing starting at the illuminated specimen, through the objective, then through a series of lenses inside the middle of the main housing into the prisms, which in turn will direct the light path into the oculars and the eyepieces, ending inside the observer's eye. In order to observe the specimen and see everything clearly, an illuminator is needed for the specimen. This separate lighting can range from a regular desk lamp through various focusable low-voltage lights attached to the microscope as well as highly specialized fiber optic cold-light systems. There are two basic ways to illuminate a specimen. Incident light will project the beam of light *onto* the specimen, while transmitted light will do the same *through* the specimen. You know those small mirrors attached to the microscope underneath the objective? When light is directed to such a mirror and the mirror is adjusted in such a way that the light bounces off its surface and enters the objective, it will go through a transparent or semitransparent specimen placed above the mirror. The result will be a bright-field illumination of the specimen—bright-field because the specimen is visible on the bright area illuminated by light bounced off the mirror. In some situations this type of illumination can produce too much interference; in order to solve that problem manufacturers of microscopes have produced illumination that is similar but with polarized filters inserted right below the place for the specimen. When two polarized filters are properly aligned, the brightness of the background will diminish and the specimen will appear to be illuminated, but with a dark field as its background. When use arises, incident light can be combined with transmitted bright- or dark-field illumination. The whole microscope is attached to the stand in a vertical position. Most of the time the stand includes the transmitted-light illuminator or at least some kind of flat surface onto which the specimen can be placed. Besides the usual stands, there are stands without that built-in surface. Such stands will allow the microscope to swing horizontally over the table, as well as allowing for much greater vertical adjustment of the instrument. The table becomes the place on which the specimen is placed. If the subject is very three-dimensional and relatively large, a stand that is free-swinging and clamped to the table is a must, as otherwise the distance between the objective and the built-in base is too small for such a specimen. A microscope that contains two basic components, the objective and oculars, is usually known as a *dissecting microscope*. In such a microscope the objective cannot be separated from the oculars, therefore camera lucida can only be attached to one of the eyepieces. Other types of microscopes allow for inclusion of various substages into the main body of the instrument, therefore camera lucida can become an integral part of the tool, allowing the illustrator to obtain a better drawing and faster production.

What is camera lucida and how does it work? The camera lucida known as a drawing tube or tracing apparatus is based on simple principle. It is a combination of a mirror

Figure 4-28. *Phascolarctos cinereus*, TMM 41106-17, a right M^2 shown in crown (B), labial (A), and lingual (C) views. Pencil and graphite powder on Strathmore drawing paper. Drawing by Zbigniew T. Jastrzębski, Field Museum of Natural History, Chicago. Published in Ernest L. Lundelius, Jr., and William D. Turnbull, "The Mammalian Fauna of Madura Cave, Western Australia, Part V: Diprotodonta," *Fieldiana*, Geology, New Series No. 11, by Field Museum of Natural History, Chicago. Drawn in 1980.

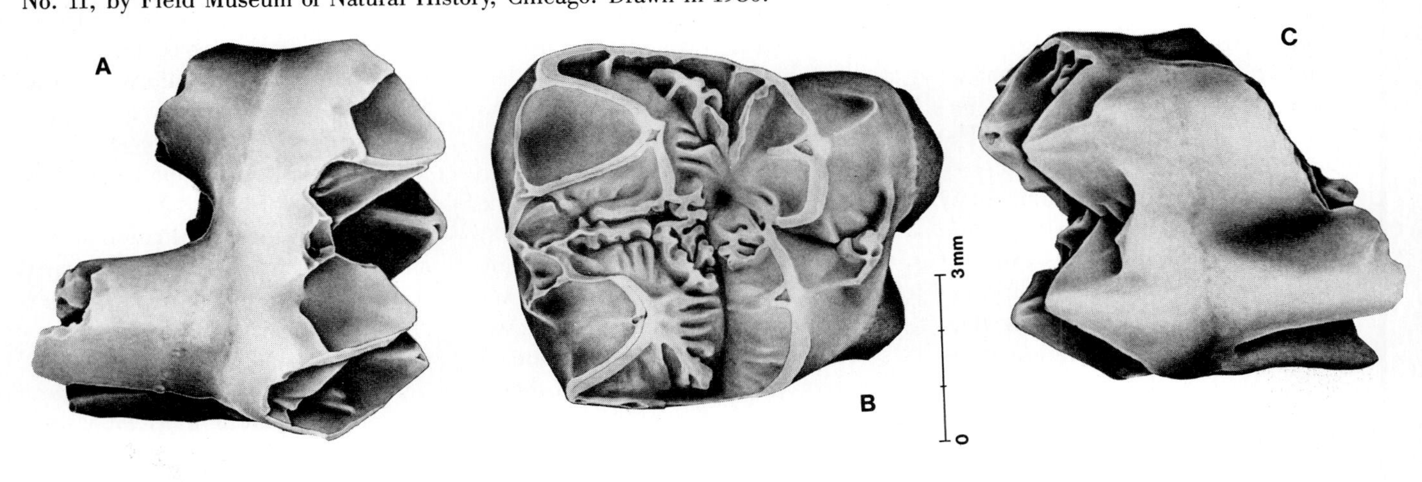

and a prism. The mirror extends outside the body of the microscope at a certain angle that allows reflection of a portion of the drawing table, with the help of the prism, into the ocular or the main body of the microscope. The result is a reflection of the drawing area, through the apparatus overlapping the specimen, viewed through the microscope. This means that the illustrator can see simultaneously the specimen and the drawing of that specimen. The light bounces off the paper, travels upward until it reaches the surface of the mirror, and then because of the mirror's angular position is directed into the microscope's optical system. A variety of camera lucidas does exist on the market, ranging from toys through very sophisticated, focusable units. The structure of the microscope will dictate the choice. When using a standard dissecting microscope, be it monocular or stereoscopic, you will have to use the type of drawing tube that clamps onto the eyepiece. You can encounter numerous versions of this device. Some will allow for viewing the specimen while drawing it; others will project the image from the microscope onto the drawing surface. As you know, in order to project any image onto some kind of surface, a strong light is needed. Your microscope had better be equipped with a powerful illumination system while using an internally reflected image. There are many combinations of microscopes and drawing tubes. The simplest and most portable professional camera lucida is produced by Zeiss Optical Instruments. While this particular model will fit in your pocket, that does not mean it is the best. What is the best is determined by the type of project you are working on and the viewing equipment available to you. In any case, while using a camera lucida in conjunction with a stereo microscope you will have to look through both eyepieces simultaneously. One eye will see the reflected drawing surface overlapping the specimen and the other just the specimen. A focusable camera lucida will give a better image; unfortunately, all focusable drawing tubes cannot be clamped on the eyepiece. All models will have to become an integral part of the microscope. This means that you must have a special type of microscope. Such a machine may have a drawing tube built in by the manufacturer or produced in such a way that the camera lucida can be placed by you into the main body of the microscope. Such microscopes come in sections. Let's take the microscope I use as an example. It is produced by Wild Heerbrugg Co. It is composed of two basic parts, the objective and the oculars, secured with locking screws. Between two main parts other parts can be fitted in. Above the objective a camera lucida is placed; above the camera lucida a part called a double-iris diaphragm is secured by the next set of screws; and the last, uppermost part is the oculars with two inserted eyepieces. All four removable sections have to be aligned properly, as otherwise the specimen will appear as a double image. Because of interchangeability of parts or sections of the microscope, the camera lucida can be placed over or underneath the double-iris diaphragm. Such a microscope will accept other additions such as special illuminators. Those can become part of the main body or can be clamped under the objective.

Is the double-iris diaphragm part of the microscope? Whenever observing a three-dimensional structure through the microscope, it is impossible to focus on and therefore see clearly the whole structure. In order to help with the depth of field, the double-iris diaphragm can regulate the input light reaching the viewer's eye. The very same exists in every photographic camera. Right behind the lens of every camera is a structure that regulates the input of light by changing the opening of the iris. The human eye has a similar mechanism. In order to see sections of the specimen that are out of focus, the diaphragm has to be closed. Less light reaching the eye allows for a sharper image. The double-iris diaphragm for the stereo microscope has two simultaneously regulated diaphragms because the microscope has two oculars. How to operate such a microscope? First, before doing anything else, make sure that you check the clearance between the objective of the microscope and the base onto which the specimen will be placed. Remember that while focusing the microscope, the housing containing all the optics will travel up and down along a designated track, while the base with the specimen will remain stationary. When looking through a microscope our naturally learned feel for distances is altered by magnification of the subject, so it is possible to crush the specimen between the objective and the base. You must slowly acquire a new sense of judgment for this distance. Remember, higher magnification will produce less space between the objective and the base. To start, be sure that you have plenty of space for the object. Instead of a specimen, you can place a piece of paper with a cross drawn on its surface under the objective of the microscope. Then the illumination system should be switched on. You need the light in order to see. The next step is to adjust the interpupillary distance. Remember that your head must be parallel to the oculars and the pupils of your eyes must look directly at the centers of the eyepieces. Slowly rotate the focusing knob until the specimen or the paper is visible. If there is nothing to see, make sure that whatever you are looking at is placed directly under the objective. Do not forget to change the objective to its lowest magnification power by rotating the magnification selector ring mounted on the objective. Once you see the subject or drawn cross, look carefully for a double image. The double image may be very obvious or may appear as a slight fuzziness. There are two reasons for this. The first is an improper alignment of separate parts of the microscope; the other is an improperly turned magnification selector. If one of the lenses within the objective is not aligned exactly as it is supposed to be, doubling of the image will result. Make sure that when changing magnification you stop in an appropriate place. The place is marked on the housing just like the magnification is clearly engraved or printed on the objective. The magnification will be designated by a number followed by an ×. In order to make sure that a particular lens is in its designated position, listen for a click and the feel of a little pressure while turning the changer. Look through the oculars and rotate the objective, seeing the image of the subject double and then return to singular form. It will be obvious to you what has caused doubling of the image. If the objective is in the proper position, the fault lies in the alignment of the segments of the microscope body. If you have the double-iris diaphragm mounted into the housing, make sure that it is open completely. At this point you do need a clear picture, therefore lots of light must reach your eyes. Once this problem is settled, look again through the oculars. One of the eyepieces will have a separate focusing device. Because usually our eyes are of unequal strength, that separate focusing device is of great help. Again looking through both oculars, focus the whole microscope on the section of the subject or on the drawn cross. While looking through the microscope close one eye, the one that is looking through the eyepiece *without* the focusing device. Keeping that eye closed, focus the other eyepiece by rotating the focusing ring. When you have finished, open the closed eye and see if everything appears sharp. In order to make sure that the best focus has been attained, close the eye matching the focusable eyepiece and refocus the whole microscope using the main focusing knob, then repeat the previously described process again.

In order to learn how to adjust a particular microscope, you will have to practice with the instrument. As numerous types of microscopes are available to the scientist and therefore to the illustrator, let the scientist introduce you to his type of microscope. During that introductory period be very careful to observe and remember necessary procedures. Ask questions; if you like, make notes. Ask the scientist if it would be permissible to practice using some kind of specimen. In most cases you will be permitted to spend some time poring over the microscope. It would be best to acquire some kind of basic experience before your encounter with any very sophisticated microscope. The reasons are many; the most important is experience. Any scientist will feel more secure if he knows that you can differentiate the focusing knob from the objective. Whenever working with a microscope, remember that it is easy to crush the specimen with the objective. Check the space between the base and the objective and make sure that during the focusing process you are gentle rather than powerful. This will be especially important when working with high magnifications. The higher the magnification, the less space between the objective and the base. Truthfully speaking, for three-dimensional objects such as a fish, plant stem, or tooth, the maximum workable magnification will be 50 times the original size of the subject. In order to obtain higher magnification with good resolution, another type of microscope, called a compound microscope, has to be used. This is a very sophisticated and complex instrument used for magnification over 100 times. Some models will have camera lucida built in, while others will not accept any additions. The tolerance between the base and objective is so small that specimens for observation are prepared in glass slides.

Usually for the type of work you will do, the dissecting microscope will be used. You may assume that present technology is sufficient for production of perfect lenses. I will have to disappoint you. Optics in every microscope will distort the image slightly. The perfect image can only be viewed directly through the center of the microscope objective. Anything to the sides is slightly distorted. But surprise! When using camera lucida, the image traced across the opening of the lens will also be slightly distorted. Depending on the magnification factor as well as the size of the specimen, such distortion can grow to unacceptable proportions. Such a traced drawing will have to be produced in more or less the same way as the photographs of the planet's surface. Your eye will see a small section of the subject, then the subject will have to be moved under the objective of the microscope in order to scrutinize its structures. The tracing of the whole subject will have to be spliced together from small sections of visible areas. Because of this, any distortion will be magnified, as every tracing of the small surface area of the subject will add more distortion. Unfortunately, this is unavoidable but easily correctable. When a specimen is of a large size—that is, when it is larger than five centimeters and when the microscope enlargement will exceed 12×, you must measure the subject's basic proportions using a proportional divider. The length and width will be established properly, roughly drawn on tracing tissue, and later easily matched with the camera lucida traced image. Such comparison will allow you to keep distortion of the subject to minimum. Depth of field will also become a problem with three-dimensional structures. You know that it is impossible to focus the microscope on the whole structure at the same time. The double-iris diaphragm will help, you say. Not really. This device will help in observation but not much in drawing. All will depend on the structural aspects of the specimen. The double-iris diaphragm will cut too much light off from the field of observation through which, besides the specimen, the tip of the pencil and parts of the traced material are visible. Remember, camera lucida reflects the surface into one of the oculars, overlapping the image of the specimen with the tracing. Less light reaching your eyes while looking through both oculars means improved depth of field, but it also means that less of the traced image will be visible. As a matter of fact the balance between illumination of specimen and illumination of the drawing surface is precarious. Excluding the double-iris diaphragm and being just concerned with keeping the specimen and the traced image simultaneously visible, you will find out that in order to see the drawn line more strongly and better, more light will be required on the drawing surface. At the same time, the illumination of the specimen should be minimal. When a specimen is not illuminated well, you cannot see what is under the microscope. The happy medium will vary depending on the need of the moment as well as on the type of specimen. In any case, you will find yourself continually adjusting the strength of light that illuminates the specimen. Usually a desk lamp will be used for illuminating the drawing area, and as such, cannot be readily adjusted.

With the specimen under the objective of the microscope and tracing tissue positioned under the camera lucida, the next problem is to find out how much you are magnifying that particular subject. The magnification factor may be clearly marked by a number on the objective of the microscope, but actual magnification of the specimen in its drawn state will differ from the designated one on the microscope. In life, everything is judged in terms of proportions. This is especially true in scientific illustration. In order to make a judgment, we have to compare two or more things and then maybe we will reach a correct decision. The proportions between the distance of the objective and the specimen on one side and the distance between the camera lucida's mirror and the surface used for drawing will affect the actual magnification of the specimen during drawing. The only sure way to know a precise magnification factor is to measure it. Naturally we do know that it will be, under normal circumstances, approximately as much as is marked on the objective, but to be sure, place a metric ruler under the objective and then trace the distance of one millimeter using the camera lucida. When the traced distance is measured and divided into millimeters, we will know how much more or less we differ from what is marked on the objective magnification factor. It is possible, by placing additional equipment such as the double-iris diaphragm above the camera lucida, to see larger but to draw a bit smaller. Additional optical components tend to enlarge the subject. The distance between the mirror of the camera lucida and the surface used for drawing can be altered by raising or lowering that surface. When a large magnification—that is, 30× or 50× is specified by the scientist, in actuality the traced image will differ; it may be 33.5× larger than what the scientist had asked for. In such a situation, make sure to note the enlargement carefully and *ask* the scientist if it is OK to use such magnification. If it is not, then you will have to play with the distances, especially with the distance between the mirror of the camera lucida and the drawing surface, until you get the magnification to what it should be.

How can you illuminate the specimen correctly? This is impossible to answer completely. It all depends on the type of specimen, the type of microscope you use, the type of lights available to you, and the requirements of the project. One thing is for sure: Do not produce shadows that you may not notice. All shadows *must* be noted by you, as none will be drawn. The accidental inclusion of shadows will destroy your illustration. Be very aware of what is happening under the microscope. Make sure that during all light adjustments, you check often by looking through the microscope. While looking through oculars, make sure that light or separate lights are moved back and forth, allowing you to see the structures separately from the shadows produced by other structures. It is very easy to make mistakes in such a situation. There are a number of shadowless illuminators on the market. Such illuminators can be attached to the microscope in one way or another, but unfortunately they will not produce strong enough light for clear observation of a small three-dimensional specimen. The other important problem evolving from the use of such illuminators is the fact that none of them will produce any shadows. For the scientific illustrator this is a problem. We use shadows produced by the variations of the surface structures as a means of recognizing surface ups and downs. Naturally we draw what we know, what we have learned through observation rather than presenting darks and lights superficially produced by illumination. You do have to remember that shadows, when included in an illustration, can be misread by the reader-viewer of the illustration. Such accidental misrepresentation can lead to serious misunderstandings.

Lights produce a quantity of heat. Because of this, you must be careful and not injure the specimen. It is easy to burn fragile tissue and it is easy to melt reconstructed specimens. Specimens, when originally stored in liquid, should be protected by liquid during drawing procedures, be it ethyl alcohol, water, or glycerine; otherwise tissues will dry up and simply burn under the heat of illuminators. Be very careful with paleontological specimens. Many such specimens are glued together with wax, which will melt under the heat of lamps. When looking through a microscope you will see such specimens disintegrating in front of your eyes. This is a tragedy. The work of a lot of scientists or preparators will go to waste. Remember that it is an art to manage a small three- to five-millimeter specimen made of minuscule particles. Ask the scientist if his specimen has been glued together with wax. Usually he or she will warn you about the fragility of a specimen before you receive it for drawing, although at times this may be forgotten. When a cold light source is used for illumination, this problem is eliminated. The cold light source employs fiber optics for transmitting light from a voltage regulator. The main source of heat, the lamp, is placed far away from the specimen, in actuality being an integral component of the voltage regulator. From this source of illumination and heat, the light travels through flexible cables made of glass fibers ending as a "cold" light under the microscope. Use of such a cold light source does not prevent the specimen from drying up. This is a problem you will have to deal with when drawing wet subjects such as fish or soft tissues.

When you have a camera lucida attached to the microscope while drawing, you have two choices. It is possible to render a finished illustration directly underneath the camera's reflecting mirror or produce preparatory materials for the final sketch. The decision will depend on the quality of equipment in use, the type of specimen, the technique used for drawing, and the need to meet a deadline. Now you must remember that producing an illustration without a preparatory sketch will call for advance thinking. If parts of the subject do not align properly, the drawing will be ruined. Remember that you will see only a small section of a relatively large specimen. Also remember the distortion produced by a microscope's optics. Draw an outline of the subject on tracing paper using camera lucida. Such a preliminary sketch, when placed on top of the partially rendered illustration, will help you in alignment. The only safe drawing technique is pencil. It is very hard to use pen and ink, stipple and line during such direct drawing. Pencil can be erased, while ink will remain on the surface of the drawing. When drawing, you must exercise great caution and watch for the following:

1. You must check the alignment of what has been drawn with the visible image.
2. You must render the illustration while looking at the drawing.
3. The microscope will be used for viewing the subject, finding things, and tracing the structures.
4. You must exercise great restraint and not press your pencil into the paper while using camera lucida.
5. You must move the drawing in conjunction with the specimen, otherwise images will not match.
6. You will have to restrain from placing your hand on the drawing paper. Sweat and grease will damage the surface, making it unsuitable for drawing.

Solid structures such as bones, teeth, or skulls are suitable for such a drawing procedure. The gain is a speedy production, as the final sketch is entirely omitted. In order to draw without the final sketch, you must be trusted by the scientist. Without such trust, it is impossible to proceed from the beginning to the final product without an intermediate step. Such a procedure can happen only after a lengthy working relationship with the scientist has been established. All in all, whenever working with a microscope, make sure to:

1. Use both eyes when looking through stereoscopic microscope. Failure will result in improper perspective of viewing and therefore exclusion of details on the side of the closed eye.
2. Be careful not to damage the specimen during positioning and drawing.
3. Adjust the microscope to your eyes. The interpupillary distance, alignment of sections of the microscope, and focusing will have to be checked and will have to fit your vision.
4. Do not scratch the lens of the microsocpe or camera lucida's mirror with a pencil or teasing needle. Damage will interfere with observation and drawing, and repairs will be very costly.

5. When not familiar with a particular microscope, ask questions. Bashfulness may result in more unpleasantness or loss of job.
6. Do not spill any liquids into optical components. Damage can result from an accidental splash of coffee. Keep liquids away from the upper structure of the microscope.
7. Adjust and readjust the density and position of light, illuminating the specimen. This will help you with observation and therefore with drawing.
8. Do not let the microscope fall off the table. Some microscope stands will slide off an inclined surface. Use a flat surface for microscope work.

Figure 4-29. White-crested Spadebill, *Platyrinchus platyrhynchos*. Watercolor on illustration board. Painting by Dr. John W. Fitzpatrick, Field Museum of Natural History, Chicago. Published in Melvin A. Traylor, Jr., and John W. Fitzpatrick, "A Survey of the Tyrant Flycatchers, The Living Bird," Nineteenth Annual, 1980–81, Cornell Laboratory of Ornithology, 1982. Painted in 1981.

The subject has been traced and deciphered and the preparatory sketch assembled. Such a sketch will be made of numerous small pieces of tracing tissue with the image of the specimen drawn on their surface. Because parts of the drawn structures have to overlap in order to assure proper positioning, a lot of layers of tracing tissue can pile up in some sections of the whole preparatory sketch. Such a sketch usually is too large to be considered for rendering. First of all, it would take a lot of time to draw and second, when the finished rendered image is reduced for printing, most of the well-drawn details will disappear or run together, making such an illustration unreadable. In order to reduce the overall dimensions, you will have to use a viewing reducing-enlarging camera and trace the image by hand or photostat your initially drawn sketch. Usually the viewing cameras are more readily available than photostats. A viewing camera is based on the very same principle as any photo camera, with two exceptions: You take the place of the film, and the camera is large. The construction of the viewing camera is actually the same as the construction of any large photo cameras. There are basically two types: One which can be or is permanently inclined, and the other, which is vertical. The bed of the camera is movable. In other words, it is possible to change the distance of the bed in relation to the lens. The lens has four stops and is mounted at the tip of the bellows. The bellows are also movable. The only integral structure that remains in a stationary position is the frame of the bellows at the other side of the lens. A plate of glass secured within this frame will serve as the surface for drawing.

The operation of the viewing camera is simple. Two separately located cranks control the distances of movable portions. One controls the distance of the glass plate; the other controls the distance from the bed to the lens. The relation of both distances produces reduction or enlargement of the subject placed on the bed. The bed is illuminated with attached lights. The construction of various viewing cameras may differ, but the basics will remain the same. In order to reduce your large sketch, place it on the movable bed and secure it with masking tape or thumbtacks. Place a sheet of tracing tissue on the glass and by turning the handles of both cranking mechanisms, see the image being enlarged or reduced. Some viewing cameras have calibrated proportion indicators built into the system. Nevertheless, the safest way to obtain proper dimension is to draw a premeasured straight line on the sheet of tracing tissue and simply adjust the image to the length of the drawn line. The only unfortunate interference during the process of adjusting proportions and tracing the image will come from light outside the camera. A sunny room is not the best place for a viewing camera. A dark room is much better because the image received through the system will be too weak to compete with lots of outside light.

The viewing camera can be used for reduction or enlargement of sketches and can also be used for tracing some specimens. Actually, such a camera is a poor-quality microscope, but it has a large area suitable for drawing. The specimen can be placed on the bed, enlarged accordingly, and traced through the glass back of the camera. The specimen must be secured well to the movable bed, otherwise it will fall off. Again, depth of field may become a serious problem, although this can be alleviated by proper adjustment of the f-stop. The smaller opening of the iris diaphragm behind the lens will give you more depth but less visibility. The viewing camera is an excellent tool for speedy production of sketches of large specimens. Whole vessels can be placed on the bed and quickly traced all to the same size. Large bones will be handled easily in the very same fashion. Photostat cameras are more complicated but basically similar to viewing cameras. The main difference is the quality of illumination of the bed as well as into the cover for the plate glass. Naturally, knowing the exposure time is the secret to success. A Photostat camera must be placed in a darkroom similar to a regular photographic darkroom. The negatives will have to be developed in the proper solution, washed, and dried. The whole process is the same as photography. A drawing, a sketch, or the object itself is placed on the movable bed and is photographed on a suitable paper negative. Again numerous versions of Photostat cameras exist on the market, as well as different negative and positive films and paper. One thing you have to remember is that pencil or any continuous tone will not be picked up on high-contrast paper negative. In order to produce a good picture of any continuous tone image, continuous-tone negative must be used. The exposure setting will differ according to situation as well as to the type of Photostat camera.

After all of this you may wonder if it would not be simpler to take a photograph of the specimen, enlarge it, and trace the image. Whenever it is suitable a photograph will be used, I can assure you. Problems with photographs are many. First of all, the camera cannot distinguish important features from nonimportant ones. Only you can do this. A photograph of a badly damaged specimen is useless for you. Second, in research institutions photography is usually handled by a separate department. It is not handled by scientific illustrators. But let's assume that you want to take a photo of the specimen under a microscope. In order to do this you need specialized equipment. Let's say that you have adapters, cameras, and all necessary equipment. Take a photograph of a fish and find out what is visible. Do not be surprised if it is not much. The camera will record everything: all the discolorations, all the reflections, and mangled and disfigured fins. In order to make any sense of the photographs, a lot of time will have to go into the very basics, the observation of the specimen through the microscope. In most instances it is faster and easier to draw directly from the specimen than from photographs of the object. Think about the illumination of the subject. How are you going to adjust the lights so that most of the features are captured with minimum effort and time? The result will be known after the development of negatives and printing of the photographs. This may take days and still not be correct or explanatory enough. Drawing bypasses the unknown because you can see the result immediately. Directed by your thinking, the process of elimination of unwanted features can take place immediately.

EXAMPLE 1. A requisition for drawings depicting squirrel anatomy as well as the specimens is in your hands. The specimens are all preserved in glycerine; some are very small. The scientist has asked you to enlarge all specimens by the same factor. As you observe the smallest and the largest, the conclusion seems inevitable: The smallest are too small to be meaningfully presented when enlarged 25 times, while the largest are a bit too large for the next higher enlarging factor. Your microscope can only magnify in steps—first by 6×, then 12×, again by 25×, and a maximum of 50×. You know that the best magnification for this project would be approximately 35 and certainly no more than 40 times. You know this, but cannot get it. Technically, of course it is possible to find whatever magnification is necessary. There are reducing or enlarging lenses for your microscope. A different set of eyepieces magnifying instead of the standard 10× will solve the problem easily. The fact is that you cannot get those eyepieces because they take time to order. It takes time to get approvals. It takes time to place those things into the budget. Yes, you have been told, if we will place those items in next year's budget then you can order whatever is necessary. Next year is of no importance because you need them now. Do not despair, but put your mind to work. After all, the greatest discoveries have been made because of circumstances. Think—a lens is a lens. Maybe better or worse, but it is a lens. Do you have a spare lens? Sure you do. It is called a reducing lens. You have been using it during renderings to check the tones on the drawing. Take it off the holder and using masking tape, attach the lens to the bottom of the objective. Take a metric ruler and using camera lucida, trace a distance of one millimeter while your miscroscope is adjusted for 50×. See by how much this lens will reduce the image. The image will not be so clearly visible, but you can live with this. For observation, take the lens off to see better; but for drawing, tape it back.

Need more reduction from the microscpe's 50×? Use two reducing lenses, taped together if necessary, and produce the drawings in the proper size.

EXAMPLE 2. Discussing a new assignment with the botanist, you find out that a series of drawings will have to be made from glass slides. Each object will have to be magnified 400×. This is beyond the powers of your microscope. You mention this to the scientist and he suggests that you should use a compound microscope in his office. Says he, "The only problem is that this compound microscope does not have a camera lucida attached into its structure." He asks you if you can draw with precision without tracing. How is it possible to do this? Proportions between structures visible on the specimen will become the guidelines. It is possible to measure any distance if an eyepiece with a graticule is used instead of the regularly used eyepiece. A graticule is a linear proportion-measuring device. It consists of a line with clearly marked distances, just like a ruler except it is built into the optical system of an eyepiece. Knowing how many parts of the distance marked on the eyepiece will fit into a desired distance on the specimen will help you to draw a grid on the tracing tissue. Once proportions between the specimen and drawing are established, you can proceed safely and speedily by looking into the microscope and adding details into the sketch. If by any chance such an eyepiece is not available, establish proportions between distant points by eye. You can easily see how many bumps on the surface of the pollen will fit into its diameter. Establish an arbitrary size for the chosen bump on the tracing tissue and then, by adding horizontal necessary distance, you will find the pollen's diameter. When drawing visual representation in such a manner, the enlargement factor may be unpredictable and not precise, although the drawing will be a truthful representation of the specimen. The scientist usually knows the size of the specimen and therefore it will be easy to calculate the magnification ratio. Just remember to divide the drawn distance by the actual size of the specimen, then you will know how many times the specimen has been magnified. If the scientist needs precisely defined magnification, calculate the diameter or the distance between the farthest points on the specimen using the scientist's information. Draw a defined, straight line. Then take the sketch to the viewing camera and reduce or enlarge it, matching that particular distance. The result will be a drawing representing precisely given specifications.

EXAMPLE 3. You have been good and thoughtful. When working with soft tissue under the microscope, you did everything right. The specimen is submerged in ethyl alcohol and it is not damaged. It took a lot of effort to decipher structures of the small piece of muscular tissue, but it is done. The scientist is glad and you feel fine. The final sketch has already been prepared, approved, and transferred to the drawing paper. During rendering of the illustration the specimen is placed under the microscope for observation and reference. While you are working the scientist calls your attention to some previously finished illustrations. "It will take a minute," he says. "I just want to ask your advice on how to arrange a plate." Leaving the drawing area and the microscope, you discuss the arrangement of illustrations. After a while, coming back to your area of work, you continue the rendering. Soon a need for reference arises, so you look through the microscope. Surprise! The tissue is all dried up and partially burned. How can this be? You have left enough alcohol in the petri dish to preserve the specimen. Can it be that this short conversation with the scientist was long enough to allow evaporation of the alcohol from the petri dish? Remember, it does not take a long time for alcohol to evaporate when exposed to strong microscope lights. Next time you leave a specimen under the microscope, make sure to switch off the voltage regulator. Make sure that another petri dish or a flat piece of glass or Plexiglas is placed on top of the dish containing the specimen. Make such procedure a habit, for covering the specimen will stop alcohol evaporation. Microscope illuminating lights are strong and produce a lot of heat. Switch those lights off whenever you leave your work area. Damaging a specimen is bad for you and is also very bad for the scientist, and therefore for his research.

The most important thing you should learn is how to operate the microscope. This may be a problem because you may not have access to one. You may not be able to practice with a camera lucida attached to the microscope. Remember that the general operation of both instruments is the same regardless of the built-in complexities of the instruments.

Exercise 1. Obtain a toy camera lucida. There is one called the Magic Art See and Draw Copier. A large toy store will have it. Examine the toy drawing tube carefully; find out where to look and how to operate this simple device. While looking through the small hole on the top of the box, point one open end toward a subject. Choose a large object; make sure that the surface under the device is black. Use black construction paper to cover an area on the drawing table. The blackness of the paper will help you to see your reflected hand with the pencil better than when the area is illuminated. Use white pencil to draw on the black surface of construction paper. Find out how the following will affect your drawing:

1. Alignment of drawn image with actual subject
2. Relations between the vertical distance of the device to the drawing surface and the horizontal distance to the subject
3. Position of your head in relation to the device

Exercise 2. Sign up for a science course where microscopes are used. Explain your purpose to the teacher. He or she should be helpful. Besides learning something, you may also have a chance to establish contact with a researcher in need of drawings.

Exercise 3. Obtain a catalog from a large art-supply store and find a photograph of a viewing camera. In such catalogs, three or four different models of viewing cameras are usually presented. While looking at the photograph, find out where the bed of the camera is and what type of illuminating system is attached to the bed. Observe the construction; locate the adjusting mechanism. It is not much of an experience to look at a photograph, but it is the cheapest way to learn and will give you some idea of the construction as well as operating techniques.

Exercise 4. If you have some spare cash, buy yourself a microscope. With a bit of looking you will find a microscope that will fit your pocket and needs. The microscope does not have to be very complicated; it can be of simple monocular construction. The most important feature you should look for is a flexibility of the adaptation of the microscope for your needs. The distance between the objective and the base should be considerable. If the microscope you are interested in does not have a reflected-light system built into it, it will be better for you. The little mirror under the objective does not have to be a part of the construction of the microscope. With a little ingenuity you can build a small stand with a Plexiglas mirror inside serving the very same function as the little mirror attached to the microscope. The next step may be more expensive. A portable camera lucida that is easily attachable to the ocular is produced by Zeiss. Call the company and inquire about prices. If it is too much for you, then try to adapt the toy drawing tube to your microscope. With a budget ranging from $100 to $500 you will be able to obtain a reasonable tool for observing and drawing.

Failures in Drawing

Not all illustrations achieve desired results. The readability of the image as well as the context of the illustration may suffer considerably because of poor rendering, bad planning, and improper printing. Basically, all failures can be grouped in three categories:

1. Failures from the scientist's point of view
2. Failures from the reader-viewer's point of view
3. Failures from the artist's point of view

The most important are the first two categories. If the scientist thinks that visual explanatory material is of poor quality, it is bad news for the illustrator, but most of the time it is a really terrible experience for the scientist. How is it possible to produce a bad illustration under the watchful eye of the researcher? It is really not that difficult. The wrongness within a drawn image is not necessarily caused by an inability to communicate with the illustrator, nor is it necessarily produced by technical inabilities of the artist. A change in the research, a new discovery, or simply the fact that another scientist had described the very same thing will change an otherwise good illustration into a bad one. In most cases a failure of the illustration will be noticed by the scientist, and such an illustration will never reach the printing stage—it will remain in the drawer or in the garbage can. Anytime the scientist considers the drawn image not truthfully representative of the specimen or not explanatory enough, it will be failure of mutual effort. It does sound pretty unpleasant, but in reality plenty of illustrations are corrected before they are submitted for review and publishing. It is a routine rather than an exceptional incident. Naturally, in many situations the communication between the illustrator and the scientist is to be blamed. Usually the illustrator will not ask enough questions about the subject, and the scientist will not forward all detailed information. As an expert in the field, he or she will not think about the most obviously basic problems which the illustrator may encounter. It is up to the illustrator to express his lack of knowledge about the subject rather than bashfully withhold his questions. Some scientists are extremely explicit in their explanations. For beginning illustrators this may be an overwhelming experience, with too much material to digest at once. The illustrator should make sure that if too much information is presented to him, he should stop the scientist and explain his problems.

The worst type of failures are those that are noticed by the reader/viewer of scientific journals. Those can be grouped in two categories:

1. Inability of illustrator to render the illustration in such a manner that it will clearly represent particulars
2. Insistence by the scientists on putting too much information into a given illustration; inability of the scientist to realize that reduction of the drawn image would destroy the context

In both cases fault lies with the illustrator, who did not perform his duties correctly. Technical ability is a must for the illustrator. There is no excuse on his part when the illustration is overrendered or if the technique chosen by him for representation of a particular subject did not work well. The illustrator should have anticipated possible loss of the drawn image caused by platemaking or by extraordinary reduction of the image.

his responsibility to know materials. The only way to acquire such information is to practice drawing and painting and to keep an information scrapbook on the behavior of papers, pencils, paints, and pens. It is impossible to know everything, therefore the illustrator must be honest. The sentence that starts "I do not know if I can do this" should be finished with "I will try and we will see the result." That is the most important area where self-criticism should be applied. Never allow yourself to be buttered up by anybody. A comment like "Your drawing is just fine" should not work on you. A "thank you" will be the response, but in your mind do scrutinize and do criticize every drawing you make. Find all imperfections and all technical mistakes. Consciously, not impulsively, appraise the situation. If you do not know a given technique, then practice. Find illustrations produced by other artists and compare their quality. Learn from your mistakes and from other artists' mistakes. Do not be afraid to admit to yourself that you do not draw as well as you should. Remember that if you seriously try, better drawings will result. When the scientist asks for too much detail, explain the problems involved in reproduction of the illustration. Maybe a different technique or different approach will be the solution. The scientist should be more than happy to listen to you. All scientists are eager to receive a good-quality product. A less-detailed basic drawing with additional enlargements of important segments of the subject can help. In any case, discuss the problem with the scientist. Be aware of the possibility of a failure. Your logic in observation should match the logic of your verbal explanations.

A beautiful drawing does not mean readable, well-organized information. Do not try to beautify any illustration because it may be

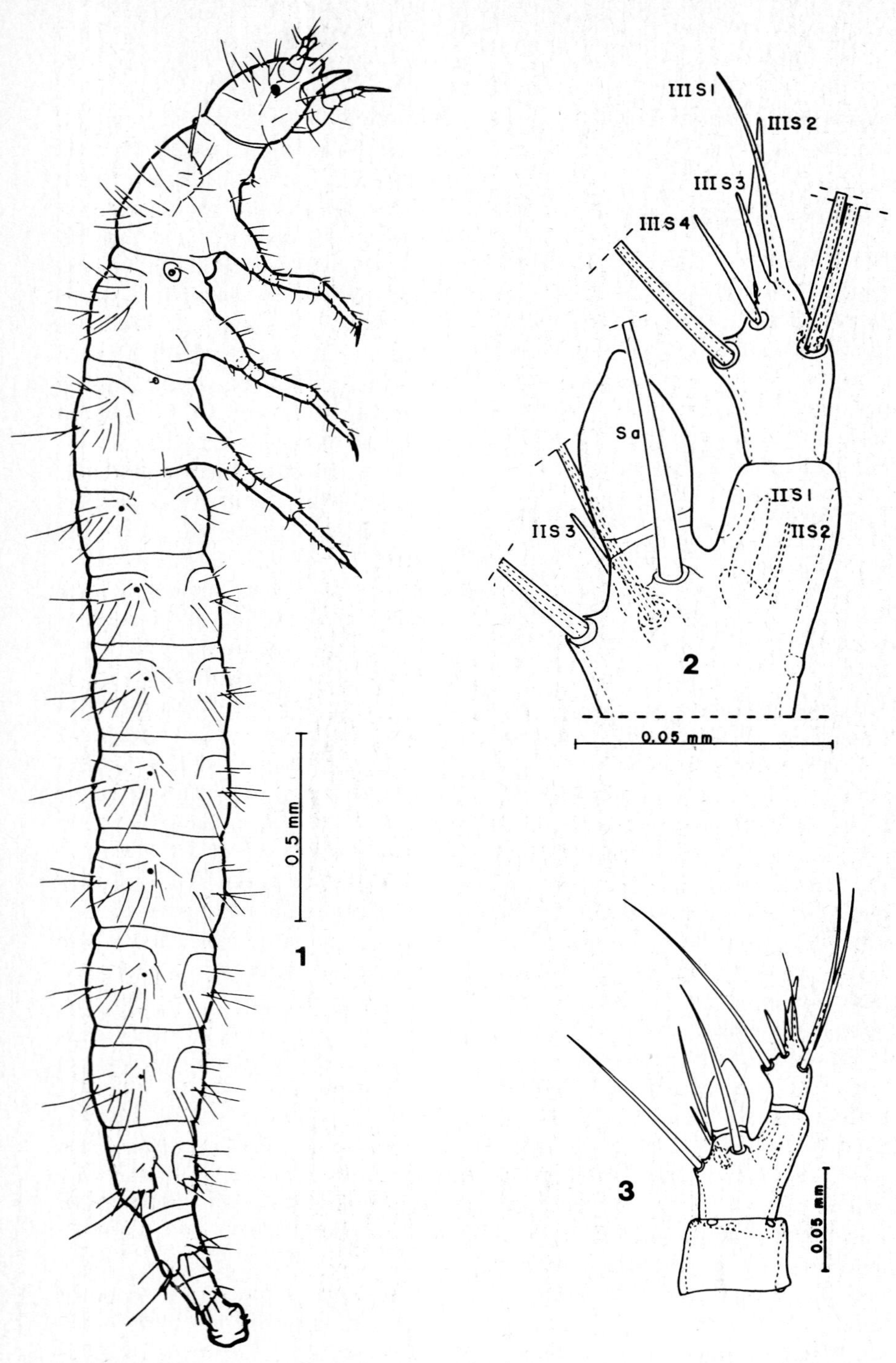

Figure 4-30. *Atheta coriaria*. Pen and ink. Drawing by Dr. Larry E. Watrous, Field Museum of Natural History, Chicago. Published in J. S. Ashe and L. E. Watrous, "Larval chaetotaxy of Aleocharinae based on description of Atheta ceriaria," *Coleopterists Bulletin*, Smithsonian Institution, Washington, D. C. Drawn in 1983.

The illustrator should be knowledgeable enough to suggest the best drawing technique to the scientist. When the scientist refuses all suggestions, the illustrator should be versatile and be able to produce a good-quality drawing under adverse conditions. The artist's responsibility is to know how to draw and paint in numerous techniques. It is needed by you for a portfolio. The main purpose of scientific illustrations is explanation of material rather than prettiness. This does not mean that lots of drawings and paintings will not be visually appealing. The subject should play the most important role in achieving a nice-looking illustration, and therefore a valuable addition to your portfolio. From your point of view, a drawing may be considered a failure when a dinosaur is drawn in outline only rather than rendered in tone as a three-dimensional structure. Actually it is more of a disappointment than anything else. The failure of a drawing or painting should be judged after the image is printed. Sometimes a very nice, well-rendered illustration will not appear the same way on the page of the publication because of drastic reduction or improperly made plates. For you, a professional scientific illustrator, this is actually a failure of the illustration. As long as the printed image is readable and easily understood by the scientist, you should be happy.

EXAMPLE 1. A well-rendered illustration of a fish is to be submitted for publication. Stipple and line have been used for the representation of shapes and tones. You, the illustrator, know that line reproduction will not be the best for this particular drawing. It is overrendered; the surface is covered by too many dots. The original looks very impressive, but when reduced and when high-contrast film is used, a lot of light tones will disappear and other, darker shades will flow together, producing unpleasant results. For that reason you suggest halftone reproduction. The idea is accepted and you ask the scientist to let you know when the drawing will go to the publisher so you can mask the image. The scientists says, "Fine, I will let you know." After some time has passed you receive a copy of the publication with a printed image of the overrendered drawing. It is printed in halftone, but first of all, an eight-by-ten photograph has been made of the original. The high contrast of the photograph has almost destroyed the drawn image. Then the image has not been masked, leaving the background covered partially by tone. The platemaker did his best to save the image. His efforts have succeeded to a certain degree. The drawing is readable, all of the features are distinguishable, and the scientist is satisfied, but you should consider such a drawing a failure. By keeping more in touch with schedules and people involved in sending the drawing for publication, it would have been possible to have the illustration printed much better.

EXAMPLE 2. A publication containing a series of pen-and-ink line drawings just came out in print. Lots of areas on the illustrations are almost invisible. Why? The drawings were small, therefore reduction of the image was not great. Drawings were kept simple for clarity and readability, but nevertheless the printed image does not match up with expectations. What type of pen did you use during drawing? What was the size of the point used for rendering? Did you check the quality of the ink? All three factors could have contributed to the failure of the drawing. In all probability the lines were too thin, although the size of the illustration did not exceed reasonable limits. Thin line will get thinner when reduced. When such an image is overetched during the platemaking process, this line will get even thinner and may disappear altogether. Perhaps the ink used for drawing was slightly diluted with water, therefore producing a line that was just too light for the camera to notice. Rather than being in despair, think back and analyze the situation. Find out why those drawings did not reproduce properly. Check the drawing ink; check the width of the point. Even make some tests and photostat those on high-contrast paper. Do not let your mistake be repeated in the future.

CHAPTER FIVE

Projects for the Portfolio

THIS selection of projects in its generality encompasses natural sciences. It is important to realize that a large variety of typical scientific illustrations cannot be drawn without appropriate specimens as well as without complicated and expensive equipment. Obtaining and storing specimens is a definite problem for the illustrator without the access to appropriate laboratories. At home, space is limited and storage cannot be safely arranged. Equipment such as a stereo microscope with camera lucida or a compound microscope is too expensive to even consider. All of this prevents us from having a portfolio full of drawings applicable to all areas of scientific research. How can you draw a mite without magnifying such subject 500 times? The projects should be considered as a base. If you like to specialize in any particular area, draw a similar subject many times. I do not think that you will have a difficult time obtaining specimens for pottery reconstruction.

Before you start any of the projects, read their description carefully. Be aware of and be prepared for unforeseen difficulties. Understand that the sketching process is as vital as rendering. Be prepared to spend at least half the drawing time preparing the sketch. All of the information is enclosed in a sketch. By sketching you will learn about the subject. The final sketch, when presented to the scientist, will represent you, as well as your ability to think and see. Make sure your sketches are properly designated in writing. In the laboratory, each sketch should contain such vital information as the magnification factor, catalog number of the specimen, and, for your information, the number designating the continuance of that sketch. When working on several projects at once, it is very easy to mix the sketches accidentally. Just imagine drawing five different fish and mixing the sketches. If designations are not marked in writing, it will take considerable time to match the sketch of a particular fin with the appropriate fish. Keep in mind that all your drawings are for reproduction through printing; they are not produced for display in a gallery, although they may eventually be matted, framed, and displayed. This means that all your drawings will be reduced during the platemaking process to, in many instances, unknown size. Be prepared for this. Use the reducing lens often to appraise the values and lines. After you finish a drawing, contact a local Photostat company and obtain two or three reductions. The different sizes of reductions will allow you to compare the final outcome. Remember that a halftone such as pencil, graphite dust, carbon pencil, or color will have to be photostatted on continuous-tone paper. High contrast will destroy the image. At the same time any line work such as stipple or line will be photostatted using high-contrast film; continuous tone is not needed. When drawing, remember that it is always better to spend more time and not to rush. The final outcome is important, as that is how you will be judged. Be sure your portfolio is prepared professionally, that it is clean and uncluttered.

It is important for you to know the requirements set by publishers for an illustration. You realize that

Figure 5-1. Fowl, comfort behavior and care of the plumage. Pen and ink. Drawing by Käthi Stutz, Zürich. Published in D. W. Folsch and K. Vestergaard, *The Behaviour of Fowl,* Birkhauserverlag, Basel, Switzerland. Drawn in 1980.

black-and-white illustrations, especially those prepared in line, are more desirable. You know that reproduction of line drawings is more economically feasible. While such requirements are placed upon you by the scientist, you should also realize that similar requirements are pressed upon the authors of published scientific materials. Scientists must comply with a publisher's requirements or their work will not be printed. Among the requirements regarding the presentation of manuscript to the publisher, a whole paragraph refers to the drawings. The *Zoological Journal* of the Linnean Society of London, England, is published by Academic Press. Under the "Instructions to Authors," the fifth paragraph refers to the preparation and presentation of the illustrations:

> 7. *Illustrations*. These normally take the form of half-tone plates reproduced from photographs, and black-and-white text-figures reproduced from ink drawings. The editors will accept illustrations in color when it is agreed that these are essential and suitable for publication, but may require a contribution to meet the extra cost.
>
> Black-and-white drawings including graphs should be in Indian ink on some smooth white card such as Bristol board, or on draftsman's tracing paper, the lines being clearly drawn and due attention paid to the consequences of reduction. Lettering and leaders may be indicated lightly in blue pencil on the drawing or in ink on a transparent overlay. Use Arabic numbers for figures and plates; capital letters for their sub-division. Original drawings should be within British Post Office size limit (610 mm length, 460 mm width).
>
> Photographs for half-tone plates are best submitted in the final size. This is especially important with electron micrographs. The print should not be marked, but labelling and a scale-line should be accurately indicated on an overlay or duplicate print clearly marked "FOR LABELLING ONLY." Indicate which side is top. The maximum dimensions available for either text-figures or plates are 20 X 13 cm.
>
> Care should be taken in stating magnification of illustrations, making it clear whether numbers quoted are before or after reduction. A scale line incorporated in the block is the most satisfactory method of indicating magnification.
>
> Authors wishing to borrow illustrations already published are asked to obtain written permission to reproduce before submitting their manuscripts. Where authors experience considerable difficulty in obtaining permission, the Society may be able to assist them, provided that copies of all correspondence between the author and the publisher are submitted.
>
> Authors may, in the first instance, submit Xerox or photographic copies of figures rather than the original drawings. If authors require the return of original artwork after publication, this should be stated clearly.[1]

As you can see, the color illustration will be printed if the author is able to arrange for expensive color separations. For comparison, let's see what another publisher of scientific materials requires. Turning our attention to *Fieldiana*, a journal published by the Field Museum of Natural History in Chicago, we read under the information for contributors to *Fieldiana*:

> *Illustrations and tables:* Original pen drawings and charts (drawn and lettered only in black ink), glossy photostats of drawings or graphs taken from other journals, and original glossy photographs for all half-tone illustrations should be supplied with the original manuscript. If practical, all lettering on glossy photographs should be done on an acetate overlay. Photos should all be cropped in margins, not cut to size. (Photocopies cannot be used for reproduction.) All illustrations should bear the author's name, address, and title of the article. The top of each illustration should be indicated. A legend should be supplied for each il-

[1]With permission from the *Zoological Journal* of the Linnean Society. Copyright, The Linnean Society of London.

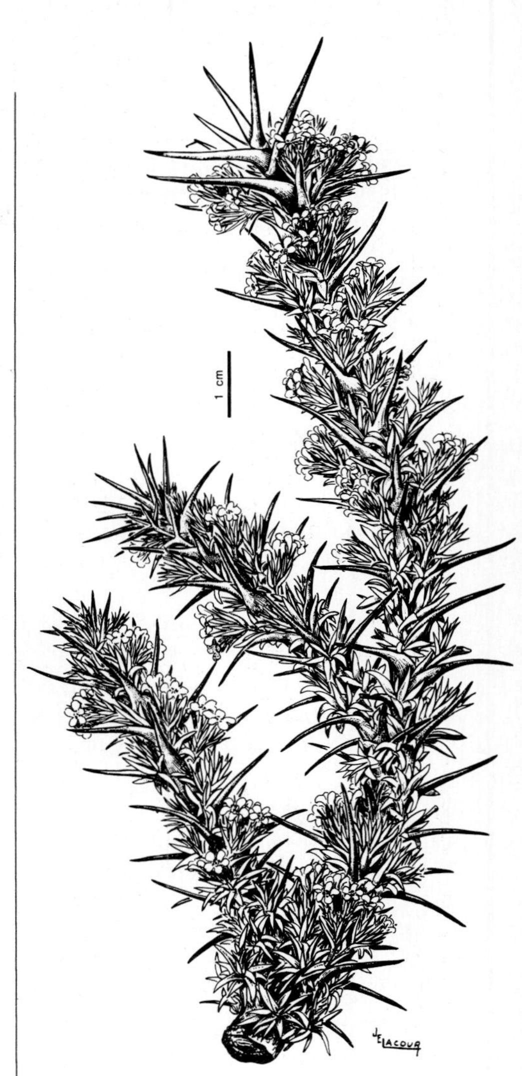

Figure 5-2. *Nassauvia axillaris*. Crow quill pen and India ink. Drawing by Josefina E. Lacour, Instituto de Botanica. Instituto Nacional de Tecnologia Agropecuaria, Castelar, Argentina. Published in M. N. Correa Flora Patagonica part 7 *Compositae*, Volume 8, by Coleccion Cientifica del I.N.T.A., Buenos Aires, Argentina, 1971. Drawn before 1971.

> lustration and all legends should be numbered consecutively. Legends must not be attached to the illustrations. Illustrations will not be returned unless a specific request is made. Tables should be kept simple and should be designed to fit a page 26 picas wide and 42 picas high.[2]

The requirements are very specific and you, as an illustrator, as well as your work will have to comply with set rules and regulations.

While drawing the sketch, a measuring instrument will be crucial. As we are, generally speaking, not using tracing devices, tools such as a compass, ruler, pointer, and proportional divider are essential. The most precise measuring tool is the proportional divider; the least precise is the ruler. Precision of measuring is relative to the type of measuring tool you are using. A ruler is hard to read and obstructs view of the subject, therefore a mistake ranging from two to four millimeters is common. This does not sound like much, but actually it is. When enlarging the specimen by a factor of two, you could be off by a whole centimeter on the very first measurement. All mistakes add up, and the drawing will be considerably out of proportion. Small subjects cannot be measured with the ruler at all. The imperfections of stamped or printed lines as well as the inflexibility of such a measuring device prohibits its usage. Even when using dividers, do not be surprised if while sketching you suddenly discover considerable imperfections in the drawing. A lot depends on the position of your hands holding the tool, as well as on the enlarging factor. A small mistake of half a millimeter will grow considerably when the enlarging factor is high. Enlarging the subject six times will produce an error of three millimeters.

[2]With permission from *Fieldiana*, Field Museum of Natural History, Chicago.

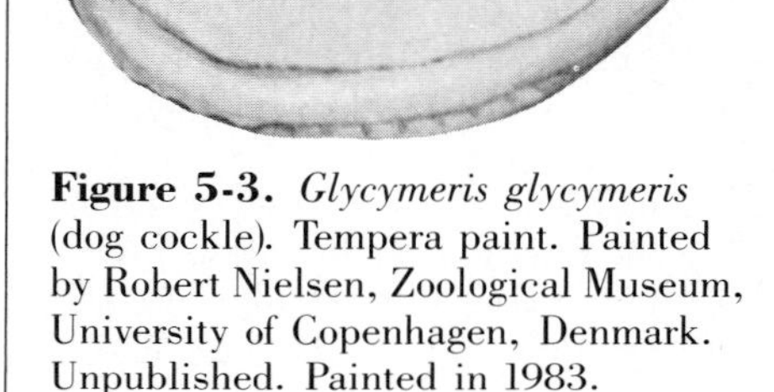

Figure 5-3. *Glycymeris glycymeris* (dog cockle). Tempera paint. Painted by Robert Nielsen, Zoological Museum, University of Copenhagen, Denmark. Unpublished. Painted in 1983.

Precision is a relative statement. Do not be obsessed with an unnecessary quest for the utmost perfect, precise drawing. This is impossible to attain. The thickness of the drawn line will prohibit such perfection. When working under a scientist's direction, precision is controlled by him. The scientist is interested in visual presentation and explanation of a set of problems; he understands the limitations created by the tools used for observation of the specimen as well as for rendering of an illustration. Be content if your enlarged, drawn image of the specimen does not exceed two or three millimeters in its imperfection. Make sure that you position the specimen correctly, as this is of great importance. Also make sure that while measuring, your head is positioned properly in relation to the specimen. Your eyes may get tired and perhaps watery—the strain is a problem. But do not try to compromise by closing one eye while measuring. I know that it is easier to align the tips of the pointers with one eye closed. I also know that the resulting measurement will be incorrect because it is taken from one side and not from directly above the subject. When you close one eye you are looking at the subject from the side of the open eye. This viewing introduces a perspective, which in most cases is an unwanted feature. When looking from one side at the subject, you cannot see the extreme parts of the other side. If you cannot see, you cannot draw. Do not be afraid to write on the sketches. All observations and comments regarding the subjects should be immediately noted. If they are not, it is very likely that you will forget them. Writing will not destroy the sketch. It will visually enhance its appearance as well as providing reference material for you. Such material proves to be invaluable during rendering, as well as being a necessary addition for a meeting with the scientist. When discussing the project with the researcher, you must be prepared for the unknown. He is the expert on the subject, not you. You, as the scientific illustrator, have a responsibility to understand the subject. In order to understand the unknown, you should ask definitive questions. How can such questions be raised if your thoughts are not recorded during initial sketching? For very practical reasons I stress a restraint from erasing. Whenever a mistake is made do not feel badly about it. Do not try to hide any wrong strokes of the pencil. Mark the wrong line, clearly stating that it is improper. Do this any way you want, but do not erase it. I use the standard typographical "delete" symbol (✗). This line can be extended from the center of the sketch

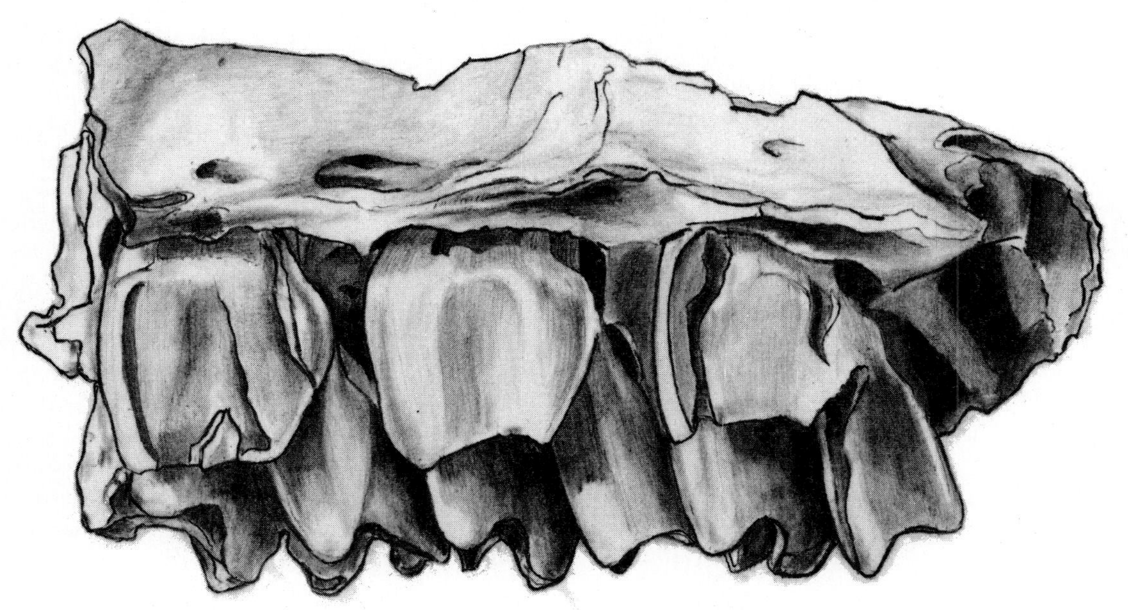

Figure 5-4. *Chaeropus ecaudatus,* TMM 41209-572, a right maxillary fragment with $M^{1\text{-}3}$ from Webb's Cave, shown in crown view. Pencil on Strathmore drawing paper. Drawing by Dr. Tibor Perenyi, Field Museum of Natural History, Chicago. Published in Ernest L. Lundelius, Jr., and William D. Turnbull, "The Mammalian Fauna of Madura Cave, Western Australia, Part IV," *Fieldiana,* Geology, New Series, No. 6, by Field Museum of Natural History, Chicago. Drawn in 1980.

to the edge of the paper without destroying the clarity of previously drawn lines. Mistakes give valid information which will help you be more precise as well as help you not to repeat the same. If it is erased, the proof of wrongdoing is lost and the new "good" line cannot be compared with the old "bad" one. It is possible to make the same mistake over and over, thinking that each new try must be better than the previous. Correction should be compared with something tangible. Your drawings are an enlargement or a reduction of the subject. Rarely is anything drawn in one-to-one scale in scientific illustration. This raises an important question: How many times should you enlarge or reduce the image? Usually this decision is left to the scientific illustrator. Remembering that the final product, the rendering, is being prepared for reproduction through printing, think about the final size of the illustration. The scientist can supply necessary information if he knows in what journal he will publish the results of his research. Many times the scientist cannot supply this information because his research is not finished and he does not know who will publish the results of his work. Be prepared for extreme reductions, and when you do not know the final size, render the illustration with this in mind. During the production of projects for the portfolio, judge the reduction or enlargement of the subject on the basis of quantity of details you would like to include in the drawing. If you would like to present a certain section of the subject in great detail, draw a separate illustration further enlarging the subject. Try not to match the size of the final rendering to the size designated for the reproduction of that illustration. It is more advisable to draw larger than the printed image. All mistakes will be reduced and therefore not be easily detectable. Always remember about the scale representing the enlargement or reduction. Usually the scale is presented for comparison with the drawn image. It is a straight line designating a certain enlarged distance. If the enlarging factor equals two, the straight line will represent an actual distance enlarged by two. The distance is measured by the metric system. Therefore, for example, the original distance of one centimeter is enlarged by two, producing a line equal to two centimeters, but it is marked as it would be in its original state—that is, from zero to one. The marking will be the same, but the length of the line will be multiplied by the enlarging or reducing factor. This does not mean that all illustrations will have a scale based on one centimeter. Depending on the size of the final rendering, the length of the line should be appropriately adjusted.

You must remember that all final renderings can be corrected, cleaned, and changed to your or the scientist's heart's desire, without affecting the printed image of the illustration. But at the same time

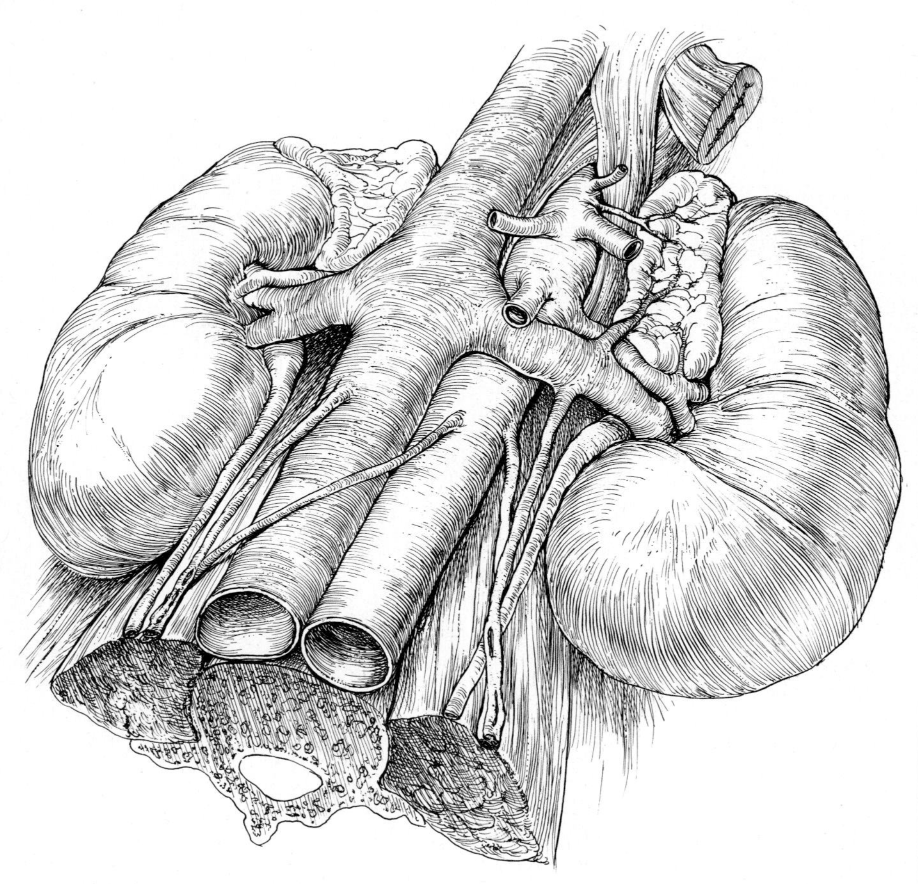

Figure 5-5. Anatomical orientation to radical nephrectomy procedure: an anatomical orientation to two surgical procedures for the removal of either the right or left kidneys. Pen (crow quill) and ink on Linkote board. Drawing by Alfred Teoli, Biocommunication Arts, University of Illinois at Chicago, Health Science Center, Chicago. To be published in Edward Beattie, M.D., and Steven Economou, M.D., *An Atlas of Advanced Surgical Techniques* by W. B. Saunders Co., Philadelphia. Drawn in 1983. In preparation.

you must be aware that all changes will be visible on the original illustration. Many times changes will have to be made because of the changes in the research, not because you have drawn the illustration incorrectly. Nevertheless, do your best, especially for your portfolio, not to inflict unnecessary damage to the illustration. This means that you should not spill coffee, water, or paints on the drawing. It can be saved for reproduction, but will look rather unpleasant for presentation. For that reason, assume a professional standing and keep unnecessary liquids, dusts, pencils, and other tools of the trade away from the drawing board. Do not forget about your hands. Keep them clean at all times. The most common cause of visual unpleasantness is finger marks. Light marks are very hard to detect during the process of drawing, but they will appear instantly when the illustration is sprayed with fixative, which brings all tones into focus. After the illustration is sprayed, it is rather hard and maybe impossible to remove any enclosed finger smudges and marks. For that and other reasons, it is best to start rendering from the one end of the drawing, progressing until it is finished. It is very hard not to smudge an illustration rendered in pencil when the drawing is full of partly finished areas. Because it is difficult to reach certain places, the drawing will get unnecessarily dirty. Another reason to work from one end of the drawing is your inability to make a decision. When we draw, we often reach a point where we face a question of what to do next. Instead of escaping into another corner of the drawing, face the question and answer it. If you do not, you will never be able to come back and answer such a question positively, because in the meantime you will be preoccupied with another set of problems, forgetting the first one.

After many hours of rendering you will like your drawing just because you have invested a lot of time in it. At this point your mind will justify all imperfections. Be very alert when you notice that the drawing looks very good. Does the drawing represent the subject? Did you "improve" on the subject a little in order to make it nicer? Are the proportions all correct? Face reality as it is; be the bystander, the viewer, when looking at your own drawing. It is much better that *you* should notice the imperfections in drawing, rendering, or counting the segments of a specimen than the viewer—especially the scientist for whom you are working.

Before doing any serious drawing or rendering, learn about the supplies. The only way to do this is to practice. Take a piece of paper or drafting film and draw something on it. See how it behaves, what kind of line it will produce, what kind of effect can be achieved. A written word cannot be a substitute for personal practice. This is a must. See if you can scratch, erase, smudge, and dry-mount. Personal experience will help you to conquer the unexpected. See if you can make an overlay; try to make a deliberate mistake in order to correct it. See if it is easy to correct a deeply embedded line in the surface of plated Strathmore drawing paper. If you are interested in a specialized portfolio, choose an appropriate assortment of projects. If you like to have many drawings of skeletal structures in pencil, find another specimen. You do not have to spend a lot of money. A single chicken bone is as good as anything. As a matter of fact, as a scientific illustrator you will draw more subjects that look like a chicken bone than elaborate portraits of a cranium. The basic steps should lead you into a routine. Do not be afraid of routine. I would venture to say that as an artist, you are already involved in many routines. You may as well learn another. The more you learn, the better control of your future you have. When signing drawings, do not become exuberant and present your signature as the most important feature. It is not. The drawing is much more important than your name. Most scientific illustrations are not signed at all. It does look strange to have fifteen signatures of an artist on one printed page. When working for a scientist, do not sign with strange symbols; try to avoid unnecessary pressure for accreditation. The scientific community is a small community. If your drawings are good, the scientific world will know about them. Any illustrations produced here may be published in France or Australia. All interested scientists will look at them, so remember that your reputation as a good scientific illustrator is in your own hands. When looking for appropriate specimens to draw, consult the local telephone book. Look under Educational Materials and Supplies. You will find a number of companies such as Fisher Scientific or Turtox-Cambosco. All of those companies publish catalogs and lists of available materials. Try natural history museums. You will find a wealth of subjects in their stores. It is not impossible—or even difficult—to find a prepared and well-preserved butterfly or an arrowhead. The real problem lies in proper positioning of the specimen, proper transfer of the image to the paper, and in the rendering.

A great many techniques of drawing as well as very important constantly encountered subjects are omitted from the selection of projects. The selection has been made according to the basic aspects of scientific illustration, being versatile and making a visually pleasant portfolio. While practicing the intricacies of scientific illustration, the techniques as well as the approach to the subject hopefully can be useful in another area of the arts.

FIRST PROJECT
Anthropology

Subject: Drawing of a pottery sherd with rim. Two drawings: frontal view and cross section.

Technique: Pen and ink on tracing tissue. Frontal view—line and stipple; cross section—line and instant-transfer tone.

Materials:

1. Tracing paper
2. Pencil, H or HB
3. Illustration board, preferably hot-press smooth surface
4. Technical pen, tips number 00 and 000 or equivalent
5. Ink, black
6. Compass or divider
7. Masking tape
8. Metric ruler
9. Metal ruler of any kind
10. Plasticine, 1 oz.
11. X-Acto knife with number 11 blades
12. Triangle
13. Stylus
14. Fixative
15. Instant-transfer tone, suggested 30% to 40%
16. Reducing glass
17. Dry-mounting tissue, removable (white)
18. Black cover-stock paper
19. Black photographic masking tape

Basic Steps:

1. Positioning of the specimen
2. Preparation of the sketch
3. Tracing the sketch into final sketch
4. Rendering
5. Dry-mounting
6. Preparation for production

PROCEDURE FOR DRAWING FRONTAL VIEW

Observation: First, observe the specimen; notice the curvature of the rim; pay attention to the type of breakage. Examine the surface for any traces of design. Make general notes, preferably in writing, to help your observations. Make sure that on all of the sketches, notes, and so on, the number found on the specimen is carefully noted. After initial observation, raise pertinent questions such as how large the drawing will be. Naturally, the final size of the illustration depends on the original size of the specimen as well as on the size of the reproduction of the illustration. Always remember that rendering very large drawings is time-consuming, but at the same time doing very small drawings prevents you from a detailed, analytical presentation.

Positioning: On the piece of illustration board draw a straight line with the pencil. Using Plasticine, position the sherd in such a way that the front, which is the outside wall

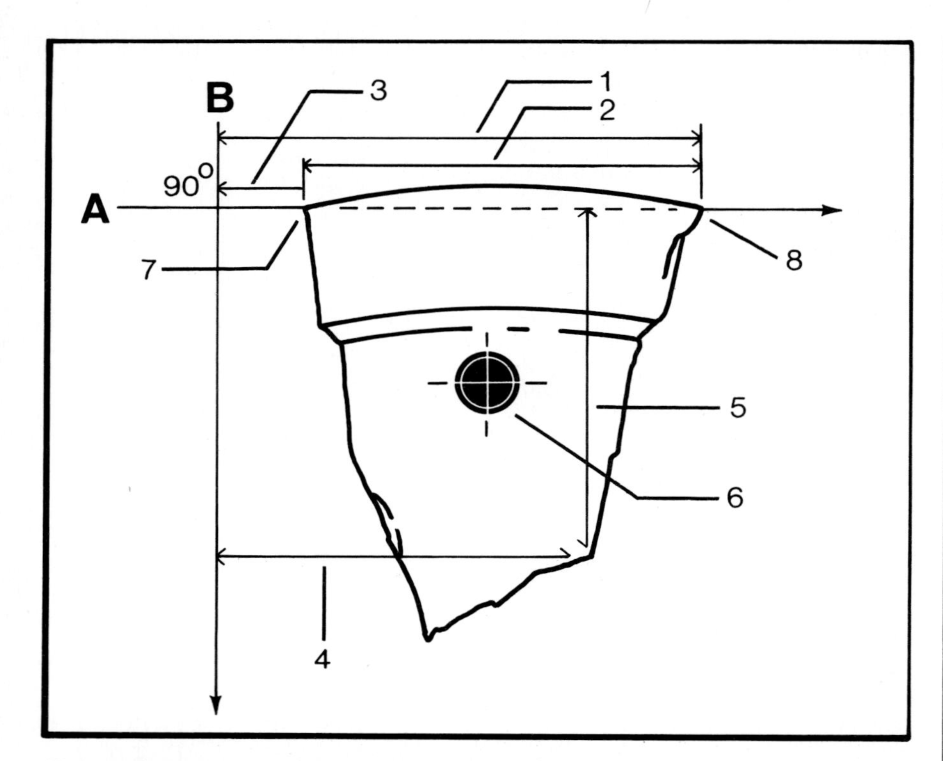

Figure 5-6. Position of the specimen in relation to the basic grid system employed in transferring the outline as well as the internal structures of the specimen. *1*—First measurement. *2*—Second measurement. *3*—Third measurement. *4*—One of the successive measurements parallel to line A. *5*—One of the successive measurements parallel to line B. *6*—Center of the specimen. *7*—Left edge of the lip of the sherd. *8*—Right edge of the lip of the sherd. *A*—Horizontal line to which the left and the right edge of the lip of the shred should be aligned when looking with both eyes onto the center of the specimen. All horizontal measurements must be parallel to Line A, and all vertical measurements should start from this line. *B*—Vertical line drawn at 90-degree angle to the horizontal line. All vertical measurements must be parallel to line B, and all horizontal measurements should start from this line.

of the pottery, will face you. Try to be careful and align the rim of the pot sherd to form a parallel line with the surface of the illustration board. Do not perch the sherd too high on a large clump of Plasticine, as the sherd may fall down and break. Specimens cannot be damaged. Move the sherd and the piece of illustration board toward yourself. With your head down look directly at the center of the sherd and align the left and right edges of the rim in such a way that both will appear to touch the drawn pencil line. Make sure the sherd will stay in this position. Check the parallel position of the line formed by the edges of the sherd. If the sherd is broken, forming a twist, try to find a middle point between the parallel alignment of the left and right edges to the surface of the illustration board of the rim and the frontal view. There is a danger that when a sherd is badly twisted because of breakage, the frontal view cannot be sufficiently visible. Do not press the sherd into Plasticine too strongly, possibly damaging the underside. Try to be gentle but firm. If the sherd moves during drawing you may encounter difficulties. Press the Plasticine firmly to the illustration board and do not obstruct the drawn line.

Drawing the Sketch: When the specimen is secured, again move the piece of illustration board with the specimen on it toward yourself. Bend over the center of the sherd and look directly at it with both eyes. Do not look at the specimen from any angle. Do not place the specimen to your right or left. It must be placed directly under you, because you are drawing the frontal view. Remember that looking from the side will create distortion, known to us as *perspective*. When looking with both eyes at the center of the specimen, mark two points on the line you drew previously. By doing this, the left and right edges of the rim of the sherd will be marked on the drawn line; in case there is any unwanted movement of the specimen you will be able to reposition the sherd. Move the sherd away from you and with the aid of the triangle draw a verticle line to the left or right of the sherd, forming a right angle with the previously drawn line. Take a larger—about 15- to 20-inch—piece of illustration board and place tracing tissue on it, attaching the corners with masking tape. Draw horizontal and vertical lines, maintaining the 90-degree angle. Remember that if the specimen is smaller than five centimeters (approximately two inches), you will have to enlarge the specimen at least two times. If the specimen has a complicated design but is larger than five centimeters, you also will have to draw the subject in enlargement. Let's assume that the subject will be enlarged twice.

Sit comfortably at the table with some space to your left and right. Move the positioned specimen toward you and look at the center of the sherd with both eyes. Holding the compass or pointers with both hands, and making sure the tips of the measuring tool are parallel to the surface of the illustration board, hold the compass or pointers above the sherd, aligning one of the tips with the vertical line and the other with the farthest edge of the rim of the sherd. Transfer the measurement, multiplying it by two horizontally onto the tracing tissue. Repeat the same measuring procedure for the edge of the rim nearest the vertical line, and transfer the point to the sketch. Now measure the distance between both edges of the rim and transfer the result to the sketch. See if it is correct—the previous measurements should coincide with the latter one. Measure the points on the edge of the broken sherd by measuring horizontally from the vertical line and vertically from the horizontal line, transferring the results onto the sketch. Do not erase mistakes. All mistakes are a valid source of information. When you find a mistake in drawing, mark it on the sketch so later you will not subconsciously repeat the same. After four or five measurements, start drawing. It is all right if your lines designating the form of the specimen are incorrect at the beginning. You have time to correct later. The design on the surface of the sherd should be transferred in a similar fashion. In order to recheck your measurements, now and then measure the specimen diagonally. Measure the distance between the left edge of the rim and lower right broken edge of the sherd. Match your measurement against the slowly forming sketch. Make sure that the enlargement is marked on the sketch in the form of a horizontal line, the scale. When the first sketch is finished, spray it with fixative and retrace on a new, clean piece of tracing tissue. Again spray the surface with fixative. At this point the sketch of the frontal view of the specimen is ready for rendering.

Rendering: Place the final sketch of the sherd on a clean piece of illustration board. Secure it with tape. Place the sheet of tracing tissue on top of the final sketch. You will see the sketching through the upper layer of tracing tissue. Make sure that the tracing tissue is of cheap quality, not to save you money, but to assure that lines and dates rendered in ink will definitely adhere to the surface of the tracing tissue. Expensive brands may have oily spots, creating problems during rendering. Prepare a technical pen. By now you should know the insides of the pen you are using. If not, take it apart and put it back together. Have the stylus handy. Move the reducing glass near you. Prepare a piece of white paper, as you will use it to check the progress of the rendering by inserting that paper between the sketch and the rendering.

First draw the outline of the specimen in line using the larger of the two pen tips. During the process of drawing, hold the surface of the tracing tissue with the stylus. While drawing, turn the illustration board around to make the drawing easier. After completing the outline in solid line, switch to the smaller tip and start applying dots to the surface of the tracing tissue. Make sure the dots are round. Hold the tissue with the stylus near the area where you stipple. Do not let the tip of the pen touch the surface of the tissue for too long—this will form larger dots. Try to be uniform. Try to stipple the general shape of the sherd. First, imagine the light coming from your viewing point. This will make all bulging areas lighter and all receding edges darker. Remember that it is hard to remove ink from any surface, especially from tracing tissue, and that you always can make the drawing darker. Add tone with premeditation. Make sure that values change smoothly and evenly. Check the process of rendering by looking at the drawing through the reducing glass. This will make the image smaller, letting you see irregularities of tone. Stand up when you use the reducing glass. Remember that it is better to have fewer dots than too many. Do not think about other things while rendering. When necessary, and with great caution, you can scrape unnecessary dots from the surface of the rendering using an X-Acto knife with a very sharp number 11 blade. I suggest, unless you have previous experience, to do this after the illustration is dry-mounted. When the rendering is finished, lay it down and protect it from dust and greasy finger marks. Start preparations for drawing a cross section of the same pottery sherd.

PROCEDURE FOR DRAWING CROSS SECTION

Observation: Remove the sherd from its previous position. Observe the broken edge. You are drawing a vertical cross section of the whole sherd, not just the broken edge. As you look directly at the broken edges, note the twists of the breakage. Choose one side of the sherd to draw. Remember, the walls of the sherd as well as all of the chips present on the edge are of no importance. Do not be distracted by an abundance of irrelevant information.

Positioning of the Specimen: Get up from your comfortable chair and move your head to the level of the table. Turn the sherd upside down. The rim of the sherd should be touching the surface of the illustration board placed on the table. Hold the sherd in one hand and position the specimen with the front or back toward you. Do not look at the rim. When the sherd is placed upside down, it simulates the whole vessel placed upside down on the level surface. You are doing the same thing as with a coffee cup in the kitchen—placing it upside down. Keeping your head lowered, look at the surface of the illustration board. Keeping the rim of the sherd stationary and touching the surface, move the sherd forward and backward. You will see a light between the sherd and the surface. You will notice that at one moment only the middle of the rim will touch the surface and both edges will be above the board. Another time, the left and right edges of the rim of the specimen will touch the board and the middle of the rim will be above the surface. You notice that light will appear at the edges or in the middle. Try to eliminate that bit of light. When all the light is eliminated, you will notice that the sherd is placed at a certain angle in relation to the board. This is the original position of the sherd, the position that at one time this sherd occupied as part of the whole vessel. The discovered angle is the same one that created the opening of the upper part of the vessel. This angle is of extreme importance, as it gives us information about the vessel; it tells us how large the opening of the vessel was. The larger the angle, the larger the opening. Secure the sherd with bits of Plasticine on the piece of illustration board in that position. Now you are ready to begin sketching.

Drawing the Sketch: Prepare the piece of tracing tissue by placing it on illustration board; secure it with tape at the corners. Take a triangle and pencil. Turn the edge of the positioned sherd toward you and bend again. Look directly at the broken edge, which should be parallel to you. Place the triangle vertically on the surface of the illustration board but with its surface parallel to you and to the broken edge of the sherd. Match the straight edge of the triangle with the most visible point of the rim. Draw a line along the edge of the triangle on the illustration board and mark that point by dot. Stand up; now, with the triangle, draw a straight line from that point toward yourself, maintaining a 90-degree angle with the previously drawn line. Draw the same combination of lines on the previously prepared tracing tissue. Make sure that the 90-degree angle is maintained. Look directly at the broken edge. Take pointers or a compass and measure the height of the sherd by placing one tip of the instrument on the board and the other aligning with the highest point of the broken bottom of the sherd. Transfer the measurement onto trac-

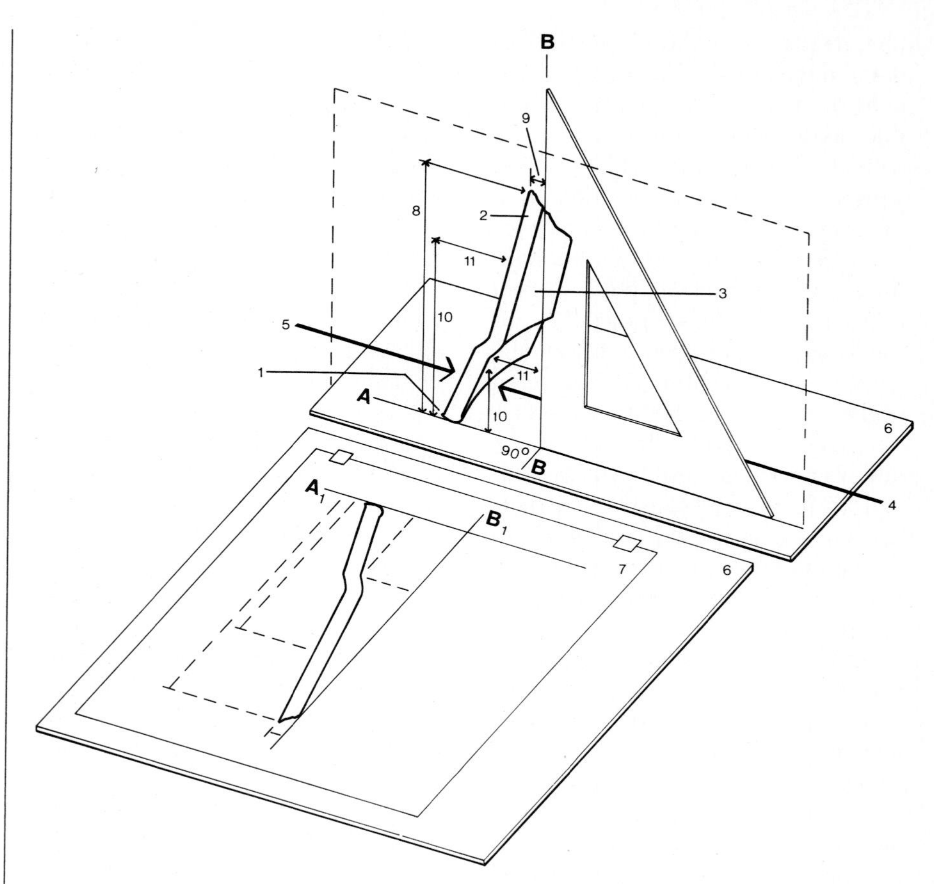

Figure 5-7. Position of the pottery sherd and the sketch while drawing the cross section. *1*—The lip of the sherd (top of the specimen). *2*—Broken edge. *3*—Inside wall (to be omitted during drawing). *4*—Path of the light stopped by wall of the properly positioned specimen. *5*—Direction of viewing and the level of the illustrator's eyes. *6*—Illustration board. *7*—Tracing tissue with the sketch. *8*—First vertical measurement. *9*—First horizontal measurement. *10*—One of the successive vertical measurements parallel to the edge of the triangle (line B). *11*—One of the successive horizontal measurements parallel to the line A. *A*—Horizontal line aligned to the broken edge of the sherd. All horizontal measurements must be parallel to this line and must be taken at the vertical plane indicated by this line. All horizontal measurements are marked on the sketch along the Line A_1. *B*—Vertical line drawn at 90-degree angle to the line A at the point indicated by the right angle of the triangle. All vertical measurements must be parallel to the edge of the triangle and must be taken along the horizontal plane indicated by this line. All vertical measurements are marked on the sketch along the line B_1.

ing tissue, remembering that you have to enlarge every measurement twice. Remember where the top of the sherd is on your sketch. The top may be facing you and the bottom may be away from you. When measuring the sherd, try to hold the tool in such a way that the imaginary line dividing both tips will be parallel to the surface of the illustration board. Make sure that the measuring tool does not obstruct your view. Take the triangle, placing it horizontal to yourself, and with the vertical side touch the previously marked point. With the other hand measure the distance between the edge of the triangle and the point of the edge of the rim. Note it accordingly on the sketch. Remember that in this situation the vertical edge of the triangle represents the vertical line drawn on the sketch. You are actually measuring the distance between that vertical line and the chosen point on the rim. Measure the distance between the very same point and the surface of the board. Note the result on the sketch by marking the point and later drawing a horizontal line through that point. That line must be parallel to the very first line you have drawn. When the vertical line drawn on the base of the vertical measurement crosses the horizontal line based on the horizontal measurement, you have transferred the measured point from the rim to the sketch.

Repeat the procedure until you have completely transferred the outline of the subject onto the sketch. Remember that all vertical measurements are noted on the sketch vertically, toward yourself. You have drawn the vertical line on your sketch for this reason. Remember that all horizontal measurements are noted horizontally, from left to right. When the sketch is finished, trace the drawn outline including the horizontal line present at the tip of the rim. This line will be of great importance during aligning the cross section with the frontal view. Spray the sketch with fixative.

Rendering: Remember that your sketch is an upside-down drawing of reality. Reverse the sketch by placing the drawn side on illustration board. Take a good look at the sketch. According to U.S. standards, the cross section must be placed on the right side of the frontal view. Remember that the outside of the rim should be facing the right side. Position the sherd and secure it with the tape. Place a clean sheet of tracing tissue over the sketch. Secure it with tape also. Take the larger pen tip and trace the outline of the sherd clearly, using a stylus to hold the tissue during drawing. Use the ruler to draw a straight line two or three centimeters long, touching the uppermost point of the inside of the rim. Wait with application of the tone until both drawings are aligned properly and are dry-mounted.

Dry-Mounting Both Illustrations: Clean the table. Prepare the hand-held iron, with the heat set low. Prepare the clean illustration board. The best is hot-press, smooth-surface board. Make sure that the dry-mounting tissue is removable. The removable kind is necessary because it is white and will give a white background to your black-and-white line drawings. Tape the tissue bearing the frontal view on the surface of illustration board, then take the cross section and place it to the right of the frontal view. Align the short, straight line on the inside of the rim in such a way that the imaginary continuation of that line will run through the left and right edges of the rim on the frontal view. Tape the tissue into position, but only from one side. Take removable dry-mounting tissue and place it under both drawings, between the illustration board and the drawings. Take care and have clean hands, clean drawings, and clean dry-mounting tissue. Everything will show after the dry-mounting procedure is completed. Always check for crumbs, and bits and pieces of matter under the dry-mounting tissue. Remove all of it with masking tape. Make sure that the illustration that is being dry-mounted touches the dry-mounting tissue directly. Take a clean piece of tracing tissue and place it on top of your drawings. Take the medium-hot iron and touch with the tip through the protecting tissue on the middle of the first and second drawings. Check that both drawings are tacked by lifting the drawings away from the illustration board. Remember to tack your drawings once and only in the middle. Take a metal ruler and X-Acto knife. Cut out both drawings, making sure that you cut through tracing and dry-mounting tissues simultaneously. Remove unwanted bits of tissues and tapes. Take the X-Acto knife and carefully separate tracing tissue, dry-mounting tissue, and illustration board from each other. This will prevent bubbles from forming. Take a clean piece of tracing tissue and place it on top of the illustrations. Carefully, with moderate pressure, starting from the middle, squeeze the air from under the illustrations with the iron. Let the dry-mounting tissue heat up, but not too much. When finished, let the drawings cool slowly.

Preparation for Production: Now it is time to draw the scale. Draw a straight line with pen equal in the proportion to the factor of enlargement. Remember that you have enlarged the specimen twice, so the two centimeters should be enlarged by two, resulting in a four-centimeter line; however, mark the line as it would have been before enlarging—that is, from zero to two centimeters. Remember always to choose round numbers for the scale. Try to avoid decimals, as they are hard to read. End the drawn horizontal line with short vertical lines. Take the sheet of instant lettering, choosing very readable numbers. Make sure that the size of the numerals is not too overwhelming and at the same time not too small. The numerals should be readable after reduction. Take the stylus, align the sheet properly, and transfer numeral 0 under the left short, vertical bar by rubbing the upper surface of the numeral with the stylus. Do the same with the number 2. Make sure that you have left space between the vertical bars and the numerals. To the right of number 2 transfer lowercase *cm*. The lower case *cm* represents the word *centimeters*. Cover the area with the backing sheet taken from the transfer lettering; burnish the numerals and letters.

For application of the tone to the cross section, take a sheet of instant-transfer tone. Place it on top of the cross section. Take the X-Acto knife. Lightly pressing the tip of the blade, cut out a rough shape of the image. Remove the sheet from the drawing, peel off the plastic layer, and place it on the drawing, burnishing it lightly. Cut the shape of the cross section with the X-Acto knife. Make sure that you cut in the middle of the black drawn line visible through the piece of cut-out instant tone. When you are finished, remove the unwanted parts of instant tone. Cover the area with protective tissue and burnish it. Turn the drawing upside down. Place a piece of tracing paper larger than the drawing on top, leaving the edge of the illustration board near you open. Make sure that the edge of the sheet of tracing paper is parallel to the edge of the illustration board. Take the masking tape and cut a piece longer than the length of the board. Hold the piece by the ends. Keeping it above the board, align the tape in such a way that when pressed down it will attach the tissue to the board. Press the ends of tape to the table. Using your hands, carefully burnish the tape to the surface of the tissue and the board. Take the black cover paper and do the same, using black photographic masking tape. Remove the illustration from the table. Place it face down on another illustration board. Using the edges of the illustration board as a guide, cut away unwanted parts of the tracing tissue and black cover paper. Turn the illustration face up. Place the metal ruler on top of the black tape, aligning the edge of the ruler with the edge of illustration board. Slightly crease the black cover paper with the X-Acto knife. Place all your tools to the side and bend the black paper along the crease. The illustration is now protected and prepared for production.

Comments: With some pottery sherds, as with few other types of specimens, the techniques of transferring the three-dimensional image onto a two-dimensional surface of paper will vary. It is possible to trace the outline of the pottery sherd in two ways. First, hold the specimen with one hand on the paper; with the other, trace the outline by following the edges of the sherd with a vertically held pencil. Such a method is crude but works with relatively flat sherds. Second, place the specimen on the bed of the reducing-enlarging copier. After reducing or enlarging the projected image, trace it onto tracing tissue placed on the glass of the copier. Both methods are used when tons of drawings are needed in a relatively short period of time. Precision is sacrificed for the speed of production. The cross section cannot be drawn an easy way. It has to be accomplished slowly by following the previously described procedure.

SECOND PROJECT

Zoology/Paleontology

Subject: Drawing of a rat skull. Lateral view.

Technique: Pencil on plated Strathmore drawing paper. Continuous tone.

Light Source: Directly from above the specimen.

Materials:

1. Pencils—H, HB, B, 2B, 4B, 6B
2. Erasers—kneaded, Stenotrace 1400 and Stenotrace 1500 series, and art gum
3. Tracing paper
4. Strathmore single-ply, plated drawing paper
5. Illustration board—hot-press, smooth-surface
6. Twin-Tak dry-mounting tissue
7. Proportional divider or pointers
8. Metric ruler
9. Metal ruler
10. Triangle
11. Masking tape
12. Stomps—very good quality, sizes small to medium
13. Fixative, Krylon Crystal-Clear
14. X-Acto knife with number 11 blades
15. Plasticine, 1 oz.
16. Q-tips
17. Watercolor brush, number 8
18. Reducing glass
19. Wooden hand supporter
20. Burnisher
21. Black cover-stock paper
22. Black photographic masking tape

Preparation of Some of the Materials:

1. Graphite from 4B or 6B pencil ground into dust, making sure that powder does not contain sharp particles.
2. Stenotrace erasers sharpened. Cut tips of erasers at about 45-degree angle with sharp X-Acto blade. Try to cut with one stroke, making sure that the edges of the erasers are without burrs and the surface of the cut is even and smooth.

Basic Steps:

1. Positioning of the specimen
2. Preparation of the sketch
3. Tracing the sketch into final sketch
4. Rendering
5. Dry-mounting
6. Preparation for production

Nomenclature:

Anterior—front

Posterior—back

Dorsal—top

Ventral—bottom

Cranium—the skull

Zygomatic arch—protruding, easily visible arch on the left and right side of this skull

Temporal bone—two large structures on the posterior section of the skull, easily visible ventrally and laterally

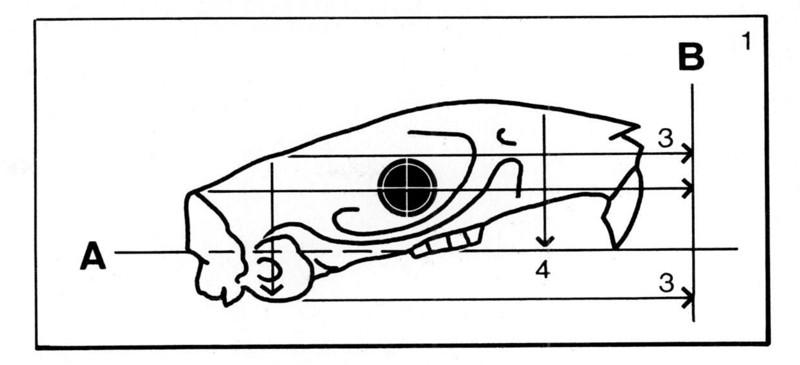

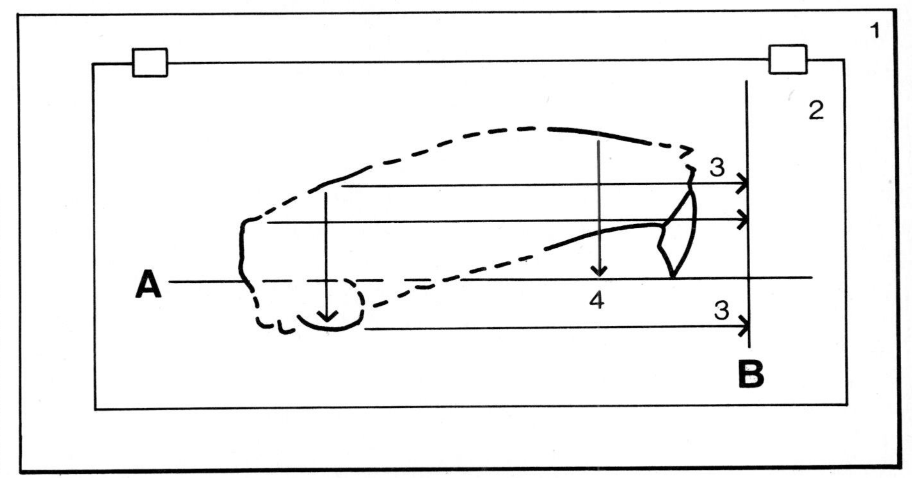

Figure 5-8. Alignment of specimen to basic grid coordinates showing initial measurements. *1*—Illustration board. *2*—Tracing tissue. *3*—Initial horizontal measurement. *4*—Initial vertical measurement. *A*—Horizontal line aligned with visible features on the front and the back of the skull. All horizontal measurements should be parallel to this line. *B*—Vertical line drawn arbitrarily at 90-degree angle to line A. All vertical measurements should be parallel to this line.

PROCEDURE FOR DRAWING

Observation: Gently take the specimen in your hands and observe it. Determine which side is lateral side. The specimen has two lateral sides—choose one for drawing. Observe the specimen from all sides. Make sure you notice a difference between discolorations and the natural form of the specimen. Notice if anything is broken. If it is, leave it as is. Do not glue anything by yourself. When you receive a specimen from the scientist, you will draw it exactly as you receive it. Pretend that is the case. While looking, note which portions of the specimen are lower and which are higher from the point of view from which you are going to draw that specimen. Note the openings of blood vessels. See what features overlap. Imagine what happens when that particular side of the specimen is illuminated from above. Make notes if necessary. Keep in mind that features closer to you will be lighter than those that are farther down. Distinguish differences in thickness of the bone. Thin bone may appear lighter, creating confusion in your mind. Not all dark areas found on the surface of the cranium are indentations.

Positioning: Take a small piece of illustration board and using a ruler, draw a straight line. Take about one-half ounce of Plasticine and place it on the illustration board near the middle of the drawn line. Secure it to the illustration board by squeezing the Plasticine onto the surface. Gently take the specimen and making sure that the lateral side faces upward, lightly press it into the Plasticine. Do no break the specimen. The zygomatic arch is fragile and can break easily. Move the whole thing toward you. Look directly at the middle of the specimen from above with both eyes; try to align a visible tip of a tooth with the drawn line. Do the same with an easily recognizable part of the posterior margin of the cranium. Get on your knees; look ventrally (the bottom) at the skull, making sure that front and back (posterior and anterior) are properly leveled. Turn the skull so the posterior (back) of the skull is facing you. Still keeping your face level with the surface of the table, make sure that the specimen is not tilted to the left or right. If it is, correct the position. Get up and look directly down from above. Again, align the tip of the same tooth with the drawn line. Do the same for the posterior margin of the cranium. When the specimen is horizontally level, vertically straight, and properly aligned, gently secure it with Plasticine. You may have to add or remove some. Remember not to clog up the openings and crevices on the lateral side against the Plasticine. Someone will have to clean the specimen and remove the Plasticine later. Take a triangle and draw a line at 90 degrees from previously drawn line, but close to the specimen. Extend that line across the illustration board. If you have a lower jaw of the specimen, place the jaw in the anatomically correct position. Remember that the lower jaw fits under the zygomatic arch. Before measuring, check the position of the specimen. It must be positioned correctly, otherwise you will not have the lateral side facing you. Make sure that the specimen will not fall. Repositioning takes time. Move the specimen toward you and with both eyes open, looking directly at the center of the specimen, mark two points on the horizontal line, one appearing to coincide with the tip of the tooth and the other with the chosen point of the back portion of the cranium. This will help you to reposition the specimen. Now you are ready to start the sketch.

Drawing the Sketch: Decide how many times you will enlarge the specimen. Such a decision is based on the actual size of the subject, the amount of detail to be presented in the drawings, and on the time available for the rendering, as well as the final size of the illustration after reproduction. The bigger the enlargement, the more time you will have to spend on the rendering of this particular subject. But at the same time, if the enlargement is not sufficient, the necessary details like separate teeth and their relative angles or features enclosed by the temporal bone cannot be drawn in detail. I suggest that you enlarge this specimen four to six times. You must remember that during the drawing procedure you will have to disregard all structures on the opposite side facing the table. This means that the partly visible teeth of the upper and lower jaws on the opposite side of the specimen will be omitted.

To start, prepare the tracing tissue and tape it to the illustration board. Draw two lines, one horizontal and the other vertical. Make sure that a 90-degree angle is maintained. Move the specimen toward you and look directly at the center of the specimen with both eyes. Take your measuring tool in both hands. Keep the tips of the tool parallel to the surface of the illustration board. Hold your tool horizontally. Do not obstruct the specimen with the measuring device, as you have to see what you are measuring. Keep the tool slightly above the specimen. Make sure your head is properly positioned. Do not close one eye. With the head positioned properly but with one eye closed, you will be looking at the specimen from one side, from the side of the open eye. In such a situation you will not see some of the structures on the side of the closed eye. Watch out for unwanted perspective. The lateral side of the specimen has to be presented with exactitude. Such exactness must be maintained through the whole sketching procedure. With measuring tool and proportional divider or pointers, measure the horizontal distance between the drawn vertical line and the farthest posterior point of the cranium. Transfer the measurements accordingly to the sketch. Remember that all horizontal measurements will be transferred horizontally to the sketch and all vertical measurements will be transferred vertically. When transferring measurements, remember that you are enlarging the image. Every measurement must be enlarged by the same factor. Remember also that when using pointers, a regular compass, or an even cruder tool, every mistake in measurement will be magnified as many times as you are enlarging the image. In order to avoid such problems use good tools, such as a proportional divider.

Repeat the same procedure with the point indicating the tip of the tooth. After two initial measurements, you have established the basic length of the subject on your sketch. Find the base of the posterior section of the zygomatic arch on the specimen. Repeat the same measuring procedure and transfer the results onto the sketch. Having established three points, remeasure the distance between the posterior point of the zygomatic arch and the tip of the tooth. Do the same with the known farthest posterior point of the cranium. If all distances coincide within a tolerance of three millimeters, you are going very well. Starting now on your sketch, establish the area for the whole structure. Use the first three points as basic guides for checking all measurements. Are you sure the first three points are correct? Make sure at all times that the measuring tool is held correctly and that your head is centered over the specimen during measurements. Remember not to erase your mistakes. Write all observations on the sketch, otherwise you will forget them. Try to use a finely sharpened pencil and after five or six measurements, draw on the sketch. When you have too many marked points on the sketch without reference lines, you will get lost, not remembering what they are. Avoid this. It is time-consuming to search for something that is already clearly marked. When lines become cluttered or when part of the sketch is drawn incorrectly, use an overlay sheet of tracing tissue to continue the drawing procedure. Draw registration marks on the first sketch and on the overlay. If needed, you will be able to separate and reposition sketches quickly. Do not get entangled in drawing extreme details. You are drawing the general shape of the specimen, but at the same time pay special attention to the shape and position of teeth—those are always of importance. Restrain yourself from shading the sketch. Write all pertinent information and connect it with arrows to relevant portions of the sketch. If necessary, on a separate sheet of paper draw the shape of a particular section of the specimen. This may help you to visualize the three-dimensionality as well as the construction of that part. Remember that during rendering you will have the actual specimen, your notes, and the clearly redrawn, retraced sketch as a reference. When the sketch is completed, spray the surface with fixative. Remember that pencil smudges easily. From now on keep your hands clean; keep your desk clean. Accidents happen, and you may transfer the graphite dust to the surface of the drawing paper by

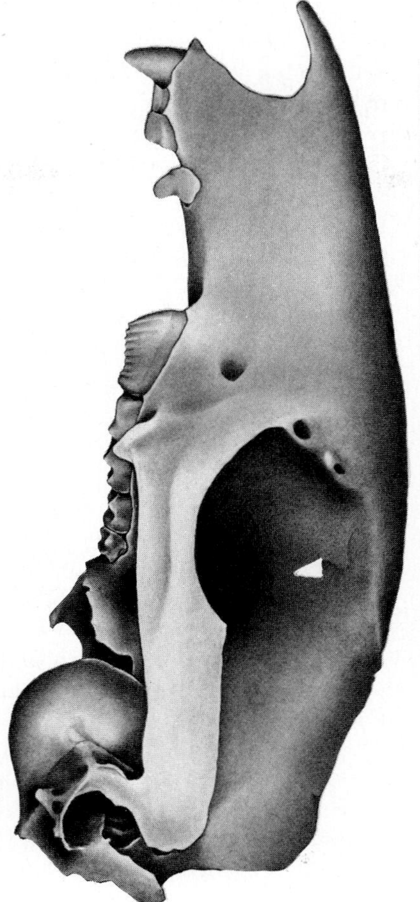

Figure 5-9. *Bettongia penicillata*, recent, FMNH 35325, skull in lateral view. Pencil on Strathmore drawing paper. Drawing by Zbigniew T. Jastrzębski, Field Museum of Natural History, Chicago. Published in William D. Turnbull, "The Mammalian Fauna of Mądura Cave, Western Australia, Part VI," *Fieldiana*, Geology, New Series No. 14 (1984) by Field Museum of Natural History, Chicago. Drawn in 1983.

mistake. With clean hands take a piece of clean, unwrinkled Strathmore plated drawing paper, preferably of one-ply thickness. Place the paper on clean illustration board. Turn the sketch drawn side to the paper and tape it at the corners. Take the whole thing to the light box and place it on top, drawing paper toward yourself. Do not tape the light box. When the paper is secured to the surface of the light box, you must accommodate yourself to the drawing situation. It is better to turn the drawing around rather than yourself. Gently, without pressing, trace the sketch onto the surface of the drawing paper. Use fine-pointed pencils (the H pencil is quite sufficient). Remember not to press hard into the surface of the paper. Any hard-pressed line is hard to erase. It destroys the surface of the drawing paper. When tracing is completed, separate the sketch from the drawing paper. Place the drawing paper with traced image up on clean, very smooth illustration board and tape it at the corners. Now you are ready to start rendering.

Rendering: Be very patient while rendering. Observe the specimen and read your notes. Be very methodical. The quality of drawing is of extreme importance. If you are doing such a rendering for the first time, be prepared to spend approximately fifty working hours on the drawing.

To start, prepare your tools; do not tape the illustration board with the Strathmore paper on it to the table. You will have to turn the drawing around in order to render various parts comfortably. Make sure that your desk is flat, uncluttered, and clean. Keep your sleeves rolled up. Prepare the necessary tools: stomps, pencils (H and 6B), Stenotrace erasers, X-Acto knife, watercolor brush number 8, wooden hand supporter, and kneaded eraser. Keep all tools away from the drawing but within easy comfortable reach.

Take a small piece of hot-press illustration board and tape it to the surface of the desk, away from the drawing. Take a 4B or 6B pencil and very generously rub the graphite on the surface of the piece of illustration board. If you have a mortar, take 4B or 6B graphite and crush it into small pieces, then grind the graphite into powder. Make sure that the resulting dust is not ground too finely nor too coarsely. If it is ground too finely, it will become very slippery, creating problems during rendering. If it is too coarse it will contain bits of unground graphite, also endangering your rendering. Place the ground-up dust in a small bottle, leaving a bit of dust to use. Do not use Microtomic brand pencils. The graphite from those pencils will not smear properly. Do not use grease pencils. Use Venus or a similar brand. Take Stenotrace erasers and sharpen the tips. The 1500 series will be used for smearing the graphite, and the 1400 series will be used for erasing fine white lines on the smeared surface. After the tips are sharpened, cut both tips at about a 45-degree angle, creating a smooth oval ending. Be sure that the X-Acto blade is sharp, and that the cut is smooth and even.

Take the 1500 Stenotrace eraser and work the cut surface into the graphite dust until it becomes very dirty. Keep the 1400 Stenotrace eraser away from the graphite dust. This eraser must be very clean at all times. Take the stomps and work their surface into the graphite dust. Make sure that the sides of the tips of the stomps are covered by the dust. Take a kneaded eraser and knead it in your hands until it is soft, then touch the graphite dust with the kneaded eraser and again knead the eraser in your hands. Bits of graphite will help to separate particles of rubber, helping later in picking up very small sections of tones. Now make sure that your hands are clean. Take a finely sharpened H pencil and, pressing moderately hard, draw the outermost *outline* of the traced drawing. Do not draw on the inside of the outlined structure. Doing the actual rendering will consist of smearing and touching up. Make sure that your touch is gentle. The plated Strathmore drawing paper has a very sensitive surface. Your object is to place the graphite dust on the surface of the paper with minimal damage to the fiber structure. When a change in tone is required and the fibers of the paper are damaged, it is impossible to erase and repair the mistake. Such an area, be it small or large, will remain darker than necessary, thereby forcing you to adjust the tone to the mistake. At such a moment you are not a controlling force, the paper and the tools are.

Do not apply tone in various parts of the drawing. It is easy to accidentally smear graphite over the paper. Start rendering in one area and finish it completely. Use a watercolor brush to remove unwanted particles from the paper. Place the arm support over the drawing and rest your hands on it. Avoid touching the paper with your hands. Start applying tone in the anterior or posterior part of the cranium. Take the 1500 Stenotrace eraser and, using moderate pressure, touch the paper on the inside of the outline and glide the eraser toward the center of the drawing for a short distance. Do not press very hard. The application of the necessary dark tone will be accomplished by repeated application of graphite dust. Do this, covering an area of about four or five centimeters. The tone will not appear smooth. Use stomps of appropriate thickness, *gently* smoothing the tone. Again, the tone will not appear very smooth. Take the kneaded eraser and roll part of it into a point. With your hands resting on the arm support, pick up the darker spots with the tip of the eraser, then retouch the area with the point of the stomp or the point of the H pencil. Take your time; do not work fast; be very methodical. When using pencil on the surface of this paper, do not press. The weight of the pencil itself is sufficient to transfer the bit of tone. Move the tip of the pencil in small circles, not in lines. Avoid cross-hatching. Any horizontal or vertical motion will creat uneven tone. The lighter spots should be eliminated by addition of tone, and the darker by elimination. From the start of rendering you will face constant addition and elimination of bits of tone. For lighter general areas, use stomps or Q-tips. Feel free to use your finger when necessary. For dark to very dark tones, use the Stenotrace 1500 eraser successively over one place as well as using softer pencil with moderately stronger pressure. Remember that is easier to add than to remove. Make sure that the zygomatic arch, one of the highest areas on the specimen, remains white. It is very important not to overrender this part, because the zygomatic arch overlies part of the lower jaw, which in turn overlies structures underneath. This is the area where you have to produce three distinctly separate values, from the darkness of the lower section to the whiteness of the arch. If proper contrast is not maintained, the three-dimensionality of the drawing will suffer. For the same reason, be sure that the bulging posterior part of the skull remains very light. Look at the drawing often through the reducing lens. Stand up, hold the reducing lens with your arm outstretched, and view the drawing through it. This will help you to estimate the values and contrasts.

Make sure that teeth appear as structures stuck into the main portion of the jaws, and not pasted on them. In order to achieve the desired result, make sure that a thin light or white line is left on the inside of the jaw. Make sure that on the side of the teeth a black line delineating the contact area between the tooth and the jaw is transformed into tone. Avoid outlines as much as possible. Let the tone end in a sharp, crisp line without that line being visible as a separate entity. It is understandable that in some areas this will be impossible to achieve, as general tone will be too light and the basic outline of the skull must be kept very dark and crisp for the sake of good reproduction. The portion of the lower jaw which is located under the zygomatic arch will have to be separated with a strip of darker tone from the zygomatic arch and with the white or light line from structures visible under that section of the jawbone. Contrast between light and dark has to be maintained during the drawing process.

Make sure that your drawing does not appear flat. As long as the back position of the skull and the zygomatic arch are not overrendered, you are on the right track. Remember that the light must appear as if it is coming from the viewer's position. This means that all the higher portions of the structure receive less tone than the lower sections. Do not pay attention to shadows; they are of no importance. Remember that you are presenting a three-dimensional form and the light from your desk lamp is strictly for observation; it should *not* be the source of illumination for the rendering. During the process of drawing, keep smoothing the tone patiently until the desired effect is achieved. Remember to keep the background clean—no fingerprints or smudges. Some areas will have to be back-lit—that means in order to properly render a particular structure, the imaginary light source will illuminate that structure from a different angle. It will appear as if it is coming from a different direction from the generally accepted vertical light. This should be done for all openings of nerves and blood vessels. Only one edge of such an opening should have a white line placed on the side of the main structure. The other should be left darker than the background. The same applies to some of the structures on the temporal bone. In this vital area the ridge swooping from the back portion of the skull comes to meet the zygomatic arch. On the ventral side this ridge should be quite dark, while the dorsal wall will remain light to white, thereby creating the illusion of a sharp edge sticking out from the paper toward the viewer's eye. Toward the ventral side the temporal bone turns sharply down. This section should be quite dark, resulting in pronounced changes of value in that area. From that ridge, ventrally, you should have successively the white portion of the ridge, the dark area under the ridge, the light section of the temporal bone coming to the dense darkness meeting the outline of the drawing. Draw all sutures in a crisp, sharp line that varies in value. This is easy to achieve when using a sharply pointed B pencil. The 1400 Stenotrace eraser should be used during removal of tone from the areas where fine white or light lines are to be drawn. Remember to cut the tip of the eraser after each use. The dirty tip will not erase, but will smudge the paper. After the whole drawing is rendered you may use a tip of an X-Acto blade to scratch very fine white lines near some of the parts of the sutures. This irreversibly damages the paper and should be reserved for the very last moment. When everything is finished, when the background has been cleaned out and is white, spray the Krylon Crystal-Clear fixative on the surface and wait for it to dry. Do this two or three times just to make sure that graphite dust will not transfer accidentally. At this point you are ready to dry-mount the illustration.

Dry-Mounting: First, make sure your table is clean and uncluttered. Take a clean sheet of illustration board. Prepare a few sheets of waxy paper used as backing sheets for instant lettering. Keep those handy nearby. Take a sheet of Twin-Tak dry-mounting adhesive larger than the drawing. With the tip of the X-Acto knife, separate *one* of the backing sheets from the adhesive and carefully remove it. Remember to be very careful. Twin-Tak is a strong adhesive. If by accident you touch the surface of the adhesive with any other surface such as a paper, the table, and especially tracing tissue, it will be permanently attached to the adhesive. Keeping the sheet of Twin-Tak in both hands, place it adhesive side up on the illustration board. Take your drawing with both hands and place the shorter side of the drawing on top of the adhesive. Be very careful—if you make a mistake, you will ruin the drawing. Keeping the rest of the drawing in the air, take the waxy backing sheets from the instant lettering and place them on top of the adhesive between your drawing and the adhesive. Those sheets will *not* adhere to the adhesive. With one hand, lower the rest of the drawing on top of the waxy sheets, making sure that the adhesive will not touch the drawing. Take plastic burnisher and another piece of waxy sheet. Place this sheet on top of the drawing and with the burnisher slowly burnish the part of the drawing along the attached side, removing air bubbles from underneath. Pick up the unattached part of the drawing and move the waxy sheets to the side for a distance of no more than two inches. Holding the drawing with one hand, slowly place it back on the adhesive while burnishing the newly attached side with the other hand. Again, burnish the newly attached portion, using the waxy sheet as a protection. Repeat the process until the whole drawing is attached to the adhesive. Remember not to make any mistakes. Once the adhesive touches the paper, it stays on it. Do not destroy your drawing by rushing. When finished, using a metal ruler, cut through the drawing and adhesive and remove the unwanted pieces of paper and Twin-Tak. Turn the drawing back side up and burnish it through the back with the plastic burnisher.

Prepare the clean sheet of illustration board to be used for the mounting of the illustration. Measure the length of one side of your newly cut drawing. Mark the position for the drawing on the illustration board with a line, preferably of blue pencil, making sure that you do not run out of space. Take the waxy sheets and place them on the illustration board. Leave about one inch of space for the edge of the drawing to be attached to the surface. Take your drawing and carefully remove the second backing sheet from the adhesive attached to your drawing. Take the drawing in both hands. Align the edge of the drawing with the drawn line. Place the drawing, adhesive side toward the illustration board and, holding the drawing with one hand, burnish the adhering edge with the other. Making sure that the waxy sheets are in the proper position, place the whole drawing on the illustration board. Repeat the process of removal of waxy sheets and burnishing until your drawing is mounted. Place the waxy sheet on top of the drawing and burnish the whole drawing with the plastic burnisher. Now you have a dry-mounted illustration.

Preparation for Production: Turn the top side of the drawing toward yourself. Take a clean sheet of tracing tissue and place it on top of the drawing, leaving about one inch of space between the tissue and the edge. Take the masking tape and tape the tissue to the illustration board. Take black cover-stock paper and place it on the top, leaving about half an inch of space between the edge of the paper and the edge of the board. Take the black photographic masking tape and tape the paper to the board. If the black tape overlaps a bit over the edge of the illustration board, bend the tape down and attach it to the back of the illustration board. Turn the whole thing back side up and place it on top of a sheet of illustration board. Take your X-Acto knife and, using the edges of the board as a guide, cut through the tracing tissue and black paper. Remove unwanted pieces of paper from the table and turn the drawing front side up. Take the metal ruler and place it on top of the black paper, covering the drawing and aligning it with the upper taped edge of paper. Using the back side of an X-Acto blade, crease the paper. Remove the ruler and bend the paper along the creased line. Bend the tracing tissue also. Now your illustration is prepared for production. For the final packaging effect you may paste a white label in the lower right corner.

Comments: Naturally the scale, a horizontal black line representing a measurable unit of proportion, should be on the drawing. You have to remember that use of the metric system is a must. The scale, as well as any other designation like labeling anatomical parts of the skull, does not have to be placed directly on the drawing. The overlay of transparent acetate attached by masking tape to the upper section of the drawing can serve as the base for black-and-white designations. If you wish to have the halftone image of the subject appear very clearly after it is printed, spearate the image from the background by masking. For the proper effect, use sheets of Rubylith. In order to separate the background from the image, place the sheet of Rubylith on top of the drawing; after cutting through the thin layer of the sheet, peel the strip off. Then, using masking tape, attach the transparent backing sheet of Rubylith to the illustration board, again forming an overlay. Remember that when more than one overlay is attached to the illustration, the masking tape holding the overlays should be on opposite sides of the board, creating free-flopping overlays. Overlays cannot interfere with each other. When Rubylith is properly attached, using a fine X-Acto knife, cut the outline of the image and peel off the background. The image should be covered by red film. The platemaker using high-contrast photography will be able to remove the background, but because red photographs black, will retain the outlined shape of the image, later photgraphed through a fine screen. In the final, printed stage the drawing but not the background will appear to be composed of fine dots. The background will be as white as the paper on which the image is printed. If line, scale, type, arrows, and so on are not placed on the acetate overlay but directly on the illustration, a separate mask has to be prepared. This mask does not have to be precise as long as *all* the line is covered by red and as long as both masks in crucial areas are separated by a very thin white space.

THIRD PROJECT
Entomology

Subject: An illustration of a butterfly, two views. The first, dorsal presentation of the specimen showing all wings unfolded. The second, dorsal presentation of one pair of wings and all legs attached to the thorax.

Technique: Pencil and watercolor on illustration board.

Materials:

1. Pencils—M, HB, B, 2B, 4B
2. Erasers—kneaded, Stenotrace 1400, and Stenotrace 1500 series, art gum
3. Tracing paper
4. Illustration board, hot-press
5. Illustration board, cold-press
6. Measuring tool, pointers or compass
7. Grid paper
8. Metric ruler
9. Metal ruler
10. Triangle or T square
11. Masking tape
12. Stomps
13. Fixative, workable
14. Fixative, Krylon Crystal-Clear
15. X-Acto knife with number 11 blades
16. Pin(s), very thin
17. Plasticine, 1 oz.
18. Reducing glass
19. Black cover-stock paper
20. Black "photographic" masking tape
21. Watercolors
22. Watercolor brushes, numbers 6 and 8
23. Carbon paper

Basic Steps:

1. Positioning of the specimen
2. Preparation of the sketch
3. Tracing the sketch into the final sketch
4. Tracing the final sketch on the illustration board
5. Rendering in pencil
6. Painting in watercolors
7. Preparation for production

Nomenclature:

Anterior—front

Posterior—back

Dorsal—top

Ventral—bottom

Thorax—easily visible, separate large structure behind the head of the insect, composed of three separate parts with a pair of legs attached to each part

Abdomen—easily visible, separate, sometimes segmented structure behind the thorax

Fore wing—the first wing

Hind wing—the second wing

Antennae—easily visible long protuberance attached to the head of the insect

Coxa—the first section of the insect leg

Femur—the second section of the insect leg

Tibia—the third section of the insect leg

Tarsus—the fourth section of the insect leg, made of two to five segments

Preparation of the Specimen: Make sure your specimen is mounted on a very thin, long pin. This is necessary for handling the specimen during observation as well as positioning it for drawing. Take an approximately four or five centimeter long thin pin and pierce the specimen in the middle of the anterior part of the abdomen or in the middle section of the posterior part of the thorax. Make sure that the pin has pierced the specimen completely and that about three centimeters of the pin has emerged from the ventral side. Leave enough of the pin sticking from the dorsal side to allow you to move the specimen by handling the pin, *not* the specimen. During the preparation make sure that you are careful enough not to crush or damage any parts of the insect. If damage occurs, make sure that you know from where it came. Take the broken part or parts and place them in separate small bottles. If you do not have small bottles, prepare a folder or folders from the paper. Make sure to label the broken parts. For example: antennae from the left side, or third leg from the right side. Do not lose anything. You may need that part for observation or for drawing. The best method is to be careful and not to damage anything.

PROCEDURE FOR DRAWING

Observation: Be very careful while handling the specimen. Insects are very fragile and can be damaged easily. First, observe and note all damaged parts. Then make note of the position of separate sections of the butterfly. Observe the head and see how much it is bent toward the ventral side, and how much it is twisted. You will have to draw the whole head and you will have to eliminate the twist. Observe the abdomen and see if you recognize all sections, making notes so you will not forget. See the position of the legs. All legs are attached to the thorax. See how badly coxa, femur, tibia, and tarsus are twisted. You will have to straighten out the accidental imperfections. Note how many veins are on the wings. Make a quick sketch and note your findings. Make sure to notice all coloration and in your mind, translate it into values of light and dark. See where the wings are attached. See by how much they overlap. Note the density and the length of hairs on the insect's body. When you have satisfied your curiosity and found the answers to the basic anatomical questions, you are ready to position the specimen.

Positioning of the Specimen: Cut a small piece of illustration board slightly larger than the specimen. Draw horizontal and vertical lines at a 90-degree angle to each other. Place the Plasticine on the illustration board away from lines. Remember that the specimen will be aligned with the drawn lines. Secure the Plasticine to the surface of the illustration board and form a cone or a relatively high (about three to four centimeters) lump. Take the specimen by the head of the pin and stick the other end of the pin into Plasticine. Look with both eyes at the center of the specimen and align the edge of one of the wings with one of the lines. Make sure that the horizontal plane of the wings is parallel to the surface of the illustration board. With insects, some anatomical features are often bent or twisted, therefore you will have to reposition the specimen while sketching in order to have a proper view of a particular part of the butterfly. The head has to be viewed directly from above. The abdomen, sections of antennae, and sections of the separate legs will have to be positioned in accordance with the plane of the wings. During handling be careful not to damage the specimen. Handle the specimen by the head of the pin.

Drawing the Sketch: Take clean illustration board and place the sheet of tracing paper on it. Secure the paper with tape. Draw two lines at 90 degrees to each other. Remember that during sketching you will enlarge the subject. You will enlarge all measurements by the same factor. The enlargement depends on the size of the specimen and the complexity of details. When preparing the sketch of the butterfly, you do not have to draw the whole subject. You will have to concentrate on the left or right side only. Your first sketch will present either the left or right part of the specimen.

To start drawing, take a measuring tool, pointers, or a compass, and keeping the tool parallel with the surface of the illustration board, hold the tool with both hands over the specimen and measure the distance between the vertical line drawn on the illustration board and the tip of the fore wing. Remember to look at the specimen with both eyes and keep your head positioned properly over the specimen. Transfer the measurement to the tracing tissue, enlarging it accordingly. Measure the distance between the vertically drawn line on the illustration board and the center of the subject at the same horizontal line as the previous measurement. Transfer this to the tracing tissue. Measure the distance between the horizontal line drawn on the illustration board and the point where the thorax touches the abdomen of the insect. Through the same type of measurements, find the point lying at the middle of the abdomen. When you have three points on the sketch aligned in a vertical fashion, draw a straight vertical line through them. This line will separate the left side from the right side of the butterfly. On your sketch you will be concerned with the side of the butterfly which is designated by the first measurement.

Repeating the procedure, draw the outlines of the fore and hind wings. Make sure that all veins and all coloration is faithfully transferred to the sketch by *line* only—do not shade on the sketch. Write, draw arrows, designate the tones and colors any way you like, as long as the markings are readable to you. Make sure you distinguish values created by colors.

When finished, check if the abdomen is in a completely horizontal position. If not, reposition the specimen accordingly. Measure the distance between the place where the thorax touches the abdomen. Do you remember your third measurement? Did you mark the point clearly? The distance between that point and the end of the abdomen should be measured accurately. The width of the abdomen, actually one half the width of the abdomen, is drawn by measuring the horizontal distance between the center of the abdomen and the outlying edge. Make sure that the specimen is placed correctly and the abdomen is horizontally parallel with the illustration board and is not twisted to the left or right. Measure and draw each section of abdomen.

After finishing drawing the abdomen, turn your attention to the thorax. You will probably have to reposition the specimen again in order to view that part correctly. Make sure the thorax is parallel to the surface of the illustration board and is not twisted to the left or right. If you cannot position the whole part of the butterfly properly, try to do this with just a section of that part, rotating or turning the specimen as needed. Assuming that the thorax on your specimen can be positioned correctly as a whole unit, measure the distance between the middle of the thorax and the points lying on the outside of the outline of the subject. Make sure that during measurements you are looking directly over the middle of the specimen and that your measuring tool is kept in a parallel position to the illustration board. Pay special attention to the corners of the thorax. Make sure the true shape is measured and drawn.

Again, you will have to reposition the specimen when starting to draw the head. Make sure that the head is tilted backward, exposing the frontal parts of the head and making them clearly visible from above. Transfer all necessary measurements on the sketch. Make sure the angle, position, and size of the eye is correctly drawn. Mark the base for the antenna and draw the first segment of the antenna. Do not draw a whole antenna immediately. Most likely you will have to reposition the specimen again. After the head is properly measured and drawn in enlargement, start observing the segments of antenna. Re-

Figure 5-10. *Charaxes jasius septentrionalis*. Watercolor and gouache on matte-finished cardboard. Painting by Ricardo Abad Rodriguez, Spanish Institute of Entomology, Consejo Superior de Investigaciones Científicas, Madrid, Spain. Published in various publications.

position the specimen if necessary and measure each of the segments separately, transferring the results to the sketch. It may be advisable to draw the antenna on separate tracing tissue overlaying the main sketch. During drawing you will have to arrange the segments of the antennae, presenting the insertions of one part into another and making the whole structure appear clearly visible. You cannot present the antennae in their preserved, dry state. If they are very badly twisted and bent, place a few drops of water on them and cover the specimen tightly overnight. The structures will absorb the water and will straighten themselves.

When finished, reposition the specimen again and prepare to draw the legs. You may have to repeat the process of repositioning for each segment of the leg. Remember that legs cannot be visible from above, as wings will obstruct the structures. The position and the state of each of the segments—coxa, femur, tibia, and tarsus—will have to be drawn separately, creating a separate illustration. Use the same technique as you did for drawing the antennae. When separate sketches of the three legs are finished, paste each sketch onto the main sketch. Make sure that sketches of the legs are attached to the main sketch on the opposite side from the wing. Now you will have a drawing of the butterfly with one pair of wings on one side and the three legs on the other.

Turn your attention to the hairs and draw them accordingly. Pay special attention to the length and the direction of the hair. Make sure that hairs do not obstruct the main shape. Do not draw all of the hairs. When finished, spray the sketch with fixative.

Carefully trace the main sketch, a half a butterfly with one pair of wings, onto a new, clean tracing tissue. Do not forget to trace the vertical line running through the middle of the subject. Fold the tracing tissue along that line and again trace the visible image, then unfold the tissue. Now you have a whole butterfly drawn. Prepare a new sheet of tracing tissue for the tracing of the legs. Trace half of the butterfly as drawn on the first main sketch. Fold the tracing in the middle and trace the center of the body. Do not trace the wings. Unfold the tissue. You will have a body of a butterfly with one antenna and one pair of legs. Make sure that their positions are correct and the sketch is uncluttered. When finished, you will have two separate sketches: one depicting a whole butterfly and the other half of a butterfly with legs not covered by the wing. Take carbon paper used for typing and a clean sheet of cold-press illustration board. Place the sketch of the whole butterfly on the board and tape it on one edge. Place the carbon paper under the sketch. Trace the sketch with 2H pencil through the carbon paper onto the surface of the illustration board. Do the same with the other sketch. Now you are ready for rendering.

Rendering: First of all clean up the drawing table and make room for your elbows, the transferred drawing, necessary sketches, the specimen, and drawing tools. You will need the sketches, your notes and the specimen to refer to during the rendering. Take the drawing of the dorsal view with all four wings and place it in front of you. Take an H or HB pencil and start with the outline. Make sure the pencils are properly sharpened.

Draw the outline of the butterfly, pressing the pencil reasonably hard into the surface of the cold-press illustration board. Be very careful and follow the outside edge of transferred lines. The outermost edge of the image must be very sharp, otherwise your drawing will not appear clearly rendered. The surface of cold-press illustration board is easy to smudge. Be very careful from the beginning and do not leave fingerprints. Make sure that during rendering the surface of the illustration board is kept white. The surface of cold-press illustration board does not accept erasure—the eraser will make a permanent mark. Such marks are of no importance to the platemaker, but they do influence the viewer's opinion about you as a professional illustrator. When finished with the outline, take softer pencils such as B or 2B and gently draw the veins on the wings. Take a stomp of appropriate size and apply the tone to the coloration areas on the wings. Work with tips of the pencils as well as stomps. Remember, do not over-render. Do not make the image too dark. Remember that later a wash of color will be applied to the same surface. The dark tones will be darker after the watercolor is applied. Make sure the light tones are in proportion to the dark areas. Check your progress through the reducing lens. Do not apply colors now. Wait until the pencil rendering is finished and sprayed with workable fixative. Be methodical; restrain yourself.

With the wings finished, concentrate on the abdomen. Render it in pencil, using stomps and pencil tips. Make sure that the abdomen appears three-dimensional. Use a reducing glass to appraise the situation. Insects are usually drawn with the source of light coming from the upper left side but this really does not apply to butterflies. The proportion between the spread of wings and the width of abdomen and thorax is too great for meaningful play of the direction of the light. Render the abdomen with the highest part being the lightest. Do the same with the thorax. Be sure that all possible segmentations are clearly visible. Do this by manipulating the tone in vital areas. Whenever a part is inserted, the area of insertion should be darker in tone than the overlapping section of the specimen. By having abrupt changes in tone, light on the upper section and dark on the lower, you will make the one part inserted into another clearly visible. Render the thorax accordingly. Make sure that the thorax is separated from the abdomen by a change of tone and a sharp line.

Turn your attention toward the head. Render the head the same way, making sure that the eyes and the first segment of the antennae are clearly visible. Render the antennae, remembering to change the tone at each section of the structure. Make sure that each structure is clearly visible as a separate entity. Make sure that you are not overcome by your imagination. Be faithful to the natural tone and the shape of the specimen. Observe the hair on the body of the specimen and be sure not to draw too many hairs. Do not let hairs destroy the rendering of the whole structure. At the same time, if there is a marked change in density of the hairs, you will have to preserve that on your drawing. Some of the hairs will stick out from the basic outline of the structure.

When finished, take a workable fixative and apply a couple of coats to the drawing. Wait until the surface is dry. Then take watercolors and a good-quality number 6 brush. Take a clean piece of white illustration board. Do you want to spill the water on your drawing? If not, make sure that water is kept away. When not using watercolors from tubes, be sure to wet all colors with water. Mix the appropriate color on white illustration board. Do not use plastic containers for mixing. Water and the pigment will bead up and you will not be able to differentiate the proportion of pigment in the water. Apply the appropriate color as a transparent wash to the surface of the pencil rendering. If not satisfied, add a little acrylic white to the color. Acrylic paint, having a plastic base, cannot be dissolved in water after it is dry; this will help you to apply the second coat of wash without removing the previous layer of pigment from the drawing. Using acrylic white, you will have to draw a thin white line with the tip of the brush along one side of the hairs in order to make the hair more visible. Make sure that the white line is thin and of consistent thinness throughout the whole body of the butterfly. The other side of the hair will have to be a dark, straight line. When rendering iridescent colors, you will have to use acrylic white and keep your brush in a semidry state during the application of color. Mix the color and add a bit of acrylic white. Dry the brush by moving the hairs back and forth on the surface of the illustration board until the tip of the brush produces numerous lines rather than a wash. Apply the pigment to the surface of the drawing by brushing it lightly. Painting the iridescence is actually using the same technique as pointillists, but in our case instead of using dots we are using lines. When finished, let the rendering dry competely. Then take the Krylon Crystal-Clear fixative and apply approximately five to eight coats. Let each coat of fixative dry before spraying the next. By coating the illustration with clear transparent layers of plastic, we are forcing the reflected light from the surface of the drawing to disperse. This makes the drawing appear fresher and makes the colors stronger and more visible.

The second illustration is rendered in the same technique. Remember to present the position of the legs properly and to differentiate the tone at the joints of coxa, femur, tibia, and tarsus.

Preparation for Production: Follow the usual procedure of securing a protective sheet of tracing tissue with the masking tape, with black cover-stock paper on top of the tissue. Make sure that the fixative is completely dry before placing the tissue on top of the illustration.

Comments: Your illustrations of the butterfly are colored drawings, rendered in the same fashion as colored book illustrations of the middle and late nineteenth century. If you can, look through *Transactions of the Entomological Society of London*, published in London by C. Roworth and Sons around 1850–51, or through any entomological publication from that period. The illustrations have been printed in black and white, lithography being the usual technique, and colored by hand. The results are very handsome.

FOURTH PROJECT

Botany

Subject: Habit of plant with enlarged leaf, section of the root, and section of the stem, all arranged into one plate, 16½ by 12 inches.

Technique: Rapidograph pen on mylar drafting film.

Materials:

1. Tracing paper
2. Pencil—H
3. Illustration board, preferably hot-press, smooth-surface
4. Technical pen, tips number 00 and 000 or equivalent
5. Ink, black
6. Compass
7. Pointers or proportional divider
8. Masking tape
9. Metric ruler
10. Metal ruler
11. X-Acto knife with number 11 blades
12. Triangle
13. Fixative
14. Reducing glass
15. Drafting film, preferably the best quality
16. Scotch brand double-coated tape, number 666
17. Black cover-stock paper
18. Black photographic masking tape

Basic Steps:

1. Preparation of the sketches
2. Arrangement of the plate
3. Tracing the sketches into a final sketch
4. Rendering
5. Preparation for production

Nomenclature:

calyx—green outside part of the flower

corolla—part of flower made of petals

stamens—thin structure growing out of the middle of the flower between the petals

pistil—the ovary of the flower, located directly in the center of the flower

petiole—the base of the leaf that turns into the middle vein, dividing the leaf symmetrically

blade—the edge of the leaf

habit—the whole plant with leaves, roots, etc., presenting a typical structure of that plant

Preparation of the Specimen: If you have access to an herbarium collection and can obtain a dried, prepared specimen, then you are ready to draw, but very likely you will have to prepare your own specimen. Collect a small plant, making sure that the roots are intact and the plant is not badly damaged. Take a sheet of newspaper and after removing the earth from the roots, place the plant between two sheets of newspaper. Newspaper is an excellent in-between step during the preparation. Arrange the plant, trying to preserve the natural composition or state of the specimen. Do not twist the leaves in directions unnatural to this particular plant, but at the same time make sure that the undersides of a couple of leaves are exposed. Take a piece of cardboard or illustration board and place it under the closed sheets of newspaper. Take another sheet of board and place it on top of the folded newspaper. Remember that in this arrangement you have the plant between newspaper sheets with the cardboards on the bottom and the top. Place something heavy on the top, trying to distribute the weight evenly over the surface of the upper board, and place the whole thing in a warm spot. Wait one week for the specimen to dry up.

After your specimen has dried, take thin, light-colored cardboard and cut a 24 by 16½ inch rectangle. Crease the rectangle in the middle and fold it, creating a folder measuring 12 by 16½ inches. Now you have a standard-size herbarium sheet. Carefully arrange the specimen; if your dry plant is too large for the sheet, break the stem in such a way that the plant will fit in the designated area. Take a transparent glue such as Elmer's and, squeezing out a bit of substance, glue the plant to the surface of the herbarium sheet. Do not dip the whole plant in the glue, and do not apply glue separately to the plant and the herbarium sheet. Use your judgment and glue the vital parts of the plant such as the stem, a section of the root, petiole, and some tips of the leaves. Your objective is to secure the plant for future use without obstructing vital parts. Remember that you will have to study the structures of the specimen. After applying glue, do not press the plant, and do not cover the specimen with anything. Let the glue dry completely. Make small pockets or envelopes to hold all broken parts or sections of the plant. Secure the envelopes to the herbarium sheet with a drop of glue. Do not forget to write the generic and specific (scientific) name of the plant in the lower right corner. Make sure that on this label you include all pertinent information regarding the place and time of collection. Note by whom this plant was collected. By including a number designating the specimen and keeping a catalog of prepared material, you are starting an herbarium collection. When all preparations are finished you will have a dried, mounted specimen just like the one a scientific illustrator receives for drawing.

PROCEDURE FOR DRAWING

Observation: Appraising the project, take into account the size of the plant, size of the leaf, and thickness of the stem. It is probable that you will have to reduce the whole plant and reduce or enlarge the leaf; it is definite that you will have to enlarge the stem. As you are choosing the proportions, be sure that all will fit into the designated area. Looking closely, make sure that by noticing all details you will definitely produce an illustration faithful to the subject. See the major veins on the leaf; count them. Make an appropriate note of your findings. Check if the underside of the leaf is different from the front. See if the stem has small hairs. Check if the leaves alternate. Make sure you know how many petals are on the flower. Note how it is attached to the stem. Make sure that you know the subject.

Positioning: For sketching the habit of the plant from the herbarium sheet, you do not need positioning. The specimen is already positioned. Do not remove the specimen from its original place. During the process of sketching sections which are supposed to be enlarged, use a thin string as a vertical and horizontal reference. Do not draw anything on the herbarium sheet. When positioning the thin string over the specimen, first tape the corners of the whole herbarium sheet to the table or to the larger illustration board. Then, using a triangle as a reference, stretch one string over the whole sheet, making sure it is near the pertinent part of the subject. Tape the ends of the string to the table or illustration board. Stretch the second string, again using a triangle as a reference, across the first one, maintaining a 90-degree angle between them. Tape the ends of the string to the surface. Do not tape the string to the herbarium sheet. Make sure that the herbarium sheet is not damaged and the mounted plant not broken or crushed. Two strings stretched at right angles to each other will give you basic vertical and horizontal reference lines.

Drawing the Sketch: First concentrate on the presentation of the habit of the plant. There are two ways to prepare the first basic sketch. One, the faster, consists of placing tracing tissue over the entire specimen and looking through the semitransparent tissue, tracing everything visible. The other is the grid technique used in previous projects. Naturally, the grid technique is much slower. As during preparation of the drawing of the habit, the precision is of a different magnitude. Usually it is not necessary to use the grid technique. The object is to draw an example of the plant with all relevant characteristics, therefore, a quick tracing is quite sufficient. Make sure that overlapping sections of various structures do not interfere with each other. It is necessary to present the density of the leaves as well as to attach them in the proper place.

When the first general sketch is finished, tape it to the illustration board and place another sheet of tracing tissue over it. Now you can refine the basic image. Make sure that veins on the leaf are properly depicted. You may have to use the grid technique in order to do this. It is sufficient to faithfully transfer the information from one or two leaves. Be observant and make sure that differences between young and old leaves are properly drawn. Pay attention to the shape of the leaf as well as to the number of veins. Make sure that the direction or the relative angle between petiole and other veins is maintained. This is of extreme importance. Make sure that if leaves have a texture, that texture is noted on your sketch. Watch for the blade of the leaf. At the same time, do not draw discoloration or do not draw cracks present on the structure. Actually, you are drawing a slightly idealized presentation of the whole specimen. When retouching the final sketch, do not exaggerate. Transfer the shape of the calyx faithfully, first by tracing and later adding any necessary correction using the grid technique. Do not forget about the roots. When your plant is too large for the final designated area of the drawing, you may remove a part of the lower portion of the stem. This may be necessary because the whole plant must be

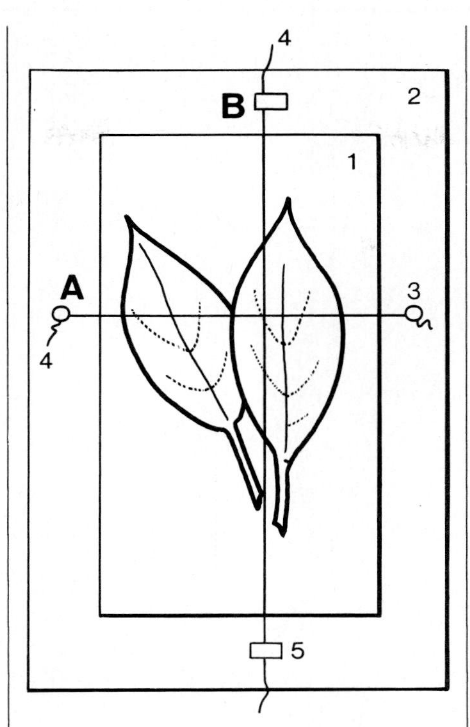

Figure 5-11. Position and attachments of coordinates of basic grid system. *1*—Herbarium sheet. *2*—Table. *3*—Pins. *4*—String. *5*—Masking tape. *A*—Horizontal line aligned to the visible features and placed over the herbarium sheet. All horizontal measurements should be parallel to this line. *B*—Vertical line drawn arbitrarily over the herbarium sheet but at 90 degrees to the line A. All vertical measurements should be parallel to this line.

shown. When doing this, make sure that the removed part does not interfere with presentation of characteristics of the plant. Leave an open space between the upper section and the lower section, but make sure the stem is properly aligned. Naturally, when your plant is very large, reduction is a must. When finished with a habit of the plant, concentrate on details—the section of the stem and a leaf. Take a separate sheet of tracing tissue. Do not draw on the first sketch, as later, while arranging drawings for the final sketch, you will need flexibility. If leaves on your specimen are large, you may trace the outline and the most visible veins by placing tracing tissue over the specimen and following the shape with the tip of a pencil. If leaves are small, you will have to use the grid technique and magnify the subject. For precision, use pointers or a proportional divider. When preparing a separate drawing of a leaf, include more details than you have done during the drawing of a habit. Make sure that the image of the leaf is not twisted; it must be presented perfectly flat. Make a sketch of the underside of a leaf in similar manner. The differences between both sides will probably be pronounced. The veins on the underside will be raised, while on the front they will appear as indentations. Pay special attention to the surface. The front may be very smooth, while the underside is quite hairy.

For sketching the stem you will need a grid as well as a separate sheet of tracing tissue. Make sure that measurements are transferred correctly. Always hold the measuring tool parallel to the surface of the herbarium sheet. Use stretched string as a reference while measuring the specimen. Do not forget to draw horizontal and vertical lines at right angles to each other on the tracing tissue. Those lines respectively represent the horizontally and vertically stretched strings over the specimen. During measuring, keep your head properly positioned over the specimen. As in previous projects, looking down at the subject, align one tip of the measuring tool with the string and the other with the relevant point on the specimen. Transfer the result onto the sketch. Do not forget to maintain the same magnification factor when transferring measurements. When necessary parts are correctly sketched, take a piece of illustration board larger than the herbarium sheet. Take a piece of tracing tissue also larger than the herbarium sheet and

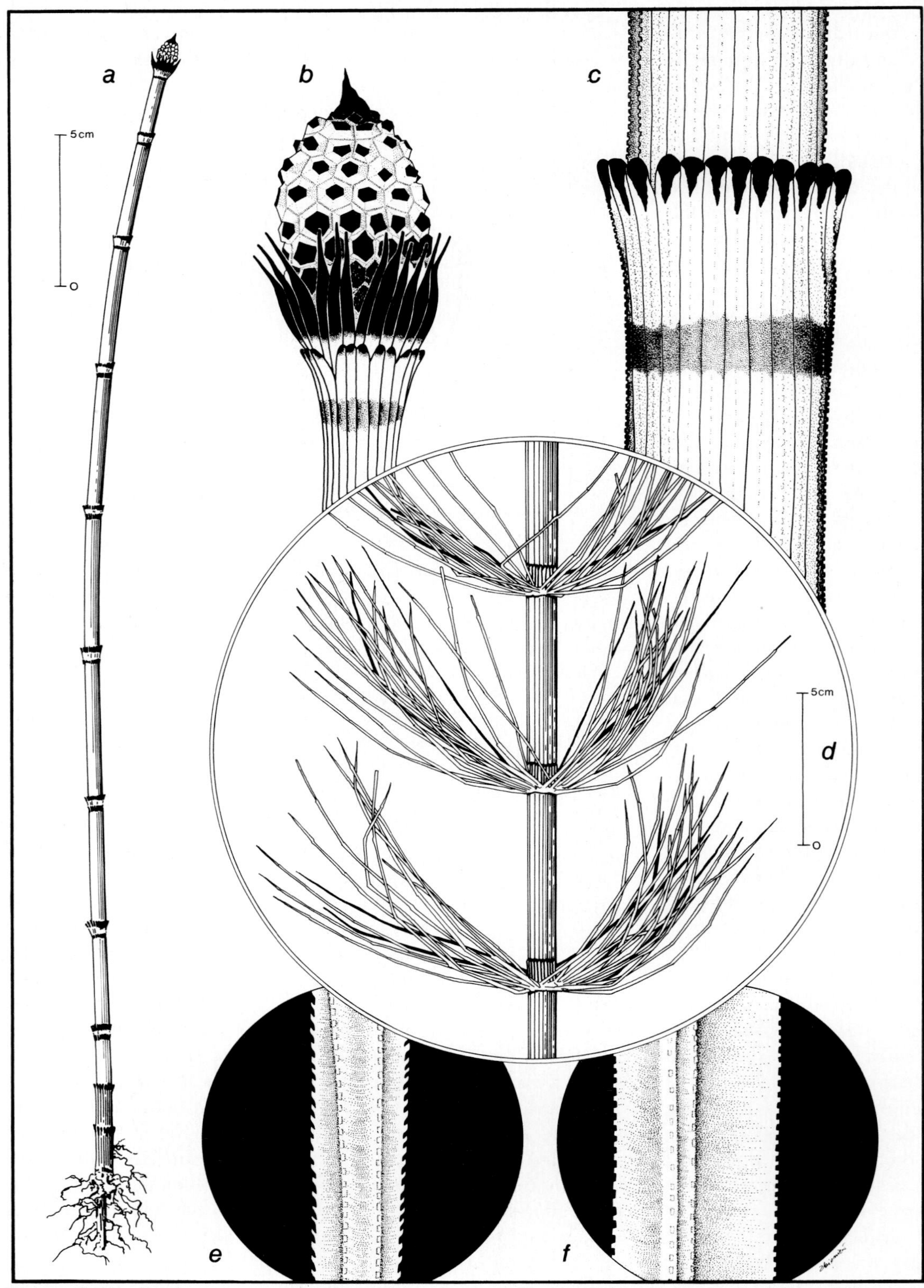

Figure 5-12. *Equisetum*. a–c, *Equisetum hyemale* var. *affine:* a, habit, ½ ×; b, strobilus, 4 ½ ×; c, stem sheath, 18 ×. d–e, *Equisetum myriochaetum:* d, portion of main stem with whorled branches, ½ ×; e, portion of branch with ridge teeth in profile, 19 ×; f, *Equisetum giganteum*, portion of branch with ridge teeth in profile, 19 ×. Pen and ink on mylar drafting film. Drawing by Zbigniew T. Jastrzębski, Field Museum of Natural History, Chicago. Published in Robert G. Stolze, "Ferns and Fern Allies of Guatemala, part III," *Fieldiana*, Botany, New Series No. 12, by Field Museum of Natural History, Chicago. Drawn in 1980.

tape it to the illustration board. Using the triangle and ruler, draw a rectangle equal in size to the herbarium sheet on the tracing tissue. Spray the sketches with fixative. When the sketches are dry, take the drawing of the habit and place it under the newly prepared tracing tissue. Place the other sketches on top of the tissue, organizing a composition that is pleasing to the eye. When finished, tape the sketch of the habit to the illustration board. Remove the sketches of the details from the top and place them under the tracing tissue in appropriate positions. Tape the sketches to the illustration board. You can draw circles or squares around the sketches for added charm or if you feel that some sketches should be separated from the background or the habit of the plant. Remember that you are free to organize the space as you like as long as the habit of the plant is presented correctly. Do not obstruct that main sketch. Do not turn the habit sideways or upside down unless this was the original direction of growth of the plant. If you like to present yourself as a professional, trace all sketches onto one sheet of tracing tissue. Spray the final sketch with fixative. Remove the first sketches from the illustration board. When finished, you are ready for rendering.

Rendering: Take a sheet of good-quality mylar drafting film produced for use with ink. Tape the drafting film over the final sketch. Prepare the technical pens. Take the metal ruler and the larger-point (00) technical pen. Draw the rectangle, making sure that the lines are straight. If you have decided to enclose some of the subjects in circles or squares, draw these first. For circles you can use a compass with the technical pen taped to one of the sides, thus avoiding additional expense. Start drawing the main subject. Use your judgment as to the size of the pen, but try to be consistent. Avoid unnecessary combination of the smaller point with the larger. When drawing the stem, make sure the line on one side is slightly thicker than on the other—this helps to create a feeling of three-dimensionality after the drawing has been reduced and printed. For better judgment use the reducing lens. Do not press the pen into the drafting film, as this clogs the tip of the pen and unnecessarily scratches the surface of the film. If you make a mistake, leave it for later correction. Do not overrender the drawing. Remember, you have drawn the enlargements for presentation of details. Remember that during reduction spaces close up, possibly blackening the image. Pay special attention to the veins. If necessary, use a French curve. When necessary, stipple to add texture or three-dimensionality. Remember that the drawing must be easy to read and easy to understand by the viewer. If you feel that it is necessary to use black background, do so but be careful not to overwhelm the whole illustration by large black areas. The plant is more important than visual effects. When stippling, use a stylus to keep the drafting film from bouncing. You will have better and rounder dots if the surface of the film is held by some kind of object before you apply the ink. While drawing, insert a white paper between the sketch and the drafting film in order to check the tones and lines. Be sure that lines are black. If lines appear gray, you have water in the pen. Clean your pen completely and dry it completely, then refill with new ink. Mistakes can be removed by scratching the surface of drafting film with the sharp tip of an X-Acto number 11 blade. Do not make a hole in the film. Be gentle and do not rush. Hold the X-Acto knife at about a 30-degree angle and scratch with the tip. For larger ink spills, use Q-tips dipped in alcohol. Make sure that all mistakes are removed—otherwise they will be printed.

Dry-Mounting: It is very hard to dry-mount mylar or acetate to the surface of illustration board. It is also unnecessary, as the illustration will be attached by a tape to the board and the whiteness of the board will be visible through the transparent material. But if you insist, you can dry-mount mylar drafting film using a combination of spray adhesive and Seal removable white dry-mounting tissue. Take your illustration and place it, image side down, on the large sheet of illustration board. Spray the back of the mylar with spray adhesive, being careful not to get the adhesive on the front of the illustration. Be careful not to overspray. The coat of adhesive should be light, and should settle as a mist. When the spray adhesive is reasonably dry, gently remove the illustration and place it on a very, very clean sheet of removable white dry-mounting tissue. Place the protective sheet of tracing paper over the front and gently burnish the drawing. Take a clean sheet of hot-press, smooth-surface illustration board and place the whole thing on its surface. Using a clean protective sheet of tracing tissue, tack the illustration at the center, using a hand-held iron with the temperature set on low. Take the metal ruler and a sharp number 11 X-Acto blade and cut through the drafting film and dry-mounting tissue simultaneously. Make sure your first cut goes through, as recutting will expose previous cuts. The edges of the mylar film and the dry-mounting tissue must meet with the line cut on illustration board. Take new, clean protective tissue and place it on the surface of the illustration. Starting at the center, applying little pressure and allowing the film and dry-mounting tissue to absorb heat, move the iron toward the edges of the drawing. It will take more time than dry-mounting paper but will work. Actually, it will be safer and better to attach mylar drafting film with one strip of Scotch number 666 tape to the surface of the illustration board. The Scotch number 666 tape is a double-stick tape with a white plastic backing, which can be used for pleasant visual effect. Do not forget to trim the tape at the edges of the film as the drafting film should be easily raised from the surface of the board.

Preparation for Production: Follow the usual procedure of securing a protective sheet of tracing tissue with the masking tape; add a black cover-stock paper for better presentation. Try to avoid the flaps of paper hanging on the back of the illustration board.

Comments: Xerox copies of the plant mounted on the herbarium sheet may be of some use during preparation of the first sketch. Sometimes, depending on the plant and the copier, some details like veins on the leaves will become more visible on Xerox copies. For small sections of the plant, use camera lucida, which will allow you to produce a faithful representation of a subject.

FIFTH PROJECT

Ichthyology

Subject: Lateral view of the fish.

Technique: Pen and ink on tracing tissue, line and stipple.

Materials:

1. Pencils—2H, H
2. Erasers—kneaded, Pink Pearl
3. White tracing paper of poor quality
4. Illustration board—hot-press, smooth-surface
5. Seal removable dry-mounting tissue
6. Technical pens, tips number 00 and 000 or equivalent
7. Pointers or proportional divider
8. French curve FC 341
9. Ethyl alcohol, vodka, or rubbing alcohol
10. Glass jar
11. Flat tray, preferably with wax on the bottom, or dissecting tray
12. Pins, thin but strong, stainless steel
13. Metal ruler
14. Metric ruler
15. Triangle
16. Tight Plexiglas cover for the tray
17. Binder clips
18. X-Acto knife with number 11 blades
19. Fixative
20. Ink, black
21. Masking tape
22. Reducing glass
23. Teasing needle
24. Paper towel
25. Stylus
26. Black cover-stock paper
27. Black photographic masking tape

Caution: Ethyl Alcohol Is Flammable. Do not smoke cigarettes near the liquid. Do not keep an open flame near the specimen submerged in the ethyl alcohol. Do not place the jar with the specimen submerged in the ethyl alcohol on the windowsill. Be very cautious. Do not set fire to your apartment. Do not drink it; it is poison. Keep your pets away from it. Maintain proper ventilation.

Preparation of Some of Materials: Make sure that 100 percent ethyl alcohol is diluted to 70 or 75 percent of the original strength, using distilled water for dilution if you plan to keep the specimen for a long period of time. It is better to buy rubbing alcohol. There are two brands of rubbing alcohol on the market, isopropyl and ethyl. Either is good, but if you have a choice, buy the ethyl rubbing alcohol.

Preparation of the Specimen: Obtain a fish from the fish market. Make sure that *all* fins and *all* scales are on the fish. Carefully clean the fish and place it in a sufficiently large glass jar, tail first. Fill the jar with ethyl alcohol and close it. Find out what kind of fish you have and write the information on a small piece of paper. Place the prepared label in the jar with the alcohol. The letters will not dissolve.

Basic Steps:

1. Positioning of the specimen
2. Preparation of the sketch
3. Tracing the sketch into a final sketch
4. Rendering
5. Dry-mounting
6. Preparation for production

Nomenclature:

dorsal fin—the first fin, counting from the head, on the top of the fish; sometimes separated into two structures

adipose fin—the second fin, counting from the head, on the top of the fish; usually small and fleshy, located next to the tail

caudal fin—the tail of the fish

pectoral fin—the fin located near the head, usually on the left and right sides of the fish

pelvic fin—the first fin, counting from the head, on the belly of the fish

anal fin—the second fin, counting from the head, on the belly of the fish

lateral line—the horizontal line, usually very visible, stretching from the gill to the tail

Figure 5-13. *Hypleurochilus aequipinnis* (FMNH 93833), male, 53.6 mm SL. Pen and ink on tracing tissue. Drawing by Zbigniew T. Jastrzębski, Field Museum of Natural History, Chicago. Published in David W. Greenfield and Robert Karl Johnson, "The Blennioid Fishes of Belize and Honduras, Central America, with Comments on their Systematics, Ecology, and Distribution," *Fieldiana*, Zoology, New Series, No. 8 by Field Museum of Natural History, Chicago. Drawn in 1979.

maxillary bone—the bone located on the side of the upper jaw

lateral—side

dorsal—top

ventral—bottom

anterior—front

posterior—back

ray—sharply ending, segmented, very visible integral part of the fin

spine—sharply ending, not segmented, very visible integral part of the fin

pores—opening(s) located on the frontal section of the head of the fish

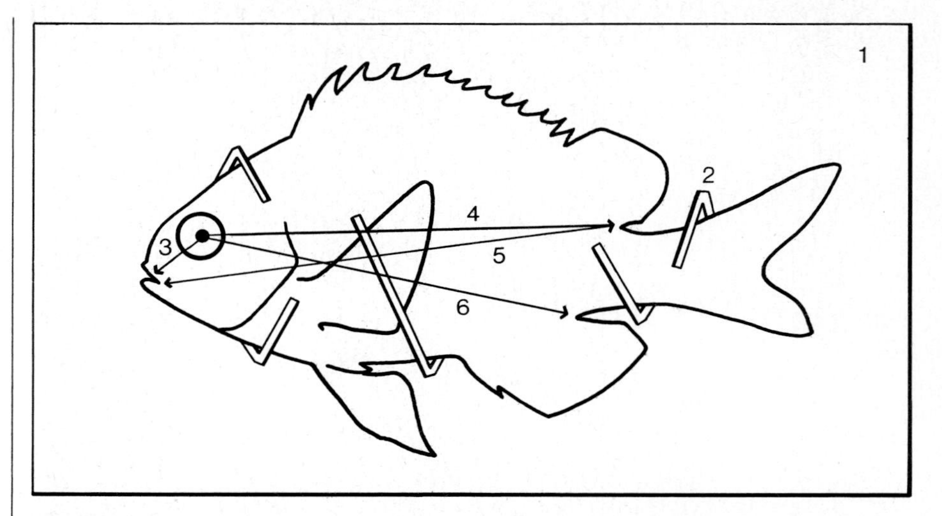

Figure 5-14. Position of the specimen in the tray, position of pins and initial measurements. *1*—Tray. *2*—Metal pins. *3*—First measurement. *4*—Second measurement. *5*—Third measurement. *6*—One of the successive measurements which must be rechecked with first three coordinates.

PROCEDURE FOR DRAWING

Observation Take the specimen out of the bottle and place it on the tray. Remember that during the whole process of drawing, the fish cannot dry out. It must be kept moist at all times as drying out causes damage to the tissue. The tissue will crack and the scales will fall out. During observation, do not submerge the specimen in the alcohol unless the subject is very small. Keep some of the alcohol nearby and sparingly use it to moisten the whole surface of the subject.

Be careful while handling the specimen. If it is to be left unattended for longer periods of time, place it back into the bottle. When doing this, remember to put the caudal fin first. In the liquid, the dorsal, pelvic, and anal fins will stretch out slightly. Such stretching may interfere with removal of the specimen from the bottle. The spines and rays may get stuck on the rim of the bottle.

When looking at the subject, keep paper and pencil handy. Write your observations. Pay attention to the coloration of the surface structures. Remember that real coloration—that is, patches of red, blue, and other colors—will fade in ethyl or regular alcohol. You are interested in values, the darks and whites. Look for definite coloration, large patches of dark value. Note this in your mind and on the paper. Closely examine all fins. Note how many spines and rays are in each fin. Write the appropriate number near the name of the fin, otherwise you will forget it. For example: pectoral fin, 2 spines, 10 rays, 1 spine; or pectoral fin, 14 rays; or anal fin, 3 spines, 18 rays, 4 spines. Remember that spines are those without segmentation; only rays are segmented. Note which of the rays (usually the first next to the spine) are singular and which are divided. Write down the information. In order to see the spines and rays better, pick up the fish and let sunlight shine through the fin. If you have a light box, use it. Place the fish on the surface of the light box, gently spread the fin, and count the spines and rays. Observe where the maxillary bone is. Note this in your memory or on the freehand quick sketch. Note where the gill is. Does your specimen have one gill opening? Make sure that you know the answer. Ask yourself questions and find answers; do not assume anything. If your specimen does not have scales, it is a catfish. In such a case, be preoccupied with coloration. Note where there are large patches of missing scales. Turn your attention to the lateral line. It is located about three quarters of the way up from the belly on the side of the fish. If you can, count all separate sections of the lateral line. In all, try to know everything possible about the subject. Remember that you are drawing *that* particular specimen, *not* an idealized representation of the family to which your specimen belongs.

Positioning of the Specimen: There are many variables when drawing fish. Positioning is only one of them. A lot depends on the size of the fish, the tools you have, and whether you are allergic to ethyl alcohol. If you happen to be allergic, then ethyl alcohol has to be ruled out as the medium with which the specimen is moistened during drawing. You must substitute distilled water, wetting the subject with it to keep the tissue from drying out. You will have to remove the specimen from the tray often and place it back into the jar or bottle full of ethyl alcohol. Such disturbance of the position of the fish does not allow you to pin the creature more permanently, interfering with measuring as well as with observation. Yes, you could wear surgical gloves, but again, not for a long period of time. Your hands will sweat and will get red. If the specimen is very small, approximately three to five centimeters, you cannot pin the fish without damaging fins or tissue. In such a situation the quantity of ethyl alcohol is minimal and you need a microscope to see details.

Let's assume you have received a reasonably large specimen, a fish approximately 18 to 25 centimeters long, from the scientist to draw. Let's also assume that you are not set up for keeping a large quantity of ethyl alcohol in the open tray. Remember that the neighbors have sensitive noses. The smell of ethyl alcohol will penetrate the whole building. First prepare the dissecting tray. Take the specimen out of the storage jar and place it on the wax that covers the dissecting tray. As you will be drawing the left side of the fish, make sure that your subject is lying on its right side. The left side is the one on the left if you are looking in the same direction the fish is looking. It is on your right side if you are looking directly at the fish's nose. Why do we draw the left side? Because in the scientific community it is the standard side for all fishes to be drawn from.

Take small pliers and thin, strong, stainless steel pins. With the pliers you will bend separate pins according to the place of insertion. You must remember that you cannot pin the fish through the tissue; you cannot push a pin through the fish's body; you cannot damage the specimen. Be very careful. Take a couple of pins and with the pliers bend the top section, forming a 90-degree angle. Hold the fish with one hand and insert one pin right next to the body of the fish into the wax in the vicinity of the back portion of the head on the dorsal side of the subject. Make sure that the bent upper section of the pin holds down the fish. Still holding the specimen, take another already-prepared pin and insert it into the wax right next to the body of the fish in the vicinity of the opening of the gill on the ventral side of the subject. Again make sure that the bent upper section of the pin holds down the body of the fish. Trying not to disturb the position of the subject, release your hold and take another pin. Find where the dorsal fin ends. Hold the fish by pressing and insert the pin right next to the body of the fish. Make sure that the pin is located exactly in the corner of the back portion of the dorsal fin and the main portion of the body of the fish. Make sure that the bent section of the pin is holding the body of the fish. Do not release your hold. Be careful if the subject jumps away from its present position. The third pin may damage the tissue and the dorsal fin. Pressing the fish firmly but not too strongly, take a pin and insert it right into the corner formed by the back portion of the pelvic fin and the main body of the fish. Four pins will not hold down the specimen. Repeat the process of pinning until you are sure that the subject is stationary. Do not overdo it. Remember that all pins will interfere with the measuring process. When the main body is immobilized, turn your attention to the fins. Start with the dorsal fin. Gently but firmly stretch the first few spines toward the left. Do not apply a lot of pressure. If the whole subject jumps away from the pinned position, the pins will rip the tissue. If the dorsal fin is stubborn and cannot be stretched at once, take your time. When the first two spines are slightly moved away from the rest of spines and rays, insert the pin into the wax right next to the inside edge of the tip of the spine. Make sure that the tissue between spines is not damaged. Try to stretch a few other spines and rays. Make sure that the bent tips of the pins will hold down that part of the subject. When finished, do the same with the other fins with the exception of the pectoral. Be extra careful with the caudal, as the first long spines are usually directly grown to the rays. Rays are very soft and can be damaged very easily; they can be broken by the slightest pressure. The tissue between rays is also very delicate. If, in your opinion, it is too dangerous to stretch the fins, do not stretch them. You must be the judge, as you are handling the specimen.

When everything is finished, pour some alcohol over the body of the fish. Make sure that all parts are wet. You do not have to submerge the whole specimen in the alcohol. For storage, use a towel dipped in the solution and cover the fish with it. Do not, if you can avoid it, move the specimen from its pinned position. You will have a harder time drawing and measuring if the position is changed. It is impossible to pin the subject for a second time exactly as you have done the first time. When taking breaks, place a wet towel on the surface. For longer periods of time, take a sheet of Plexiglas about the same size as the dissecting tray and, using binder clips, secure the cover. Make sure that the towel covering the specimen is very wet and that you have enough binder clips to secure the Plexiglas lid tightly. You must stop evaporation of ethyl alcohol, otherwise tissue and scales will become brittle and will be damaged. In an ideal laboratory situation you should submerge the whole fish in ethyl alcohol. The subject should be covered

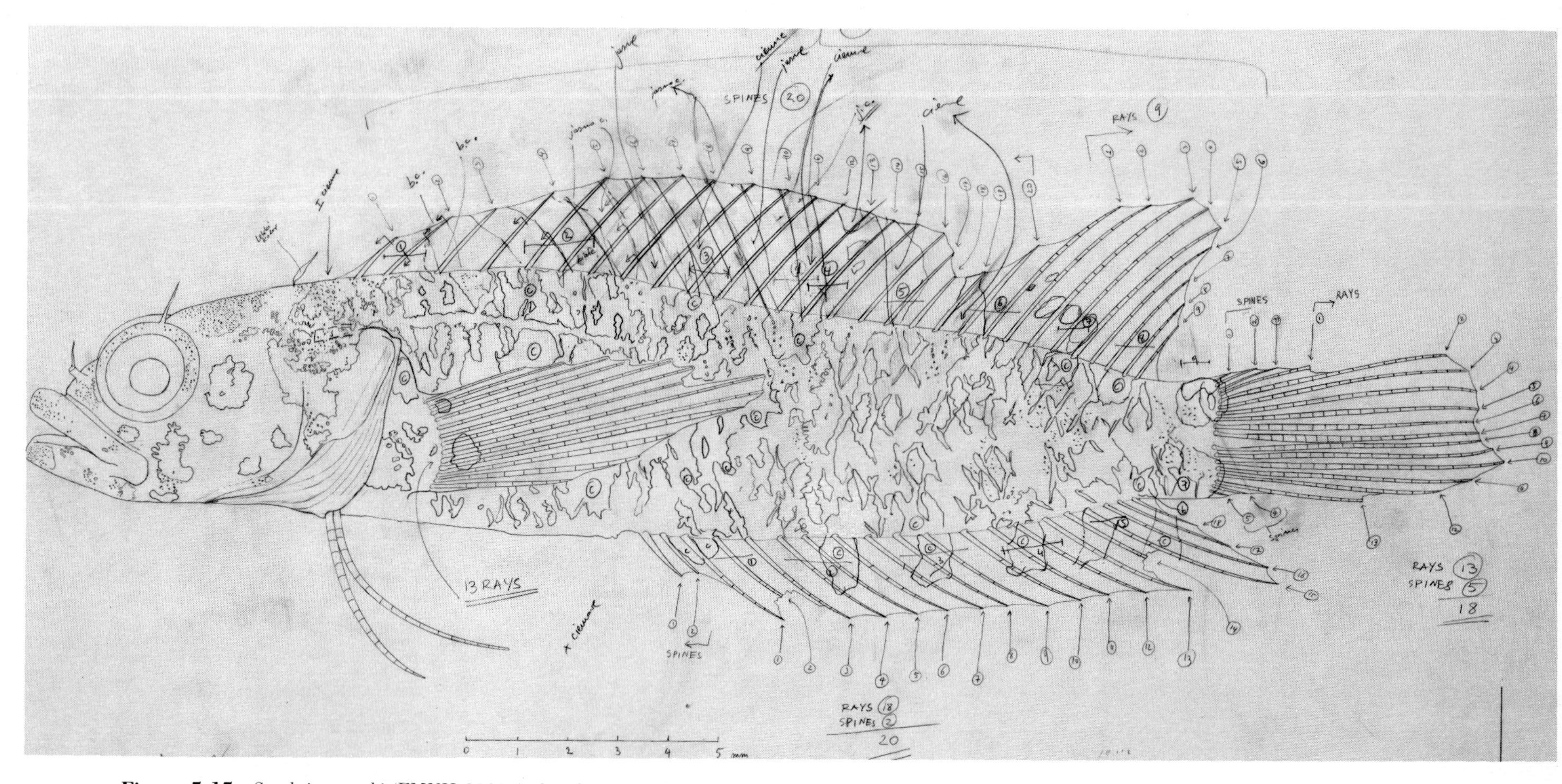

Figure 5-15. *Starksia starcki* (FMNH 93836), female, 23.1 mm SL. Pencil sketch (final sketch) on tracing tissue for figure. Drawing by Zbigniew T. Jastrzębski, Field Museum of Natural History, Chicago. Drawn in 1979.

completely by the solution, definitely assuring preservation.

Drawing the Sketch: When drawing a lateral view of a fish, you must be prepared for a long project. As we are not using tracing devices to obtain the proportions, outline, and details, but will measure everything by hand, the project will be even longer. If this is your first drawing of such a subject, be prepared to spend approximately 45 to 55 working hours on the sketch alone. Naturally, the amount of time you spend on the sketch will vary in proportion to the complexity of the specimen you are drawing. During the process of sketching you will try to obtain information pertaining to the proportions of the subject, the coloration pattern if such exists on your specimen, the correct shape, and the placement of all the fins. You will do your best to make sure that each fin has the appropriate number of spines and rays, that the length of spines and rays is correct. You will find all breathing pores and will count scales on the major portion of the body. You will also depict and count all sections of the lateral line. If scales are missing, do not draw them. Scarred tissue will be marked as it is on the specimen, and visible internal structures will be omitted. There are two exceptions to the rule. If your subject has a great many very small scales, I strongly advise you not to get involved in counting each separate scale. Without a microscope you may be helpless. And, of course, catfish do not have scales, and as you cannot improvise you cannot draw them.

First, establish an enlargement factor. That depends on the size of the specimen, the quantity of details, and the final size of reproduction of the finished illustration. Take a sheet of illustration board and a sheet of tracing tissue. You do not have to have tracing tissue as large as the enlargement of the subject will be. The first sketch will be retraced anyway into a final, clean sketch, so the small pieces will not interfere during the rendering. When you have decided on the enlargement, make an arbitrary mark in the form of a cross on the middle of the left side of the tracing tissue. Let this mark represent the middle of the pupil of the fish's eyes. Take your measuring tool and keeping the tool parallel to the surface of the table, look down directly on the specimen and measure the distance between the center of the pupil and the farthest tip of the fish's nose. If you are using a proportional divider, you will be much more sure of the precision of the enlarged measurement. When using pointers, be extra careful. Transfer the measurement to the paper and draw a straight line. Measure that line with a metal ruler. Mark the measurement on the surface of the ruler. Now take the ruler and align one mark with the center of the previously drawn cross, then mark the other point of the measurement on the tracing tissue. Move the ruler in a semicircle, repeating the process. You will have a semicircle of marks on the tracing tissue. Naturally, if you have a large compass, you will escape that monotonous process. Do you have a beam compass attached to a beam bar? Maybe you can make such a tool by attaching a long, strong pin and a pencil to a metal ruler. Having one definite point, the center of the pupil, and establishing a distance to the other, the tip of the fish's nose, still leaves you in the dark. Measure the distance between the center of the pupil and the back part of the dorsal fin. Repeat the process of enlarging and transferring the initial measurement. Measure the distance between the back part of the dorsal fin and the tip of the nose. Enlarge and transfer the measurement. You know that the point on the nose is located someplace on the semicircle to the left of the pupil, and the point designating the back portion of the dorsal fin is located some place on the semicircle to the right of the pupil. Having measured and enlarged that distance, match it with both semicircles. Mark it. Now remeasure both distances to make sure you are right. Remember the enlarging factor; you must maintain the same enlarging factor throughout the whole process of sketching. When those three points are secured and known to you, measure the distance from the center of the pupil to the ventral base of the pectoral fin. Be sure to concentrate on precision. Keep the measuring tool parallel to the surface of the table. Do not forget to moisten the specimen at regular intervals. Recheck the fourth measurement against the previously established points.

By now you have a basic knowledge of measuring. Establish the outline of the body of the fish through this method. Do not forget the gill opening. Do you really know the shape of the maxillary bone? Did you check and note breathing pores? Look for them at the area above the upper jaw and around or near the eye. If teeth are pronounced, especially if they are present on the lower edge of the upper jaw, stretching from the front of the jaw all the way to the back of the maxillary bone, include them in your sketch. Do *not* include the outlines of the fins on the sketch. All fins will be drawn separately. But during the sketching of the basic shape of the subject, be sure that the anterior and posterior base of each fin is clearly established and indicated. Where is the beginning of the caudal fin? Look closely and find the places of insertion of rays into the main body. This will be the beginning of the tail. You may need to use a teasing needle. With a sharp steel point, touch each spine and each ray and quickly you will be able to differentiate between hard substance and soft tissue. All spines and rays are of cartilaginous matter. Remember to be careful and not to damage the specimen. Make sure that all of the points of growth of the spines and rays are indicated on the basic outline. Number them. Make sure that the distance between each of the rays and spines is measured. Clearly mark spaces between spines and rays. Later it will be easy to misalign those structures. Make sure that you exclude the right pelvic fin from the sketch. You must remember that you are drawing a left lateral view, and everything on the opposite side is of no importance.

When you are positive that the main body of the subject is correctly drawn, turn your attention to the dorsal fin. Take a separate clean sheet of tracing tissue and tape it to the illustration board. Prepare the tools: teasing needle, pencil, and pointers or proportional divider. Remember the enlarging factor. By now you know how many spines and rays are in the dorsal fin, right? Count them again. Pay attention to the last ray of the posterior section of the fin. The last ray may be a double structure—that is, it may grow out as a singular cartilaginous thin structure, but right after emerging from the body divides itself into two separate, segmented protuberances. It is important to know this and to draw this. The same will apply to the pelvic or anal fin. The last anterior ray may be a double structure. When you know which protuberance is a spine or a ray and which is a singular or double structure, take a measuring tool and measure the length of the first spine of the dorsal fin. Measure from the base to the tip. Enlarge the measurement accordingly, and transfer the measurement to the tracing tissue in the form of a straight line. Clearly mark the beginning and the end of that spine. Now you have a straight line depicting the length of the first structure of the dorsal fin. Measure the base of that first spine. Note this on the drawn line in the appropriate place. Make sure that you have labeled what fin you are drawing. Make sure that the appropriate number designating the order of the structures, spines, and rays is noted by the drawn straight line. Write on the tracing tissue "dorsal fin," and next to the drawn straight line a number 1. Write an explanation next to the number 1 whether the fin's structure is a spine or a ray. Measure the second structure and again draw a straight line on the tracing tissue. Mark the measurement in writing. If you do not write you will get lost, you will not know what is where and in what order it is to be presented. Continue the process of measuring and enlarging. When you get to the first ray, make sure that you note on the tracing tissue if that structure has a singular tip or is divided into two or more thin structures. Usually the first ray has a singular tip, the second is double, and the rest are divided into many. The singular and the double tips should be clearly depicted on the final drawing (at present just note them on your first sketch). The endings of rays divided into many structures do not have to be depicted with exactness, as long as it is apparent that there are many thin, segmented structures. Continue until you have all spines and rays measured and represented on your first sketch as straight lines. Make sure that you note the horizontal distance between spines and rays. Some may be growing right next to each other; others may be separated considerably. Notice the density of segmentation on the rays. Usually you do not have to count the segmentation. It is important to know at which point segmentation starts on the ray. Is it next to the body of the fish? Or perhaps it starts midway between the point of growth and the tip of the ray? Note the density, as the density will have to be depicted correctly.

Repeat the process of measuring, enlarging, and transferring for all the fins. Make sure that each first sketch of the separate fin is on a separate piece of tracing tissue. This will make your life easier when the final sketch of the separate fin is drawn. When drawing the first sketch of the caudal fin, pay special attention to the dorsal and ventral sides of that fin. Right where the fin begins, still appearing as a part of the main body of the fish, you may find very small spines. Usually those structures are not easily visible. Take a teasing needle and carefully turn the tip from right to left; check if spines are present. The tip of the pin will encounter a hard, very small, cartilaginous structure. This is a spine. Do not mistake a scale for a spine. Make sure that you count the spines and rays on the caudal fin. The caudal fin is not symmetrical. Find the dividing space surrounded by the two shortest rays. Dorsally you will have more cartilaginous structures.

When all first sketches of the fins are ready, prepare yourself for the process of drawing the second series of sketches. Take a piece of tracing tissue and place it over the main sketch of the fish. Trace the area designating one of the fins. Let's start with the dorsal fin. On the tracing include a portion of the outline of the fish. Trace the beginning and the end of the fin. Trace the places designating the insertion of the separate spines and rays. Number them. Be sure that you will not confuse the areas designating the growth of spines and rays with the separating spaces. It is a good idea to mark the width of the spines and rays at the place of their emergence with a horizontal line. Do not forget to number each structure. Remove the newly traced image and tape it to the illustration board, taping one side of the tracing tissue. You will have to lift that sketch from the illustration board. Take the first sketch of the dorsal fin consisting of the straight lines. Place it under the newly traced image. Align the *last* ray of the dorsal fin with the designated space on the newly traced image. Make sure that you leave enough space between the outline designating the main body of the fish and that ray. This is of great importance, as the last ray cannot interfere with the outline of the fish. Yes, this means that the whole fin will appear as if it is stretched toward the head of the fish; yes, it is artificial. But it is very necessary, as it allows visible presentation of all the structures. At the same time, do not overstretch—remember, you cannot draw the fin tilted forward. When the straight line representing the last ray of the dorsal fin is aligned with its designated space, take a French curve and use it to draw a semicurved line depicting the structure. Remember not to overdo it. Look at the specimen and compare. Remember that this ray is markedly divided; be sure this is visible. Do not exceed the truthful length of the ray. Do the same with the rest of the rays and spines, making sure that *all* are parallel to each other. Remember the spaces between the structures. The resulting sketch will give you the shape and the structure of the fin. Mark the segmentations on the rays. Be sure to notice from where the tissue found between spines and rays grows. Does it grow directly from the tips of the spines and rays? Make sure that you note the curvature of the upper edge of that tissue. Draw the other fins, with the exception of the caudal, in the same way. Remember to start with the last structure, leaving a reasonable space between the body of the fish and that structure. Remember to consult the specimen, and to moisten the specimen at regular intervals. In the laboratory you have the scientist to consult with; at home it is only yourself. Be extra observant.

The second sketch of the caudal fin you will start drawing from the middle. Remember where the dividing space is? Did you mark that clearly? Start with the ray closest to the dividing space and proceed dorsally. As in previous second sketches, align the straight line representing the length of that ray with the appropriate place designating its growth. Take the French curve and align the edge of the tool in such a way that the tip of the ray is above the dividing space and the drawn line curves away from you. Remember about the width of the tissue between the first middle ray on the ventral side and the one you are drawing. Repeat the process of aligning and drawing until you get to the first ray before the series of spines on the dorsal side of the fin. Check the width of the tissue found between the spines. Here some of the spines may actually be attached to the last ray. Making sure that you are correct, continue until the rest of the spines are appropriately drawn. Probably you will have to expand or stretch that portion of the caudal fin. Make sure that you do not expand the area from which the spines and rays grow, but if your specimen has a lot (35 to 50) of spines and rays on the caudal fin and if the anterior area of the main portion of the body of the specimen is very thin, then for the purpose of readability the back section is expanded and all rays and spines are slightly separated. Draw the ventral part of the caudal fin the same way as the dorsal. Make sure that the rays and spines will curve toward you, in the opposite direction to the curvature of rays and spines of the dorsal part of the caudal fin. Make sure that you have counted the spines and rays correctly. Make sure that segmentations are marked on the rays.

When all fins have been clearly reworked in the second series of sketches, trim the tracing tissue near the drawn fins. Take the sketches and align each of them in its appropriate place on the basic, main sketch of the whole fish. Tape each sketch securely to the main sketch. Do not obstruct your previous drawing with the tape. Try to avoid application of the tape within the main body of the fish. Try to apply the tape to the sketches securely but away from previously drawn images. When finished, turn your attention to the scales. Examine the surface of the fish. See if and where larger patches of the scales are missing. See how small the scales are. Compare their size with the relative size of your enlarged image of the fish. Are the scales on your specimen too small to be meaningfully visible on your drawing? Remember that the finished product, the rendering of that specimen, is designated for reproduction. Reproduction means that your drawing will be reduced and then printed. After enlargement, are the scales of your subject about two or three millimeters in width or length? Make a decision. Without a drawing tube (camera lucida) you will spend lots of time measuring the position and outline of the scales. It is necessary to know this if the drawing is to be used to illustrate somebody's research. At the same time, if the scales are much too small, after the enlargement, to appear as separate countable images when the finished drawing is reduced, do not count the scales. Concentrate on the general pattern created by the rows of scales. In such a case you must show the relative size of groups of scales. When your specimen has large-enough scales, I strongly advise you to count all of those within the main portion of the body of the fish. The only exceptions are scales on the fins. The scales on the fins should represent the pattern visible on appropriate fins. Take a new sheet of tracing paper and affix the sheet to the illustration board over the main sketch. Tape this sheet on one edge only, making an overlay. Prepare your measuring tool and teasing needle. Using the teasing needle find the outlines of the larger scales next to the opening of the gill. Establish the location and the outline of that scale by measuring the distance between the scale and already known points. Be sure that you are accurate. Repeat the process until all scales are drawn on the overlying sheet of tracing tissue. Count all of them. Make sure you have located the lateral line. Make sure that the darker markings, the attachment points of scales, of the lateral line are counted. This is very important. The quantity of those markings is essential information for classification of the fish. When finished, look at the coloration. Make sure the patches of light and dark are correctly drawn. Make notes on the sketch. Spray both sketches with fixative. Remove the sketches from the illustration board and place both layers on the light box. Do not tape the sketches to the light box; tape them to each other. It is easier to turn the sketches around while tracing than to turn yourself. Be as comfortable as possible. Your drawings will be better and your hand will not shake from tiredness. If you do not have a light box, use a window. Then you have to tape the sketches and a new, clean sheet of tracing tissue to the window. Do not try to hold the whole thing with one hand and draw with the other. It will not work. Trace the whole image: the outline of the fish, all the fins, all scales and coloration markings. Write all pertinent information on the new sketch. Do *not* trace the scales found under the pectoral fin. Remember that this fin with all its components must be very visible. When you are finished tracing, spray the new sketch with fixative. This is a final sketch, which is presented to the scientist for corrections. It is to your benefit to have a clean sketch for a professional presentation. Do not forget to moisten the specimen during the process of drawing, and do not remove the specimen from the original pinned position. When the final sketch is correct, take a new sheet of illustration board and tape this sketch to the surface. Be careful not to tear the tracing tissue. Take a new, clean sheet of tracing tissue and place it over the final sketch. Tape it to the board. Now you are ready for rendering.

Rendering: For the time being, keep the specimen moistened and covered tightly. Clean the table to make room for rendering. Have enough free space to turn the illustration board around. Prepare the technical pens, making sure they have been cleaned and the flow of ink is not interrupted. Make sure the ink is not diluted with water—drawn lines must be black. Prepare a sheet of white opaque paper. While drawing you will have to insert this sheet between the rendering and the sketch in order to check the density and the quality of dots and lines. Have the reducing lens handy, and do not forget about the stylus, which you will need to keep the tracing tissue down during application of dots and lines.

First, take the technical pen with the larger point—I suggest 00 or equivalent—and draw the outermost outline of the subject. Do not draw fins. If parts of scales are the outline, treat them as the outline. Use the stylus to keep the tracing tissue down. Move your hand steadily. Do not try to draw long lines in a flamboyant fashion. Divide the long spaces into shorter ones. Keep your elbows on the table—you will have better support and your hands will not get tired quickly. Try to relax your muscles. Draw reasonably short strokes, but not too short; join them carefully. Remember that it is difficult to correct the drawing by scratching out imperfections. It can be done, but very, very carefully. The tissue is very thin, and it is easy to make a hole in it. Draw the outside edge of the eye socket with a solid line. Draw the opening(s) of the gill with a solid line. If the maxillary bone is definitely separated from the tissue and you can lift this structure with a teasing needle, draw the base in solid line. Do not shake your pen above the drawing; accidents happen. When shaking the pen, wrap the tip in Kleenex or a paper towel, then shake. You do not want a blob of ink on the drawing.

Take the French curve and start with the dorsal fin. You will have to draw all spines and all rays with solid lines unless they are covered with very heavy, nontransparent tissue. I assume that the spines and rays are not covered with very heavy tissue. Start with the first line to the left. Align the French curve carefully. Press the curve with one hand and, keeping the tip of the pen away from the edge of the French curve, draw a solid line starting from left to right if you are a right-handed person. At the junction of the spine with the body you will have to use a freehand line. Do not remove the French curve from its position. Wait. After the ink has dried, move the curve slightly and draw a solid line for a second time, overlapping the first. You will have a heavier line than the outline. You will do this to all left edges of the spines and rays on the adipose, dorsal, pelvic, and anal fins. Do not tape pennies to the bottom of the French curve. It is not

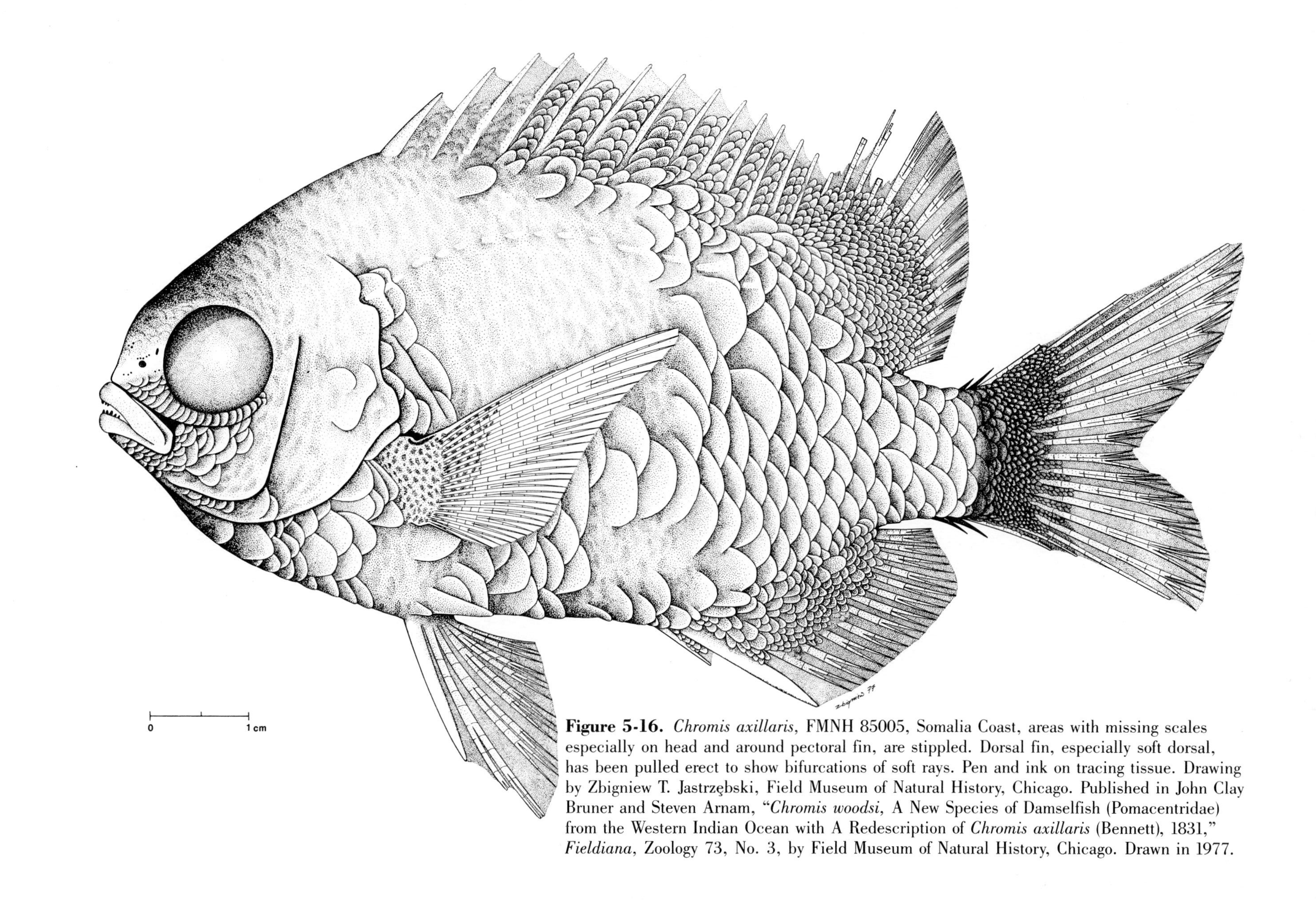

Figure 5-16. *Chromis axillaris*, FMNH 85005, Somalia Coast, areas with missing scales especially on head and around pectoral fin, are stippled. Dorsal fin, especially soft dorsal, has been pulled erect to show bifurcations of soft rays. Pen and ink on tracing tissue. Drawing by Zbigniew T. Jastrzębski, Field Museum of Natural History, Chicago. Published in John Clay Bruner and Steven Arnam, "*Chromis woodsi*, A New Species of Damselfish (Pomacentridae) from the Western Indian Ocean with A Redescription of *Chromis axillaris* (Bennett), 1831," *Fieldiana*, Zoology 73, No. 3, by Field Museum of Natural History, Chicago. Drawn in 1977.

necessary, and it is more cumbersome. The technical pens do not have long-enough tips of the points to allow for the raised surface of the French curve. I advise against large sizes of pen tips, such as number 2 or larger. Usually the resulting line is not continuously straight on both sides of the line unless the pen is kept in a vertical position. Such a position interferes with the type of drawing you are doing. The right side of the first dorsal spine is to be drawn with a smaller tip than the one used for drawing the left side. Use tip number 000 or equivalent. Repeat that process of drawing for all spines, but make sure that successive lines depicting the spines and rays do not end abruptly at the places where those structures join the main body of the fish. Remember that the scales are overlapping spines and rays. Yes, the scales are transparent, and the spines and rays are visible through the surface of the scales, but do *not* draw the structures under the scales. The scales must be visible as entities in themselves. When drawing rays, make sure that the tips of the rays are depicted properly. See the working sketches and the specimen for information. Do not forget about segmentation. On the surface of the rays draw short but curved lines. Make sure that the curvature of those lines is uniform. The thicker left line and the thinner right one will help the spines and rays appear more three-dimensional after the reduction of the illustration.

When you have finished drawing outlines of the dorsal, adipose, pelvic, and anal fins, turn your attention to the pectoral. Draw thicker lines starting from the dorsal side of the pectoral fin. When you reach the last ray on the pectoral fin, be sure that both sides of that particular ray are drawn with a heavier line. The whole fin must be visibly separated from the surrounding body of the fish. The adipose fin, if your specimen has one, will be drawn with the same weight of line as the basic outline of the fish. Be sure that your lines are clean. Be sure that all spines and rays are parallel to each other. When drawing the caudal fin, start from the middle. Remember the dividing space? Start from the last ray of the upper section of the fin. The inside line will be the thinner one, the outside thicker. When you reach the last spine located on the dorsal edge of the caudal fin, make a decision. If those spines are very small, you will have to fill in the space designated by the outline of the spines with black ink. Otherwise there is a possibility that after the reduction of your drawing such small structures will not be clearly visible. Yes, those structures may be very transparent on the specimen, but at the same time they must be very visible. After the illustration is reproduced, they should be visible enough to be easily counted by those who read the illustrated research paper. Draw the edge of the tissue connecting the rays and spines with a thinner line. Turn your attention to the scales. Draw all scales without the French curve. Use the stylus to hold the tracing tissue. As you draw, turn the drawing around for your convenience. Use one thickness of line. Use the smaller tip of the pen. Do not forget about teeth. If the teeth are very small you will have to fill the spaces with black to make them visible after the reduction. When all lines are drawn, begin applying tone. Start with the head of the fish. Tone will be represented by separate dots. Do not rush; be very methodical. When drawing, think about drawing. It is better to work in one area and finish it rather than to jump around the drawing with your pen. If you do this, you are not in control of the situation. The reason for not finishing one area and going to another is your indecision. You are avoiding making a decision. Face the facts; do not avoid them. Start with the eye of the fish. Carefully, holding the tracing tissue with the stylus, apply dots to the surface. Remember about coloration of the eye. Remember that the center of the eye is white. Make the eye appear as three-dimensional as possible. Do not overrender. The density of the application depends on the coloration of the specimen. If your specimen has distinct coloration—that is, a lot of contrast—make sure your drawing will represent the truth, but at the same time do not overrender.

Do not forget that after reduction, the very dark areas will become darker, maybe too dark. Remember that the edge of each scale must be visible. Always leave the space near the inside edge of the scale white. Start applying tone to the scale from the outside edge of the overlapping scale. Make sure that your dots are dots, not small lines. Do not be sloppy. Your sloppiness will be more visible to the viewer than to yourself. Use the reducing lens often to appraise the darkness of the drawn tone. Take a piece of white opaque paper and insert the paper between the sketch and the rendering. Stand up, take the reducing lens in your hand, stretch your arm, and then look at the drawing. The farther you are from the drawn image, the smaller it will appear. The printed size of your illustration may range from three to six inches. Make your drawing appear approximately that size. See what happens with the values and contrasts. Do not try to render your subject very three-dimensionally. Remember that dark value, needed for three-dimensionality, may be mistaken by the scientist for coloration of the subject. When your fish is very dark, almost black, adjust the tone of the whole subject in relation to the lightest area. You do not have to have a pitch-black drawing to convey the darkness of the subject. Make sure that you leave the edges of the scales white. Scales must be countable after reduction and printing.

Do not forget about breathing pores. Fill the opening in black. Do not mistake the breathing pores with the nasal opening. Nasal openings are much larger; some will have a flap of tissue growing from the anterior side of the opening. Leave a light to white tone around breathing pores. Make sure that flaps of tissue near the nasal opening are left white for better contrast with the background. Do not forget the lateral line. Dark or light segments forming the line must be very visible. Whenever the scales are missing do not draw them; draw the scarred tissue. The scarred tissue will have a pattern; try to maintain that pattern. Such scarred areas must be visibly different from the sections of the body covered by scales. During drawing check with your notes, the working sketches, and the specimen. Do not let your imagination run free. Do not become enchanted by the forms and the drawing process. Remember that you are drawing a very particular subject and that you have to draw that subject as it is in its natural form. When finished, carefully remove the specimen from its pinned position and place it, tail first, back into the container.

Dry-Mounting: Be very careful when dry-mounting. You have invested a lot of time in this drawing—do not mess it up at the last minute. Think twice before you do anything. Take a very clean sheet of hot-press illustration board and a very clean sheet of Seal removable white dry-mounting tissue. You may have to buy a whole roll. If it is too expensive, then try to splice two sheets together. The edges of two separate sheets will be slightly noticeable. Remember to cut the drawing after you have tacked it to the illustration board. Remember to separate the edges of tracing tissue and the dry-mounting tissue. Remember to pick up all crumbs and lint from the illustration board and dry-mounting tissue with masking tape. Remember to keep your hands clean, as fingerprints will show. Remember to place protective tissue over the drawing when applying heat. Remember to set the iron properly. Too much heat will create bubbles. It is better to have a cooler iron than one that is too hot. Take your time. Follow the dry-mounting procedure described in the First Project.

Preparation for Production: Carefully tape the protective tissue to the top of the illustration board. Do the same with black cover paper using black photographic masking tape. Do not forget to crease the black cover-stock paper. Remember to trim the edges of the tracing tissue and black paper. Add those final touches for professional presentation. Do not forget about the scale, a straight line designating the enlargement factor.

Comments: Why not take a photograph of a fish and escape the pains of drawing? A photograph will not show what is desired by the scientist. It will present a tangled, hard-to-decipher image, an image where everything is noted indiscriminately, torn tissues as well as reflections of light. It will be impossible to distinguish separate scales. The spines and rays will not be presented clearly. A multitude of selected information will not be available to the researcher. For that reason, careful analytical drawings are needed. In some cases, especially when working with deep-sea fishes, a reconstruction of fins may be needed. But in most cases, drawing does present the specimen faithfully, depicting all missing rays, spines, and scales. A drawing offers better readability than a photograph.

While drawing small subjects, a microscope is necessary. When fish is about four to five centimeters long and the drawing must be produced quickly and very accurately, a camera lucida attached to the microscope is used to assure the accuracy and speed up the production of the sketch. Nevertheless, the rendering will still be drawn the same way as you have done it.

Some scientific illustrators draw the eye of the fish differently than I have described. The eye may be rendered black with a reflection of light in the corner. I strongly feel that such rendering may misinform the reader about the truthful coloration of the eye.

The tracing tissue used for pen-and-ink rendering should be of poor quality, because such tissue will accept ink better and more uniformly than good-quality vellum. Now and then it happens that because of a slightly greasy surface, a section of good-quality vellum tracing tissue will repel ink. I do not want that to happen to anyone, especially when most of the rendering is finished. This will create serious difficulties which can be overcome only by cutting out a "bad" section of the tissue and splicing in a new sheet. Such treatment of the tissue will be visible to the viewer, although not to the camera. Reproduction will not suffer, but the presentation of the drawing will be affected. Why using tracing tissue at all? Just to save time and money. Not every scientific illustrator has unlimited resources and unlimited time. Deadlines have to be met. By using tracing tissue for a rendering, you save time, as you do not have to retrace the final sketch onto any surface. Yes, you can use mylar drafting film instead of tracing tissue. But again, I should remind you that this costs more than tracing tissue. If you are using drafting film, try to obtain the best quality. It will affect your drawing. Also watch out for the thicknesses of the lines and dots. Ink will spread out slightly on any drafting film.

SIXTH PROJECT

Amphibians and Reptiles

Subject: Dorsal view of a grass frog, *Rama pipiens*, stressing coloration patterns.

Technique: Pen and ink, pencil, graphite dust on mylar drafting film.

Light Source: Upper left side.

Materials:

1. Pencils—2H, H, B, 2B
2. Erasers—kneaded, Stenotrace series 1400 and 1500
3. Tracing tissue
4. Stomps
5. Technical pen with tip number 00 or equivalent
6. Ink, black
7. Magic Art See and Draw Copier (toy camera lucida)
8. Pointers or proportional divider
9. Glass jar
10. Ethyl alcohol, vodka, or rubbing alcohol
11. Flat tray, preferably with wax on the bottom, or dissecting tray
12. Long pins, thin but strong, stainless steel
13. Metal ruler
14. Metric ruler
15. Tight Plexiglas cover for the tray
16. Binder clips
17. X-Acto knife with number 11 blades
18. Ground graphite
19. Fixative
20. Masking tape
21. Reducing glass
22. Paper towel
23. Stylus
24. Two small, separate table lamps
25. Black cover-stock paper
26. Black photographic masking tape

Caution: Ethyl Alcohol Is Flammable. Do not smoke cigarettes near the liquid. Do not keep an open flame near the specimen submerged in ethyl alcohol. Do not place the jar with ethyl alcohol on the windowsill. Be very cautious. Do not set fire to your apartment. Do not drink it; it is poison. Keep your pets away from it. Maintain proper ventilation.

Preparation of Some of Materials:

1. Make sure that 100 percent ethyl alcohol is diluted to 70 or 75 percent of the original strength, using distilled water for dilution if you plan to keep the specimen for a long period of time. It is better to buy rubbing alcohol. There are two brands of rubbing alcohol on the market, isopropyl and ethyl. Either is good, but if you have a choice, buy the ethyl rubbing alcohol.

2. Graphite from 2B or 4B pencil ground into dust, making sure that the powder does not contain sharp particles.

3. The Stenotrace eraser should be sharpened. Cut the tips of erasers at about a 45-degree angle with a sharp X-Acto blade. Try to cut with one stroke, making sure that the edges of erasers are without burrs, and that the surface of the cut is even and smooth.

Preparation of the Specimen: Buy the specimen from any company selling educational materials. It will be prepared for you.

Basic Steps:

1. Positioning of the specimen
2. Preparation of the sketch
3. Tracing the sketch into a final sketch
4. Rendering
5. Preparation for production

Nomenclature:

Dorsal—top

Ventral—bottom

Lateral—side

Anterior—front

Posterior—back

Tympanum—visible structure located slightly behind the eye on the left and right sides

Eye mask—large, visible structure enclosing the eye

Dorsolateral fold—long, visible, thin structure located on the left and right sides of the frog, stretching from the tympanum to the hind leg

Webbing—thin tissue connecting the toes of the hind feet

PROCEDURE FOR DRAWING

Observation Prepare the tray with the layer of wax on the bottom. If you have a dissecting tray, use it; it is better. Take the specimen from the container and place it on the tray. Keep a jar of alcohol handy—you will need it to moisten the specimen. Without alcohol the tissue will dry out and the specimen will be damaged. If you are allergic to the alcohol, use distilled water. Keep a pencil and paper at your side. When looking at the specimen, observe the coloration pattern. The assignment specifically stresses that aspect. Disregard torn tissue. Make sure you notice major anatomical features: the front and hind legs, the dorsolateral fold, how long the dorsolateral fold is, how big the eye mask is. Look at the specimen from the dorsal side and from the top. What do you see? Do you see all of the coloration marks on the edges of the hind legs? Can you distinguish and differentiate the tympanum from the general coloration? Check if the legs can be moved relatively easily. Do not damage the specimen; the scientist would not like this. Manipulate the legs carefully. With your fingers, try to stretch out the feet of the hind legs. Do not damage the tissue stretched between the toes. Turn the subject on the back. Examine the belly of the frog. Pay special attention to the patches of coloration which will be foreshortened when looking at the subject from the dorsal side. Pay attention to the sides and see where and how the coloration pattern merges or disappears. Know the subject. The better you know the subject the better and more truthful drawing you will produce. Do not forget to moisten the specimen at intervals.

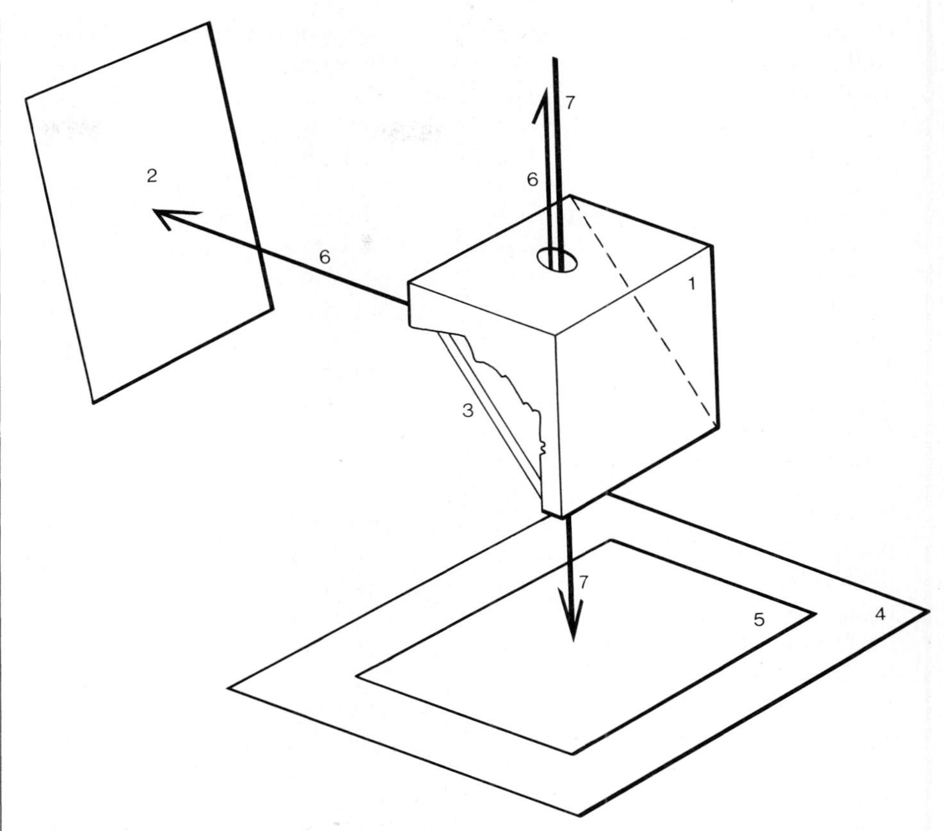

Figure 5-17. Position of the Magic Art See and Draw Copier (toy camera lucida) in relation to the specimen, the drawing paper and the eye. *1*—Magic Art See and Draw Copier. *2*—Positioned specimen. *3*—Glass splitting the beam of light. *4*—Table. *5*—Black paper. *6*—Beam of light reflected from the specimen. *7*—Direction of viewing the beam of light reflected off the glass and the black paper.

Positioning of the Specimen: You will need a pair of pliers to bend the upper section of the pins. Place the specimen on the ventral side. Take two long, thin, stainless steel pins and bend the upper section to form a 90-degree angle. Holding the specimen, insert one pin into the wax at the posterior corner formed by the front leg and the body. Make sure that the pin is well inserted and that the bent section of the pin is holding the body of the subject. The upper section of the pin should overlap the dorsolateral fold. Still holding the specimen, take another pin and insert it into the wax on the horizontally opposite side of the first, at the posterior corner formed by the front leg and the body. Do not pin the specimen through the tissue, as this causes permanent damage. Do not exert extra pressure; be firm but gentle. Take two more pins and bend their upper sections with pliers. While holding the specimen with one hand, insert the pins in the anterior corner formed by the hind legs and the body. Make sure that the bent sections of the pins will overlap the posterior part of the dorsolateral fold, holding the posterior section of the body well. Make sure that the subject will not move and that all four pins are holding well. Prepare two more pins, bending both of them about in half. Stretch the left hind leg, moving it toward the head of the frog. Make sure that the first section, the thigh, will be at about a 50-degree angle in relation to the main body of the subject. Insert the pin in the junction formed by the thigh and the calf. Make sure that the bent section of the pin overlaps the leg and will hold the specimen down. Do the same with the right hind leg. At this moment the pins holding the posterior section of the frog are exerting considerable pressure. You may need to add a couple of pins in the anterior area in order to stabilize the situation and keep the specimen down. Pin the calves of both hind legs in a similar fashion, making sure that the coloration presented on the posterior margin of the thighs is well visible. This means that the position of the calves will be slightly exaggerated. Make sure that you see the coloration, but do not pin the legs in an extremely unnatural position. Turn your attention to the feet. You will have to twist the hind feet in such a manner that all toes are clearly visible. Make sure the feet will stay in their pinned position. Do the same thing with the front legs. Do not stretch the front legs laterally. The toes of the front legs should be visible also. Make sure that during the process of securing, you continue to apply moisture to the specimen. At present you have a pinned subject, but one still not positioned properly.

You will be using a Magic Art See and Draw Copier, a toy camera lucida of relatively good quality, for transferring the image to the paper. In order to make use of it, take the tray with the pinned specimen and securely place it in front of you, making sure that it is set vertically. The specimen will be directly in front of you. Take a small metric ruler and attach it to the wax on the tray near the specimen. Make sure the tray will not fall down. Illuminate the subject, using the desk lamp. Make sure that you do not forget to moisten the specimen. Clean the table and prepare for sketching.

Drawing the Sketch: Take the Magic Art See and Draw Copier and find a small, round hole on one of its sides. Look through that round opening while pointing the copier at the subject. Be sure the mirror is facing the specimen. Do you see anything? Place the black paper on the surface of the table; make sure the subject is moistened and illuminated. Look again through the round opening, keeping the copier above the black paper. You will see an image of the specimen appear on the flat surface of the black paper. While keeping the copier in one hand over the paper, you will notice that your hand as well as the subject is visible. Take a pencil and place it directly under the copier, on the surface of the black paper. The pencil will be clearly visible. The image of the subject is being reflected by a combination of mirrors; a regular mirror bounces the image to the surface of the "one-way" mirror, which in turn reflects it to your eye through the round opening. You are seeing a reflected image overlapping the actual surface of the table. Replace the black paper with the tracing tissue taped to the white illustration board and take another look through the copier. The reflected image of the specimen will be harder to see because of the whiteness of the paper. Adjust the illumination to the specimen in proportion to the visibility of the reflected image. More light on the specimen and more light on the board will not help. Try different proportions of illumination. Generally, the more light on the specimen, the worse you will see the pencil. The more light on the paper and the less on the specimen, the more you will see of the pencil and the drawn line, but the less of the specimen. Find a happy medium. You will see enough to trace the subject by following the reflected shapes with the pencil. Make sure that the copier is at a reasonable distance from the surface of the table and is secured in the desired position. Do not tilt the tray holding the specimen. This will produce a distortion known to us as perspective. Remember that you are drawing a dorsal side of the subject. Do notice that the enlarging factor is controlled by the distance between the specimen and the copier. The metric ruler affixed next to the subject will be projected, also allowing you to draw the enlarged measuring unit, the centimeter. Very possibly you will not be able to see the whole specimen at once unless it is placed at a considerable distance from the copier. In such a situation the resulting reflected image may be too small for meaningful rendering of details. Keep the subject relatively close to the copier, the distance being designated by the details on the specimen as well as the size of the resulting tracing. The closer the specimen, the larger the tracing. The farther the specimen from the copier, the smaller the image and the tracing.

While tracing, pay special attention to the coloration, which is obstructed by the shadows resulting from the illustration of the specimen. Make sure that the basic outline of the subject is correctly traced—the head, eyes, legs, and toes. When keeping the copier a short distance from the specimen, you will have to move the specimen and the tracing tissue together in a horizontal or vertical direction. This will be necessary for alignment of sections of the subject temporarily outside the range of the copier. Do not move the copier. By moving the copier you will endanger the proportions as well as the alignment of parts of the image during the drawing. Do not move the specimen forward or backward—again, you will endanger the proportions of the specimen in relation to the already drawn sections of the subject.

Be careful. Think. Do not forget to moisten the specimen, nor to trace the relative scale. When finished tracing, take the copier off the table. Take the tray with the specimen and place it near the first sketch. Examine the specimen and the first sketch. Look for discrepancies; correct them. When you are not sure about the placement of a particular segment of the coloration on the first sketch, remeasure that portion of the specimen with pointers or the proportional divider. Find a point that is easily visible to you—for example, a back portion of the eye. Measure on the specimen the horizontal distance between that point and the one you are trying to establish. Make sure that the measuring tool is kept parallel to the surface of the bottom of the tray. Establish the enlargement factor by measuring the traced centimeter and dividing it by the actual centimeter size found on the metric ruler. Remember that one centimeter has ten millimeters. Enlarge the measurement by the established enlarging factor and transfer it to the first sketch. Repeat the same process of measuring, enlarging the distance from known points on the specimen to the points found in the questioned area until the problem is solved. Make sure that you designate all values in writing on the sketch.

Do not render on tracing tissue, as it is not suitable for tone rendering when pencil is used. Any rendering will destroy the readability of the information contained in the first sketch. Restrain yourself from erasing. All mistakes are a valid form of information to you. Make corrections clearly and you will not repeat the same mistake. Because the copier may have distorted the image or you may not have aligned the vertical position of the specimen with the copier, check the general proportions of measuring. When you are sure about the correctness of the first sketch, spray it with fixative. After a while take another sheet of tracing tissue and place it over the first sketch. Secure it with tape. Then trace the subject and all coloration areas in outline.

Try to be clean and professional. Do not spill coffee on your sketch. Do not forget about the specimen. When the specimen is being observed, moisten it periodically. When it is not needed, place a wet paper towel or thin gauze over the specimen. Keep the specimen under the Plexiglas cover with binder clips holding the edges of the cover very tightly. After the second, final sketch is traced, take the new, clean illustration board and tape the final sketch to the surface. Clean the table and prepare for rendering.

Rendering: Take the technical pen with the 00 tip or equivalent. Make sure that it is in working condition. Refill the pen with black ink. Prepare a paper towel or Kleenex. While shaking the pen, a paper towel or a Kleenex should be wrapped around the tip of the pen to serve as a protection against accidental spillage of ink. Take a sheet of mylar drafting film and tape the sheet to the illustration board. Secure the sheet with the tape on one side only, allowing for the insertion of white opaque paper between the final sketch and the drafting film. Prepare a piece of white illustration board and secure it with the tape to the table at a reasonable distance from the drawing. Pour a bit of ground graphite on the small piece of illustration board. Prepare the stomps and place them in the vicinity of the ground graphite. Make sure your hands are clean—wash them even if you think they are already clean. Sit comfortably with your elbows resting on the table. Take the technical pen and the stylus; bring the illustration board with the final sketch and the mylar drafting film toward you. Carefully, without rushing, trace the outline of the subject in ink. During tracing, make sure that the lines are applied smoothly. Do not attempt to draw a long line. This will add to the tension, resulting in shaking or trembling of your hands. Draw small sections of the outline. Make sure that the small sections of the line are joined properly by slightly overlapping the previously drawn section of the line with the new line. Press the drafting film with the stylus, holding down the surface of the film. Do not outline the coloration patterns, the dorsolateral fold, nor the tympanum. But *do* draw the outline designating the eye. Also make sure that the structures of the toes are outlined. Because of their narrowness, make sure that the ink lines designating the edges of the upper structures of the toes are separated, and white space between the edges of those structures is maintained. Draw the outside edge of webbing. During the process of drawing, turn the rendering conveniently around. Remember that you are in control of the rendering. If you are comfortable while drawing, the results will be better. Remember that if you make a mistake while drawing the outline, you cannot scratch out that mistake.

During the application of the graphite all scratches and especially the edges of the scratched areas will become darker than the undamaged surface of the drafting film. After the ink has dried out, take the smaller and larger stomps and dip the angled surface of the tip into the ground graphite. Smear the graphite with those stomps on the surface of the illustration board, working the particles of graphite into the surface of the stomps. Try the density of the tone produced by the already-prepared stomp on the small white piece of illustration board. After being sure about the value produced by the stomps, start rendering the general shape of the subject. Apply the tone with the stomps. Hold them at a 45-degree angle to the surface of the drawing. Use the tips of the stomps in the areas near the outline. Make sure that you use right size stomp—smaller for narrow areas and larger for the main sections of the subject. Press the stomps reasonably hard in order to achieve darker values. Remember that

Figure 5-18. Blotched or pied pattern of rear of thigh in frogs of the *Amolops jerboa* species group. Outline in pen and ink, tone in graphite dust on mylar drafting film. Drawing by Zbigniew T. Jastrzębski, Field Museum of Natural History, Chicago. Published in Robert F. Inger and Paul A. Gritis, "Variation in Bornean Frogs of the *Amolops jerboa* Species Group with Description of Two New Species," *Fieldiana*, Zoology, New Series No. 19, by Field Museum of Natural History, Chicago. Drawn in 1982.

graphite dust will not stick to mylar drafting film as it does to paper or illustration board. All tones, from darkest to lightest, will appear semitransparent when the drafting film is raised above the surface of the illustration board. For the darkest tones, use pencil. Carefully, with the well-sharpened tip of the pencil, apply the graphite, moving the tip of the pencil in a circular manner. Do not start rendering in various parts of the drawing. Be methodical. Start from the head and proceed toward the bottom. Leave the eyes white. Make sure that you do not forget the dorsolateral fold. Toward both sides of the subject, starting from the dorsolateral fold, the values will become darker. This will introduce three-dimensionality to the subject. Near the joints between the legs and the body, use pencil. Pencil is the only tool that will bring the appropriate darkness. Do not overrender the upper area. If the value is too dark in the upper area, you will not be able to achieve three-dimensionality. Drafting film will accept a limited quantity of graphite. The coloration of the subject will be rendered after the first tone is applied.

Take the kneaded eraser and roll it into a pointed cone. With that cone, by touching, remove the graphite from the surface of the drafting film. Touching the graphite with a kneaded eraser will produce a light to very light tone. Apply moderate pressure. If needed, move the tip of the kneaded eraser across the surface, making a line. After each time, knead the eraser in your hands and form a new, clean tip. When the tip of the eraser is dirty it will not pick up a sufficient amount of graphite. If it is too dirty it may leave a dark mark on the surface of the drafting film.

Whenever the coloration of the subject is darker than the general tone, use pencil or a Stenotrace eraser to apply more graphite in that area. Use the specimen, your notes, and the first sketch for reference. Do not forget to moisten the specimen periodically. Keep your hands clean, as any dirt will show easily on mylar film. You want the drawing to be clean and professionally finished. Let the ink outlines be visible. The black lines will not distract from the overall rendering.

When drawing, do not forget about the tympanum. It is too regular in its shape to be confused with general coloration. Do not be extremely precise. The object of the drawing is to present the distinct coloration of the subject, not to reproduce that particular pattern faithfully. At the same time, do not take liberties and do not introduce improvements into the natural patterns. Check the drawing through the reducing lens. In order to improve visibility, place a sheet of opaque white paper between the drafting film and the sketch. If you make a major mistake, such as smearing graphite across the illustration, do not panic; slowly remove the smear with a kneaded eraser. Tone applied with pencil will not be as easy to remove as the tone gently laid on the surface of the drafting film with the stomps. Stenotrace eraser may be the answer. If you have to resort to scratching, do this last, right before the application of fixative. When the rendering is finished, apply a thin coat of Krylon Crystal-Clear fixative. Do not overdo it. A thick coat of fixative may stick to the protective tissue and will stick to an acetate overlay. Do not forget to carefully remove the pins from the specimen and place it in the container full of alcohol.

Dry-Mounting: Try not to dry-mount mylar drafting film to the surface of the illustration board. Dry-mounting is unnecessary and can ruin your drawing. The illustration should be attached to the illustration board by a strip of masking tape at one edge. The upper section of the drawing is an appropriate place for the tape. The whiteness of the board will be visible through the semitransparent film, serving as a white background for all of the tones. For the clean, professional look, use Scotch double-stick tape number 666. Leave the white plastic backing on the tape for pleasant visual effect. Do not forget to trim the edges of the tape with an X-Acto knife. The illustration should be easily lifted from the surface of the board. If you insist on dry-mounting, do it before the fixative is applied to the surface of the illustration, or use a backup sheet supplied with instant lettering as a protective from the heat. The light blue greasy-feeling paper will not stick to the fixative. The tracing tissue or other paper will destroy the drawing. The fixative will melt into the surface of the protective paper. Use only removable dry-mounting tissue, first spraying the back side of the illustration with spray adhesive.

Preparation for Production: Place a protective sheet of tracing tissue over the drawing and the black cover-stock paper over the tracing tissue. Tape respectively the tracing tissue with regular masking tape and the black cover-stock paper with black photographic tape. Do not forget about creasing the black paper and folding it back.

Comments: When your first, rough sketch is too large for a meaningful rendering, you will have to reduce it to an appropriate size. This can be done in two ways: first, through photostatting the sketch; second, by tracing the sketch on a device called a reducing-enlarging copier. The reducing-enlarging copier is built like a large camera with bellows, lens, and movable bed. The drawing is placed on the illuminated bed and the image projected on the large ground glass at the back of the device. After adjusting the projected image to the desired size, it is traced by hand on the sheet of tracing tissue.

SEVENTH PROJECT

Botany

Subject: Habit of plant to be used for general identification.

Technique: Watercolors on Strathmore plated drawing paper.

Materials:

1. Pencils—H, HB, B
2. Erasers—kneaded, art gum
3. Tracing paper
4. Strathmore single-ply, plated drawing paper
5. Illustration board, hot-press
6. Pointers
7. Metal ruler
8. Metric ruler
9. Masking tape
10. Brush number 8 with very good, pointed tip
11. Watercolors of any kind
12. Fixative, Krylon Crystal-Clear
13. X-Acto knife with number 11 blades
14. Reducing glass
15. Black cover-stock paper
16. Black photographic masking tape

Basic Steps:

1. Preparation of the sketches
2. Preparation of the final sketch
3. Tracing the final sketch on the Strathmore plated paper
4. Painting
5. Preparation for production

Nomenclature:

calyx—green outside parts of the flower

corolla—part of the flower made of the petals

stamens—thin structures growing out of the middle of the flower, found between the petals

pistil—the ovary of the flower, located directly in the center of the flower

petiole—the base of the leaf forming the middle vein, dividing the leaf symmetrically

blade—the edge of the leaf

habit—the whole plant with leaves, roots, and so on, presenting a typical structure of that plant

Leaf Types:

entire—a leaf that has a smooth, unbroken blade, slightly oval in shape

toothed—a leaf with its blade interrupted by many sharply ending irregularities similar in shape to *entire* leaf, and having petiole deeply inserted at the base

lobed—a leaf with numerous rounded lobes and a smooth blade

linear—long, thin leaf with slightly toothed blade

lanceolate—a leaf similar in shape to the *entire* leaf, but slender, with slightly toothed blade

ovate—a leaf characterized by profound similarity to the letter *O*, with smoothly toothed blade

Leaf Arrangement:

clasping—leaves attached directly to the main stem of the plant in clasping fashion *alternate*—leaves with *petiole* attached to the main stem of the plant in alternating fashion

opposite—leaves, usually with short *petiole* attached to the main stem of the plant directly opposite to each other

basal—leaves attached only to the base of the plant

whorled—clusters of leaves, usually with short *petiole* attached to the main stem of the plant directly opposite to each other with alternating direction of the clusters

perfoliate—leaves attached directly to the main stem of the plant with the main stem perforating the base of the leaf.

Preparation of the Specimen: Buy the specimen from any store that sells potted flowers. The specimen should be fresh, with stems, leaves, roots, and perhaps flowers. Try to obtain a single green plant. Do not buy very colorful and complicated plants. Complicated coloration on the leaves and flowers is pleasant to look at but may present difficulties during painting. Keep your plant potted until ready to draw the roots.

PROCEDURE FOR ILLUSTRATING

Observation First take note pad or a sketchbook and prepare B pencils. Place the plant in front of you and thoroughly examine the subject. Keep in mind that you are about to prepare an illustration representing the general characteristics of the specimen. The viewer should be able to compare the finished illustration with what exists in nature and recognize *that* plant. Keep in mind that in this project you are not required to draw the specimen from a specific side. You are free to choose any view, as long as the drawn image presents *that* specimen. At the same time, realize that foreshortening is not allowed. A three-quarter view is prohibited. Perspective cannot be implemented, and presenting the plant from above or below will not meet required standards. With all of your freedom, the plant must be presented in lateral view.

Examine the structures carefully and make notes. First note the coloration. As the illustration will represent the subject in color, and the coloration may change rapidly due to the handling of the specimen as well as to other natural causes, make sure that the description of colors of the leaves, petals, and the stem is noted immediately. Describe the colors any way you like. Draw a quick sketch of a leaf and write. Mix the right color and paint a small patch, writing next to it an explanatory note. Make sure that you remember from what part of the plant that sample came. Do not investigate the roots at the beginnning. Leave this for last. Look at the shapes of the leaves; notice how the leaves grow. Find the proportion between the new and old leaves. See which are the small ones. Are they on the top of the plant? Where are the large leaves? How many are there? Make sure to examine the surface of the leaves. How much does the front differ from the back surface? Make sure to notice the differences in color and texture. See the veins. Are they sunk into the front surface? Do they become a three-dimensional configuration on the back of the leaf? How regularly does this particular feature appear throughout the plant? Look at the bases of branches and see how they are attached to the plant. See how the petiole is attached to the leaf as well as to the branch. Find the leaf arrangement. Does it fit any of the three basic categories? Do the leaves grow on opposite sides from each other? Do they grow at the base of the plant? Do the leaves alternate? Find the pattern of alternation. Make sure that you are aware of the leaf arrangement on the stem. Count the main veins on several leaves and see if the number of the veins remains the same. If not, ask yourself why not. Is it because of the age of the leaves? Did you find some regularly appearing irregularities in the structure of the plant? Pay attention to the thickness of the main stems supporting leaves. Separate the stems carefully and find out from where they are growing. Find out if the stems overlap each other. Find out how they are overlapping.

Be sure you are examining one plant. Are you looking at one plant or a couple of plants planted together? It is common for store-bought potted plants to consist of two or three planted in one pot. If your subject has a flower, examine it, paying attention to the pattern of the petals. Make sure to notice if fresh buds are present. During the observation take notes in writing, recording the coloration and characteristics of the structures. Pay attention to the flexibility of the stems; note how flexible the structures of the plant are.

Remember that the sketch you are about to prepare will represent the plant, but will be composed of selected features. You will be the judge. You will have to include in the sketch the information pertaining to the general proportions, size of the leaves, veins on the leaves, flexibility of the stems, buds, flowers, structure of the stem(s) and roots.

Drawing the Sketch: The sketching procedure has many variants. All depend on the size and complexity of the subject. Generally speaking, you will present a synthesis of the subject. But if your plant is a simple, small structure, there is nothing to eliminate. On the other hand, if your subject is very complicated, you will have to select certain features. Why, you may ask, should

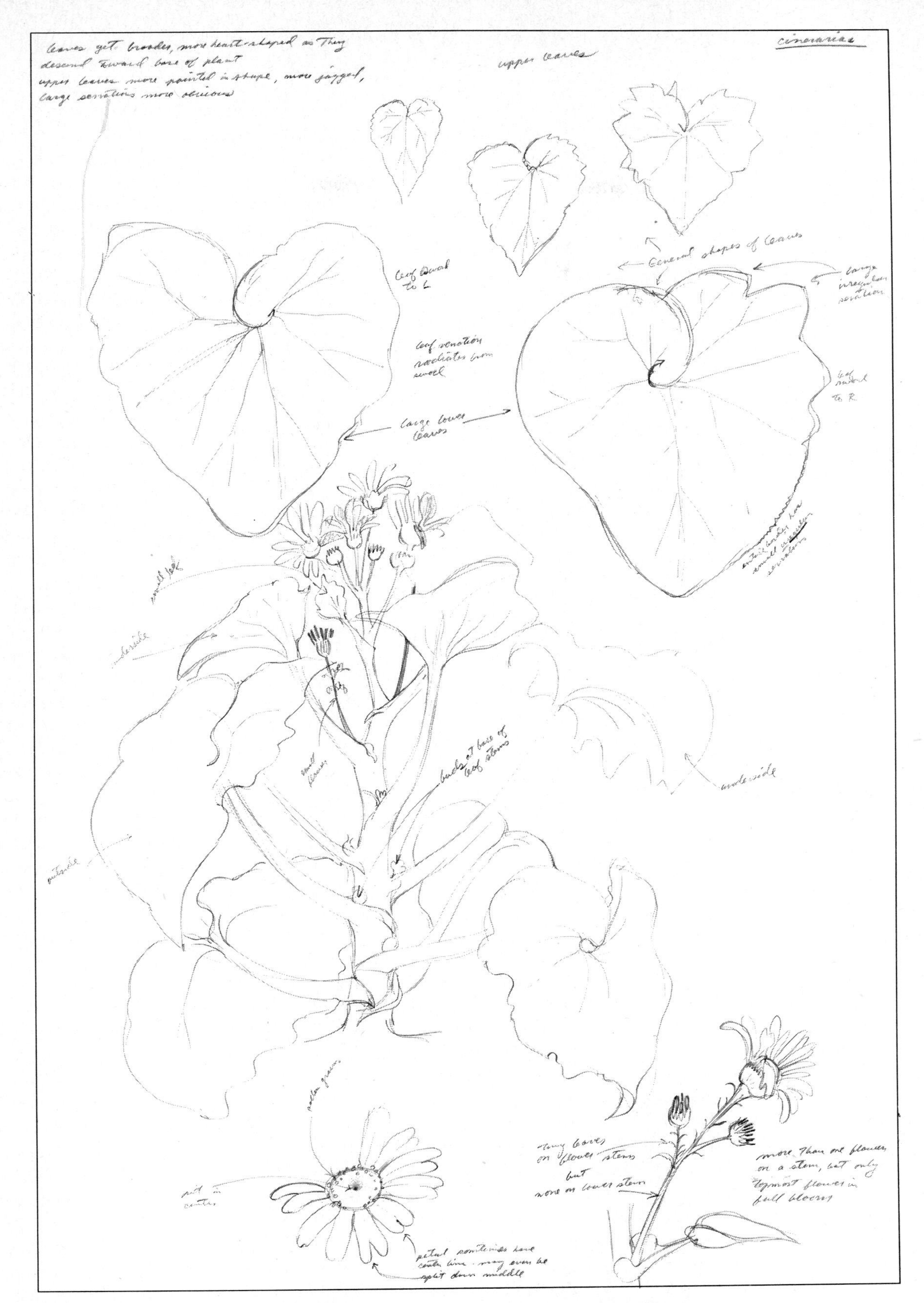

Figure 5-19. *Senecio cruentus*, preparatory sketch. Pencil on Strathmore drawing paper. Drawing by Sarah Forbes Woodward, Chicago. Unpublished. Drawn in 1979.

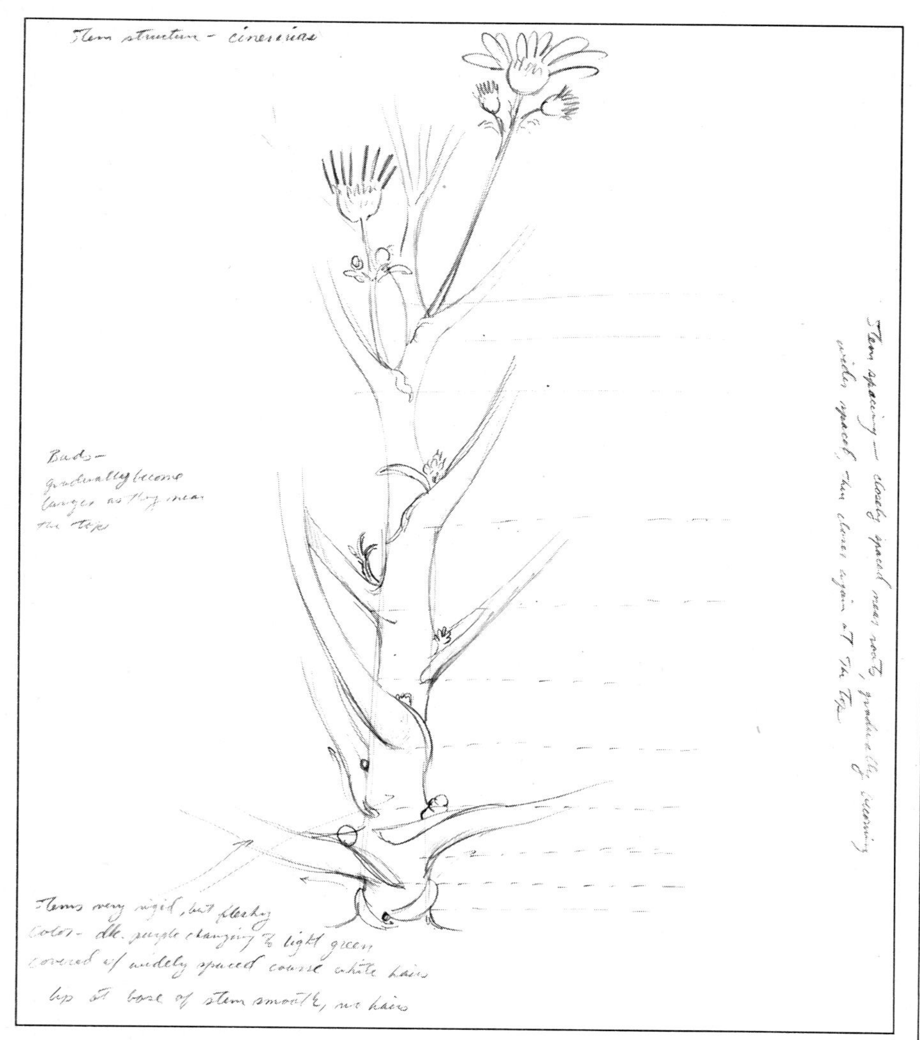

Figure 5-20. *Senecio cruentus*, preparatory sketch. Pencil on Strathmore drawing paper. Drawing by Sarah Forbes Woodward, Chicago. Unpublished. Drawn in 1979.

you choose some things and not others? Why not draw and paint the plant as it is? Readability is the answer. The final rendering will have to present clearly the structure and the patterns characteristic to this plant. Here lies the problem. Let's say that your subject is a plant with a complicated stem structure. The stems grow from the root but overlap each other. The plant has lots of leaves. It is very dense—so dense that it is impossible to deduce the intricacies of the structures just by looking at the plant. In such a situation you have to be very selective, but at the same time you cannot forget the basic characteristics: bushiness; lots of stems and leaves. Such a structure is much harder to draw correctly than a simple blade of grass.

Let's assume that you are facing a plant with complicated structure. First, before picking up the pencil, clean up your table. Make room for your tools and your elbows. Take a sheet of illustration board and tape the tracing tissue to the surface. With the pointers, ruler, or some other measuring tool, measure the height and width of the structure. Find out if you will have to enlarge or reduce the subject during sketching. As always, note the enlarging or reducing factor. Using a ruler and triangle, draw a rectangle on the tracing paper. The borders of the rectangle denote the furthermost edges of the subject. The bottom line represents the earth from which your potted plant is growing. The roots of the plant will later be placed under this line. Mark this sketch as the first sketch. Do not draw anything else on this sheet of tracing tissue. Take another sheet of tracing tissue and place it over the first. Tape it to the illustration board, not to the first sheet of tissue. With the ruler, draw a straight line, vertically dividing the rectangle into halves. Examine the plant and find a stem that is the most basic to the plant; measure its length. Mark its length on the vertical line. Take a triangle or a ruler and place it at front of the plant in the vertical position, matching its straight edge with the main stem. Measure the horizontal distance between the straight edge of the ruler and the upper section of the stem. Note this on the second sketch. Measure the thickness of the stem at ground level and mark it on the sketch. Measure the vertical distances between the petioles supporting the leaves. Consult with your notes and find the average. Note this on your second sketch. If your plant has branching secondary stems, measure their length as well as their relative angles. Transfer the information to the sketch in the form of a line. Slowly, you are drawing a "stick figure" of the plant. You are transferring the basic information to the tracing tissue. Find out where the leaves are attached and note this with some kind of marks on the sketch. Do not draw the leaves as yet. Drawing leaves will call for a separate set of sketches. Measure the length of other basic stems and note the information on a separate sheet of tracing tissue for later reference. Take a new sheet of illustration board and a new sheet of tracing tissue. Remembering the enlarging or reducing factor, concentrate on the base of the plant. See which stems obstruct the view of others. Measure the width of the stems and draw a simple sketch presenting the construction of this section of the plant. Place a new sheet of tracing tissue over the first sketch of the base of the plant. Through tracing, clean up the image and eliminate unnecessary stems by ending the structure with a horizontal line or by complete removal.

Do not shade. Do not apply any tone to the images or the tracing tissue. If you feel it is necessary, draw an arrow pointing to a pertinent place and write a description of the tone. Write a note if the stem is convex or concave, adding any other comments regarding the structure of this section of the plant. The surface of the tracing tissue does not accept pencil tones properly. By shading you will obstruct the information enclosed in the sketch. Remove the sketches of the base of the plant from the illustration board and affix a new sheet of tracing tissue. Examine the leaves on the plant and at the same time consult your notes. Select a typical leaf. If you can spare it, remove the leaf from the plant, straighten the leaf's surface and place it under the clean sheet of tracing tissue. Holding the leaf in a stationary position, trace the outline of the blade onto the tracing tissue. Make sure that the leaf is flattened and that the tracing represents the actuality.

If you do not want to damage the specimen or if you cannot inflict any damage because of restrictions imposed on you, transfer the outline of the leaf by the grid technique. Take pointers and measure the length of the leaf. Transfer this measurement to the tracing tissue in a form of a straight line. Measure the distance from the base of the leaf to about one quarter of the actual length of this leaf. Note where this point falls on the surface of the leaf. Mark the measurement on the sketch. Measure horizontally the width of the leaf at one-quarter length. Be sure to maintain a 90-degree angle between vertical (length) and horizontal (width) measurements. Again mark the measured width of the leaf on the sketch. Proceed with measuring and transferring the results to the tracing tissue until the basic outline of the leaf is completed. Then check the blade of the leaf scrupulously. Determine the type of the leaf. If the blade is toothed, note the characteristics of the irregularities and count the separate teeth. Transfer the relevant information to the sketch. Make sure that you draw both the front and the back side of the leaf. Count the basic, most visible veins; carefully note their angle relative to the main vein. The main middle vein is an extension of the base of the leaf, the petiole. Note the texture of the leaf and describe it in writing. Do the same with coloration, describing the coloration in writing. It is important that you begin to notice the values presented by the coloration. Describe the values the same way as you did when examining a black-and-white subject. Write "This section is very light," or "This place is very dark compared to the base of the leaf." When finished, you will have a complete sketch of the structure.

The next step is to draw smaller and larger leaves, naturally in proportion to the original sizes of the leaves found on the plant. Determine the different sizes of the leaves by measuring the specimen and by consulting your previous notes. There are only two possibilities: one, that all leaves are virtually the same size; two, the plant has three to four distinctly different sizes of leaves. Facing the second possibility and knowing the relative size of the small, medium, and large leaves, prepare separate small sheets of tracing tissue. Take a sheet of clean tracing tissue and place it over the basic sketch, actually the complete sketch of the leaf. Tape the tissue securely and with the H pencil, draw the middle vein. With the same pencil draw on the inside of the first image a line parallel to that depicting the blade of the leaf. Make sure that the distance between those lines is relative to the actual difference of size between the leaf presented in the first sketch and the small or smallest leaf you are drawing. The tips of the veins close to the blade will be cut by the second, smaller image, but their number as well as the relative angle will be preserved. The irregularities of the blade will also be preserved because the outline of the smaller leaf is represented by the parallel line running along the outline representing the blade of the larger leaf. When drawing the large leaves, larger than the first-sketched image, do the same but draw the outline presenting the blade on the outside. When you have three or four flat leaves drawn, examine the curvature and the twisting of the leaves on the plant. Examine the possible foreshortenings. If your plant has leaves that twist and/or bend, find the exemplary ones on the plant and estimate how much two-dimensional space those structures will take. Do not be concerned at this point with three-dimensionality of the leaves, but pay special attention to the two-dimensional area, represented by some kind of rectangle, occupied by the leaf you are observing and sketching. Determine this area by measuring. Look directly at the leaf and with the ruler, pointers, or any other measuring tool, measure the structure vertically and horizontally.

Draw the rectangle on a new sheet of tracing tissue. Take the sketch of the leaf and place it under the sheet with the rectangle drawn on it. Naturally the sketch of the leaf, being drawn flat, will be larger than the rectangle. Take a new sheet of tracing tissue and place it over the drawn rectangle. You will now have three layers of tracing tissue: the first with the sketch of the flat leaf, the second with the rectangle, and the third, clean one, ready to be drawn on. On the third, clean sheet of tracing tissue, within the boundaries of the rectangle, find and make the tip as well as the base of the leaf. The foreshortened image that will be drawn cannot exceed the boundaries of the rectangle. Carefully observe the middle vein on the leaf you are about to draw. Make sure that you understand the curvature of the surface of the leaf. Within the rectangle, draw the line from the tip to the base, curving the line and simulating what you have observed in nature. If you do not succeed on the first try, you must not go out of the rectangle borders. Do it again. As you know, some parts of this curved line will later be obstructed by the surface of the leaf and not be visible. At present it is important that the whole line representing the middle vein be clearly and visibly drawn. Draw horizontal straight lines depicting the width of the leaf, crossing the middle vein at right angles at about the same places you drew on the very first sketch, the basic sketch of the leaf. Again, be careful not to wander out of the designated boundaries of the rectangle. Paying no attention to the irregularities of the blade, connect the tips of the horizontal lines, following the curvature presented by the line that depicts the middle vein. By doing this, you are drawing the outline of the leaf. When finished, take a new sheet of tracing tissue and place it over the newly drawn rough sketch of the foreshortened leaf. Remove the sketch representing the flat leaf. Trace the new image, clearly implementing the irregularities of the blade and omitting sections of the drawn image that are obviously obscured by overlapping parts of the leaf. In order to add a desired twist to the drawn structure, redraw or actually retrace the new image, moving the base or the tip of the leaf to the left or to the right. Be extra careful *not* to align the tip of the leaf with the blade. This must not happen. It is very difficult to paint or draw a three-dimensional image when the sharp tip touches the outline. It is very difficult to present the feeling of space, of distance between the tip of the leaf and the surface under the tip, because the tip and the blade will tend to merge together in the viewer's eyes. The reason lies in not enough contrast between the shapes.

When four or five leaves are presented in their final form, clean the drawing table and take a new sheet of illustration board. Take the very first sketch of the plant, the sketch representing the plant in "stick figure" fashion. Tape it to the board, taping the upper edge of the tissue. Take a sheet of clean tracing tissue large enough for the sketch of the whole plant. Again, tape this sheet to the illustration board with one piece of tape at the upper edge of the tissue. Make sure you will be able to insert previous sketches under both sheets of tracing tissue with ease. Now you will prepare the five detailed sketches of the whole plant. It will be a composite of all the other sketches. Take the sketch representing the bottom portion of the plant and place it under both sheets of the tracing tissue at the appropriate place, as designated by the "stick figure" representation of the plant. Trace the image onto the clean sheet of tracing tissue. When finished, remove it. Continue drawing stems, consulting with the specimen and your notes. On the "stick figure" of the plant you have previously marked places where leaves should be attached. Insert the appropriate sketches of the leaves and trace them. Remember that if you would like to have a leaf pointing in the opposite direction, all you have to do is turn the very same sketch face down, drawn side down, and you will achieve the desired result. Continue matching the separate sketches with appropriate areas and tracing them. Try not to overload the new sketch with extensive information. If your plant has flowers, leave a couple of areas blank for the additon of flowers. When arranging or composing the first complete sketch of the whole plant, pay special attention to the clear presentation of information. Try to restrain yourself from erasing. Draw with smooth, clean lines. Do not move your hand in back-and-forth, short strokes. Such motion of the hand produces multiple lines that are hard to read. If a mistake is made, mark it with a clean line leading from the wrong information to the edge of tracing tissue and end the line with a wiggle, just as you have always done. Do your best to avoid sharp edges of the images from touching other sharp edges. When finished, turn your attention to the flowers.

The coloration of the flowers has been already described by words as well as by samples of color. Therefore, you can devote yourself to the study of the structure of the flower. Start with a simple preliminary sketch presenting the general proportions. On the overlying sheet of tracing tissue, develop the details of the construction. Make sure that petals are correctly drawn and that the calyx is presented visibly even if it is not that easily noticed on the plant. When finished, insert the final sketch of the flower under the almost-finished first sketch of the whole plant and trace it. Do not overload the drawing with too many flowers, an exception being a specimen that would be characterized by numerous flowers. In this case you will have to draw enough flowers to preserve the feeling of the plant's basic character. When the first sketch of the whole structure is finished, spray the sketch with fixative. Remove all other sketches from the illustration board, take a new sheet of tracing tissue, and place it over that sketch. Tape it to the board. Clearly retrace the image, implementing all necessary corrections. When finished, clean the table and take a couple of newspaper sheets. Place the newspaper on the table and carefully remove the plant from the pot. Clean the roots under running water, removing clumps and particles of earth that obstruct the view. Place the plant on the newspaper and prepare the measuring tools. Take the illustration board and place a clean sheet of tracing tissue on it, taping it to the surface of the illustration board. Study the root system, look at it and measure it. Draw a rough sketch. Be sure that the coloration and texture are noted properly. Do not draw every segment; be selective. General information is sufficient. Do not exaggerate even if you think that such would make the illustration more outstanding. When you have finished drawing and have noted all necessary information, place the plant back into the pot and clean the drawing table. You do not want any dirt on the sketches. Retrace the basic sketch of the root system, making the image more coherent, and spray the sketch with fixative. Take the last sketch of the whole plant and insert the clean version of the root system under the tissue. Make sure that the sketch is aligned properly. Trace the image; when finished, remove the sketch of the root system and spray the sketch of the whole plant with the fixative. When the fixative has dried, take another large sheet of tracing tissue and place it on top of the sketch of the whole plant.

Before retracing the first sketch of the whole plant, examine the image carefully. Look for areas that are hard to read, such as overlapping leaves, stems, and flowers. Remember that your illustration is being produced for reproduction and is an explanatory depiction of the structural aspects of that plant. Check foreshortening of the leaves and look for exaggerations. If you find any, implement necessary corrections. During the final stage, refine the sketch; later you will not have a chance for changes. Carefully trace the corrected image and spray the final sketch with fixative. Spray both sides of the sketch in order to avoid accidental transfer of graphite to the surface of Strathmore paper. Wait until the fixative has dried. Take a sheet of one-ply, smooth-surface, plated Strathmore drawing paper and prepare for the transfer of the image to that paper. The transfer of the final sketch will be accomplished by tracing. The easiest way to trace the image to opaque paper is by using a light box. Take the sheet of Strathmore paper and turn the sketch drawn side toward the paper. Tape the sketch to the paper. Take the taped sketch with the paper and place it on the light box, paper side up. Do not tape the paper or the sketch to the light box. While tracing, you will have to turn around the paper with the taped sketch, correlating the direction of the line about to be traced with the motion of your hand. This will result in a better and cleaner transfer of the image. Do not fight with the physical limitations of your anatomical structure. Be relaxed; do not exert unnecessary strain on your muscles. When working in a stationary position it is very hard to place your body at the correct angle to the part of the drawing to be traced. It is much easier to adjust the drawing to your stationary body. As an added benefit, it will be easier to reach to certain parts of the drawing without smearing previously drawn lines, and as your hands will not get tired as quickly, the drawn lines will be much more under your control and therefore much more stable. Do not exert too much pressure on the paper during tracing. The image should be readable but not very dark. If the traced lines designating the image are very dark, you are asking for trouble during painting, as at least some of these lines will have to be covered with paint. Even when watercolors are mixed with acrylic white it may not help. The lines should be thin and visible. When there is no access to a light box, use carbon paper, the very same kind you use to make copies while typing. Here the process of tracing is different. First take an illustration board and tape the Strathmore paper to its surface. Then place the sketch, with the image facing you, on the top. Tape one side of the sketch to the board. Do not tape it to the paper, as during the removal of the tape the surface of the Strathmore paper may be damaged. Place the carbon paper under the sketch, with the carbon side facing the paper. Carefully trace the image, continually checking the tracing for accidentally omitted sections of the image. When finished, remove the sketch and carbon paper. During the final tracing be sure your hands are clean, and be sure that the surface of the Strathmore paper is kept clean. Remember that painting is still to be done and there is plenty of time for possible accidental damage. When the transfer of the image from the final sketch to the surface of the

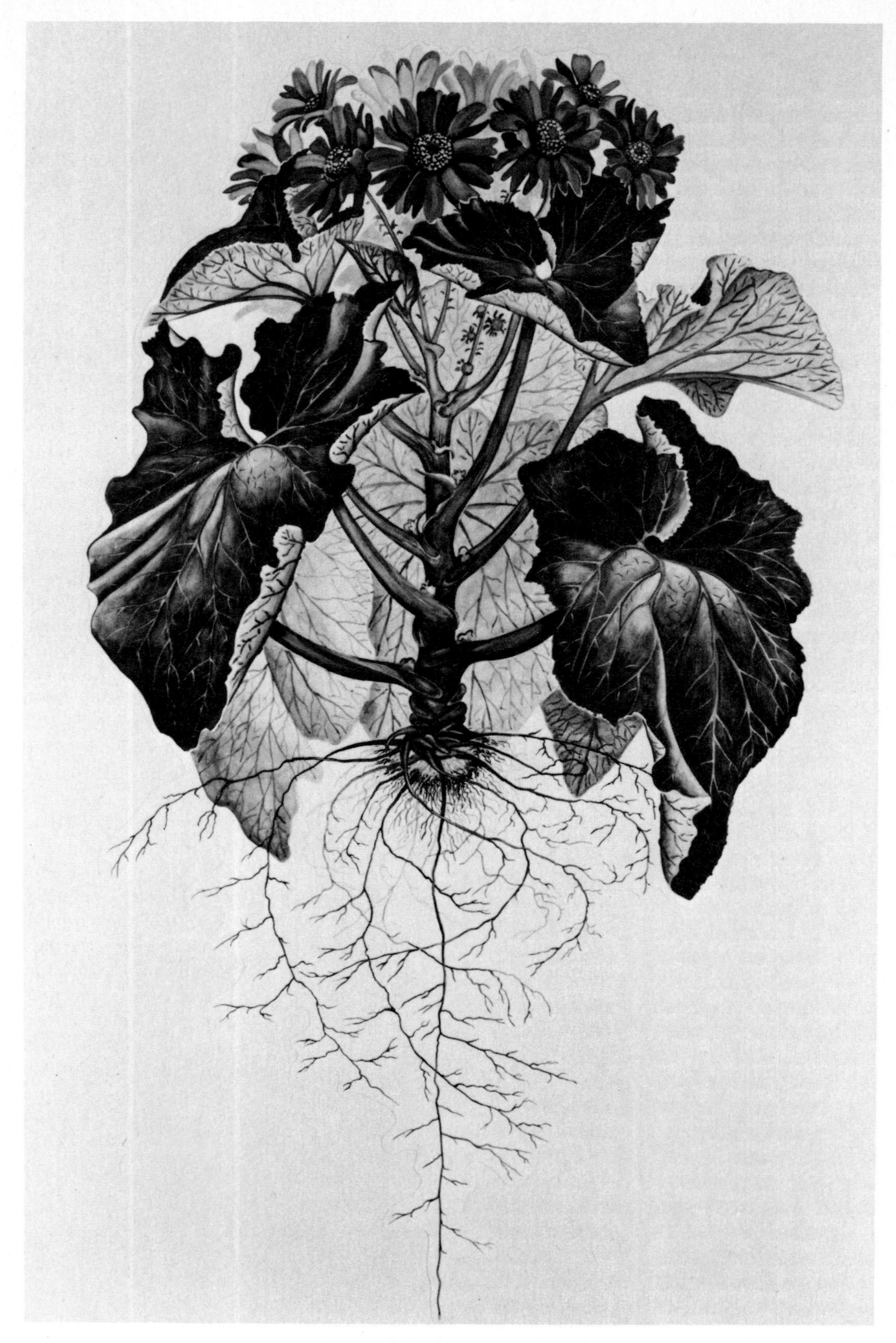

Figure 5-21. *Senecio cruentus*. Watercolor with addition of acrylic white on Strathmore one-ply drawing paper. Painting by Sarah Forbes Woodward, Chicago. Unpublished. Painted in 1979.

Strathmore paper is finished, clean the table of unnecessary instruments and sketches and prepare for painting.

Rendering: Prepare the watercolors and brush or brushes. Watercolors do not have to be of the best quality—as a matter of fact, expensive watercolors are not very good for painting on a smooth, more-absorbent surface. Good-quality watercolors are manufactured for use with watercolor paper, which is very absorbent. This means that the pigment in such paints is ground to an extremely fine powder and when applied to the surface of absorbent paper, will embed itself into the fiber structure of that paper. In our painting, we are looking for the opposite: The pigment should stay on the surface of the paper with a minimum of absorption. Such a situation will allow us to remove unwanted tone relatively early, without much damage to the surface, as well as without changing the future color or the tone applied to the washed area. During preparation of our illustration we are *not* interested in perfection of watercolor technique, but we do want the final results to be correct. Use any watercolors, but you will get better results if you will use a simple, cheap, and poorly manufactured set. The combination of the smooth, plated surface of the paper and the relatively coarsely ground pigment goes hand in hand.

Why use such a combination? The answer lies in the reality of scientific illustration. Most scientists have Strathmore paper on hand because most illustrations are drawn in pen and ink. This paper has an excellent surface for the pen-and-ink technique and is widely used, therefore it is most probable that it is what you will have to paint on. On the other hand, because of the paper's fiber structure, it is resistant to the pigment. The result is that any mistake you make is correctable.

Prepare the brush. You do not need a variety of brushes, but you need a very good-quality brush with a very finely pointed tip. Avoid brushes with round tips, as they will not produce a fine line. During rendering, you will need a brush that can hold a reasonable quantity of water but at the same time can produce thin lines. Avoid very small brushes, those with two or three hairs. Such brushes are capable of producing a thin line, but because of the small quantity of water they will retain, they will produce a very short, thin line. It is hard to connect short lines into longer ones. I strongly suggest a number 6 or 8 brush of any make, as long as it ends in a finely pointed tip.

You will need one tube of acrylic white paint. The plastic-base paint will be mixed with watercolors, producing a nonwashable surface, in case you need an overpainted opaque surface; be it as a preparation for further application of colors or as a final statement. Mix colors on a white surface; the illustration board is the most suitable. Try to avoid the plastic containers made for the mixing of paint; they will cause water to bead up, preventing you from making a proper estimate of the proportions between pigment and water and

Figure 5-22. *Laminaria japonica*. Watercolor. Painting by Feng Minghua, Institute of Oceanology, Chinese Academy of Sciences, Quingdao, People's Republic of China. Unpublished.

therefore preventing you from knowing the quality of the tone. When all necessary painting supplies are ready, make sure that some kind of delicate tissue of high-absorbent quality is within your reach. Absorbent tissue such as Kleenex or toilet tissue will be needed throughout the process of rendering; it will be used for soaking excess water from the brush as well as from the paper, and also for removal of the paint (pigment) from large areas as well as while painting the veins on the leaves.

During the process of painting you *must* differentiate between colors and values. Remember that light colors can have different values. Thinking in black and white may help you to solve this important problem. The painting, any realistic painting, first presents the differentiation of values just like a black-and-white drawing. The closest association between painting and drawing is presented in a continuous-tone pencil drawing depicting a tangible object. It may be very helpful to look at seventeenth and eighteenth century Dutch still-life paintings. When analyzing these paintings, pay special attention to the ever-changing values and to the direction of the light. During painting you will have to be very conscious about light and dark areas, irrelevant of the color. It is advisable to exaggerate the contrasts between light and dark, thereby producing a more three-dimensional-looking image. The distance or the depth is enclosed by such contrasts. For a three-dimensional effect it really does not matter where the darks and lights are placed. The dark can be all the way in the back of the image or it can be right up front. More important is the change of values. The edge of a dark area must be met by the edge of a light area. When values are the same, the image will appear flat. Some time ago one of my teachers of painting told me that in order to paint well I should draw well. He said that through the knowledge of realistic drawing, the path to realistic painting would be open. He was right. In actuality, painting is drawing in color. Keep a large container filled with water nearby, but not too close to the illustration and certainly away from your elbows. Restrain yourself from placing a jar filled with water near the edge of the illustration board. During painting you will turn around the illustration board with the Strathmore paper taped to its surface. When turning the board around, it would be very easy to knock over the water jar and destroy the rendering. Make sure that while painting, dirty water will be disposed of and changed frequently. Do not keep your brush, bristles down, in the container for too long. By resting the tip of the brush on the hard bottom of the container you will damage that tip. Wash the brush in the water and touch the soft, absorbent tissues with the bristles. The tissues will remove excess water. Then roll the bristles of the brush on the clean piece of illustration board in order to align the bristles properly, forming the pointed tip again. Sometimes a cloth may be used for the same end, and the palm of your hand will also serve the purpose quite well. Skin, being very sensitive to pressure, will allow you to feel the tip of the brush. The only drawback will be if your hand is dirty—if you do this, remember to wash your hands frequently. When a brush is to be stored for a period of time, it should be cleaned with water and soap. Wash it, but not to the extreme, on a very thin layer of soap. This is necessary in order to keep the bristles of the brush together; roll it on the palm of your hand and place it, wooden end first, into an empty container. Avoid placing plastic sleeves over the bristles—it is virtually impossible to place such a device over the bristles without one or two bristles being damaged by being bent backward.

Be professional and avoid removing excess water from the brush by the means of a strong arm and the gravitational force. This may be fun, but your apratment will become a pigpen. Start the procedure of painting by wetting all paints in the container with water. Make sure that the water dissolves the pigment. Check the density of the value of the dissolved pigment on a piece of white illustration board.

Before starting the actual application of the paint to the traced image, think about the general direction of light that will appear to illuminate the subject. For example, if a leaf with its tip bent forward has a dark value applied to the frontal section of the bent tip and a light value placed toward the back and under the leaf, on its folded part, such a combination of values will produce an appearance of illumination coming from below the leaf. During painting you will be forced to back-light some areas; nevertheless, a general impression of a strong uniform direction of light should be apparent to the viewer. Do not forget to use the reducing glass to appraise the values during painting. The value you start painting will produce a chain reaction, forcing you to apply light and dark values according to this first step. Do not confuse values with color. Each color has a great many values, therefore do not assume that the green you want has to be taken directly from the factory-produced green. You will have to mix and prepare the correct green, adding other colors and controlling the density of pigment with water. For that reason it is important that the water used for cleaning brushes should be changed often. By having clean water you are assured that the proper value of a particular color will be correct. Impurities in water will affect your painting.

You have two ways of applying paint to the paper. One is to paint the whole designated area using a uniform, flat wash, waiting until that area dries, and removing the pigment from lighter parts as well as adding more pigment to the areas which should be darker. This type of painting is more difficult because the underpainted surface requires more careful judgment and more careful work. Constant addition and subtraction of pigment must be performed carefully, as the surface of the paper cannot be damaged too much. Use of this technique will produce a more stiff-looking presentation of the image. In some cases, depending on the size of the area to be painted, you will not be able to produce a desired effect quickly. The most important danger exists in rushing to apply dark values which later may have to be removed completely or in part.

The second technique uses accidentally applied tones as well as blobs of water for the benefit of the painting. Here the flat areas of underpainting cannot be used. Tone and color are applied to small sections of the larger area and are left to dry. When the pigment has dried, some of the accidents are removed by washing. If you do not have experience in painting, I strongly advise you to use the second technique of application of the pigment. Start with the leaves. Assume that the frontal section of one leaf will be lighter in value than the back section of the same leaf. Leaves are green; therefore, mix a dark green with a touch of red and a little bit of blue. This will produce a brownish dark green. Check the value and the color on the white illustration board. If it is too brown, add a little green or a touch of yellow. Remember that the back portion of the subject must be markedly darker than the front, but at the same time you cannot mix a completely different color. It must be, generally speaking, green—brown-green, bluish-green, it really does not matter. Later you will implement necessary corrections. Have a generous amount of water on your brush and start painting from the tip of the leaf if the leaf is small, or from the middle if it is large, moving the brush toward its base. Let the brush rest at the base of the leaf for a second. The prolonged stay of the brush will allow the water to flow to the surface of the paper, transporting the pigment and depositing it in a larger quantity than a quick movement would do. Remove the brush and let the whole thing dry. Do not attempt to cover a large leaf with one stroke of the brush, because there is a great probability that you will go out of designated boundaries. Apply the pigment with deliberation and thought. Do not rush. Your object is a well-painted plant.

Having placed the first stroke, you have a smudge of color ranging in value from light to dark. Mix the same color and repeat the procedure of application of the pigment to the surface of the paper. Do not worry about inconsistency of the tones, and do not worry about darker lines created by those two overlapping brush strokes. Make sure that the back of the leaf, the area near the stem, is dark in value. For the front section of the same leaf, prepare a lighter tone. Again apply the paint starting from the tip of the leaf (if the tip of the leaf is pointing toward the viewer). Let the whole thing dry. Assuming that you have drawn another leaf under the first painted one, make sure that the values on that leaf are markedly different. This means that the dark value of the first leaf must coincide with a light value on the second, and the light value of the second should border a dark value on the first. The play of values is of utmost importance and must be carried throughout the whole painting, as these changing values will produce the illusion of three-dimensionality. The stem of the plant placed among the leaves should also vary in value according to overlapping and underlying tones. If you find it necessary, and that depends on the type of plant you are painting, use some acrylic white in mixing colors. This will allow you to place the tone—which is lighter because of the presence of white and not easily washable—making it easier to overpaint. To remove unwanted spots of dark value, use a wet brush and absorbent tissue. Roll the tissue into a thinly pointed cone. Wet the desired area with a brush, quickly touching the wet area with absorbent tissue. The water with pigment will be soaked up by the tissue. Then apply the desired correction—you can do this when the area is either wet or dry. In both cases you will succeed, but with different results. If the tone is applied to a wet area, the newly applied tone will spread rapidly. When the area is dry, you will have to brush the pigment much more carefully when matching the edges of the values. Do not be sloppy; make sure that the background remains white during the whole process of the painting. Because Strathmore-plated drawing paper does not absorb pigment very well into its fiber structure, you will be able to wash out accidental splotches of paint from the background. In order to completely remove an unwanted feature, prepare clean water, a clean brush, and a quantity of absorbent tissue. With the clean brush dipped in clean water, wash the surface well; quickly blot the wet area with absorbent tissue. Repeat this until all the pigment has been removed, being cautious not to damage the surface of the paper very badly. Do the same thing when removing unwanted tone from the painted area, being careful not to destroy the rest of the image.

The veins on the leaves can be painted in two ways, depending on the characteristics of the vein system. If the veins are thin, use the tip of the brush to wash the pigment previously placed by drawing a line with a clean wet brush over the vein; then with a tip of rolled absorbent tissue, carefully remove the dissolved pigment, creating a thin line of lighter value than the background. When the veins are pronounced, mix a color of darker value than the background and draw a line with the tip of the brush on one side of the vein. Wait until the newly placed pigment has dried; then, with a clean brush, wash the desired thickness of the vein without disturbing the drawn darker line, and remove the unwanted dissolved pigment with tissue or another brush.

In order to lighten in value an area that is of dark value but thinly painted, add a wash of pure yellow. The pure yellow will sufficiently change the quality of the value without disturbing the basic coloration. The primary difficulty encountered when using a combination of nonabsorbent paper and watercolor is the instability of the pigment. This can be overcome by patience or by a light coat of workable fixative. Unfortunately, it is harder to remove the pigment when correcting or changing if the illustration has been sprayed with fixative.

The charm of such a combination of materials lies in the ability to correct mistakes easily. When slowly rendering and using more opaque layers of pigment, a cracking on the surface of the paint may occur. Do not be alarmed with this, as the finished painting will be sprayed with a protective coat of Krylon Crystal-Clear fixative. Repair the cracked area by retouching and continue painting. When the appearance of texture must be rendered, use a semidry brush. Mix the appropriate color and brush the tip of the brush on the surface of the illustration board until most of the water is removed from it and the pointed tip is slightly flattened, producing numerous thin lines. Apply the paint to the surface of the desired area in short strokes. Such application will cancel the unwanted differences of value present in areas where two brush strokes overlapped, making the area smoother as well as adding desired texture. When finished painting, leave the illustration taped to the illustration board and apply the Krylon Crystal-Clear fixative in a well-ventilated area. First spray the illustration lightly in order to preserve the pigment. After dry-mounting, spray the image five to eight times, drying the illustration well between sprays. Besides protecting the image, fixative applied in numerous layers will enhance the coloration by making it appear darker, deeper, and brighter.

Dry-Mounting: You must be very careful during dry-mounting. In order to protect the surface of the illustration, use only the bluish backup sheet from Letraset instant lettering. This tissue will not adhere to the fixative sprayed on the illustration. Follow the usual procedure: Take a clean sheet of hot-press illustration board, position the illustration on it, tape two corners of the illustration to the board, insert Seal removable dry-mounting tissue between the illustration and the board, place a protective sheet on top of the illustration, and tack the illustration in one spot only. Do not forget to trim the illustration together with the dry-mounting tissue and separate the Strathmore paper from the dry-mounting tissue and the board. Then, applying moderate pressure, dry-mount the illustration. Do not use tracing tissue or any kind of paper for protection of the illustration during dry-mounting, because the coat of fixative will absorb the fiber structure of such protective papers, causing them to stick to the surface of the illustration. When pulled back by force, the regular paper will remove the pigment from the surface of the Strathmore paper, destroying the illustration. Make sure that the backup sheets from Letraset instant lettering are used as a protection. After dry-mounting is finished, spray many coats of Krylon Crystal-Clear fixative on the painting.

Preparation for Production: Tape a sheet of tracing tissue to the top of the illustration board with regular masking tape, and the black cover-stock paper with black photographic tape. Crease the upper section of the black cover-stock paper after trimming the unwanted parts of tracing tissue and cover-stock paper, using the edge of the illustration board as a ruler. Remember to keep the X-Acto knife straight, with the back part of the blade close to the edge of the table. Draw the scale and designate it appropriately.

Comments: It is impossible to avoid mistakes when painting. As the final destination of our project is reproduction, we can repair mistakes very easily. The most common mistake is caused by the sloppiness of the illustrator, resulting in a dirty background. How many times have you accidentally dropped your brush on your drawing? How many times have you accidentally painted outside of the designated area, cropping up into the background? In order to remove all these unpleasant symptoms, cut the drawing out from the dirty background. Do this *only* after dry-mounting tissue has been attached (tacked) to the illustration.

That's where the biggest problem lies. You are lucky if your painting has been executed on one-ply Strathmore. One-ply paper is easy to cut through; two- or three-ply is much more difficult. When tacking the dry-mounting tissue to the illustration with the background to be removed, make sure the dry-mounting tissue is attached under the painted surface. Otherwise the illustration will be easy to remove, not being tacked, while the dirty background will remain on the illustration board. When cutting, be sure to cut through both layers simultaneously. It is hard to recut the same area. Be sure that you have sharp X-Acto blades. Do not use larger matting knives; they are awkward and cannot be used with any finesse. Use only number 11 blades. When cutting, clean up the messy edges of the image by cutting a bit into the rendering. Remember that if you cut across the board, the cutting mark will be visible. When finished, dry-mount the illustration as usual. If a part of the plant has to be repainted, do this on a separate piece of Strathmore; when finished, splice the new part, matching it well with the rest of the image, then dry mount it. Instead of fixative, feel free to use a quick-drying varnish. The result will be the same. The layer of transparent varnish will disperse the light, making the colors appear deeper.

EIGHTH PROJECT

Anthropology

Subject: Drawing of a stone tool, indicating planes and the direction of cleavage.

Technique: Quill pen and ink on mylar drafting film.

Light Source: From upper left side.

Materials:

1. Pencils—H, 2H, 4H
2. Erasers—kneaded, art gum
3. Tracing tissue
4. Quill pens
5. Pen holder
6. Ink, black
7. Mylar drafting film
8. Pointers
9. Magic Art See and Draw Copier
10. Metric ruler
11. Metal ruler
12. X-Acto knife with number 11 blades
13. Plasticine, 1 oz.
14. Reducing glass
15. Stylus
16. Small pencil-type flashlight
17. Masking tape
18. Scotch brand double-coated tape, number 666
19. Black cover-stock paper
20. Black photographic masking tape

Description of Some Materials:

1. Crow-quill pens are one of the oldest tools used by artists for drawing. As the name implies, the original crow-quill pens were made from a quill, the feather of a bird. The natural flexibility of the quill pen allows for variable thickness of the line, resulting in the quality of that line unsurpassed by other drawing devices. A common pen, the technical pen, will produce a very uniform line. They have been designed to do this, by being designed to overcome the difficulties associated with the flexible tips of quill pens. A technical pen can be successfully used by a relatively unskilled hand with good results. The quill pen requires practice, and practice is burdensome and time-consuming, therefore not very feasible for our fast-rolling civilization. Nevertheless, the beauty of a line produced by the quill pen cannot be surpassed. At present, a variety of metal quill pens is marketed. Again, most of these pens are manufactured with ease of use as a priority and are designed to perform specific tasks in the art of calligraphy. Pens such as Graphos, Osmoroid, or Speedball are in such a category; they cannot be used in our project as their tips range from flat to round. We need a pen with a pointed tip. Such tip is to be found on both Gillott and Hunt artists' pens. Usually the Gillott or Hunt metal quill pens are easily available in small sizes, which do not hold lots of ink, thus preventing us from drawing relatively long lines, although they can be diversified in thickness. Naturally, your choice of the size of pen depends entirely on the size of the drawing. For a drawing that does not exceed six inches in length, I suggest Gillott number 604EF, "Double Elastic"; "Spencerian" number 1 produced by Ivison Phinney Co.; Hunt's 513EF or "Ex-Fine" number 512. For smaller drawings, use the Gillott "Mapping" pen number 291, Gillott crow quill number 659, or Harper "Extra Fine." The problem facing you is the choice of the size and flexibility of the point of the pen relative to your muscular system and drawing habits. It is important to find the right pen for you, because the thickness of the line is produced by the pressure exerted with the hand on the tip of the pen. According to the pressure, the tip will spread apart slightly or close together, producing the desired line. I strongly advise that you buy a variety of pens and try all of them until you feel secure that you have some kind of control of the qualities of line without too much stiffness. The tip of the pen should not scratch the surface used for drawing. It should glide smoothly, without sudden interruptions, allowing for an even

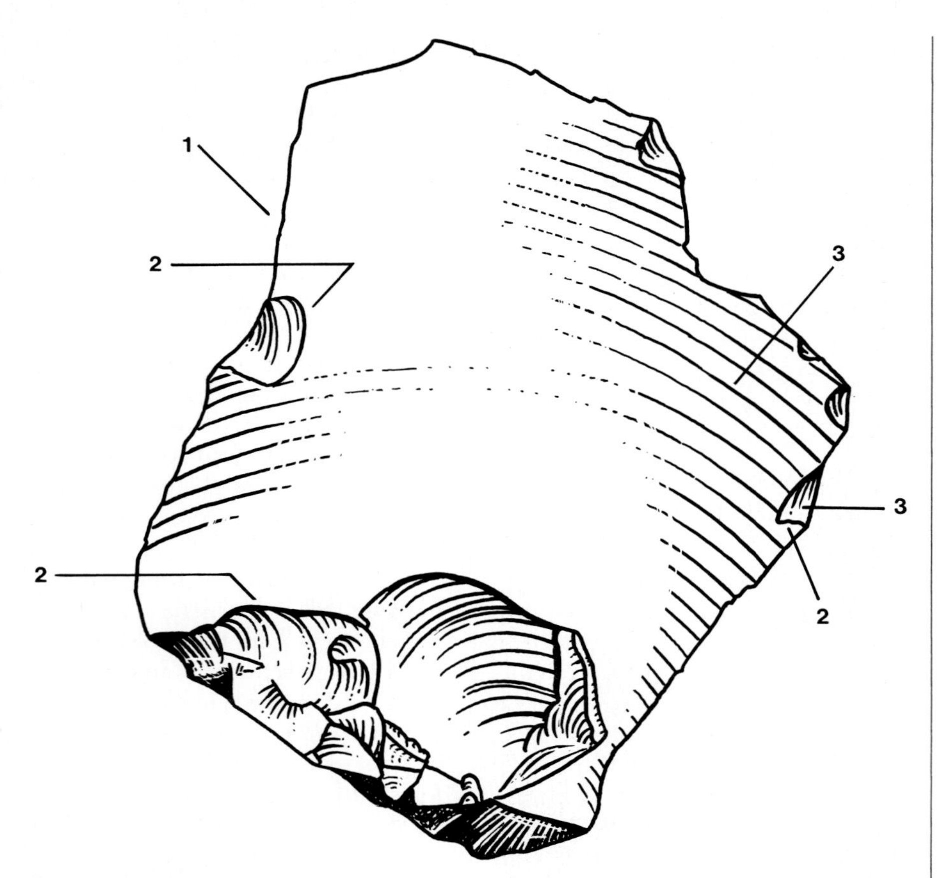

Figure 5-23. Diagrammatic representation of the stone tool. *1*—General shape of the subject. *2*—Edge of a chip. *3*—Direction of the breakage of the chip.

flow of ink. All new quill pens have to be washed with soap and water in order to remove a protective greasy coating from their surface. If pens are not washed, the ink will not adhere properly to the undersurface of the tip and will bead up in the upper section of the pen. The result is your aggravation as well as a sudden, unpredictable downsurge of ink, producing a nonobjective blob instead of a well-defined line.

If you really enjoy the process of drawing, make yourself a real quill pen. In order to do this, you need three or four large quills from a bird. Not all birds have proper feathers for a quill pen, but you can make quite a good pen from any large feather—one from a pigeon, turkey, or crow. You can find an abundance of feathers waiting for you in parks and forests. First disinfect the feather with vodka or rubbing alcohol, then dip the quill in boiling water for a couple of seconds. Prepare a sharp X-Acto number 11 blade and cut the tip vertically, the same way as you would if you were sharpening a pencil. Try to make a smooth, even cut. If the inner structure of the quill is visible, remove it carefully. Remember that burrs will interfere with the drawing, therefore eliminate all burrs by shaving the edges with a sharp blade. The quill may have a tendency to remain slightly twisted after the first cut. Be sure that the twist has been properly eliminated. Place the quill with the uncut side down on illustration board, and make a straight vertical cut directly in the middle of the uncut wall of the quill. Be very careful not to damage the structure, and make sure that this cut will divide the point into two halves. Eliminate all burrs; cut the extreme point of the tip horizontally. Make sure that only the very last part of the tip is eliminated. The quill pen should produce a thick line. The initial thickness of the drawn line will be proportional to the width of the tip.

2. The other very important tool is the holder for the metal quill pen. There is a variety of holders on the market. Try your best to obtain a holder that is not very thick at the lower area. A very thick holder will interfere with drawing. If you buy a holder with cork at the place where your fingers will hold it, remove the cork. The proper thickness should not exceed the thickness of a pencil. Avoid sculptured holders at all times. Holders that have variable thickness are supposedly designed for better grasp. Unfortunately, those holders presuppose the position of your fingers on the holder. Remember that the holder of crow quill pens is designed for your fingers; your fingers are not designed for the holder.

Preparation of the Specimen:

1. Obtain the specimen from any shop that sells arrowheads. If you are lucky, you can find an arrowhead in your own backyard.

2. Buy a piece of obsidian from any store that sells minerals. Obsidian is a volcanic glass from which an abundance of stone tools were made. Break it with a hammer into small pieces. *Caution:* Be very careful, as obsidian chips produce very sharp flakes that can cut on touch. Wrap an old towel around the piece before striking it with the hammer. Do not cut yourself, and definitely watch for small, flying particles. Do not endanger your eyes while breaking the piece into small fragments.

After breaking the large piece, select a small fragment about three to four centimeters long.

Basic Steps:

1. Observation
2. Positioning of the specimen
3. Preparation of the sketch, depicting the outline of the subject and the major areas of flaking
4. Tracing the sketch into a final sketch
5. Searching for indication of the smaller areas of flaking and the direction of cleavage
6. Preparation of a series of sketches presenting the direction of cleavage
7. Preparation of the sketch presenting a coherent presentation of all pertinent informations
8. Rendering
9. Preparation for production

PROCEDURE FOR DRAWING

Observation: As always, examine the specimen from all sides. During examination, remember that you are looking for two definite bits of information: the cleavage of the chipped areas and the direction of the cleavage. The three-dimensionality of the subject is secondary to those two prime areas of interest.

Cleavage is an area with more or less definitive boundaries obtained by splitting a rock. It is the area from which the flakes of the rock have been chipped away by the maker of the stone tool. In order to shape a tool from the rough stone, the toolmaker has to strike that stone with forceful skill, cleaving away bits and pieces. During such process the small pieces of the rock are knocked out, marking their previous location with edges. One of the first things you must do is to find those edges and therefore the shape of separate cleavages. As the traces of the direction of the breakage are also evident, the next step is to find the direction of the cleavage. It sounds simple, but as a matter of fact this project is very difficult; for this reason, be extra careful during observation as well as during the sketching process. Check and recheck your analytical approach, because scientific expertise is not available to you. During observation disregard all superficial scratches and cracks found on the surface of the specimen. Disregard all external and internal discolorations. Most edges designating the boundaries of the separate cleavages will be relatively easy to find; however, depending on the type of material from which the tool was produced, as well as the preservation of the specimen, deduction of the edges may become a complicated procedure. When examining a tool made from obsidian, make sure that transparency of the volcanic glass does not influence your judgment. All the visible internal structure such as bubbles or traces of other minerals should be disregarded. When drawing a flint tool, the internal structures will not be visible because the material is opaque. Prepare to have a pencil-type flashlight ready, as you will need it at all times. The best way to find the cleavage and its direction of breaking is to illuminate the particular area of the tool at a very sharp angle and observe the resulting shadows. Take the tool in one hand and the flashlight in the other. Shine the light on the tool at a level almost parallel to the area of interest. The ridges will be illuminated. Then slowly turn the tool back and forth and try to locate the semicircular undulations. Those undulations, wavelike apparitions similar to the ripples of the water created by the toss of a pebble, are the directional lines of the cleavage. Sometimes a great concentration of such markings exists in a particular area, while in other places only one or two may be found. Take your time observing; make sure that before sketching you will be able to differentiate between surface imperfections and directional markings. Remember that directional markings will be more rather than less parallel to each other. This means that if in a particular area you find a semicircular wavelike formation as well as sharp-looking but extremely shallow lines forming a parallel or starlike formation running against the undulations, the sharp lines are surface imperfections. Most of the time such lines will be of rugged construction and as such, are relatively easy to differentiate from the directional lines. Directional markings in any of the areas will always go in *only* one direction. Never will you find the directional lines of the cleavage crossing each other. Whenever you notice such a situation, remember that one set of the markings is an irrelevant surface imperfection. On closer examination, using the flashlight, you will find the characteristic differences. This means that within an area designated by the sharp ridges, the directional markings will be confined to one basic direction, even if in some of the smaller sections of that area are other short, linelike structures appearing to contradict the basic information. If in your judgment such formations appear as a truthful directional line, look for the edge or edges of the cleavage, as those edges must exist around such linear structures. Some of the chips can be extremely shallow and therefore the boundaries can be easily overlooked. Manipulate the light, illuminating the subject in order to notice everything. The most difficult areas to decipher are those where the boundaries of the cleavage were damaged by the natural forces of erosion. This is especially evident on the surface of flint tools such as arrowheads. The edges of the cleavage will not be as clearly visible as on obsidian tools, especially the tools which you may obtain for the purpose of drawing. Such freshly broken pieces of obsidian offer a good choice for exercise because of the freshness of the breakage. At the same time a drawing of an arrowhead may be more visually pleasing than the irregular split of the obsidian.

In situations where it is impossible to detect any traces of the directional lines, the lines are not drawn. As always in scientific illustration, without evidence an illustrator cannot invent things. In order to make the observations a little bit easier, use white chalk, rubbing it gently over the surface of the tool. Some of the cleavage ridges may become more visible, helping you to extract the pertinent information. Restrain yourself from scratching the surface of the tool with sharp objects such as needle of the compass. Such scratching will destroy the surface of the subject, making it unfit for research. If by any chance a soft foreign matter is present on the surface, remove it by soaking the tool in water and washing it.

During observation, choose one side of the specimen for drawing. Avoid three-quarter views—such views are dramatic, but not accepted by science. Choose one of the flat sides for drawing. When drawing for an archaeologist, the position of the subject will be chosen by him; all major decisions pertinent to the project will be made by him. You will have a great many questions regarding the truthful shape of the cleavage and the direction of the breakage. In this difficult project you will have to rely on your clear thinking and observation. Remember that you are interested in extracting two basic pieces informations from the subject: the areas of cleavage and the direction of breaking of the separate chips. Remember that with the exception of old flint tools, sharp or very prominent edges designate the cleavage. Remember that the directional lines are much softer and not as prominent as the ridges. Remember that the directional lines resemble undulations similar to ripples on the water. Make sure that you notice the difference between superficial markings and sudden deep crevices. Do not be overly concerned about the changes of three-dimensional structure of the subject; if you cannot deduce any information, leave that area unfinished. Make sure that if by any chance you encounter an unbroken surface, mark such an area with an imitation of the texture found on the subject. Make sure that you practice illuminating the subject from different angles for a while before making any valid observations.

Positioning: When you have chosen one side of the subject, you are ready to position the subject properly. Take a small piece of white illustration board and prepare a little Plasticine. Place the Plasticine on the illustration board and form a small cone, making sure that it is properly secured by squeezing the lower portion of the cone to the illustration board. Do not make the cone too tall. Try to be reasonable and keep the specimen close to the illustration board. Take the specimen and place it horizontally on the Plasticine. Make sure that the proper side of the specimen is facing you when you are looking directly at the center of the specimen. In order to assure exactness of position, move the board with the specimen placed on the top of the Plasticine toward you so you can look directly at the surface of the specimen when your head is lowered over the drawing table. Gently push the subject into the Plasticine, maneuvering it into the desired position. Make sure that the specimen is not crushed against the illustration board and at the same time will stay securely in place. During preparation of the sketch you will have to remove the specimen from its position for further study to assure that all cleavages and the directional lines will be drawn correctly. As you may find yourself in need of repositioning the specimen, it would be wise to mark its original position on the illustration board. Move the mounted specimen toward you. Make sure that its position on the table is directly under your eyes. Do not look at the specimen from left or right, but make sure that with both eyes open, looking directly at the specimen, you can clearly see the sharp points of the outline. Take the sharply pointed pencil and while looking at the center of the subject, mark on the illustration board the most outstanding features of the outline. Make sure that you will remember the features you have marked. If the specimen has a very complicated outlying edge, feel free to write explanatory comments on the illustration board. When finished with positioning, you are ready for the first basic sketch of the subject.

Drawing the Sketch: The very first task awaiting you is the translation of the contours of the specimen onto the tracing tissue. As you know, there are two basic techniques that can be used to do this with desired accuracy: tracing and the grid technique. The grid technique takes more time than tracing, therefore you will use that later; we are all aware that time means money. Prepare the Magic Art See and Draw Copier, a piece of illustration board, and a sheet of tracing tissue. The Magic Art See and Draw Copier is a toy drawing tube or camera lucida. It is very cheap and at the same time will give very good results. It is a boxlike, small structure with two consecutive sides open, with a combination of a mirror and a small plate of glass installed inside. Above the mirrors a small, round opening permits the viewer to look inside the device with one eye. When the device is directed toward the desired subject, it is possible to see the reflection of that subject through the round opening superimposed on the tip of a pencil placed underneath the device. In order to become acquainted with the principle of the device, look through the round opening, directing the copier toward a brightly illuminated subject like your window; place the black or very dark paper under the device. When you see the reflected image, place your hand on the paper and move it back and forth until you are able to see the reflected image and your hand superimposed over each other. When you feel secure with the dark background, try white paper. Now you will notice a marked decline in the readability of the subject. The reason for this is an improper balance between the illumination or the brightness of the subject and the reflection of the light from the white surface. In order to see the image better you will have to adjust the difference of illumination. With the white paper used as a surface for drawing, you will never be able to see the subject very clearly, just well enough to trace the important elements, thereby saving yourself considerable time.

To proceed, take the positioned subject and mount the piece of illustration board with the specimen on it vertically. Make sure that a 90-degree angle is maintained between the surface of the table and the subject. Take the metric ruler and draw a line one centimeter long next to the subject. You will need this measurement as a reference when drawing a scale. The enlargement or reduction of the subject depends on the proportion of horizontal distance between the device and the subject and the vertical distance between the device and the surface prepared for drawing. Prepare two separate lights. One is needed for the illumination of the subject and the other for possible illumination of the tracing tissue.

Position the specimen as well as the device. Make sure that the right angle is maintained, otherwise your drawing will be a distorted (perspective) representation of the subject. Place the tracing tissue taped to the illustration board under the device and first trace the line depicting one centimeter. As later you will have to remove the specimen, and use the pointers to measure a few distances by hand, make sure that from the very beginning you are aware of the enlarging factor. It will make your life easier when the enlarging factor is represented by a round number. Your sketch should not exceed fifteen centimeters or so. It may be necessary to adjust the distances in order to obtain the proper enlarging factor. When the enlarging factor is correct, trace the outline of the subject as well as all visible ridges, using a well-sharpened H pencil. While tracing, do your best to maintain stability and readability of the line. Do not erase your mistakes; mark the wrong lines, but preserve them on the sketch as evidence, thereby making sure that this mistake will not be repeated.

If you do have access to other reducing/enlarging devices such as an Artograph, Goodkin, or Art-Tec viewer or an opaque projector, use it. If this is impossible and you cannot find the Magic Art See and Draw Copier, use the centuries-old grid technique in order to obtain the outline of the subject. Why do I not suggest photography? Try it and find out. Photography is more time-consuming. It is harder to capture the information needed for our project through photography. It is almost impossible to correctly illuminate the stone tool and snap a picture of the details. Such a task is especially difficult when the subject is obsidian.

When an outline of the specimen and all ridges visible on the surface of the subject are correctly drawn on this first sketch, take the specimen from its vertical position. Place the specimen in front of you and using the flashlight, illuminate the surface and determine if all the cleavage has been observed and drawn. As it is hard to notice all the details when using a drawing tube, you will have to add necessary details. In order to be very correct, use a vertical and a horizontal line drawn at right angles as a reference in measuring and transposing additional details. Draw using simple line; remember that erasing mistakes will not help in overall presentation. This basic sketch will be refined later. When finished, remove the specimen from its established position and start drawing the directions of the cleavages.

Before drawing further, retrace the first sketch. Place a new tracing tissue over the first sketch and carefully trace the image. Remove the first sketch from the illustration board and affix the second sketch, using masking tape. Place a clean sheet of tracing tissue over the second sketch and tape it to the board. Without retracing the image visible through the tracing tissue, draw the directional lines of the cleavage. Do not draw many lines in one particular area. Keep this informational sketch readable. End lines with arrows indicating their direction. Remember that the indications of the directional lines can be found when moving the specimen slowly back and forth with the light shining on the surface at a low angle, illuminating each cleavage separately. Hold the specimen in your hands and

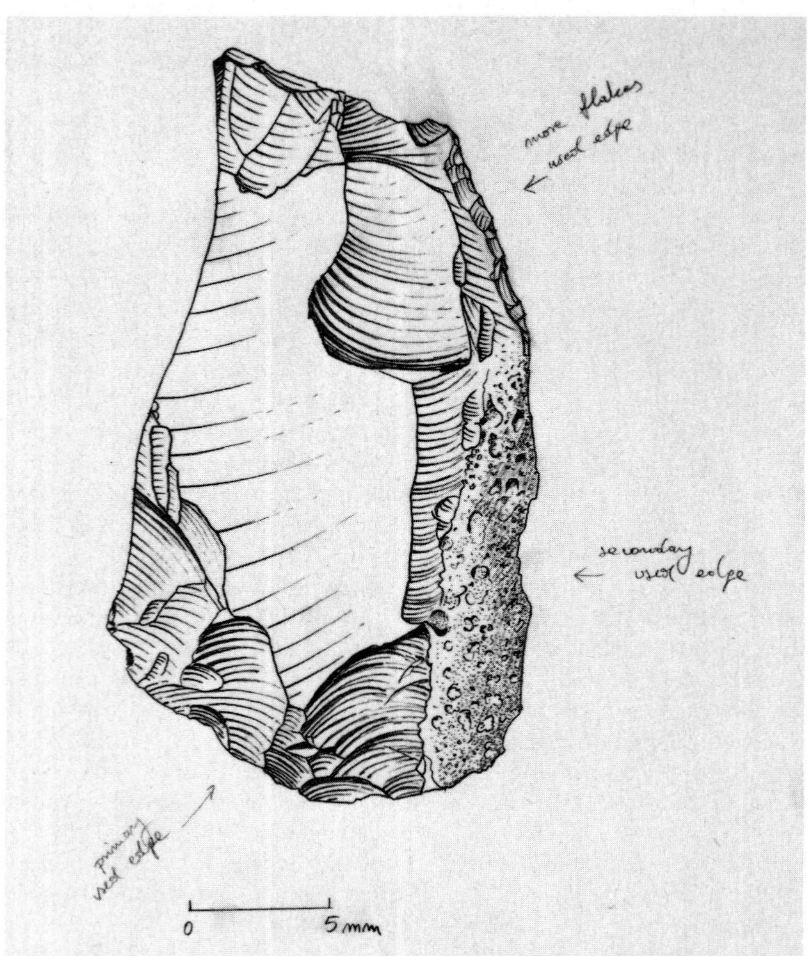

Figure 5-24. Obsidian implement from Sumatra, TP A6, final sketch. Pencil on tracing tissue. Drawing by Zbigniew T. Jastrzębski, Field Museum of Natural History, Chicago. Unpublished. Drawn in 1979.

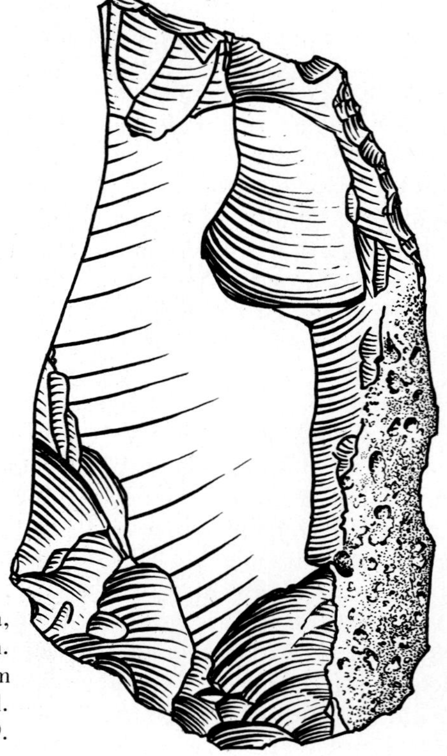

Figure 5-25. Obsidian implement from Sumatra, TP A6. Crow quill on mylar drafting film. Drawing by Zbigniew T. Jastrzębski, Field Museum of Natural History, Chicago. Unpublished. Drawn in 1979.

make sure that the illuminating light barely skips the surface. Watch for glimmers of slightly curved lines, more or less parallel to each other. Such lines do not have sharp ridges and look more like undulations. Some areas may have an abundance of such markings, making the directional lines easy to observe and therefore to draw. Others may have very few, and some none at all. Make sure that a particular area of cleavage is illuminated from different directions, especially when observing an obsidian tool. Disregard surface imperfections such as cracks and scratches. Concentrate on small sections of the subject, slowly rotating the specimen under the light. Draw what you see and make sure that you write all of your thoughts immediately. This will help you in the process of elimination of false information. Remember that you do not have to draw on one sheet of tracing tissue. As a matter of fact, it may be better to have several sheets with notes rather than one overworked sketch. Do not get discouraged, as the stone implement you are drawing may be hard to decipher. Arrowheads made of flint have different surface characteristics from those made from obsidian.

Have some white chalk handy. Gently rub the surface of the flint implement, moving the stick of white chalk horizontally. The chalk will catch the upper surfaces of ridges and undulations, making them more visible. If your tool is quite flat and relatively large, five centimeters or longer, use ground graphite rubbed over a thin layer of smooth tissue. One layer of Kleenex or toilet tissue is the best. Keep the tisue tightly wrapped around the subject with one hand and using your finger, rub powdered graphite over a section of the tissue covering the specimen. While doing this, make sure that you apply the powdered graphite in smooth, one-motion strokes, maintaining the same direction through one application. Repeat the procedure as many times as necessary, using a new, clean piece of tissue. The results will depend on the surface characteristics of the specimen, not on your skill. Throughout the observation and drawing, remember your goal. Do not get involved in too much experimentation with rubbings and sketchs. All of this material is supplementary. Your main problem lies in deciphering the surface characteristics of the implement in question. Make sure that your drawn lines are legible. Make sure that the angle, the curvature of the directional characteristics, is properly presented. This is of extreme importance. Because of the small size of many areas of cleavage, especially near the edges of the tool or near the major ridges, you will not be able to measure the direction of cleavage. Observe the direction of the line in relation to the outlying areas, then make a guess. After the initial transfer of the observation, go back to the same area and recheck whether you were correct. Naturally all mistakes will be corrected by retracing, not by erasing. Draw simple lines; do not shade the drawn object—tonal shading does not help. It destroys the clarity of the information. Remember that the whole drawing has to be drawn in line only. Later, through manipulation of density and the thickness of the lines, you will establish some feeling of overall value. With the thickness and density of the lines, the presentation of the direction of illumination as well as the three-dimensionality of the whole subject will be achieved.

Do not forget to draw registration marks on all sketches. Registration marks—a cross within a circle—should first be drawn on the bottom sketch and then traced on the overlaying tissue. Such marks will help you to align both images easily when necessary.

Make sure that your observations are correct. Think before you draw; remember that you do not have to restrict yourself to one sheet of tracing tissue. When satisfied with your results, take a new sheet of tracing tissue and place it on the top of the two previous ones. Remember that for the viewer, it should appear that a light is illuminating the specimen from the upper left side. Your drawing must show this clearly.

Draw the outline, using a softer B or 2B pencil. A softer pencil will allow you to simulate the quality of the crow-quill line. The outline and all the ridges should be drawn in a uniform but thicker line than the directional lines. First trace the outline of the subject and all lines representing the ridges. The thickness of the line should be approximately one millimeter if your drawing is about fifteen centimeters long. Determine the dark sides—they will be on the right and the illuminated portions on the left. All depends on the actual three-dimensionality of the subject you are drawing. If your subject is very flat, do not exaggerate. Such a subject must appear flat. If it is very three-dimensional, then your drawing should appear accordingly three-dimensional. Remember that the most important part of the drawing is a presentation of the information pertaining to the cleavage and its direction of breaking. Start with the left edge and continue this concept throughout the whole drawing. Avoid having dark touching dark. You must have continual change between dark and light in order to make the ridges stick out toward the viewer. Therefore, in well-illuminated areas use few lines; in darker areas, use more. When drawing a directional line from the left, be sure that such a line starts from the outline or ridge and continues, *narrowing* in width just as the line made by a crow-quill pen would, and stops without touching the opposite outline. This means that in any well-illuminated area you will have a directional line starting from an outline and proceeding to about half of the designated space, leaving a white area between its end and the other outline. The right sides of cleavages should be back-lit in order to be easily differentiated. This means that in the sections which theoretically should not be illuminated at all, the right side or the lower right side will remain white, suggesting a light coming from another direction than the designated upper left. Remember that the only way to control the values is by the density and the thickness of the directional lines.

Do not attempt to represent the form of the surfaces of the specimen in any conventional way. When drawing very sudden drops of the structure, the thickness of the line representing the cleavage border as well as the density of the lines representing the direction of the cleavage can increase. The lines are symbolic and designate the direction of the breakage. Remember that only by the subtle play of the densities of those symbolic lines is it possible to formulate a *general* feeling of the three-dimensionality of the separate cleavages and the whole specimen. You will probably have to draw and redraw the last sketch a number of times before you grasp the concept. Do not get discouraged. Use the reducing lens and look at your drawing from a distance. Remember that what really matters is the readability of the information after the reduction and publication of the drawing. Remember that the reduction will be in the vicinity of 300 percent!

When finished, spray the sketch with the fixative, wait until it dries, and carefully trace it. From this final sketch your rendering in ink will be produced. Take a new, clean illustration board and tape the finished final sketch to the surface. Make sure the sketch is fixed and the board is clean. Wash your hands, clean the drawing table, and prepare for the rendering.

Rendering: The key to drawing with a crow-quill pen is the steady hand of the artist and the proper combination of the particular pen with the surface to be drawn on. Naturally, practice will help you to obtain the necessary skill. If you have not used the crow-quill pen previously, spend a considerable time practicing. Use a variety of papers, experimenting with different pens. Make sure that all the papers have a smooth surface. The pen cannot scratch but has to glide on the surface of the paper. Remember that the fiber structure of papers differs. By practice I do not necessarily mean drawing straight lines, although such procedure is beneficial, especially if you remember to vary the thickness of the drawn line. Consider the first ink tracing of the final pencil sketch as a necessary practice. Be prepared that in order to obtain proper results you will redraw—or I should say *retrace*—that sketch a number of times. Remember that the size of the pen is proportional to the size of the drawing, the details to be presented, and to the size of the final reduced product, the printed reproduction of that drawing.

To start, clean the drawing table, prepare a bottle of ink, insert the crow-quill pen into the holder, and place a sheet of good-quality mylar drafting film on top of the final sketch. Make sure that the drafting film is secured only on one side with masking tape. Make sure that the new metal crow-quill pen has been washed with soap and water before it is used. Cut a small sheet of white opaque paper and keep it handy. While drawing, insert that paper periodically between the mylar drafting film and the final sketch, giving the white background needed for proper appraisal of lines and tones. When arranging and preparing the necessary tools, be sure the drawing ink is placed at some distance from the drawing. You do not want to knock over an open bottle of ink with the edge of the illustration board. Do not forget about the reducing lens. The lens is necessary for proper appraisal of values produced by the thickness and density of lines *during* the process of drawing. As you have the information as well as the values preplanned and clearly presented on the final sketch, the rendering in ink is in actuality a tracing of that sketch. Sit yourself comfortably and place your elbows on the drawing table. Place the illustration board with the final sketch covered by mylar near you but not too close. Remember that you must feel comfortable and at ease in order to control the line. Take the crow-quill pen in one hand. Bend slightly over the table, moving your chest forward. Without removing your elbow from the table, and keeping it in a stationary position, move your hand over the surface of the table. Notice the direction the tip of the pen will make: It is a small semicircle. During drawing you will have to use this inherent ability, aligning each directional line with the natural motion of your hand. Remember to hold the pen approximately three-quarters to one inch above the tip. Find the best, most comfortable place to hold it. At the same time make sure that the pen is positioned at about a 45- to 50-degree angle in relation to the surface of the drafting film. The best way to hold the pen is to let the holder rest on the edge of the middle finger in the area of the lower portion of the fingernail, keeping it between the thumb and index finger. The thumb and index finger will provide the rotation of the pen; the up-and-down motion of the wrist will provide the necessary pressure. Dip the crow-quill pen in the ink and take a look at how much ink you have on the pen. The pen cannot drip but at the same time cannot be too dry. Take the pen, dip it in the ink, and slightly extending the fifth finger to the side, draw the outline of the implement. Press down the surface of the mylar with the stylus while drawing, making sure that it will not bounce under the pressure of the pen. The outline should be uniform, being slightly thicker than the width of the middle sections of the lines representing the direction of the cleavage. Depending on the complexity of the image, the outline can be considerably thicker than all of the directional lines. The width of that line is proportional to the quantity of details depicted near the outline as well as the size of the whole drawing. When drawing, make sure that you turn the illustration around on the table as needed. Do your best to connect separate strokes. Try to continue the line, drawing it in one direction. This helps when joining separate, shorter lines. Make sure that you do not spill ink on the surface of the illustration. Watch for overflow of ink when connecting two wet lines. When there is too much ink on the tip of the pen and too much ink on an already-drawn line, during the connection an ink blob can easily form. Such accident occurs more easily if the pen is held incorrectly. Do not hold the pen in too horizontal a position. Remember that the angle between the pen and surface of the illustration should be between 40 and 60 degrees. When mistakes happen, do not panic. Wait until all of the ink has dried and then appraise the situation. If a lot of ink has been spilled, it may be better to start from the beginning; but if you have missed connecting two strokes or a small blob has formed, you can repair the damage later. Mistakes can be corrected relatively easily after the ink has dried out by scratching unwanted sections of the line with the tip of a sharp X-Acto blade. It is always better to leave all mistakes intact until the whole drawing is finished and then correct them. During scratching, the surface of the mylar film will be damaged. Depending on the type of mylar film, you may encounter difficulties drawing on the scratched surface. Move the pen toward yourself rather than away. It takes some experience to move the crow-quill pen against the natural movement of the arm. Remember that the thickness of the line is controlled by the pressure exerted on the tip of the pen. The more pressure, the thicker the line. When drawing the outline, be sure that the pressure is uniformly constant, otherwise the line will vary in thickness. Be sure not to smear wet portions of the drawing. Before placing your hand on the drawing, always look for small wet sections of previously drawn lines. Be aware that lines drawn with a crow-quill pen will not have a uniform layer of ink. This is a natural result of the characteristics of that pen. The ink flow from the tip of the pen to the surface will vary, resulting in differences in drying. Make sure that all small changes of the outline are carefully presented.

When the outline is finished and the ink has dried out, proceed to draw the ridges representing the boundaries of separate cleavages. Keep those lines more or less uniform in width. Remember that the boundaries should be clearly visible to the viewer after the drawing has been finished and published. Use the reducing lens for looking at the drawing during the production of the illustration. Take the previously prepared sheet of opaque white paper and insert that paper between the mylar drafting film and the final sketch, thereby making the freshly drawn ink image clearly visible. When drawing the cleavages, make sure that all small chips, so carefully presented on the final sketch, are properly transferred to the ink rendering. This may call for a variation of the thickness of the line, as in the smaller areas you may have to use a thinner but uniformly drawn line. The lines representing the direction of the cleavage must be executed with great care. Beside the informational context, those lines will present the three-dimensionality of the implement. Trace the designated lines from the final sketch, making sure that the width of the ink line will vary. Start with the left side and start drawing from the outline of the whole tool or from the outline of the cleavage. Do not get carried away and do not change or improve the sketch. Remember that at present you are concentrating on the quality of the line. You are preoccupied with the control of the drawing tool. Pressure on the crow-quill pen will result in a heavier, thicker line, but the thickness of that line will depend on the type of the pen you are using. Press relatively hard if the pen is not very flexible, releasing the pressure as you continue drawing the same line. Make sure that the drawing is positioned properly and that the curvature of the line falls within the natural movement of your hand. If some lines are long and cannot be drawn in one stroke, do not force the impossible. Draw such lines in sections, making sure that the short strokes are connected correctly. When having difficulty connecting the short strokes, do not connect them; leave a bit of space between. Just make sure that at that point the thickness of the new section of the line is about the same as the previous, further narrowing the newly drawn line. On the dark side, the right side, make sure that backlighting of the cleavages is presented clearly. Without this the separate chips will not be clearly visible. When portraying a sudden drop of the surface, thicken the area considerably, making the drop visibly pronounced. When finished, wait for the ink to dry completely. Then take the new X-Acto blade and scratch out all mistakes. Insert the white paper under the drafting film and observe the surface of the film. Look for light smears, fingerprints, and other unwanted features. Remove the larger smears by washing the area with a Q-tip dipped in rubbing alcohol. Do not touch the drawn lines with the wet Q-tip, as this will produce more smears. Take the drawing off and place it on the new, clean piece of illustration board. Compare the final sketch with the rendering. Use the reducing lens to appraise the tonal values. When some areas appear darker, lighten them by carefully scratching some of the darkness. Remember that directional lines must be clearly visible. At the same time you can scratch across those lines, cutting the longer lines into small sections and lightening the value of that particular area.

Among your sketches find information pertaining to the enlargement of the subject. Tape the drawing to the illustration board and draw the scale under the image of the stone tool. Do this using a ruler and the Rapidograph pen or thin lettering pen. Regular crow quill is not very suitable for drawing straight lines with a ruler. Make sure that the proportion between the length of the straight horizontal line and the enlargement factor is maintained. For example, if the drawing is five times larger than the specimen, the enlarging factor is five. This means that every centimeter has been enlarged five times. In this case the scale, a horizontal straight line equaling five centimeters, is an enlarged representation of one centimeter. Therefore such a line should be designated as one centimeter. On the left side place a zero, and on the right side a number one, using instant lettering, a lettering guide, or a scriber. Make sure that numbers are visible and easy to read. Designate the measurement in small lowercase letters. The word *centimeters* is customarily shortened to two letters—*cm*. When everything is finished, clean up the drawing table, remove the ink to a safe place, and get ready for dry-mounting.

Dry-Mounting: Do not dry-mount mylar drafting film with Seal dry-mounting tissue or Twin-Tak. Use tape to secure the finished illustration. Take the clean sheet of hot-press illustration board and place the drawing on its surface. Position the drawing correctly and tape the upper edge of the drafting film to the illustration board. For elegant results, use Scotch double-coated tape number 666, taking advantage of its white plastic liner. Naturally any tape, masking or transparent, can be used as long as the illustration is attached to the board only at one edge. Remember that the white background must be visible through semitransparent drafting film in order to assure the proper black-and-white reproduction. A color background is not advisable, as it distracts from the information. The illustration must be attached in such a way as to produce a free, easily raised flap. This is necessary because during the initial steps of the reproduction process the drawing will be placed on a vacuum table. The suction of the vacuum will flatten the drawing, and the camera will be able to register the proper, undistorted image.

Preparation for Production: Cover the illustration with tracing tissue, taping the tissue along the upper section of the illustration board. Over this place a sheet of black cover-stock paper, remembering to use one-ply paper, and again taping this sheet along the upper section of the illustration board. Use the black photographic masking tape for a handsome presentation. If needed, be prepared to designate the final reduction. This can be achieved in various ways. Place the illustration in front of you on the drawing table and tape the board in two places to the table, assuring the needed stability. You will have to know something about the final printed size of the illustration. You will have to know at least one dimension, horizontal or vertical. If you do know both dimensions, vertical and horizontal, proceed the same way as when you calculated the enlargement of the specimen. This time the *illustration* will be reduced to a certain size. If you do know only one dimension—usually the horizontal one, such as the width of the page or width of the printed column—calculate the reduction of the width of the illustration only and let the vertical dimension "hang out." You can mark the reduction by percentage or by directly writing the size of horizontal dimension in inches. I advise you to use inches because in case the platemaker makes a mistake, this will add more weight to your possible argument. If necessary, you can use a ruler and instantly prove to anybody that the reduction is incorrect rather than hassling about perccentages. When you write a capital *R* designating reduction and clearly indicate three inches as a final size, you surely cannot find a better proof to your statement. If you want to, use a reducing wheel, a device made of two circles with desginated distances marked in proportion to each other. The larger wheel gives the size of reproduction, while the smaller gives the size of the original artwork. A small window is cut for two additional readings, the percentage of the original size of the artwork and the number of times of reduction. Such a device can be obtained from an art-supply store. When marking the reduction of the illustration, open the black cover and the protective tissue. Use a good-quality metal ruler and the triangle. Align the ruler with the drawn scale, assuring the horizontal position of the designating line. Using the *opposite* edge of the ruler, draw a horizontal line with non-reproducing blue pencil. Do not press very hard, but make sure that the line is visible. Align the edge of the ruler with that line and using the triangle, mark the farthest point of the drawing on that line. Make sure that those marks are short, vertical lines and that they are visible. Draw an arrow on the horizontal line on the inside of those vertical markings pointing to those markings. Under the line write capital *R* or the word *Reduction* and the designating number, using percentage or inches.

Comments: In order to become proficient in crow-quill technique, do practice. Use smooth-surfaced paper. Remember that you will have to draw a lot, so obtain cheap paper. Brown wrapping paper is a reasonably good paper for practice. Draw straight lines cautiously, varying the lines in width. When drawing straight lines, move your whole body along the edge of the table rather than the arm only. Stand firmly on both legs, centering yourself at the middle of the table. Start from the left if you are right-handed and from the right if you use the left hand. Keep your arm above the table but in a relatively stationary position. Remember that the line you draw will be directed by the movement of your torso. Move slowly, with concentration; do not rush. You have to learn to be deliberate and patient. Keep your eyes on the line you are drawing. Draw lines close to each other. When the pen runs out of ink, dip the pen in ink and continue the same line. Another good exercise is drawing the form of an apple. Draw the outline of an apple, then draw the curved surface of that object, making sure that the lines starting from the left side are thinner than the same lines on the right side. Repeat this process of drawing numerous times. Remember, the more you draw, the better you will become.

NINTH PROJECT
Zoology and Paleontology

Subject: Drawing of a vertebra, dorsal and right lateral view.

Technique: Scratch board, line only.

Light Source: From upper left side.

Materials:

1. Pencils—2H, H, HB
2. Tracing tissue
3. Illustration board, hot-press, smooth-surface
4. White scratch board number 00
5. X-Acto knife with number 11 blade
6. Scratch-board steel graver, 25 lines
7. Scratch knife number 20
8. Crow-quill pen, Gillott mapping pen number 291 or equivalent
9. Pen holder
10. Ink, black
11. Brush number 6
12. Metric ruler
13. Metal ruler
14. Plasticine, 1 oz.
15. Reducing glass
16. Stylus
17. Carbon paper
18. Masking tape
19. Black cover-stock paper
20. Black photographic masking tape

Basic Steps:

1. Positioning of the specimen
2. Preparation of the sketch
3. Tracing the sketch into final sketch
4. Rendering
5. Preparation for production

Nomenclature:

Lateral—side

Dorsal—top

Ventral—bottom

Anterior—front

Posterior—back

Body—the main ventral portion of the vertebra

Spinous process—large, easy-to-see dorsal portion of the vertebra, positioned posteriorly at the angle in relation to the body

Demi-facet—concave area located in side of ventral section of the body

Transverse processes—dorsally located, easily visible protuberances

Articular process—easily visible protuberances located ventrally from the transverse processes

Tubercles—a pronounced irregularity of the transverse process found at the upper border, lower border, and one externally

Description of Some of Materials:

1. A scratch-board graver is not essential for this project, although it can be used very effectively. The number of lines, in actuality separate blades, per inch designates such gravers. Therefore, a 25-line graver will produce 25 parallel scratches; the 35-line graver will produce 35 parallel scratches. The density of the lines may not be suitable for relatively small draw-

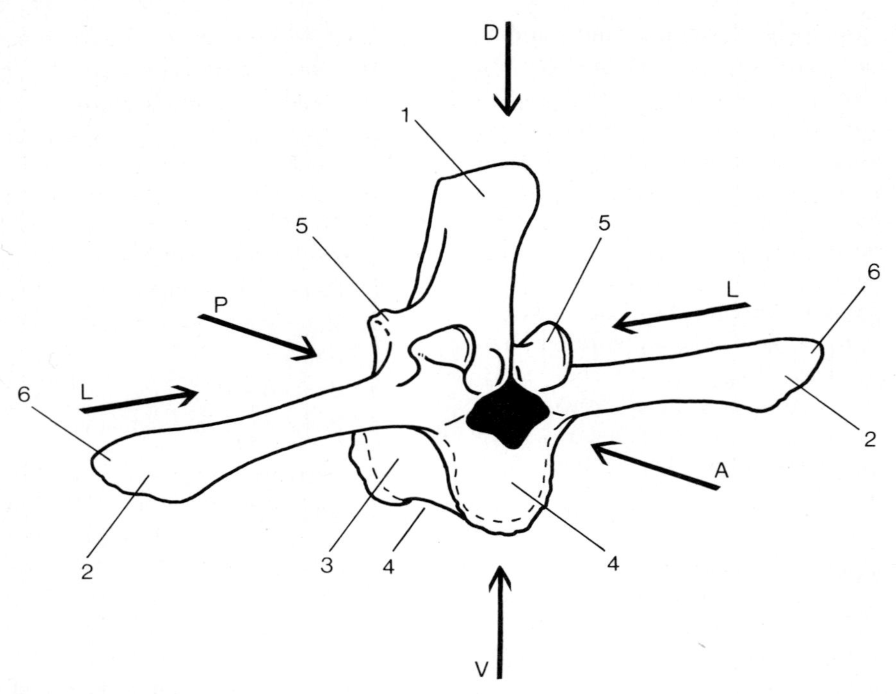

Figure 5-26. Diagrammatic representation of lumbar vertebra, front view. *A*—Anterior; *D*—Dorsal; *P*—Posterior; *L*—Lateral; *1*—Spinous process; *2*—Transverse process; *3*—Body; *4*—Demi-facet; *5*—Articular process; *6*—Tubercles.

ings. At all times the number of lines per inch must be in proportion to the size of the drawing.

Do you remember why you are preparing the drawing? It is to be reduced and printed, helping to illustrate the scientist's research. If for a larger drawing, let's say an illustration eight inches by five inches, 40-line graver is used, there is a good possibility that the final product, the reduced reproduction of that illustration, will not be readable. Why? In order to meaningfully answer this question, perform a small experiment. Draw a straight line five inches long. Divide this straight line into 200 equal segments (40 lines per inch multiplied by five inches). Draw horizontal lines at right angles to the five-inch line. You will have 200 lines drawn very close together. Take this drawing to the Photostat company and ask for one positive line print, reducing the five-inch line to two inches, and see the results. The 200 horizontal lines will in parts appear as uneven black, "closing" themselves too much. The effect is the same when an illustration is overrendered or overdrawn, when too much details in proportion to the reproduction size has been presented. Keep in mind that the drawing must be readable *after* it is reduced and printed, as it will be judged by the recipients not in its original size but only in its printed state. Overrendered drawings may look good in their original size, but are unreadable after reproduction. For this reason, use a scratch-board graver with caution. At all times make sure that the number of lines produced by the graver is in proportion to the final printed size of the illustration.

2. Make sure that you have a sufficient quantity of number 11 X-Acto blades. This is of importance, as during use the tips of the blades will break often and the blade, although still good for cutting, cannot be used for scratching.

3. Carbon paper is the same as that used for making copies during typing. It is very helpful in transferring the sketch to the scratch board.

Preparation of the Specimen: Obtaining a large-enough specimen may be a problem. First contact commercial companies selling educational materials. Their names, telephone numbers, and addresses can be found easily in the telephone book. Call or write and ask for a catalog of educational or biological supplies. Look under the Skeletons—Biology in order to find a vertebra. Secondly, check the neighborhood butcher shop. There you will be able to obtain a section of a lamb's or pig's spine. Take the specimen home and carefully clean the tissue without damaging the bone structure. Do not scrape; just cut off the larger sections of tissue. Do not break the segments of vertebrae during handling. Prepare a pot and boil the remaining tissue in water until virtually all of the tissue has fallen off the vertebrae. Remove the specimen from the pot and carefully remove every remaining bit of tissue from the surface of the specimen. Separate the vertebrae from each other and let the specimen dry. Prepare yourself for drawing.

PROCEDURE FOR DRAWING

Observation Take the prepared specimen in your hands and observe its shape and its surface. Make sure that you understand which side is dorsal and which side is lateral. Hold the specimen in such a way that you will be able to see a large round opening clearly. When looking at the opening you are observing an anterior (front) or a posterior (back) end of the vertebra. On one side of the opening you will be able to notice an oval shape, and at the opposite end there will be a pronounced growth of three protuberances. Two of those will be situated at an angle in relation to the opening. The third is positioned in between at the central location. Observe the third protuberance. It is the spinous process. We know that the spinous process is located on the dorsal portion of the vertebra. Turn the specimen around along the vertical axis of the spinous process and look directly at the top (dorsal) of that structure. If the outermost section of that structure is small and slightly rounded, the vertebra belongs to the upper section of the spinal column. If the same section is elongated and flat, the vertebra you are looking at came from the lower portion of the spinal column. Look closely at the top of the spinous process. If the end, or actually the flat top of that structure, forms a slight triangular shape, the sharp tip of that triangle indicates the anterior (front) of the vertebra. The opposite end, the broader end, indicates the posterior (back) of the specimen. When the specimen you are looking at comes from the upper section of the spinal column, turn the vertebra sideways. The spinous process will be very pronounced, being rather long and relatively narrow. Hold the vertebra by the opposite end of the spinous process. From this chunk of the bone all of the protuberances grow out. Those are the tubercles. The opposite side of the spinous process is the main body of the vertebra. When holding the subject by its posterior and anterior ends, make sure that the body of the vertebra is positioned vertically in relation to you. You will easily notice that the spinous process is located at the angle to the body. The direction in which the spinous process is pointing indicates the posterior (back) of the vertebra. Naturally, the opposite end is the anterior end of the specimen. When holding the specimen as you are, you are looking at the lateral side. One of the drawings will be of the lateral side of the specimen. The other, dorsal view will present the spinous process sticking out toward you.

After becoming familiar with the positions and the general designations of the top, bottom, front, and back of the specimen, pay attention to the surface irregularities. Turn the specimen with the ventral side toward you. This is the side which is opposite to the spinous process, being the bottom of the body of the vertebra. Observe the anterior and posterior ends of the vertebra. To the left and right you should find concave areas. Those are designated as demi-facets and indicate the attachment of the ribs. Observe the concavity of the body. Turn the specimen around to see how much of the observed structures you can see from the lateral side. On the vertebrae from the posterior section of the spinal column the articular process will be located near or on the lateral side of the dorsal section of the body. At present you are looking directly at this tubercle. Dorsally, away from the body of the vertebrae, a large concave area will give the beginning to the transverse process. The transverse process, when viewed from lateral side, partially obstructs the spinous process. Observe the changes of the surface, as starting from the ventral edge of the body you find a rather dramatic change in the configuration of that surface. The three-dimensionality of the drawing of that specimen will depend on the correct presentation of dark and light values. Behind the spinous process you should see a part of the transverse process on the other lateral side of the specimen. Still observing the vertebrae from the posterior section of the spine, turn the specimen with the dorsal view facing you. The top of the spinous process is at the center, while transverse processes are to the left and right. Below the transverse processes you should see the partially obstructed articular processes. Observe and pay attention to the angle of the posterior and anterior sections of the articular processes. The surface structure of the articular processes changes very rapidly. When looking at the vertebrae from the anterior section of the spine, you will notice that the spinous process is greatly elongated. While drawing the dorsal view, you will have to make sure that the length of the spinous process appears properly foreshortened. By not being aware of the changes in the surface structures of the specimen, it is possible to make a serious mistake by not presenting that process properly on the drawing. Please notice that when looking from the dorsal side the articular processes are at front of the transverse processes and do not present great difficulties during rendering. On the other hand, the depth of field enclosed between the uppermost part of the spinous process and the articulate processes is difficult to portray. The foreshortening of the spinous process does present a challenge.

In order to correctly draw the subject, you must know that subject very well. I suggest that a vertebra from a back section of the spinal column should be the subject of the first attempt. Such a vertebra has more volume, more mass, and therefore is easier to draw. The foreshortenings of the spinous and transverse processes are not as overpowering as on the vertebrae from the frontal part of the spinal column. When observing the specimen, do not be distracted by discoloration of the surface. Differentiate between darker areas caused by the thinness of the bone and an actual surface indentation.

Remember that the specimen you are observing has been prepared by you and therefore has not been bleached. Make sure that all bits and pieces of the dried, unremoved tissue are observed. Little things like that can alter the truthful shape of the subject. Watch for sharp ridges, small openings in the surface of the bone, and pay attention to the shadows cast by tubercles. The direction of the shadows depends on the placement of the source of light. Usually the placement of the lamp is accidental, resulting in a certain direction of illumination that is not questioned by the observer's mind. Make sure that you know what is in the shadows. If it helps you, draw quick sketches of the structures you observe. Explanatory notes combined with such quick sketches will be of great help.

Positioning of the Specimen: Start with the right lateral view. The lateral view is easier to draw than the dorsal. Prepare a bit of Plasticine. Draw a horizontal line on the illustration board, using ruler and an H pencil. Make sure that the pencil is properly sharpened and the line is thin. Move the board toward you. Make sure that you are able to look directly at the area near the line. During positioning, use both eyes—do not close one eye. Viewing the subject with one eye gives false information. Place some of the Plasticine on the illustration board, slightly below the drawn line. Secure the Plasticine to the illustration board by pressing the edges into the board and slightly smearing the Plasticine on the surface. Make sure that you have enough space between the top part of the small mound of Plasticine and the board to accommodate your specimen. Remember that you cannot crush the specimen when pressing from above. At the same time, the specimen cannot be too far from the board. Find the right lateral side on your specimen. In order to do this, look at the spinous and transverse processes from the dorsal side. The edges of transverse processes are similar to the letter *V*. When the narrow section of the *V*-shaped structure is directly toward you, the front of the specimen is away from you and the right lateral side is to your right. Turn the specimen sideways, making sure that the spinous process is facing you. As you know, the body of the vertebra is located exactly opposite the spinous process. Observe the outermost edge of the body. The left and right ends are very clearly visible. Place the specimen on the prepared Plasticine, pressing firmly but gently. Align the left and right ends of the body of the vertebra with the drawn line. Get up from the chair and look at the vertebra along the surface of the drawing table. Turn the illustration board with the specimen so that the spinous process will be to your left or right. Notice the angle of the spinous process in relation to the surface of the table. The spinous process must be exactly parallel to that surface. Gently adjust the specimen as much as necessary. Again look at the subject from above, making sure your head is positioned correctly over the specimen. Check if the ventral edge, the outermost edge, of the body is aligned with the drawn line. Repeat the process of positioning until the left and right ends of the body are aligned exactly with the drawn line and the spinous process is parallel with the surface of the table. After satisfactory positioning, start the sketch.

When the sketch of the right lateral view is completed, remove the specimen from its position and reshape the Plasticine in order to position the dorsal side easily and correctly. Make sure that the posterior of the vertebra is facing you. Place the specimen on the Plasticine and press it firmly but gently. Find the transverse processes. Align the left and right outmost edges of the transverse processes with the drawn line. Look directly at the posterior of the specimen. Check if the spinous process is placed exactly in a vertical position. If not, readjust the specimen. Do not forget to recheck the dorsal alignment. If you find that the specimen is positioned slightly diagonally, leave it that way; it is all right. The outermost left and right edges of the transverse processes are used as reference points during sketching. The sketch of the dorsal view of the vertebra may be drawn diagonally in relation to the horizontal reference line. Do not worry, as the final sketches of both views will be aligned before the rendering will start. The object of positioning the specimen is to obtain a definite relationship between the horizontal reference line and two points on the specimen that are distinct but separated by some distance. During the positioning of the dorsal view, be aware of the fact that the depth of field of the specimen, as viewed from above, is greater than from the lateral side. This is caused by the natural dimension of the specimen. If the specimen positioned for the lateral view is located too close to the illustration board, the sketch of that view will have slightly different overall dimensions than the sketch of the dorsal view, making precise alignment of both sketches an impossible task. Make sure that the center of the opening visible from the anterior or posterior side is at the same distance from the surface of the illustration board during the positioning of the specimen for right lateral and dorsal view. This will assure the correct presentation of the depth of field. In order to do this, look at the center of this opening when positioning the specimen.

In both cases you will have to get up from the chair and look directly along the surface of the table into the posterior or anterior section of the specimen. If you want to measure, use a small metric ruler or pointers. Make sure that the ruler is placed at right angles to the surface of the illustration board and that the pointers are held correctly.

Drawing the Sketch: The enlargement or reduction of the subject depends on the actual size of the specimen. As you know, a rendered drawing should be larger than the size of the printed image of that drawing. This does not mean that the size of the initial or first sketch should be limited to the dimensions of the rendered illustration. The first sketch can be reduced or enlarged through various means such as grid, Photostat, or opaque projector to the final size. Naturally, the initial as well as the final size of the sketch is dictated by the quantity of details to be presented as well as the technicalities involved in preparation of that sketch. When drawing a vertebra larger than two centimeters, be prepared to face a one-to-one ratio. This means there is a good probability that you will be asked to draw such specimen in its actual size. I assume that you have a subject larger than two centimeters; therefore, do not produce too large a sketch. For practical reasons, your final sketch should not be larger than five centimeters in its horizontal dimension.

First, establish the enlargement or reduction factor. As the specimen is positioned with the right lateral side toward you, take the metric ruler and measure the distance between the outermost dorsal edge of the spinous process and the outermost ventral edge of the vertebra body. This initial measurement will help you to establish the proper enlargement or reduction factor. When drawing transverse and superior articular processes, be prepared for a struggle with presentation of the foreshortening of those structures. To start, take a sheet of illustration board and a sheet of tracing tissue. Tape the tracing tissue to the illustration board. Draw a straight horizontal line on the tracing tissue, using a ruler and a well-sharpened H pencil. Take a small triangle and the positioned specimen. On the illustration board, near the left or right side of the specimen, draw a vertical line. Make sure that a right angle is maintained between the previously drawn horizontal and the new vertical lines. Do not touch the specimen while drawing the vertical line. Do not draw this line too far to the left or right. The distance between the specimen and this line should not exceed one centimeter.

Look directly at the center of the positioned specimen and, with the pointers, measure the distance between the anterior and posterior edges of the body of the vertebra. Enlarge or reduce the first measurement according to your previous decision. Mark the anterior and posterior tips of the body on the previously drawn horizontal line. Remember to use both eyes when measuring, and to hold the measuring tool, the pointers, with both hands, making sure that the tool is parallel to the surface of the table. Do not hold the measuring tool vertically. Such a position obstructs your view and causes great inconvenience. Make sure the measuring tool is held about three quarters up from the lowest section of the specimen. Do not measure distances at the level of the table. The three-dimensionality of the specimen does not allow for this. In order to compensate for distortion caused by the depth of field, you must take all measurements at more or less the same level. The anterior portion of the articular and transverse processes located on the right lateral side are closest to you, while the transverse process on the opposite side, the left lateral side, is one of the farthest, most visible structures. As you can see, you have to compensate for the depth of field by keeping the measuring tool relatively high, but not above the specimen.

Having the first measurement assured, clearly marking the anterior and posterior ends of the body, measure the distance between the anterior end of the body and the vertical line. Enlarge the measurement as needed and mark the measurement on the tracing tissue. Then measure the distance between the posterior end of the body and the vertical line, again transferring the obtained information to the tracing tissue. Those two measurements should end in one place, the place being a transferred location of the vertical line to the tracing tissue. By having two coordinates, the horizontal and the vertical, you will be able to find any point on the subject as well as be able to transfer such a point to the tracing tissue with great accuracy.

In order to define the shape of the subject, measure the distance between the anterior edge of the body and the anterior edge of the spinous process. This is a vertical measurement. Hold the pointers with both hands on the left side of the vertebra, making sure that the measuring tool is positioned parallel in relation to the surface of the table. If the anterior portion of the spinous process on your subject is at the angle, do not measure that distance at the angle. Make sure that both tips of the measuring tool are at a right angle in relation to the horizontal line drawn on the illustration board under the specimen. This means that one of the tips of the pointer, the one away from you, will be very close to the anterior edge of the body, while the other, the one closer to you, will be pretty much away from the anterior edge of the spinous process.

You will have to extend the dorsal edge of the spinous process to the left in your imagination when aligning the tip of the pointers which is closer to you. When doing this, be careful not to touch the specimen with the measuring tool, as you may knock the specimen out of its position, thereby making the first measurements inaccurate. If by any chance you find such measuring burdensome, turn the specimen anterior side toward you and then measure. I do not recommend a lot of turning of the specimen during measuring, as there is a possibility for inappropriate positioning of your head over the specimen. Again, this may result in inaccuracies, as some small structures may become invisible by being obstructed by larger ones. The inexperienced eye will not notice such discrepancies.

Transfer this last measurement to the sketch, marking the distance vertically. Then measure the distance between the vertical line drawn next to the specimen and the very same anterior edge of spinous process. This is a horizontal measurement and as such, should be transferred horizontally to the sketch. Having the anterior edge of the spinous process indicated on the sketch, do the same with the posterior edge of the same process. Here, the farther tip of the measuring tool will be away from the posterior edge of the body, while the one closer to you will almost touch the posterior edge of the process. Transfer this vertical measurement to the sketch, then take the horizontal one. Measure the distance between the posterior edge of the spinous process and the vertical line drawn next to the specimen; transfer this measurement to the sketch. Notice all irregularities on the dorsal edge of the spinous process and carefully find the pertinent points on this structure by measuring each point twice—once vertically from the horizontal line and once horizontally from the vertical line.

Having all these data properly transferred to the sketch, you can correct those points, drawing in fact the dorsal edge of the spinous process. Having this vital information, use it to measure similar irregularities found on the outside, ventral edge of the body. As you are very familiar with the dorsal edge of the spinous process, having marked its irregular edge on the sketch, use this information to find the very same on the ventral side of the body of the vertebra. When finished, correct the points and make sure that the central section of the body is presented properly. Concentrate your attention on the transverse process of the right lateral side. This is the edge that sticks upward, slightly at an angle next to the spinous process. Pay attention to more pronounced irregularities of the outside edge of this process as well as to the fact that some parts of this edge, notably the anterior section, display a certain thickness. Measure this edge just like the previous ones, but be sure that you clearly define the posteriorly located point from which the wall of the next process, the transverse process, begins. This point may be or may not be clearly visible. Also pay special attention to the left side of that process, which actually ends in a rounded protuberance. Watch carefully for discoloration caused by improper preparation; do not confuse the discoloration with the inside, ventral, rounded surface. Watch for shadows, and disregard all shadows cast by irregularities of the surface.

Again, after the measurements are taken and transferred to the sketch, connect pertinent points, drawing the outside edge of that process. Turn your attention to the transverse process located on the left lateral side. This process may not be completely visible; only the anterior or posterior section of the transverse process may appear in view, because most of this process is obstructed by the spinous process. The view of most of this process is obstructed by the spinous process. The distance, as viewed from above, between the edge of the right lateral and left lateral transverse process, can be minimal. Be sure your measurements are correct. After finishing, recheck your measurements. When sketching the superior articular process, keep in mind that you are looking directly on something that is sticking straight up. This means that this process does have two edges presenting thickness and in some parts, especially in the posterior section, a slightly twisted surface. For delineation of the articular process, you need to know both the dorsal and ventral edges. Because you are viewing this structure directly from above, be extra careful about knowing what you are measuring. The left side, the anterior part of this process, is relatively round but clearly visible in parts. The right side, the posterior section, may have a very sharp, thin blade coming directly at you. In order to notice this feature, look at the process from the side. Make sure that light does not interfere with your observations, as they are the base for the measurements.

Do not erase anything. Mark your mistakes or probable mistakes clearly. Use well-sharpened pencils during the drawing. Avoid heavy lines—a heavy line produced by soft lead such as B, 2B, or softer will easily be wider than the thickness of the blade found on the posterior section of the articular process. The anterior section of this process hangs over the portion of the body of the vertebrae. Make a distinct note on your sketch and be sure the whole anterior section of the vertebrae is measured correctly. Pay attention to the fact, depending on the vertebrae, that a portion of the opening may be fully visible when you look directly at the anterior or posterior end of the subject. From underneath the right side, the posterior side of the articular process, a sloping bony structure connects itself with the transverse process. The curvature of this structure is important, therefore make sure that you make a note of that curvature on the sketch. A crude, explanatory cross section of the subject should be prepared. Do not shade on the sketch—shading always destroys the readability of the information presented in the form of lines. The whole posterior, right side of the vertebra should be measured and transferred accordingly to the sketch. During preparation of the sketch, make sure that you do *not* trust yourself. Always check and recheck measurements.

When the first sketch becomes too cluttered with lines and notes, take another sheet of tracing tissue and place it over the first one. Tape the second sheet in position and continue drawing. When the working sketch of the right lateral view of the vertebra is finished, spray the sketch with fixative. Make sure that the surface of the tracing tissue is dry before placing a new sheet over the working sketch. Retrace the first sketch cleanly and carefully on the new sheet of tracing tissue, making sure that the curvature of the various parts of the subject is marked by writing or by drawing arrows presenting the surface. Again, this is information that will be needed during the rendering. Spray the final sketch with fixative and prepare to sketch the dorsal view of the same subject.

After positioning the subject correctly, again draw a vertical line next to the specimen, making sure that the right angle between the horizontal and newly drawn vertical line is maintained. Prepare a new sheet of illustration board and a new sheet of tracing tissue. Draw horizontal and vertical lines, using a ruler, triangle, and well-sharpened H pencil, making sure that both lines are at right angles to each other. Look at the transverse processes. With the pointers, measure the distance between the horizontal line and the clearly visible point on the posterior section of the left transverse process. Transfer this measurement to the sketch. The posterior sections of the transverse process tend to turn down, presenting continuous downward curvature toward the posterior end of the spinous process. There are no sharp ends, therefore you must pick a point that is clearly distinguishable located on that section of the process and use it as a first reference. Measure the horizontal distance between the vertical line and the same point. Transfer the measurement to the sketch. Turn your attention toward the anterior section of that process and measure the horizontal distance between the vertical line drawn near the specimen and the outside edge of the front, anterior point of the transverse process. Did you position the specimen correctly? Are both outside anterior edges of the transverse process aligned with the horizontal line? If not, be sure the specimen is repositioned. Transfer the last measurement to the sketch. Remember to notice the pronounced thickness of the frontal section of this process. This means that both edges of the left anterior process will have to be designated on the sketch. Follow the irregularities of the outside edge of the left anterior process, taking appropriate measurements vertically and horizontally until the process is described in line on the sketch. Turn your attention to the right anterior process and do the same. Make sure that the inside curvature of the front and back of those structures is measured and transferred to the sketch correctly. The tolerances are very small—actually, you are measuring millimeters of a structure, therefore be very careful. The front section of the *V*-shaped left and right anterior processes turn downward and hide under the spinous process, making a very sharp edge. Any sharp edge will make your life easier, because it is tangible and therefore easy to notice and delineate.

Having the outside edges of the transverse processes drawn, turn your attention toward the articular processes clearly visible to the left and right of the transverse processes. When measuring, make sure to hold the measuring tool properly at about the level of the edges of the transverse processes. Do not jump up and down when making measurements. The subject is much too three-dimensional to do this, and the result of such haphazard measurements is disastrous. Measure the outside edges of the articular processes in the same way you did all others. Remember not to be confused by discoloration, shadows, and Plasticine. Do not forget to make sure that the body of the vertebra is visible in the anterior or the posterior section of the specimen.

Last, turn toward the spinous process. The top of the spinous process will be slightly narrower in its anterior area and wider at the posterior. Make sure that this is properly noted. The upper, dorsal part of the spinous process may be quite flat, but at the same time may have irregularities on its surface. Make sure that all of those, within reason, are noted and transferred to the sketch. The anterior section of the spinous process forms a ridge that travels downward at a certain angle. Make sure that the thickness of that ridge is marked properly. The frontal section of that process can be of complicated structure. Make measurements of all clearly visible points in order to find the truthful shape of this ridge. Make sure the measuring tool is kept parallel to the surface of the table and above the transverse processes. Mark on the sketch the curvature of the walls of separate structures in order to be familiar with the surface and to have definite information on hand. When the initial sketch is finished, do not forget to spray it with fixative before tracing the image into a clear, final sketch.

Having both final sketches ready, the drawing of the right lateral and dorsal views of the vertebra, you will have to align them both in such a way that respective structures will correspond to each other. Remove both final sketches from their places and prepare a sheet of illustration board. This time, tape the illustration board to the table. Prepare the ruler and the triangle. Take the right lateral view of the vertebra and position the sketch on the illustration board, making sure that the posterior of the vertebra is facing you and the spinous process is vertical. Tape the sketch to the board in two places, but only on the right side of the sketch. Make sure that to the left of the lateral view there is enough space for the sketch of the dorsal view. Take a new, large sheet of tracing tissue and place it on top of the sketch. Make sure that this sheet is definitely larger than the sketch. Tape the clean sheet in two places but only on the top, making an easy-to-lift flap. Take the ruler and align its edge with the outermost anterior and posterior points of the body of the vertebra; draw a straight, vertical line. Take the dorsal view and insert the sketch under the larger sheet of tracing tissue to the left of the lateral view. Move the sketch of the dorsal view close to the lateral view, leaving approximately three centimeters of space between the sketches. Take the triangle and align one edge with the vertical line drawn through the ventral side of the body, making sure that the other edge forming a right angle will touch the uppermost tip of the transverse process; draw a horizontal line. Remove the triangle and adjust the dorsal view in such a way that the right transverse process will touch that line. Make sure that both sketches are separated by some space. Take a triangle and align one of the edges with the just-drawn horizontal line and the other vertical edge going through the middle of the spinous process on the dorsal view. Draw a straight line. Align the triangle with the second straight vertical line and move it up or down until it touches the outermost tip of the right superior articular process. See if the tip of the right articular process from the dorsal view is aligned with the tip of the same process on the right lateral view. If the tips of this process are aligned, it means that very likely the rest of the features are aligned. If it is not, try to move the sketch of the dorsal view up or down until it is aligned. If you find this impossible, align the tip of the right superior articular process on the dorsal view as close as possible to the same on the lateral view. Recheck all major points in this way.

If you have a major problem it means that one of the sketches is smaller than the other, and that you are facing a typical problem caused by the depth of field. This also means that one of the views has been measured at a lower level than the other. Most likely this will be the lateral view, because it is flatter than the dorsal. Perhaps for the drawing of one of the views the specimen was positioned on too large a clump of Plasticine and thus was raised too high in relation to the other view. If you find a great discrepancy, start the dorsal view from the beginning. Make sure that the overall dimension of the important lengths and widths has been measured and compared with the sketch of the lateral side. If you do have access to a reducing/enlarging camera, place the dorsal view on the bed, reduce it to the proper size, and trace the image. When all parts are aligned properly, hold the untaped sketch of the dorsal view with one hand through the larger sheet of tracing tissue. Prepare a couple of pieces of masking tape. Lift the edge of the larger sheet of tracing tissue without disturbing the position of the dorsal view and tape the loose sketch to the board or to the other sketch. Make sure that you do not apply the tape over the drawn image. Remove the larger sheet of tracing tissue completely, as you will not use it any more. Turn your attention to the overlapping sheets of tracing tissue. Secure the relative position of both sketches by taping them together and remove any obstructing leftover parts of tracing tissue by tearing them off. Do not cut—it is very hard to cut through one sheet of tissue without damaging the other underneath. Place a metal ruler on top of the tissue to be removed and press the ruler down firmly. Grasp the unwanted tissue in the other hand and *slowly* pull the tissue toward you, pressing against the edge of the ruler. When tearing, pull the tissue close to the surface of the table, making sure that it is moved toward the edge of the ruler. Remove the masking tape from the sketch of the lateral side and turn both sketches image side down.

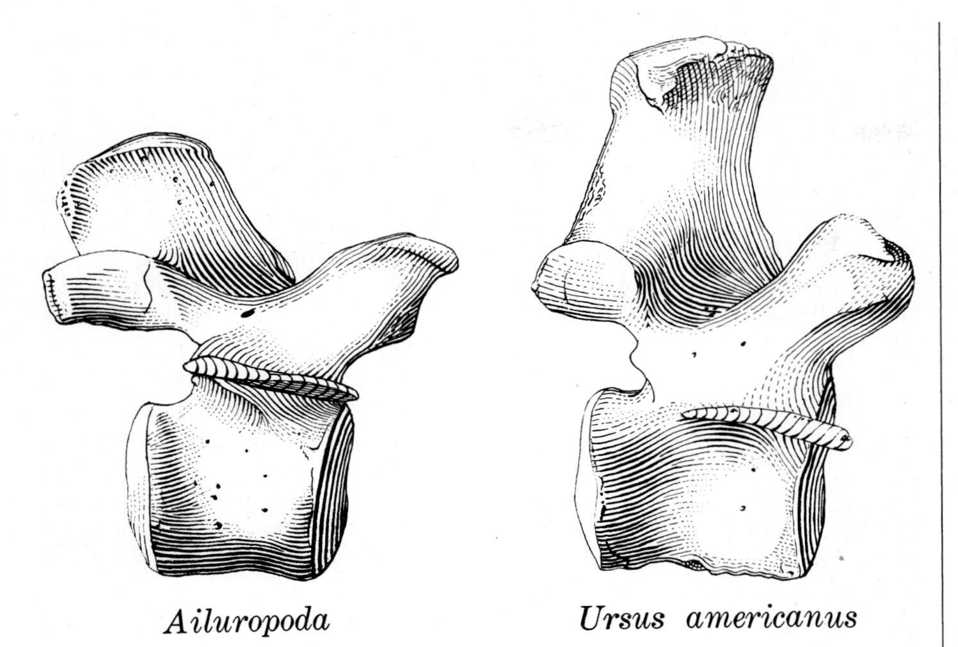

Figure 5-27. Third lumbar vertebra of *Ailuropoda* and fourth lumbar of *Ursus*, seen from left. Crow quill on two-ply bristol board. Drawing by John C. Hansen, Chicago Natural History Museum (Field Museum of Natural History), Chicago. Published in D. Dwight Davis, "The Giant Panda, A Morphological Study of Evolutionary Mechanisms," *Fieldiana*, Zoology Memoirs, Vol. 3, 1964, by Chicago Museum of Natural History (Field Museum of Natural History), Chicago. Drawn in 1940.

Remove any overlapping tissue obstructing the view of the drawn image. Now you are ready to transfer the sketches to the scratch board.

Remove the tape from the illustration board. Take the piece of white scratch board of appropriate size and tape its corners to the illustration board. Take the taped sketches and place on top of the scratch board, drawn side up. Gently, and using very small pieces of tape, secure the sketches in two places onto the surface of the scratch board. Take the sheet of carbon paper and insert it carbon side down between the sketches and the scratch board. Take the well-sharpened H pencil and, applying a gentle pressure, trace the outlines of both views, transferring the drawings to the surface of the scratch board. Do not trace the lines depicting the curvature of the structures. Those designations are for your information only and not for tracing. When finished, remove the carbon paper. Then carefully remove the masking tape holding the sketches. Do not tear the surface of the scratch board. Pick up the sketches and tape both of them to the illustration board. You will need both sketches as references during rendering. Before rendering, clean up the table and wash your hands. You must have clean hands, as the surface of the scratch board must be kept clean.

Rendering: Prepare the brush, ink, crow-quill pen, and a small piece of scratch board. Try the scratching technique before attempting the rendering. Take a brush, dip it into the ink container, and apply an opaque layer of black ink on the surface of the scratch board. Wait until the ink is completely dry, then take a graver and, applying moderate pressure, move its blades across the black surface. Observe the resulting white lines. Repeat the same technique, but change the amount of pressure and again look at the white lines. Your objective is to remove the upper surface of the board. The thickness of the lines should change in proportion to the applied pressure. Push the graver against the surface. You are actually engraving a number of parallel lines into the surface of the scratch board. The direction in which the graver is pushed, the angle of the graver in the relation to the surface of the board, and the amount of pressure applied are all critical. Do not gouge into the board; be gentle with the tool and the surface of the board. There are three basic ways to hold the graver. First, the crude method is similar to holding a pencil. Take the graver and let the upper part of the blade rest on the middle finger. The thumb and index finger will hold the graver from both sides. The blades can be facing any direction. Make sure that the graver is held at about a 45-degree angle in relation to the board. Apply pressure and scratch the surface of the board.

The second and third methods are similar to holding an engraving tool. Hold the graver between the thumb and index finger with the end of the handle almost parallel to the surface of the board and almost parallel to yourself. Hold the graver in such a way that all four of your fingers will be located on the opposite side of the thumb, with the thumb slightly below and actually touching the board. The distance between the tips of the blades and the surface of the board will be controlled by the location of the thumb and by the twisting movement of the wrist. The area of the palm near the fifth finger should be stretched out horizontally and you should feel some tension in the muscular system of that area. The second joint of the fifth finger should be bent and sticking upward, while at the same time the second joint of the index finger should be flat and the index finger fully extended. The first joint of the index finger should be bent, because the tip of that finger is pressing the tool against the thumb. The fifth finger will press the handle of the graver against the side of the hand near the third joint of that finger. The movement of the hand and the graver will be horizontal, with the blades pushing against the surface. The thickness of the lines will be controlled by the twisting motion of the wrist. In order to engrave lines vertically, you have to turn the drawing around, adjusting the image to the motion of the hand, or change the hold on the graver.

This brings us to the third basic technique of holding the tool. Turn the palm of your hand upward. Place the front part of the handle on the tip of the index finger making sure that sharp blades are facing upward. Automatically the middle, fourth, and fifth fingers will bend upward, preventing the tool from rolling off your hand. Continue their reflexive motion and grasp the handle of the tool with those three fingers. In order to maintain the proper position of the graver on the tip of the index finger, you have to press the handle with the thumb. Now you are holding the tool between the middle, fourth, and fifth fingers, with the thumb pressing on the side of the graver. Keep the fingers positioned as they are, but be sure that the end of the handle is firmly resting against the palm of the hand near the wrist. Holding the tool, turn the hand down with the blades of the graver facing the board. The graver is almost parallel to the surface of the board. In order to engrave or scratch, push the graver against the surface of the board. With this hold you can change the direction of the engraved lines more freely by the horizontal motion of the wrist, while the depth (this means the width) of the lines is controlled by up-and-down, vertical motion of the wrist.

When using an X-Acto knife, hold it between the thumb and index finger, letting the weight of the knife rest on the side of the index finger. The edge of the X-Aco blade should be facing your fingers. Scratching will be accomplished by horizontal motion with the side of the tip of the blade. Make sure that you have a good supply of number 11 blades, as the tips of that blade tend to break easily. You can resharpen the blades using a sharpening stone, but you may encounter difficulty in shaping the tips of the blades properly. Such blades will be good for cutting, but not for scratching the surface. Do not use large matting knives. Such knives are awkward to hold and their blades are much too large for our need.

Start rendering by using the crow-quill pen. Wash the new pen with soap and water in order to remove the greasy coating. Ink will adhere properly when the pen is clean. Dip the pen into the ink and draw the outlines of both views. Make sure the line is of the same width. Draw the outside of the processes in the outline, but leave the areas of that outline open wherever sharpness of the process is not visible. In the dorsal view the very forward section of the spinous process may have a rounded edge. The anterior section of the transverse process turns downward in the dorsal and lateral views, while at the same time is usually delineated with a sharp edge. Knowing where the lower areas are located and knowing the direction of illumination, estimate where the dark tones will be placed. As the light comes from the upper left side, the right side and the posterior section of both views will be darker than the left and frontal parts of the specimen. In the lateral view, the left transverse process—the one located underneath the spinous process—must be darker than the spinous process itself in order to preserve the feeling of three-dimensionality of the subject. In the same view the back section of that process will be even darker than the front part of that process. The left edge of the transverse process in the lateral view will be a sharp outline, while the right side of that ridge will be indicated by a very light tone. Remember that during rendering you must have sufficient contrast between the lower sections and the upper parts of the subject. Make observations as to where the dark or light tones will be on the drawing. Do not forget that the major source of light should appear as if it is coming from the upper left side, but at the same time do not forget that structures in the dark should be back-lit in order to be presented properly.

After the observations are made, take the brush and apply an even coat of ink in appropriate areas. Let's start with the lateral view. The outline is ready and it is dry. Apply an even coat of black ink to the visible surface of the left transverse process, leaving the ridge of that process white. This means that the black will touch the outline of the frontal section of the spinous process, the frontal small section of the left transverse process, and part of the outline of the right transverse process, but will *not* extend to the edge of the dorsal section of the ridge of the left transverse process (the left side of the drawing), leaving that ridge white. Make sure that the structure above the spinous process is separated by the solid blackness. The posterior, small section of the left transverse process should be painted solid black. Do not overpaint drawn outlines. Apply an even coat of black ink to the middle section of the frontal area located between the right transverse and articular processes, extending the flat blackness almost to the top of the right transverse process. The ridge of the transverse process should remain white.

Wait until the ink is completely dry. Scratch the frontal area of the left transverse process, producing vertical parallel lines. Use any of the scratching tools—the choice of tool depends on the size of that area. Naturally you cannot use a large graver if that area is small, and this is probably the case. When scratching, make sure that the distance between the scratched white lines decreases toward the ridge of the left transverse process, making the pure blackness change into a light tone. The effect of scratching should remind you of a well-executed engraving. Turn your attention toward the area located between the right transverse and superior articular processes. The darker areas should be in the lower part of the right transverse process, with decrease of tone as the surface nears the upper ridge of that process. Try your best to move the scratching tool vertically, pressing the tool deeper into the surface of the board whenever more light is desired. Make sure that the edges of the flat-painted black are broken by the scratching tool. Remember that if you make a mistake you can rescratch that area; you can also add extension to the black lines produced by the removal of the flatly painted black ink with the crow-quill pen. The general tone produced by the white and black lines should be lighter by about 50 percent as compared to the frontal section of the left articular process. Such contrast will help to produce the feeling of three-dimensionality. To the left of the ridge of the transverse process, on the surface of the spinous process, paint the flat black. Keep this blackness near the ridge of the right transverse process. This darkness, when broken by the whiteness of scratched lines, will help to separate the spinous process from the right transverse process. A few vertical crow-quill lines may be added later in order to remove the harshness of the suddenly ending dark shape. Do not forget too separate the anterior section of the spinous process from the left transverse process by scratching a white line along the angled edge of the spinous process. As you know, the anterior section of the superior articular process ventures into the anterior part of the transverse process, making an *S*-curve. A portion of the transverse process actually goes underneath the articular process, therefore you must make this section darker than the overlying structure. Apply the flat blackness of the ink, wait until it has dried out completely, and scratch white parallel lines into this surface, making sure that the area under the *S* curve is darker than the rest. Naturally, the surface above will be very light. Draw some lines with crow-quill pen on the upper surface if you find it necessary. Use a reducing lens for proper judgment of tones. Proceed toward the body of the vertebra. The articular process on its left dorsal side should be quite light in value. This means that the tone produced by the black lines should diminish as it continues upward. The front section of that process usually is formed into a rounded shape. All rounded shapes are difficult to render with line. Make sure that darkness from the lowest section of the structure found between the processes rapidly changes into light, almost disappearing, and then again changes into dark, just to be scratched back to the light depicting the top of that round tip. The posterior part usually ends in a sharp ridge. You must underline that ridge on the left illuminated side. You are forced to add dark lines either by using crow quill or by painting a black shape and scratching white lines into the surface of the board. As the right side of that process is in the dark and the light is illuminating the ridge from the upper left side, feel free to cover the whole side of the body of the vertebra with ink, using a brush. Wait until the ink is dry; then, starting from the right, ventral side, scratch white lines, slowly increasing the distance between the lines, thereby making the tone appear darker as it gets closer to the ridge. The ventral side of the body will also be painted black with white vertical lines scratched toward the right, ventral edge of the body. The result should produce a change of tone from dark to light and again to dark in the horizontal direction. The right side of the articular process will be dark, changing to light as the tone continues horizontally, then again into dark, getting darker as the body curves downward.

When drawing the dorsal view apply the very same concept, but watch for more changes in the surface structures. Always make sure that outside edges of the ridges are separated by the dark tone, especially on the right side. The right side of the right articular process should be very light. The light source is located on the left side and because of this fact, the transverse process casts a deep shadow on the articular process, but at the same time we are forced to break such darkness by back-lighting. If this is not done, the curvature of the dorsal side of the articular process cannot be presented properly. Watch for the right side of the left transverse process. This section should be much darker than the opposite similar section of the right transverse process. Lastly, do not place any tone on the left side of the spinous process, but produce about 20 percent tone on the surface of the left transverse process in order to separate this vertical structure. At the same time the right side of the spinous process should be very dark, contrasting with the upward-curving surface of the right transverse process. Using a crow-quill, mark some of the surface changes on the uppermost, flat section of the spinous process, but be very careful not to overdo this. The more tone you place on this surface, the more you will flatten the image of the specimen. When finished scratching, carefully appraise the drawings; when necessary, add lines with a crow-quill pen, taking care to join the lines correctly. Clean the table, remove the ink, and wash the brush, preparing yourself for the next step. Remember to note the enlargement factor by writing it in the corner of the drawing or by drawing appropriate scale.

Dry-Mounting: Dry-mounting is not necessary. If you do want to dry-mount the drawing, follow the usual procedure, using any dry-mounting tissue. Make sure that the drawing is covered by protective tissue during the dry-mounting. By now you realize that the surface of the scratch board is easy to damage. In order to preserve the unmounted drawing, take a sheet of tracing tissue larger than the drawing. Place the drawing on half of it and bend the remaining part of the tracing tissue, covering the drawing with it. Do not tape the tissue to the drawing; tape tracing tissue to tracing tissue.

Preparation for Production: If the drawing is not mounted and is to be stored for a long time, take a sheet of black cover-stock paper and bend the paper in half, making a folder. Keep the drawing and all sketches in such a folder. When sending the drawing for reproduction, take a sheet of illustration board and place the drawing on top. Tape the upper section of the drawing firmly to the illustration board as well as the lower left and right corners. The illustration board will serve as a protection against accidental bending of the illustration. Make sure that protective tissue is placed and taped over the illustration and that a sheet of cover-stock paper will reinforce the protection of the top of the drawing. All necessary notes pertaining to reduction of the image should be written very legibly.

Comments: Scratching is an imitation of the engraving technique, and as such has as many variations as the engraving technique itself. The scratching technique can be used on a variety of surfaces, as long as the surface does not splinter in an unpredictable fashion. The best materials to use are scratch board and mylar drafting film. Lines do not have to be drawn with a crow-quill pen; a technical pen can be used quite effectively. Also, carbon paper is not the only tool that can be used to transfer the image from the sketch to the surface, although transferring the image with carbon paper is one of the most efficient ways.

The scratched lines are actual symbols of tones, varying in their widths and producing light and dark. It is good to know the subject well, so a feeling of the surface of the subject can be presented properly. When a device such as an opaque projector is available, feel free to make use of it and place the subject inside. Trace the projected image, saving yourself a lot of time. The only drawback of such a system lies in the fact that when a beginning illustrator gets accustomed to using tracing devices, he will not be able to draw without their help.

TENTH PROJECT

Anthropology

Subject: Pictorial reconstruction of pottery, with explanatory presentation of design.

Technique: Pen and ink on tracing paper; line and stipple.

Materials:

1. tracing paper
2. Pencils—H, 4B
3. Illustration board, preferably hot-press, smooth-surface
4. Technical pen, tips number 00 and 000 or equivalent
5. Ink, black
6. Pointers or proportional divider
7. Calipers
8. Masking tape
9. Metric ruler
10. Metal ruler
11. Triangle
12. Kleenex
13. X-Acto knife with number 11 blades
14. Reducing glass
15. Plasticine, 1 oz.
16. Stylus
17. Instant-transfer tone, suggested 30 to 40 percent
18. Seal removable dry-mounting tissue
19. Black cover-stock paper
20. Black photographic masking tape
21. Seal-Lamin glossy laminating film

Preparation of Some Materials: Take a 4B graphite pencil and grind it into dust, making sure that the resulting graphite powder does not contain any large, sharp particles.

Basic Steps:

1. Preparation of sketches
2. Tracing the basic sketch into clean final sketch
3. Rendering
4. Dry-mounting
5. Preparation for production

Nomenclature:

Rim—the uppermost section of the vessel

Neck—the area underneath the rim

Shoulder—the area between the neck and the main body of the vessel

Body—the main section of the vessel

Base—the lowest section of the vessel

Preparation of the Specimen: Pictorial reconstruction is a puzzle, and such a puzzle is relatively easy to solve when you already know the answer. A helpful friend will solve this problem. Ask a friend to obtain a vessel of relatively large size, preferably handmade, covered with a three-dimensional design. Have your friend record the shape and design of the pottery through a drawing or photography, then break the vessel and select four or five small pieces for you. The drawing or photograph will be of some interest to you for a pictorial representation of the unbroken vessel for comparison after reconstruction is finished. When breaking the vessel, wrap a towel around it and smash it well, making sure the remaining pieces are of relatively small size. In order to proceed with the project, you must have a piece or pieces of the rim as well as some parts of the neck, shoulder, and body. Sections of the base are helpful but not necessary.

If you do rely only on yourself, then obtain five or six vessels, record their shapes, and break all of them. Leave the broken pieces in one container for a period of time, then pick four or five sherds. Do not forget to stash away previously recorded pictorial representations of those pots; restrain yourself from using them as a reference during the reconstruction. When making a selection, make sure that the sherds belong to the same pottery.

PROCEDURE FOR DRAWING:

Observation: Observation and measurements are vital. Prepare paper for note taking and simple calculations. Arrange the pieces of pottery on the table and observe their shapes. Look closely at the fragments of design. You do know that remains of design were once connected in a continuous way. Such a presumption sounds basic—wait! Were those bits of design *really* once connected, creating some kind of pattern? Perhaps there was a large empty space between each remaining piece. Do not assume anything. Think in terms of what you have. Possibly the design is continuous and will help you to position the sherds correctly; perhaps not.

Arrange the sherds on the table the way you think they would be arranged on the vessel. The rim, being the most obvious and most easily recognizable, will be on the top, and the rest of the sherds underneath. Do not try to recreate the three-dimensionality of the vessel at this time. Be concerned with the probable position of the sherds. Number the sherds—take white labels, write numbers on them, and paste the labels on the inside wall of the sherds. You will refer to the sherds by number. Do not paste anything on the outside of the sherds. The outside is very valuable to you, for it contains important information. Take calipers and measure the thickness of the walls of each sherd. Make two or three measurements. Write them down—for instance, "Sherd number 3, upper section 4.53 millimeters, middle section 4.54 millimeters, lower section 4.58 millimeters." A consecutive change in thickness of the wall of the sherd will indicate the direction of this particular sherd as well as indicating its position on the surface of the vessel. In most cases, the walls of handmade pottery are thinner toward the top and thicker toward the bottom. The sections of the rim may be thicker than the main portion of the body. All sherds, with the exception of the rim, will have unevenly broken edges. This means that the position of a sherd or the theoretical surface of the vessel cannot be estimated quickly. Such a sherd cannot be smashed with other sherds. It does not have a definitely established top or bottom as is the case with the sherd from the rim. The position of such a sherd is crucial to the reconstruction of the design, to the cross section of the vessel, and to the length and width of the pottery. In the horizontal direction such a sherd will have a relatively uniform thickness, while in the vertical the thickness may change drastically. Pay attention to the configuration of design and make appropriate notes, always referring to the particular sherd by number.

Make a tentative decision regarding the size of the finished rendering. You do not know the size of the vessel, therefore in actuality you do not know the size of the sketch. Be prepared for enlargement or reduction of the sketch or parts of the sketch during later stages of work. During observation, do not forget that all estimates are only a probability; this will remain so during the whole sketching process. The shape, length, width, thickness of the walls, and design are unknown to you. Base all of the deductions on facts, not on imagination. Most observations are to be done in conjunction with positioning separate sherds.

Positioning of the Specimen: Look closely at the sherds placed in front of you and prepare a data sheet. Take calipers and measure the thickness of every sherd. Note all of the measurements. As all the sherds are numbered, remember to note the number associated with the particular sherd. On the basis of the thickness of the walls of the sherd, rearrange the position of the specimens on the table. Naturally, the rim will be on the top, and the remaining sherds will fall someplace below the rim. Do not let the sherds touch each other. When measuring the thickness of the walls, make sure that three sides of the sherd are taken into consideration. Make two or three measurements per side to make sure that changes in the thickness are noticed. The vertical thickness of the walls will vary, while horizontally it will remain almost the same. Naturally, the sherds from handmade pottery will vary in thickness, while walls of commercial or mass-produced pottery will remain virtually the same in thickness. On the basis of the measurements, rearrange the position of the separate pieces. Those that are generally thicker will be placed toward the hypothetical bottom of the vessel, and the ones with the thinner walls will be toward the top. On the basis of the measurements, the horizontal and vertical position will also be established.

The next step is the decoration. Again, look closely at the sherds and in your mind try to connect the flow of the decorative pattern. Depending on the pattern, it may be a relatively easy process to deduce the possibilities.

Observe the broken edges of the sherds and see if there is a possibility of fitting separate broken sections of the vessel together. If there is, and you do discover that two pieces will match, place those sherds together. Do not exert any physical force when matching separate sherds. Make sure that edges are not broken during investigation. Before proceeding further, place a sheet of tracing tissue on the table and transfer all the sherds from the table to the tracing tissue. Mark the established positions by drawing a crude outline of each separate piece; designate each outline with the appropriate number.

Prepare to position the rim of the vessel for the sketch of the frontal view. Take a small piece of illustration board and draw a horizontal line, using a ruler and well-sharpened pencil. Take a bit of Plasticine and place it on the illustration board, making sure that the Plasticine is secured properly. Use your fingers when pressing the edges of the Plasticine to the board. Press downward and outward, making sure that the main portion of Plasticine sticks up. Place the section of the rim on top of the Plasticine. Make sure that the edges of the rim are at the same height from the illustration board. Look directly at the center of the sherd with both eyes. Align the left and right edges of the rim with the previously drawn line. Recheck the distance between the edges of the rim and the illustration board. Draw a vertical line near the left or right side of the sherd, making sure that a right angle will be maintained between both lines.

Proceed to draw the sketch of the frontal view of the sherd. When the sketch is finished, position it for curvature of the upper part of the vessel. In order to start properly, place the sherd upside down on a piece of illustration board, with either the inside or outside of the sherd facing you. Look at the sherd along the surface of the table. Make sure that the rim of the sherd, which is the rim of the vessel, is placed in such a position that light will be stopped by the edge of the sherd. To be sure, move the sherd back and forth along its edge while looking directly at it. If there is light visible underneath the left or right edge, eliminate that light by inclining the sherd. Eliminate the light between the center of the sherd and the surface of the illustration board as best you can. The less light, the closer you are to the truthful representation of the angle of the upper section of the pot. When you are satisfied, secure the sherd by applying Plasticine to the outside and the inside of the rim. Do not use anything else besides Plasticine; other products which look like Plasticine will damage the specimen because they are very difficult to remove from the surface of the object. Turn the sherd with the broken edge toward you. Look at the edge directly. Do not prop up the subject. Do not raise it above the surface of the table. Let the stable surface of the table be the main reference with which your body will have to comply. Take a plastic triangle and place it in front of the broken edge with the surface of the triangle facing you. Make sure that the surface of the triangle is parallel to the broken surface of the wall of the sherd. Draw a horizontal line on the illustration board along the edge of the triangle. Again look directly at the broken wall of the sherd and place the plastic triangle exactly on the previously drawn horizontal line, making sure that the right side of the triangle is near the outside or inside of the sherd. While looking along the surface of the table at the break, align the vertical edge of the triangle with the most visible point on the rim. Remember where the rim of the sherd is. The sherd is upside down with its rim facing the surface of the illustration board. Mark that point on the illustration board with a well-sharpened pencil. Stand up and while looking from above, draw a vertical line starting from the marked point and going toward you. Make sure that this line is at a precise right angle in relation to the previously drawn horizontal line. While sketching, disregard the visible curvature of the walls of the sherd. Remember that you are interested only in drawing a cross section of the specimen.

Drawing the Sketch: Establishing an accurate diameter of the vessel at the rim, neck, shoulder, and base, as well as the vessel's vertical dimension, are the primary functions of the initial procedure of sketching. Start with the information that is easily obtainable—that is, the diameter of the vessel at the rim. You know that vessels have round openings unless the pot is very unusual. Check the sherd or sherds belonging to the rim for proper information. If the rim section does not have an unusual configuration, you can safely assume that the rim is some kind of a circle. Take the sherd in your hands and place a ruler on the inside of the walls of that sherd. Measure the distance between the edges of the sherd at the uppermost section of the rim. Transfer this measurement to the tracing tissue by drawing a straight line. If you are enlarging or reducing, do not forget to enlarge or reduce all measurements, using the same factor. Divide this line in half. Clearly mark all points. Draw a line, using a triangle and well-sharpened pencil, crossing the previously drawn line at the middle. Make sure the new line is positioned at right angles to the other, forming a cross. One line, designated AB, represents the distance between both edges of the sherd; the other will serve as a base for marking part of the radius of the circle. The first line is a chord of the circle; the second will become the radius. In order to establish the second measurement, place the ruler on the inside of the sherd at the same position from which you took the first measurement. For your convenience, make sure that the flatness of the ruler is touching the inside edges of the sherd. Be sure that the upper edge of the ruler is aligned properly. You have only two hands, and one is needed for the second measurement. Hold the sherd and the ruler in one hand and take a pair of pointers in the other. The pointers will be used to measure a *section* of the radius of the circle created by the rim of the vessel. While holding the sherd and the ruler together, do not press very hard on the ruler. Excessive pressure will bend the ruler slightly, diminishing the distance between the ruler and the inside wall of the sherd, affecting the radius and therefore the diameter of the rim. Knowing where the *middle* of the distance between the left and right sides of the sherd is located on the ruler, measure the distance between *that* middle point and the exact middle of the inside of the wall of the sherd. Look down from above, holding the sherd and the ruler in one hand and close to you. Hold the pointers horizontally in the other hand. Make sure that the distance you are measuring is a distance between the middle of the chord, formed by the ruler, and the middle of the remaining part of the circumference of the rim. Do not measure any other points on the inside wall of the sherd except the exact middle. Those two measurements are crucial because they are the base from which the diameter of the rim can be deduced.

Just to reassure ourselves in what we know about the diameter of the rim, let's go back. First, we know the distance between the left and right edges of the sherd as measured on the *inside* of the sherd. This distance is represented on the sketch as a straight line designated AB. We know the distance between the middle of that line and the inside wall of the sherd. Draw a straight line crossing AB at the middle and extend one end of this line considerably. Make sure that the new line is at right angles to AB. At the point where the line AB crosses the new line, designate point E. Take the second measurement and place it on the new line starting from point E, and mark the end of that measurement on that line. Designate this point as C. Designate the opposite end of that line as D. Now you have a cross. The line AB is crossing the line CD at right angles at point E.

There are a few ways to calculate the radius and the diameter of the rim besides the trial-and-error method. If points AC or BC are connected with a straight line, this line forms the base of the isosceles triangle—one in which two sides are equal and the two angles are equal. This means that the angle at the point C will equal the angle at the point A or the point B. In order to establish the value for the radius of the circle, measure the angle C with a protractor and transfer the value of the angle to either point A or point B. Starting from one of those points, draw a straight line until that line intersects CD. The new point is the center of the circle. Designate this point as F. Take a ruler and measure the distance between C and F. This distance is the radius of the circle you are looking for. Take a compass and draw a proper circle. In order to directly obtain the value of the diameter, employ the principle of mean proportion. Remember that a perpendicular extending from any point on the circle to the diameter of that circle is the mean proportional between the segments of the diameter. Also remember that any point on the circle connected with the opposite ends of the diameter of that circle will form a right triangle. Because of that fact, a mean proportion exists between certain sections of such a triangle. The diameter of the circle can be deduced.

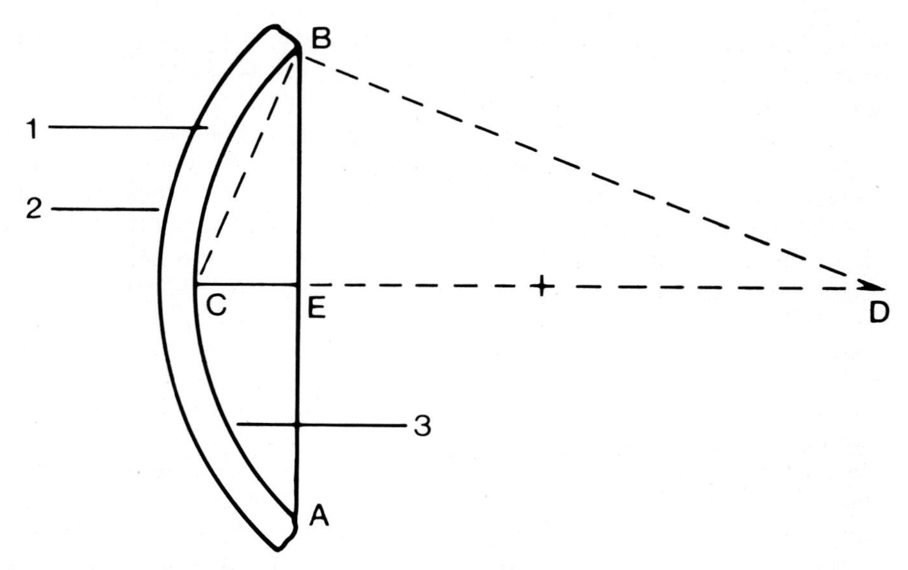

Figure 5-28. Finding the diameter of the vessel. *1*—Pottery sherd; *2*—Outside of the wall; *3*—Inside of the wall; *CE*—known value; *BE*—known value.

Let's look again at what we have. The known value that is a horizontal distance between the edges of the sherd is represented by the straight line AB; the center of that line is marked as point E. On the perpendicular line drawn from point E the other known value, the distance between the ruler and the center of the rim, is represented as EC. We do not know the diameter of the circle. Obviously, the diameter of the circle must extend from the point C to the unknown point D, lying on the extension of the perpendicular line EC, and must be the opposite of point C since point C is taken from the edge of the rim of the sherd. As soon as we learn the unknown distance ED, we will know the diameter of the rim. The fact is that a perpendicular drawn from any point on the circle to the diameter of that circle is the mean proportional between the segments of the diameter. In our situation the perpendicular extending from the point located on the circle to the diameter of that circle is represented by the line BE and is of known value. The segment of the diameter is represented by the line CE and also is of known value. Having those two values we can form an equation

$$CE/BE = BE/ED$$

with BE being the mean proportional. Therefore

$$BE^2 = (CE)ED$$

from which the unknown value ED can be extracted because

$$ED = BE^2/CE$$

As we know the value represented by CE and have learned the value represented by ED, therefore

$$ED + CE = \text{diameter of the rim}$$

The laws governing the proportions of a right triangle offer us another way to find the unknown. On our sketch we have marked point E as the middle of the chord represented by AB. Again, the distance between C and E—C being the point located on the circle and E being the point located on the diameter of that circle—offer us the following equation

$$CD/CB = CB/CE$$

CD is the unknown distance, while CB and CE are values we can measure. Therefore

$$CB^2 = (CD)CE$$

leading us to the final conclusion represented by

$$CD = CB^2/CE$$

and therefore giving us a direct answer to our question. The diameter of the rim is represented by the value enclosed in CD. All the calculations apply not only to the diameter of the rim, but also to the diameter of the walls of the vessel. This means that it is possible to know the exact value for the diameter of the vessel at any point on the body as long as the evidence of that section of the vessel in the form of a sherd is in our hands. Do not forget that if enlargement or reduction is necessary, all the measurements must be enlarged or reduced by the same factor.

Trial-and-error methods are fast and relatively accurate. As a matter of fact, trial-and-error methods are used more widely than any of the calculations, but have one built-in problem: Both trial-and-error methods are more applicable to findings of the diameter of the rim than to the diameter of the body or neck of the vessel. Having an intact and reasonably large piece of rim is essential. The curvature of the rim is matched with previously drawn circles of various diameters, and the result is instantaneous. The other trial-and-error method takes a bit longer, but is good for very small sections of the rim. The distance AB has to be measured and divided in half at the point E. A straight line has to be drawn from the point E at right angles to AB. The sherd has to be positioned upside down, matching A and B with its left and right sides. The curvature of the rim is drawn with a well-sharpened pencil by running the tip of the pencil along the edge of the sherd. During the procedure, make sure that the sherd is not damaged; also make sure that it will be held in proper position. As the center of the circle lies on the line E, someplace along that line there is a point where the compass should be inserted in order to draw a circle equal to the diameter of the rim. The problem is to find that point. Take a compass and keeping one of its points on the line E, try to find the center of the circle by drawing circles and matching the points AB as well as the curvature of the rim.

In order to learn the vertical curvature of the wall of the vessel, first you will have to place all pertinent sherds in their appropriate positions. Then you will have to deduce the diameter of the vessel from those sherds. The easiest and usually the most informative section of the vessel is a rim with part of the neck. Draw a cross section of that sherd. First position the sherd correctly. As you know, the sherd has to be placed upside down on a smooth, sturdy surface such as a piece of illustration board. It has to be angled correctly. In order to deduce the correct angle, look along the surface of the table directly at the front or back of the sherd. Make sure that the visible light seeping underneath the upside-down rim at the edges or the middle is eliminated by back-and-forth motion of the sherd. When the sherd is in its correct position, secure it with Plasticine. Make sure that the sherd will stay in its position. Turn the broken edge of the sherd toward you. Look directly at that edge and using triangle and the pencil, draw two reference lines, one horizontal and one vertical. The horizontal line should be parallel to the broken edge of the sherd, and the vertical must be at right angles to the horizontal. It is very helpful if the vertical line coincides with some easily visible feature on the edge of the sherd such as the tip of the rim or one of the wall. Repeat the same on the sketch. Draw the horizontal and vertical lines. The vertical and horizontal measurements must be taken while looking directly at the broken edge. Remember that all vertical measurements will be transferred to the vertical line, while horizontal measurements will be parallel to the horizontal line, giving two respective coordinates on the sketch for every horizontally and vertically measured point. Proceed until the thickness, the shape, and the angle of the wall are properly transferred to the sketch.

Having deduced the diameter of the rim, draw an appropriate circle on the new sheet of tracing tissue. Draw a horizontal line through the middle of the circle. Trace the first sketch of the cross section of the sherd from the rim and place it on the right side of the circle, but under the tracing tissue. Make sure that the new sketch of the diameter of the rim is drawn on a large sheet of tracing tissue and that it is attached to white illustration board with tape at the top only. When positioning the sketch of the cross section, make sure that proper alignment of that sketch in relation to the diameter of the vessel is maintained. Make sure that you know if the diameter is of the inside or outside of the rim. The inside is preferable because it is easier to align other sketches to that first basic sketch. Make sure that the outside wall of the sherd is facing the outside, right side of the circle; tape the sketch into position. You now have two layers of tracing tissue, the one with the circle placed on top of the one with the cross section. Now prepare yourself for deciphering other sherds. Design patterns have to be deciphered and the sherds will have to be carefully measured in order to pinpoint their respective positions on the walls of the vessel.

If all the sherds were measured during the initial stages of observation, utilize the collected data. First, draw outlines of the sherds or copy previously drawn outlines. Then take a roll of thin toilet paper or Kleenex tissue as well as ground graphite needed for preparation of rubbings of the three-dimensional design from the surface of the sherds. Remember that each sherd should be numbered. Very likely the design on all sherds will be quite similar in its basic structure and therefore may be difficult to assign to a particular sherd. The numbering system will help to match a particular sherd with the design after the rubbings are finished.

Take a clean piece of illustration board and pour some ground graphite on its surface. If you are using Kleenex, make sure to separate the two sheets from which Kleenex is made. Do not tear or rip them. Be very gentle, as the material you are handling is fragile. Carefully place one thin sheet of Kleenex on the surface of the sherd, slightly wrapping it around the subject. Hold the wrapped sherd in one hand, and with a finger of the other hand spread a thin layer of powdered graphite over a section of the Kleenex. Make sure that the Kleenex will not move over the surface of the sherd; also make sure that the sherd is stroked against the design. Move your graphite-dipped finger along the surface of the sherd, trying to catch all the ups and downs of the design. Try to cover a reasonable area without destroying the thin sheet of tracing tissue. Do not worry about accidental smearing. Your object is to produce the information pertaining to the design. The rubbing itself is not to be admired. If the design is complicated, make a few rubbings from the same sherd. Take a thin (.0015), glossy laminating film known as Seal-Lamin and cut a piece larger than the rubbing. Place the rubbing with the image up on illustration board. Take the prepared laminating film and place it over the rubbing, making sure that the *glossy* side of the laminating film is facing up and the matte side is facing the rubbing. The matte side will adhere to the rubbing. Tape the corners of the laminating film, making sure that film is slightly stretched. Place a sheet of tracing tissue on top as a protection and use a hand-held iron to apply heat. The manufacturer-recommended temperature is 225°F. Keep the iron moving across the protective sheet for about 15 seconds. Take the iron off, remove the protective tissue, and cut the rubbing at its edges with an X-Acto knife. Because Kleenex is delicate and transparent, the laminating film will penetrate the tissue and will tack it slightly to the board. Peel off the rubbing carefully and turn it over, making sure that it is lying on the smooth surface of the illustration board; then laminate the other side of the rubbing. The Kleenex fiber structure will almost disappear, leaving the graphite very visible and in a sense producing a transparency. Such a transparency is very hard to destroy, preserves the image very well, and can be used in an opaque projector. If many transparencies are made of the rubbings take from one sherd, a difficult design becomes clearer, especially when transparencies are placed over each other on the surface of the light box. The added benefit is the possibility of making a photographic contact print or a very credible blueprint copy directly from the rubbing.

When all necessary information pertaining to the design has been recorded and preserved, take a separate sheet of tracing tissue and draw a straight line representing the diameter of the rim. Attach this sheet to the illustration board and place the rubbings or tracings of the parts of the design underneath. Deduce the continuity of the whole design, basing your judgment on the pieces of existing evidence. The positions of separate sherds are fixed in their respective vertical places because of measurements of the diameter of the vessel, as well as because we know that thicker walls must be toward the bottom of the vessel. What is really left to deduce is the continuity of the design. You know that if a given sherd is located at a definite height, that does not mean that this particular sherd cannot be moved horizontally. The only clues to its horizontal position are the bits and pieces of design. Sketch the probable design lightly and see if it makes sense. If something does not fit right, do not force it. Move the pieces horizontally and see if you find any solid information. It all depends on the complexity of design, readability of remaining sherds, and your logical thinking. Draw outlines of known sherds on this sketch, making sure they do not obscure the deduced design. Make sure that you have not expanded the horizontal dimensions of the vessel. Fill the missing areas with design based on the evidence. For example, if a surface of the vessel is covered with indentations, by now you should know the relative density of indentations per square inch. You should know a general direction from which the force came that made the indentations. You should know the organization of the indentation on the surface of the vessel. Are those depressions aligned with each other, forming circles? Do those indentations create a linear pattern? Does the size of indentations change? Do they have regularly appearing irregularities? If such regularly appearing irregularities exist, why are they irregular? What was the possible tool used to make the indentations? Was it as simple as a finger or perhaps as complicated as a specially carved wooden paddle? In which direction was the vessel turning when the design was being applied to its surface? From such questions and observations it is possible to deduce a lot of information about the process of pottery making as well as about the individual potter. It is possible to deduce if the potter was left- or right-handed. It is possible to deduce why this decoration was applied to the vessel.

When finished, place another clean sheet of tracing tissue over the deduced and sketched design. Draw a rectangle, keeping in mind that the true vertical dimension is still unknown to you. Leave the lower area of the rectangle open for possible expansion. Make sure that the rectangle encloses all important sections of deduced design. The upper edge of the rectangle is the diameter of the rim. If the rim happens to be sculptured, this must be presented all along the upper horizontal edge of the rectangle. Trace the design cleanly and do not shade. Do not apply any tone. This sketch is a final sketch of the reconstruction of the design. It will be rendered separately and will be placed to the right of the rendering depicting the shape of the vessel.

Now let's go back to the reconstruction of the shape and the cross section of the vessel. All the sketches pertaining to the design should be within easy reach, serving as references. Take the illustration board with the sketch of the diameter of the rim and cross section of that rim attached to the right side. Mark the middle of the diameter and draw a long, straight perpendicular line downward. Make sure the right angle is maintained between the horizontal line representing the diameter of the rim and the new straight line. Use a triangle to start and the ruler to finish. This line divides the vessel into two halves. You will be preoccupied with the right side of the vessel. The left side will be a mirror image of the right. Knowing the thickness of the walls of the rim is essential, because this information allows you to start proceeding downward toward the base of the vessel. Plot the radius of the inside of the vessel's wall, basing your judgment on the information gained from the other sherds. By now you do know where, vertically, each of the sherds should be placed. You do have two values pertaining to the thickness of the sherds. For each separate sherd draw two horizontal lines parallel to the line representing the diameter of the rim. The lines will represent the upper and lower edges of each sherd. Make sure that any possible confusion is eliminated by writing numbers of the sherds in their respective places. The length of those horizontal lines should be equal to the radius of the vessel at the levels of the sherds. If necessary, insert a sketch of the cross section of a particular sherd under the large, main sheet of tracing tissue. Remember that the radius of the vessel at any given level is a radius of the *inside* walls. This means that the thickness of the walls must be added in order to reconstruct accurately that particular section. All of this visual information—the horizontal lines, the cross sections of sherds, the cross section of the rim with its well-established angle, and the thickness of the walls of each sherd inserted in the appropriate place—will give a certain shape to the vessel. Naturally, this shape is mostly visible in areas where concrete evidence was available; the other areas will remain blank. All such areas will remain questionable as to their truthful thickness and shape. Are they beyond deciphering? Not at all. By now we know the thickness of the wall of the vessel in various places. We also know the shape of the vessel in the same areas. By reasoning, we can reach a conclusion that will be extremely close to the original size and shape of that vessel. Think: The walls slope downward at a certain rate. We know that if the diameter of the vessel remains the same at various levels, the walls of the vessel must be parallel to each other. This would mean that the base was probably at right angles to the walls. But if the walls curve toward the base and there is no evidence of any fluctuations of that curvature, by extending the curvature of the wall we can ascertain the shape of the base. It is much easier to find the height of the vessel if a part of the base is available, but assuming that such evidence does not exist, simply continue the two curving lines representing the thickness of the wall of the vessel. If the vessel was made by hand, try to imagine how that vessel was held by its maker. Also you know that the thickness of the wall must increase toward the base in order to withhold the pressure of the clay forming the upper section. Connect lines representing the cross sections of separate sherds with each other, therefore forming a complete outline of the right side of the vessel.

When finished, take another sheet of tracing tissue and place it on the top of the last one. Draw the straight line representing the diameter of the rim. Draw the perpendicular straight line from the middle of the diameter of the rim downward. Take a French curve and neatly trace the hand-sketched cross section of the vessel, using a hard, well-sharpened pencil—an H or 2H will be just right. Do not use 4H or harder lead because such a well-sharpened pencil will not leave a decisive mark on the tracing tissue and when pressed, will rip the tissue. When tracing, make sure the produced line is visible and of uniform thickness. The outline depicting the cross section of the right side of the vessel should include the uppermost section of the rim and should be extended all the way to the center of the base, where it should end, forming a short, straight, vertical line. Naturally this short vertical line, representing the deduced thickness of the middle section of the vessel, will be an integral part of the finished rendering.

Turn your attention to the opening of the vessel. The circle representing this opening is not represented in perspective. It is possible to look down at the opening and at the same time see the cross section of the vessel. As the diameter of the vessel is and will be represented as a horizontal distance between the left and right walls, we need not worry about the degree of foreshortening of the opening as long as it looks reasonable to the eye of the viewer. As a matter of fact, there are two standards for presentation of the opening. One calls for direct viewing of the vessel from the side, omitting the opening of the vessel completely. In the other the opening is presented as an ellipse. The decision as to how to present the opening is based on what will or should be presented. If a sherd, be it a separate piece or a more complex section of the rim, should be depicted from the front *and* back—that is, the outside as well as the inside wall should be represented—the opening should be presented as an ellipse. If this is not necessary, then the rim will be a straight line.

Remove the tape and fold the sketch along the vertical line; trace the right side carefully on folded tissue. This will give you the whole vessel when the tissue is unfolded.

When finished, carefully unfold the sketch and reposition it in the previous place. Then, on the same sketch, you will have to draw the sherds that have been used as the evidence for reconstruction. Their vertical and horizontal positions are fixed by their cross sections as well as the design pattern. As the actual vessel is three-dimensional, some sherds may have to be drawn in foreshortening. The sherds depicted near the left or right wall should be respectively foreshortened when such a decision is made. How do you reach such a conclusion? What is the basis of such a decision? It all depends on two major factors. The first is a deciphered design dictating the position of a particular sherd. The second is a bit of information pertaining to the vertical cross section of the vessel enclosed in the sherd. If one of the sherds has on its outside wall some kind of indication or resemblance of the handle, spout, or other unusual three-dimensional formation, it obviously should be depicted as a part of the cross section of the vessel or it should be positioned near one of the walls, presenting that form from three-quarter view. Such information is of great value and should always be clearly depicted. Should all sherds known to us be drawn within the parameters of the pictorially reconstructed shape of the vessel? Not necessarily. The decisive selection should be based on the overall value of that sherd to a particular project.

You may easily encounter some sherds which because of their size, irregularities, or the unreadability of their surface design are impossible to place properly. Although you may be positive that such sherds do belong to the vessel under reconstruction, and although those sherds do supply information of some value, it may be beneficial to the project as a whole to omit them from the drawing. Remember that not all sherds came from the front of the vessel. Some indeed had formed the back wall. If you do judge that one sherd is more of an obstruction and is not beneficial to the presentation of the whole vessel, remove it from the drawing. At the same time be aware that such a decision is usually left to the scientist. If he decides that a sherd you think is of no value should remain in the sketch, it simply means that *this* sherd stays in the drawing. When the right side, the slightly presented cross section of the vessel, is interrupted by a sherd, draw the cross section separately. This means that the reconstructed vessel will be presented without solid lines continuously depicting the right inside and outside edge. The only solid, hard lines will be the ones that represent the outside wall of sherds placed on the left or right side of the vessel. The general shape of the vessel in the areas between the documented sherds will be presented with dashed lines. The sherds will at all times be enclosed in a sharp outline that depicts their image clearly. The cross section of the sherd should be on the right side of the sherd. The drawing will consist of an outline of half the vessel, drawn in continuous line. Why must such a drawing be presented on the right side of the reconstructed vessel? Two standards are used by archaeologists, the American and the British. The American calls for presentation of the cross section of the vessel on the right side of the frontal view of that vessel, while the British does the opposite.

After making such decisions, drawing the sherds in their respective places, making sure that all traces of design are depicted on the surface of appropriate sherds, trace the whole sketch carefully. When finished, you will have the final sketch, to be used as a base for rendering. Make sure that if the cross section is drawn separately, the upper point of the inside of the lip will have a short horizontal line drawn from the lip to the left. This horizontal line is the only information you will have during alignment of the cross section with the frontal view of the whole vessel. Knowing now the height of the vessel and the distance from the outside wall of the bottom to the horizontal line representing the diameter of the rim, transfer the measurement to the sketch representing the reconstructed design. By doing this, you establish the lower border of the rectangle. Remember that the rectangle enclosing the reconstructed design must be of the same height, vertically, as the vessel itself. Retrace that rectangle carefully, making sure that the design is readable and clearly presented. This final sketch serves as the basis for the pen-and-ink rendering. The reconstructed design will be placed to the right of the cross section of the vessel.

Rendering: Clean the table, wash your hands, prepare the technical pen and the ink, find the French curve, and make sure that your coffee is a considerable distance from the drawing area. Take a new, clean illustration board. Take the sketch of the reconstructed vessel and spray both sides of the sketch with fixative—the front to make sure the graphite will not smear and will not transfer to the rendering, the back just in case. Place the dried sketch on the illustration board and tape the four corners with masking tape. Take a new sheet of tracing tissue. Make sure that this sheet is of poor quality, but transparent enough to allow you to see the sketch. Poor-quality tracing tissue will take the ink at all times, while more expensive may cause the ink to bead up in some places. Tape this sheet of tracing tissue on top of the sketch. Prepare the white opaque piece of

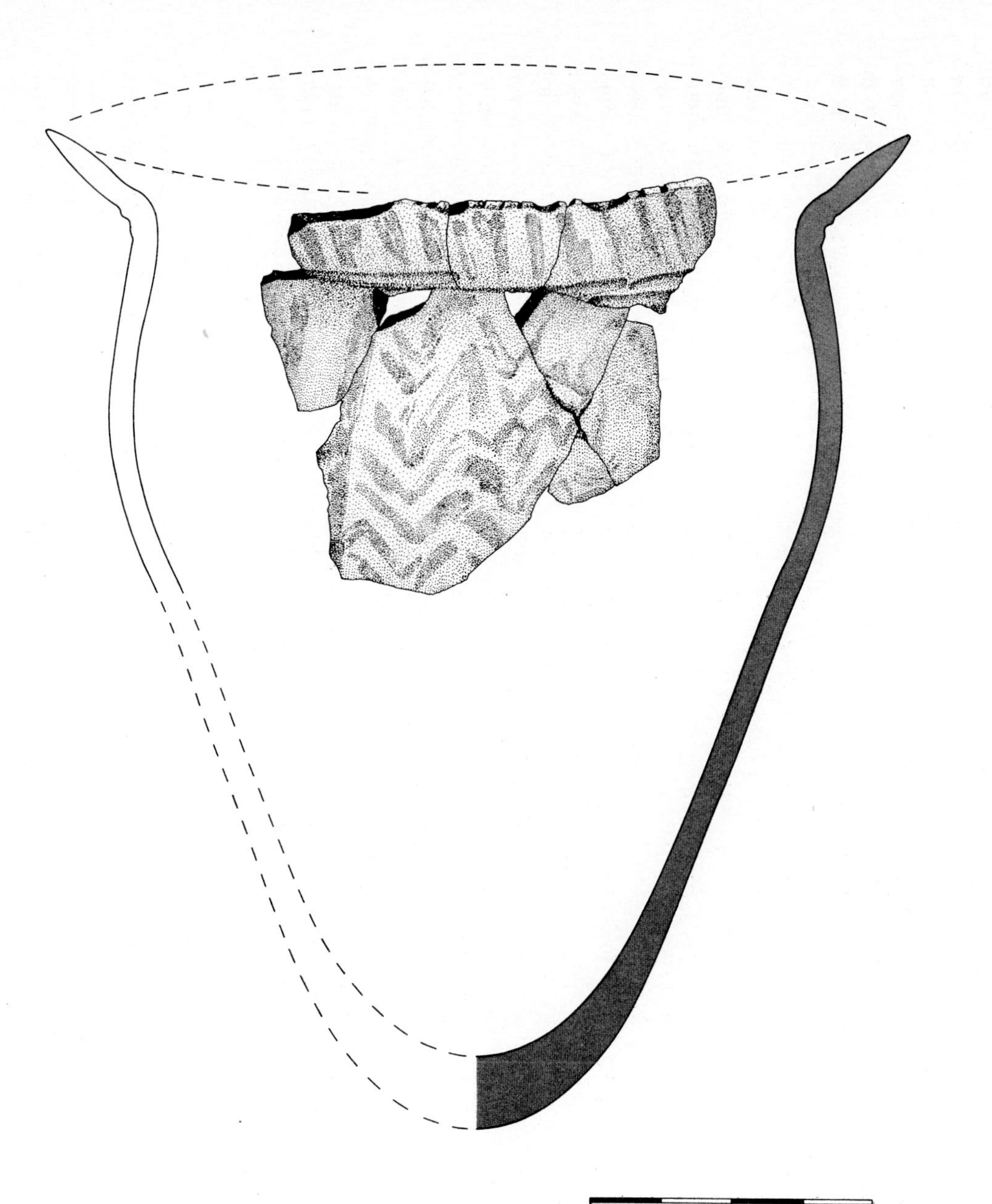

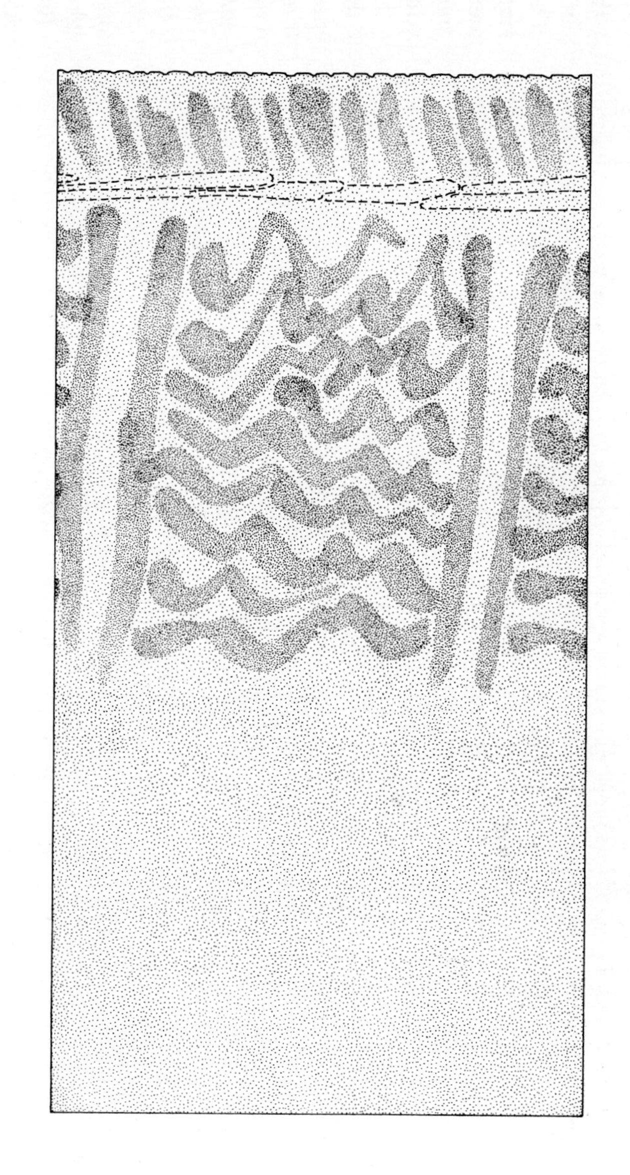

Figure 5-29. Reconstructed vessel from Bouganville, Papua, New Guinea (Solomon Islands). Pen and ink. Drawing by Zbigniew T. Jastrzębski, Field Museum of Natural History, Chicago. Unpublished. Drawn in 1978.

paper to be used as a white background when checking on the progress of the rendering. Have a reducing glass handy at all times. Depending on the size of the drawing, use a relatively larger pen point to draw outlines of sherds and the cross section. Make sure that stipples, the dots representing the tonal value, are smaller. When using point number 1 for the outline, use point number 00 for stipples. When using point number 00 for the outline, use point number 000 for the stipples. Before starting rendering, make sure that you have a stylus or something else to hold the tissue during rendering. When the tissue is not pressed down with some kind of pointed, slightly rounded tool, it will bounce up and down when the tip of the Rapidograph pen touches its surface. This bounce will cause irregularities during drawing. Holding the stylus in one hand and the pen in the other, draw all outlines of the sherds in solid black line. Do not shade the pen over the illustration. I know that it is tempting, but eventually it will result in a blob of ink in a place you do not want it. Render the tone representing the design by making dots—the more dots, the darker the tone will be. In order to check the rendering, take a white opaque sheet of paper and place it between the sketch and the rendering. Take the reducing glass and get up from the chair. Stretch your hand out and look through the glass. The image will be reduced, giving an approximation of how it will look after reduction and printing. Restrain yourself from overrendering. Such drawings do not reproduce well when reduced. The dots will merge, producing unwanted and uncontrolled black spots.

Remember that this illustration presents an overall view of the vessel, its dimensions, and the shape. The sherds are drawn because they support the visually presented theory. Use a French curve when drawing the ellipse and the cross section of the vessel. Do not stipple any tone in the cross section. Later, use an instant-transfer tone. Such commercially produced tone is much more uniform than tone produced by hand. If the cross section of the vessel is a part of the frontal view overall presentation, then it must be drawn in continuous line. When using a French curve, do not keep the tip of the pen right next to the edge of the curve. This can result in seepage of the ink under the curve, causing a smear. When connecting lines, try to be consistent. Lines that are badly connected will be very visible and will distract from the cleanliness of presentation. Take your time. Draw a small section and wait until the ink is dry before continuing. Do not draw various unconnected sections of the line. It is very hard to connect such short but separated lines. Usually this affects the thickness of walls, as the rendered cross section will not match the final sketch. Turn the drawing around as much as needed. If you are right-handed, start from the left so you can see the produced line. If you are left-handed, start from the right for the same reason. Keep the tip of the pen a little bit at an angle so the drawn line will not be next to the edge of the French curve. Make sure that the French curve is clean before placing it on top of the final rendering. It is a good idea to wash the French curve with soap and water, wiping its surface completely before using. When wiping, use a white tissue and pay special attention to the edges of the curve. Draw the opening of the vessel, the ellipse, in short, evenly spaced lines, using the French curve, with the exception of the frontal view of the sherd. The short, evenly spaced lines should not touch the cross section, nor should they touch the opposite, left wall. Leave some space between the ellipse and the walls of the vessel for better readability. Remember when drawing the design on the separate sherds not to exaggerate. The design should be presented as it is on the sherds.

A more precisely rendered depiction of the design will be drawn on the separate sheet of tracing tissue as a reconstructed design. In order to do this, take a separate illustration board and tape the final sketch of the reconstructed design at the four corners. Be sure that this sketch is also sprayed with fixative before rendering. Take a new sheet of tracing tissue and tape it on the top. Draw the outline of the rectangle in solid line. If the rim, the upper horizontal line of that rectangle, is a part of the design, make sure that this is properly represented. The rest of the design will have to be stipples, presenting the three-dimensionality of the design if such exists on the specimens. The background will be stippled in uniformly presented tone. This means that the whole rectangle will have a tone of one type or another. The presentation of the design depends entirely on the type of the design. Again, do not overrender the drawing. If the design is extremely complicated, include an enlarged section of it composed in a strategic location. In some situations, such as the presentation of figures, the whole reconstructed design should be drawn in line only, with the factual parts of the design drawn in solid line and reconstructed in stippled or dashed. When the renderings are finished, get ready for dry-mounting.

Dry-Mounting: Prepare the usual tools used for dry-mounting. Make sure the dry-mounting tissue is removable. Removable tissue is waxy and white. We need a white background for our semitransparent tracing tissue; it is essential for proper reproduction. Use a regular iron. Make sure there is no water in the iron; set the temperature of the iron for approximately 150–180°. In other words, set the iron for the lowest temperature. Prepare new illustration board and make sure all the renderings will fit on its surface.

Start with positioning. Take the overall reconstructed view and place it on the left side of the illustration board. Make sure that the diameter of the rim is parallel to the horizontal edge of the board. Tape the upper edge of the rendering to the board in two places. If you have a separate drawing of the cross section, place it to the right of the general view. Take a ruler and carefully align the upper tips of the rim with the short horizontal line drawn from the cross section. The cross section must be positioned at the correct angle. When finished, tape the upper edge of the illustration board. Do the same with the representation of the reconstructed design. Make sure that the upper edge of that drawing is correctly aligned also. Tape the upper edge in two places. Now you have two or three separate flaps attached to the illustration board. Take the clean dry-mounting tissue and place it under the illustrations. It does not matter if you have a large sheet of dry-mounting tissue or three separate small sheets. Make sure the dry-mounting tissue fits under the drawn images. Take the new sheet of tracing tissue and place it on top of the illustration or illustrations to protect the renderings. Take the iron and using its tip, tack the separate images at their centers through the protective tissue. Do not press very hard. Touch the tissue with the tip and hold it for a couple of seconds. Do this for each separate drawing. Put the iron away and remove the protective tissue. Take an X-Acto knife with a new, sharp blade and, using a metal ruler, cut the drawings through the tracing tissue and dry-mounting tissue at the same time. If you have to cut at an angle, do so. Such a situation may occur in the area located between the direct view of the whole vessel and the separate drawing of the cross section. When cutting separate illustrations, be sure to cut through both tissues at one stroke. When repeating a cut, be careful and do not leave any dry-mounting tissue on the illustration board. When finished, remove unwanted parts of the tracing tissue as well as the dry-mounting tissue. Then carefully separate the illustrations from the dry-mounting tissue and then the dry-mounting tissue from the illustration board. Do not tear your drawings. Take your time and be careful. After this, take a new clean sheet of tracing tissue and place it on top of the drawings. Take the rim and starting from the middle of the larger image, slowly squeeze the air by applying pressure and melting the dry-mounting tissue with the heat of the iron. When everything is dry-mounted, draw a scale, using the metric system, to designate the enlargement or reduction factor. The cross section of the vessel should be covered with instant-transfer tone or painted black after the illustration is dry-mounted.

Preparation for Production: Follow the usual procedure and tape a sheet of tracing tissue to the upper section of the illustration board. Place the sheet of black (or your favorite color) cover-stock paper and also tape it to the upper section of the board. Make sure that the cover stock is creased and easily bendable. Keep all the sketches in a separate folder; they may become handy. If a reduction of the illustration must be marked, do this on the white illustration board under the drawn illustrations. Use a blue pencil and be very straightforward in your instructions. Write legibly—the platemaker or photographer will have to read your handwriting.

Comments: The process of pictorial reconstruction of pottery does not produce a visually "interesting" illustration, nor should it. Such an illustration should be clearly drawn and should contain all pertinent information. You know that the purpose of such a drawing is to present a definite, explanatory, overall picture of the vessel. Reconstruction is a puzzle, and as in all puzzles, there are many unknown factors controlling the outcome. Be logical and methodical; utilize all available information. The closer you look, the better the result. The first project should be used for reference.

After the reconstruction is finished, take a look at the sketch or photograph of the shape of the vessel, made before it was broken to pieces. Compare this sketch with the finished illustration. See how much you are off. Analyze where you might have gone wrong. This will help you better to understand the process of reconstruction. It will help you to observe the evidence, the bits and pieces of the broken pot more clearly. A table of second powers is included in Appendix B for you to calculate the diameter of a vessel.

CHAPTER SIX

Other Artists' Techniques

YOU must wonder about the techniques and approaches used by other scientific illustrators. What are their procedures, materials, and tools? How are their drawings produced? Before answering such questions you must realize that techniques can differ as much as artists' personalities. Each illustrator has a different approach toward application of a particular technique. Each description reflects a lifelong struggle for perfection. Continual improvement, continual experimentation with materials is prompted by the need to meet deadlines and the need to satisfy the scientist and the illustrator alike. There are almost as many variants of any one technique as there are artists. It is important for you to have a small sample of other artists' statements in order to understand your own potential. Such material is useful as a reference and as an example. First of all, be aware that scientific illustration is a major field and is to be found in all of the corners of the world. Secondly, realize that nonartists, the scientists, are an important part of the scientific illustrator's life. This chapter contains materials written by both professional illustrators and scientists. The group, a unified body of experts in their respective fields, coexist as a working unit presenting a product which continually builds our civilization. You, the future scientific illustrator, will become a part of his actively productive entity. When looking through descriptions of techniques, materials, and tools, think of your own advancement. The examples presented here can be actively followed. If you need a camera lucida, build one. Perhaps the first try will not be entirely satisfactory, but the second will definitely be improved and useful. Did you ever try to use an airbrush? Even if you did, make an effort and follow the experts, seeing for yourself the results of your experimentation. Remember that only practice will provide the necessary improvements.

Nélida Raquel Caligaris's Rendering of Anatomy of Insects[1]

Drawing an insect involves preparation of insects, observation, sketches, and final rendering. Such a complicated procedure calls for patience as well as steady hands.

Preparation of Specimen for Drawing

1. Before boiling the specimen in sodium hydroxide (HO Na), the wings must be removed and placed directly in phenol (c_6H_5OH).[2]
2. The specimen is boiled in 10 percent solution of sodium hydroxide and after boiling is placed in phenol.
3. Specimen is dissected, using Zeiss stereoscopic microscope for viewing and dissecting needles for separation of structures.
4. All structures are placed in petri dishes containing balsam of Canada diluted in xylol and are properly positioned for drawing.

Sketching:

1. For sketching and drawing, an Olympus HB microscope equipped with camera lucida is used continually.
2. In order to assure accuracy, each separate structure is correctly placed under the objec-

[1] Facultad de Ciencias Naturales, Argentina.

[2] Warning from author—poisonous chemical.

tive of the microscope and viewed. With the help of camera lucida, the image is transferred directly to the surface of the Shoeller Hammer white, smooth-surface drawing paper.

3. If necessary, French curves are used for detailed representation of some structures.

Rendering:

1. Final rendering is achieved with 0.1-mm and 0.2-mm Rotring drawing pen. The image is produced using line and stipple. Different sizes of the pen are used for easy implementation of various thicknesses of the line.
2. Tonal range and three-dimensionality of details are based on careful observation and continual comparison of drawn image with actual structures seen through the microscope.

Nélida Raquel Caligaris's Rendering of Crustaceans[3]

For proper visual representation of crustaceans, a photograph of the specimen as well as the actual subject can serve as a source of information. The specimen's anatomical structures are carefully observed and the project is discussed in detail with the scientist. In some instances, minor modifications are allowed as long as appropriated changes will not distort nor misrepresent the subject. All modifications must be approved by the scientist before rendering begins.

The procedure of study and drawing is as follows:

1. Subject is carefully examined.
2. Decision pertaining to the position of the subject is reached through discussion with the scientist.
3. The subject is positioned in exactly the way it will be drawn.
4. Photograph is taken of the specimen. Carefully established enlargement or reduction factor will serve as the basis for the size of the future sketch. The x ratio must be noted, as it will be needed in preparation of the scale.
5. The photographed subject is traced on tracing tissue in outline, using pencil. All questionable areas are clearly marked for reference.
6. After approval of the sketch by the scientist, the image is transferred to white, smooth-surface

[3]Facultad de Ciencias Naturales, Argentina.

Figure 6-1. *Aegla neuquensis neuquensis*. Pen and ink on two-ply, medium-surface paper. Drawing by Nelida Raquel Caligaris, Universidad Nacional de La Plata, Facultad de Bellas Artes and Facultad de Ciencias Naturales y Museo, La Plata, Argentina. Published in Estela Celia Lopretto, "Fauna de Agua Dulce de la Republica Argentina," by Instituto de Limnologia "Dr. Raul Adolfo Ringuelet," Argentina. Drawn in 1982.

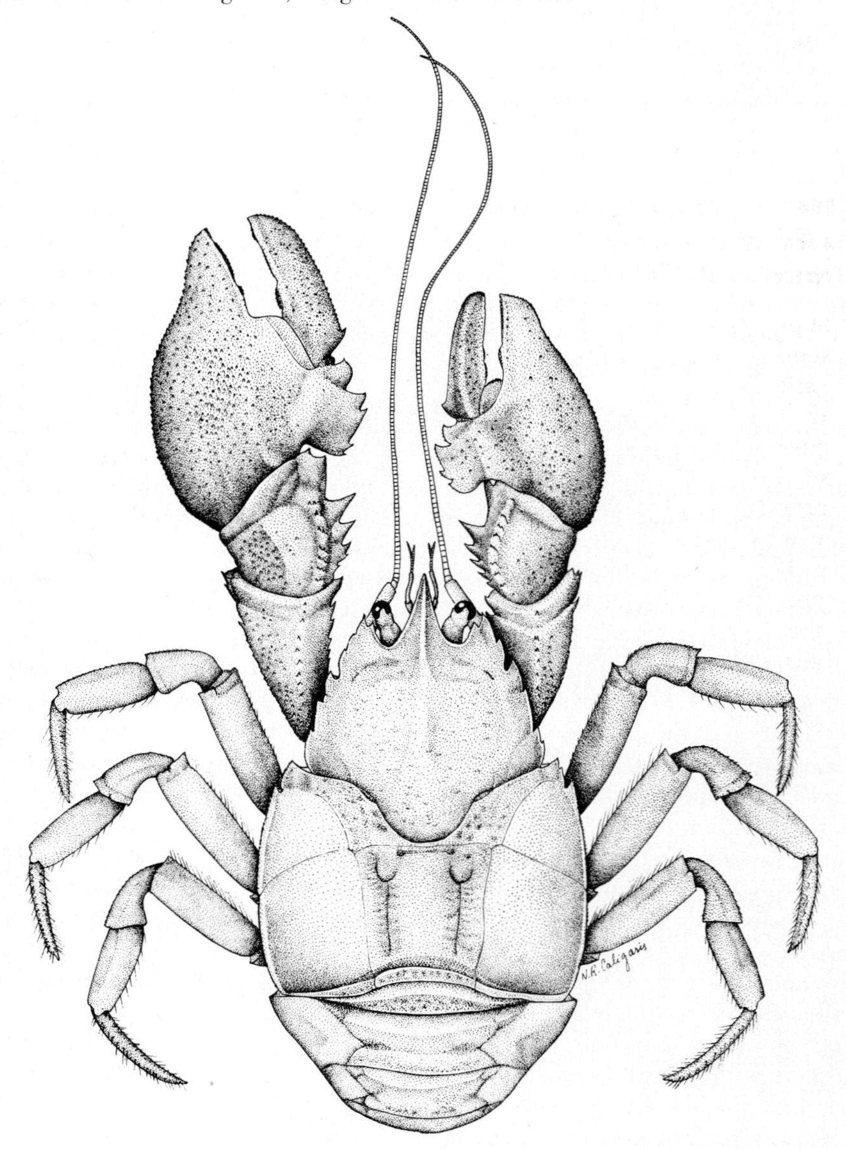

Shoeller Hammer drawing paper, making sure that the tip of the pencil does not destroy the surface of the paper.

7. Every detail of the transferred sketch is compared with the actual subject. Small details of the surface configurations are observed through the microscope. With care, necessary changes are implemented, preparing the transferred image for rendering in ink.
8. The final rendering is achieved through a combination of line and stipple using 0.1-mm Rotring drawing pen.
9. When the image has been drawn in ink, all visible pencil lines are erased from the surface of the drawing.

Yves Coineau's Indian Ink on Scratch Board, Outline Drawing of Microarthropods[4]

The preparation of ink drawings representing the anatomical structures of microarthropods is complicated and tedious. Mastering the technique will enable the illustrator to take advantage of proper graphic expression and will help in production of a correctly represented image. It is most useful when drawing microscopic anthropods (mites), but at the same time can be applied to larger anthropods. Basic principles can be transposed to other areas of specialization.

Preparation of the Sketch

The original drawing is made with the help of a camera lucida on a good-quality writing paper with a black-lead 2H pencil. Sharp and precise lines produced by the 2H pencil will not smudge under the hand. Proportions have been indicated with great care and precision; the corrections on the initial sketch are applied freehand. Observation of the subject is a must during such a procedure. Usually a higher magnification is needed for clear observation of necessary details. During this stage a softer pencil (HB) can be used for underlining some of the important details and structures. As a matter of fact, a colored pencil will be of great help. Separate colors will denote certain structures, such as the extension of sclerites, for example. Naturally, all colored lines will be rendered in India ink at a later stage.

Consequently, the original, first sketch as well as the corrected, second sketch must have every possible detail of the specimen noted. Complete data for all structures is necessary, later allowing the illustrator to concentrate more on the ink rendering than on observation. Such a sketch when stored in archives will constitute an important document, replacing the specimen itself if the subject is destroyed by dissection or clumsiness.

Drawing Surface: Choosing an appropriate scratch board is important, as not all surfaces will correctly accept the transfer of the sketch as well as the ink.

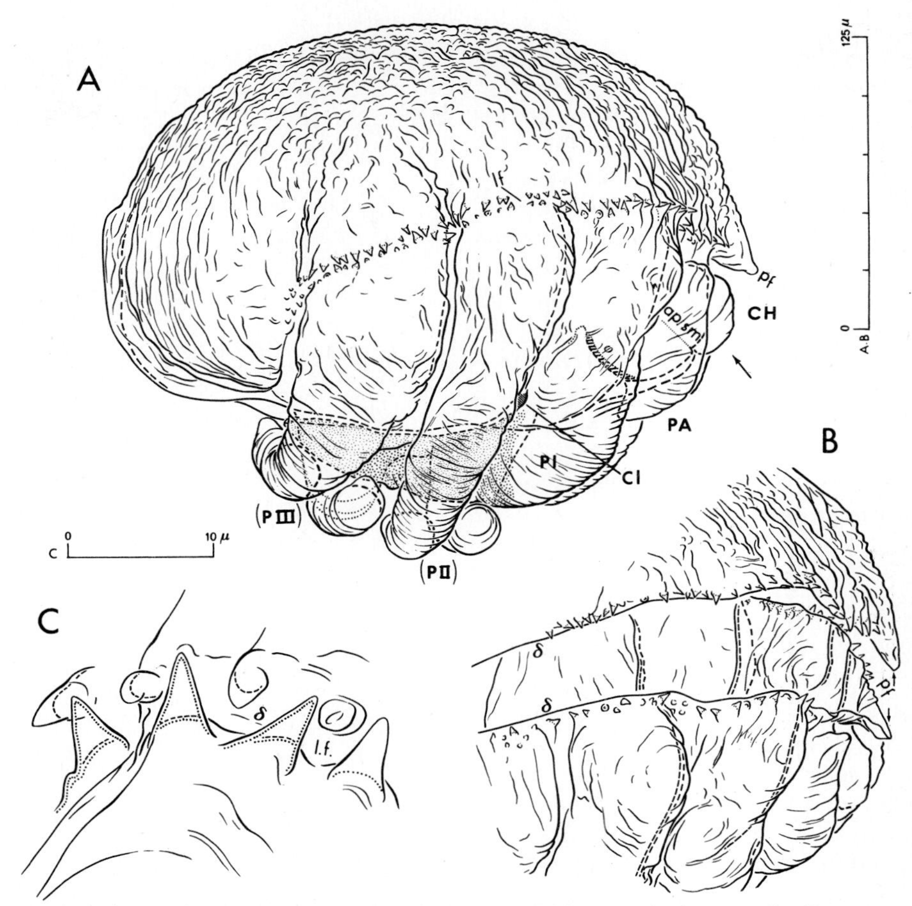

Figure 6-2. *Microcaeculus steineri delamarei*, Prelarva—A, from aside; B, detail of the anterior part after molding; C, detail of the latero frontal furrow. Crow quill on scratch board. Drawing by Dr. Yves Coineau, Museum National d'Histoire Naturelle, Paris, France. Published in Yves Coineau, "Elements pour une Monographe Morphologique, Ecologique et Biologique des *Caeculidae* (Acariens), "Memoires du Muséum National d'Histoire Naturelle," N. series, series A, Zoologie, Tome LXXXI, 1974. Drawn before 1974.

[4]Muséum National d'Histoire Naturelle, France.

The scratch board must be coated and must not be too flexible. All too-soft and glossy boards should be avoided because such surfaces will change the thickness of the drawn ink lines. When the coating on the board is too thin or too dry, the surface must be rejected. The best surface will accept ink without difficulty and will allow for scratching when implementing additions or correcting mistakes. The rigidity of the surface will insure better protection of the drawn image during the lengthy process of rendering.

Copying or Tracing the Sketch onto the Surface: Once the scratch board has been selected and the sketch approved, the transfer of the image to the board must be accomplished. In order to copy or transfer the image, a tracing method is most useful. The back of the approved sketch is blackened with 2B lead, "Progresso" (Hardtmuth), and the sketch is placed, blackened side down, onto the surface of the scratch board. Any excess lead can be removed by rubbing the back of the sketch with soft paper. Using masking tape, affix the sketch to the drawing surface, then follow the image on the sketch with a blunt needle. A hard lead pencil (5H) can be used, but with a disadvantage. Usually, extra marks and lines are added to the sketched image when the illustrator misses or goes out of the traced line. Use of the needle helps to avoid feature modifications of the important sketch. In addition to this, the scratch board is not damaged, offering a smooth, untextured surface that allows better control of the pen and ink.

Tools for Pen and Ink Rendering: During rendering I use two kinds of pens: those with flexible tips and Graphos pens. Pens with flexible tips are the classic crowquill pens, allowing for variations of the thickness of the drawn line.

Flexible-Tip Pens:

1. A pen with a long, thin tip is more flexible and allows you to make thick and thin stokes; for example, Joseph Gillott's mapping pen, number 291.
2. Pens with a thick-set tip will be more rigid and must be kept back to draw thin and regular lines, for example, Gilbert Blanzi Poure "Atome," number 423.
3. Very fine pens look like a vaccinotyle; for example, Brandauer 518.

Comment: The curve of the pen body must be identical to the curve of the penholder, because the constraints exerted on the pen body, when not suitable, give rise to a distortion at the level of the tip, and the pen will scratch the surface of the paper.

Graphos Pens (Pelikan): For some years technical pens have often been used in place of crowquill drawing pens. Technical pens are perfectly suitable for work on celluloid or plastic semitransparent surfaces, but should not be used for drawing on scratch board. The point of the technical pen constitutes a sharp edge that scratches the surface of the scratch board, which is then ruined. The line becomes thicker and the ink duct is apt to become blocked.

The Graphos fountain-pen system has a set of calibrated pens that allow better control of the drawn line. (See Yves Coineau, *Comment Réaliser vos Dessins Scientifiques*, p. 31, Fig. 3-12). The fountain pen contains a staggered duct which, because of capillary action, keeps back some of the ink, therefore allowing the illustrator more freedom in handling the drawing tool. Personally, I use a medium-flow duct number 2 with Pelikan "La Perle" or Paillard "Yang Tse A2" inks. Graphos pens consist of seven separate series, each used for precisely described applications. Two types of Graphos pens are useful for morphological illustrations:

1. Drawing pens of the series A have nine sizes, ranging from 0.1 to 0.6 millimeters. The pens are used for drawing exactly calibrated lines. When using such pens, the illustrator must keep the pen in a more vertical position than usual so that the two halves of the point will be situated exactly on both sides of the sketched line. If the line is curved, the pen must be rotated between the illustrator's fingers in order to accomplish the same result. Using this method, a curved line can be drawn freehand without hands shaking. (See Yves Coineau, *Comment Réaliser vos Dessins Scientifiques*, p. 61, Fig. 3-7.)

2. The pens belonging to series 0 have tips ending with a round shape. The 0.2-mm and 0.3-mm sizes are the most useful.

These pens have a double advantage in that they are firm enough to draw regularly calibrated lines, but subtle enough to allow modulation of line thickness.

Their spatulate tip insures a good mobility on the coating surface.

The Scraper: I use a Swann-Morton number 10 surgical blade fitted on a Martor-Scapel number 23-0-00 handle. Those preferring a rounded handle can use Martor-Boy number 31-133 with the same blade, broken in the middle of its length. Freehand scratching can be made progressively by tangential successive approaches while blowing out the dust to control the advancement of the work. In some situations it is possible to drive the back of the blade with the help of a transparent French curve and remove unwanted areas with precision.

The Reduction and the Plate: Large reductions should be avoided because the final appearance of the image in its printed form is too difficult to imagine. Only long experience can be of some help. Each element constituting the plate must be drawn from a preparatory sketch, which in turn has been prepared with the subject oriented to the appropriate reference plan. When each separate illustration is properly positioned, a precise comparison between the drawings and analogous objects is possible. If possible, the position of separate components of the plate should be aligned with their respective natural orientation, with the light coming from the upper left side at a 45-degree angle. The plate must be composed properly and the drawn images should be balanced, with lettering and appropriate scales filling the gaps.

CODES FOR VISUAL REPRESENTATION. The following codes must be observed:

1. All directly seen structures must be drawn in continuous lines.
2. Structures that are visible through transparent tissue must be drawn in dashed lines.
3. The surface plan notion often has a conventional aspect and is answerable from the morphological point of view. When one draws a transparent worm like a *Rotifera* the integument is generally indicated only at the level of the outlines and the organs are seen directly.
4. If two successive plans are seen directly, the farther must be isolated from the nearer by a clear gap indicating that there is no direct structural continuity at this level.
5. Sometimes an arthropod shows two symmetrical organs (two setae belonging to the same pair) standing out in profile above their support, one on top of the other. In order to avoid overcrowding on the plate, only the end of the second seta should be drawn shifted slightly to one side.
6. In arthropods, the mechanical qualities of the integument should be rendered by lines of different thicknesses.
7. A thin line should be used for representing joint membrane, while a scleritized area will be rendered in thick line.
8. The outlines of solid-mass structures like those of the body and legs will be represented by thick lines; details and organs will be rendered in thinner lines.

Continuity and Structure Homology: The use of visual variants in the drawing, showing the thickness of the elements or the modification of their shapes, allows a better understanding of the structure of the specimen.

For example, in mites, the outline of the ectoskeleton will be represented by a thick line divided into short dashes where the hidden parts are, and its inside wall represented by a thinner line divided into longer dashes. In order to represent the integument mass, the space between those two areas will be filled with lines. When several layers belonging to the same structure are situated in the same section plan, the lines must be of the same length and thickness and must have the same orientation. In order to avoid overcrowding on the illustration, the parallel lines must be restricted to well-planned areas of illustration, helping the readability rather than encumbering the drawing. The orientation of parallel lines is reflected by the section plan with the unifying representation of the image being of utmost importance.

Naturally, in order to underline the homologies or present the differences a different type of stipple or dashed line can be used. Special symbolic patterns such as undulated lines or other schematic representations can be applied to various structures such as tendons. It is advisable to restrain from continual repetition. If a specific detail is found all over the surface of the subject, it is enough to draw only a position rather than to clutter the image with too much detail. Relatively empty areas of the drawn image are most suitable for such treatment. Enlargements of a specific section should be drawn separately. The rendering of the volume and the three-dimensionality of the subject can be accomplished by conscious manipulation of representation of surface details.

All annotations, comments, letters, and other designations must be aligned horizontally and must be parallel to the edge of the printed page. Leaders, the lines that help point to specific structures, must be horizontal or must be placed at a 45-degree angle for uniformity.

Maria Cristina Estivariz's Procedure of Drawing[5]

During the initial stages of production of a correct image, two separate measuring techniques are used by the artist. The first is a combination of direct tracing from photography and measuring the subject. The second is based on comparisons of proportions within the subject while using the microscope.

For illustrations representing crustaceans, an actual-size photograph of the subject serves as the basis for the final sketch. The following is a routine:

1. Proportions as well as basic forms are traced from the photograph onto the tracing tissue, resulting in a one-to-one-ratio image.
2. The sketch is redefined by measuring the subject, if available, and adding more details to the final sketch.
3. The final sketch is transferred to the white drawing surface,

[5]Museo de la Plata, Argentina.

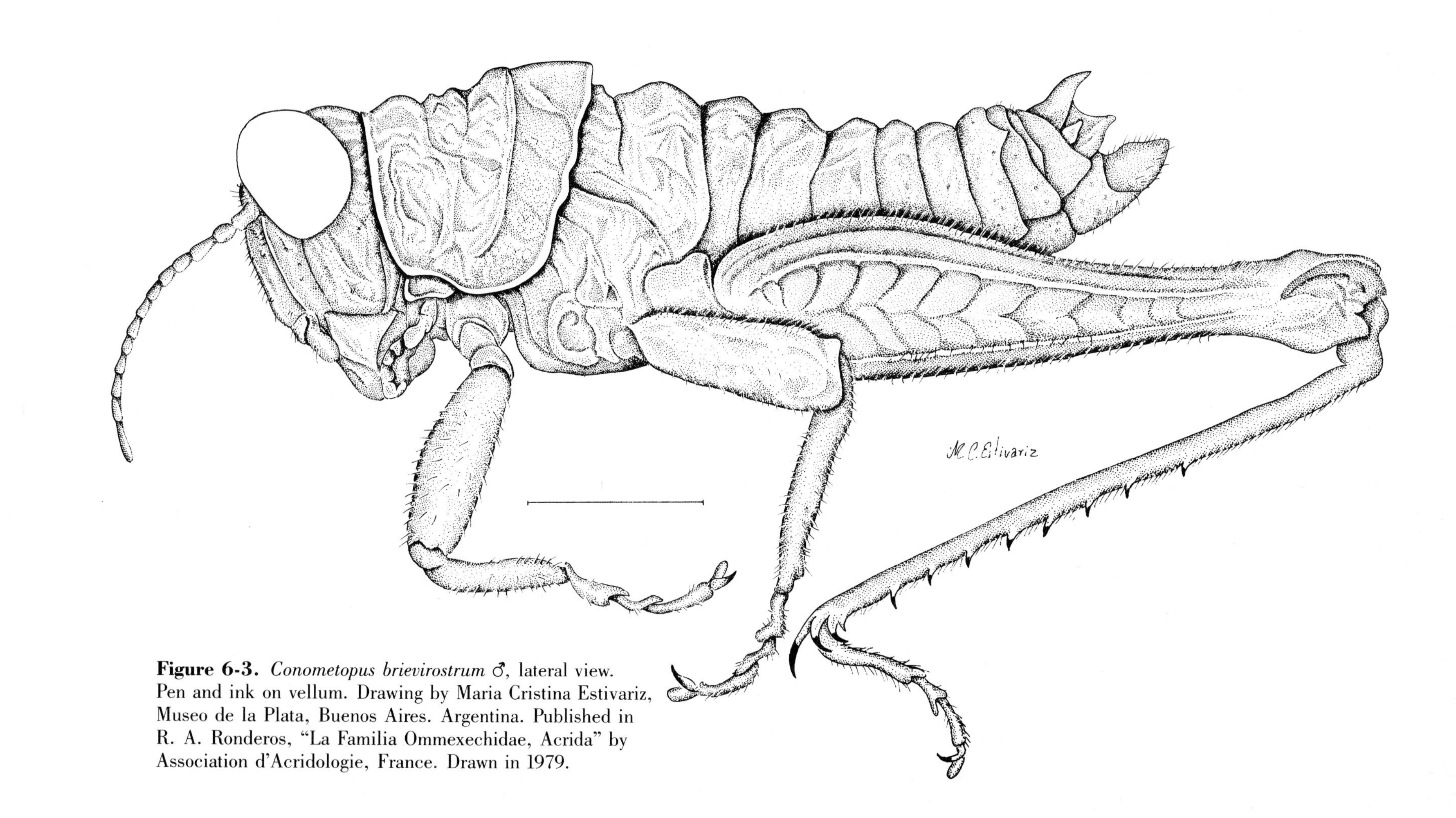

Figure 6-3. *Conometopus brievirostrum* ♂, lateral view. Pen and ink on vellum. Drawing by Maria Cristina Estivariz, Museo de la Plata, Buenos Aires. Argentina. Published in R. A. Ronderos, "La Familia Ommexechidae, Acrida" by Association d'Acridologie, France. Drawn in 1979.

consisting of a suitable smooth-surface paper.

4. Rendering is achieved by using line and stipple with 0.1-mm Rotring drawing pen. In order to obtain small points (stipples) and thin lines, the pen should have a completely empty ink tank.

During production of the illustration, information about the structures is obtained directly from the specimen.

1. The subject is correctly positioned and placed under the objective of the stereomicroscope.
2. For observational and drawing purposes, a reticle (graticule with linear or other designations) is placed into one of the oculars.
3. The proportions of separate sections of the structures are compared in size, using the reticle with linear or cross-hair designations as a means of establishing the relations between structures visible through the microscope and the sketched image.
4. The sketch of the whole subject is produced through careful observation and comparisons of proportions on the subject as well as on the sketch.
5. After the sketch is finalized, the cellularic tracing paper is used as a rendering surface.
6. The rendering is achieved with a 0.1-mm Rotring drawing pen with its ink tank empty for better control of the diameter of dots and the thickness of lines.

Maria Cristina Estivariz's Rendering of Insects[6]

Illustrating insects calls for attentiveness on the part of the illustrator. Only specimens specially chosen by the scientist are submitted for descriptive studies. The illustrator must return all specimens to the scientist in undamaged condition. All anatomical structures must be presented with care and exactness. In some instances, minor modifications of sections of legs and antennae can be implemented, provided that such modifications are relevant to diagonostical features only.

The procedure of study and drawing is as follows:

1. The specimen is carefully examined through the stereoscopic microscope.
2. One of the eyepieces of the binocular microscope contains a reticle used for measuring proportions by comparing relative sizes of the visible features.
3. The specimen is rotated under the objective of the microscope until a satisfactory position for drawing a particular anatomical structure is found.
4. The image of the specimen is sketched on grid paper. The grid on the drawing paper must correspond with designations marked on the reticle in order to establish proper proportions between structures visible through the microscope and those that are drawn.
5. When the whole insect has been sketched and the final version approved by the scientist, the rendering will start.
6. Tracing tissue and 0.1-mm Rotring technical pen are used for the final drawing. In order to obtain the finest possible line, the ink tank in the pen must be almost empty. A small drop of ink is sufficient.
7. The drawing surface is placed over the approved sketch, and dotted tonal areas are executed in parallel series of minute points. Tonality on the rendering is controlled by topography of the anatomical structures of the specimen.
8. Dots are not overlapped, presenting an overall gray tonality and soft appearance of the drawing. After finishing the rendering in the above manner, dots are added to some areas in order to underline desired features. Structures that cast shadows, such as tubercles, ridges, keels, and legs, are usually treated with extra tone.
9. Originals do not exceed 50 centimeters in order to avoid excessive reduction. The softness and gentleness of tone can be preserved better if the finished illustration is not drastically reduced. If the original pen-and-ink rendering is longer than 50 centimeters, the possibility for losing 70 percent of the work quality is great.
10. The technique of rendering requires careful but quick application of dots in order to maintain high quality of tonal ranges as well as reasonable time-work rate.

Nancy Shinn Hart's Method for Preparation of Botanical Illustration[7]

Botanical illustrations, although prepared for scientific purposes, are used as illustrative material on book covers and in exhibits. Nancy S. Hart keeps such a dual purpose in mind when preparing visual material. In order to complete a botanical drawing, the process is as follows:

1. Plant species are selected for illustration. One of the institution's botanists, taxonomists, or dendrologists accompany the artist into the arboretum's living collections to select a plant specimen. The scientist is usually chosen to accompany the artist because of his

[6]Museo de la Plata, Argentina.

[7]The Morton Arboretum, Lisle, Illinois.

Figure 6-4. *Pinus strobus* (white pine). Pen and ink. Drawing by Nancy Shinn Hart, The Morton Arboretum, Lisle, Illinois. Published in *Morton Arboretum Quarterly*, Vol. 13, no. 2, 1977. Drawn in 1977.

special knowledge of a given subject.

2. Several specimens are cut and brought into the greenhouse cold rooms to keep them as fresh as possible for as long as possible.
3. In order to learn as much as possible about the subject, the illustrator does research on the specific plant in the Sterling Morton Library. The drawing must be as close to the typical plant as possible, and be without insect holes or weather or nutrient damage.
4. Occasionally, with fragile plants or in very hot weather, the photographer will be requested to take some black-and-white photos of the specimen for use after the plant has wilted.
5. If necessary to clarify details, a few sketches are made using a microscope.
6. The living specimen is then considered for descriptive illustration and clamped into a stand, where it will remain in position until the drawing is completed. When necessary, the specimen is returned to the cold room to assure longer preservation of material.
7. The plant is measured and drawn in a one-to-one ratio. The drawing materials are simple:
   a. 5 H pencil
   b. Pink Pearl, ink, and kneaded rubber erasers
   c. Strathmore four-ply, hot-press rag board
   d. Hunt crow-quill dip pen with a variety of points, with the number 104 point as the most useful
   e. Pelikan black India ink
   f. Technical pens, occasionally
8. A complete pencil sketch is drawn on the Strathmore board.
9. The sketch is corrected by the scientists.
10. The approved pencil sketch is then inked, using a variety of rendering techniques, including strokes and points. Crow-quill dip pens are preferred because of the sensitivity of lines achievable with such drawing tool, reflecting the complex surfaces and structures of plants.
11. The drawing is checked after completion by the scientists, then it goes to the printer.

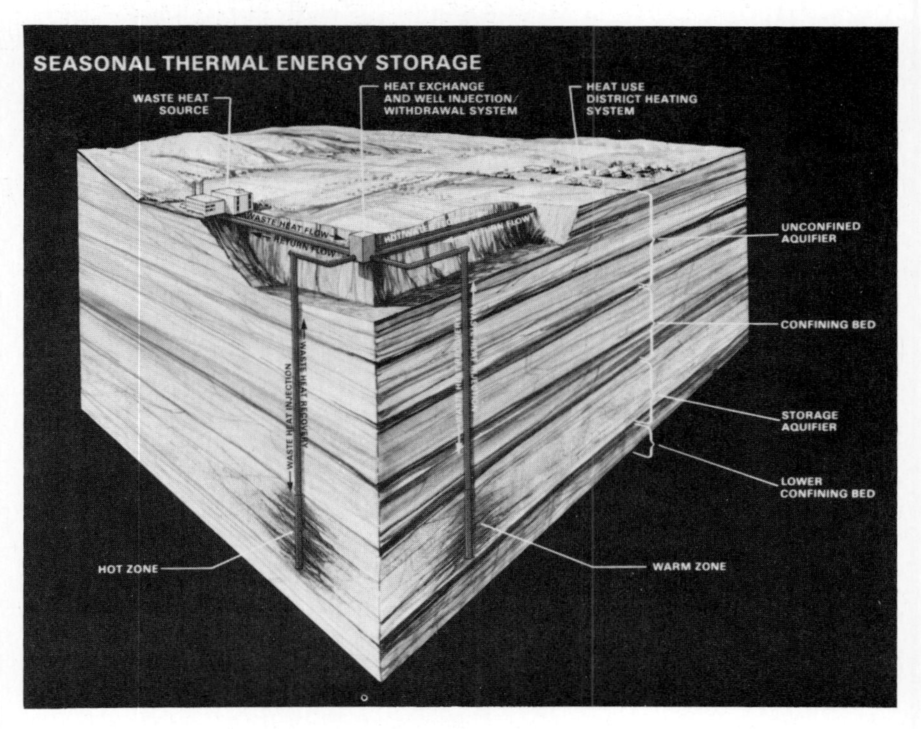

Figure 6-5. Seasonal Thermal Energy Storage. Brush and acrylic on illustration board. Painting by Mark P. Hutton, Battelle Memorial Institute, Pacific Northwest Laboratories, Richland, Washington. Produced for 35-mm slides and 8-by-10-inch viewgraphs (transparencies). Painted in 1981.

Mark Hutton's Acrylic on Illustration Board[8]

1. Prepare the sketch appropriately with requirements in pencil on vellum tracing paper. All necessary details must be implemented and approved.
2. Final version of the sketch must be rendered in ink, using technical pen (Rapidograph) on new sheet of tracing paper or on the final sketch, providing that all pencil lines will be erased, leaving a clean, inked image.
3. From the inked version of PMT (photo-mechanical transfer), a print in matte finish must be made. The PMT print is a direct positive image obtained through a new version of photostatting techniques developed by the Kodak Company. During the PMT procedures, the drawn image can and should be enlarged if the original pen-and-ink drawing is too small for meaningful rendering.
4. The PMT print, properly washed, must be affixed to Crescent illustration board using double-stick adhesive such as Twin-Tak. Caution must be exercised because improper handling of double-stick adhesive can produce air bubbles or can destroy the PMT print.
5. Light coat of gesso must be placed over the dry-mounted PMT print. The gesso coat should allow for good visibility of the image, therefore it must be transparent.
6. After a coat of gesso has dried, application of acrylic paints can

[8]Battelle Memorial Institute, Pacific Northwest Laboratories, Richland, Washington.

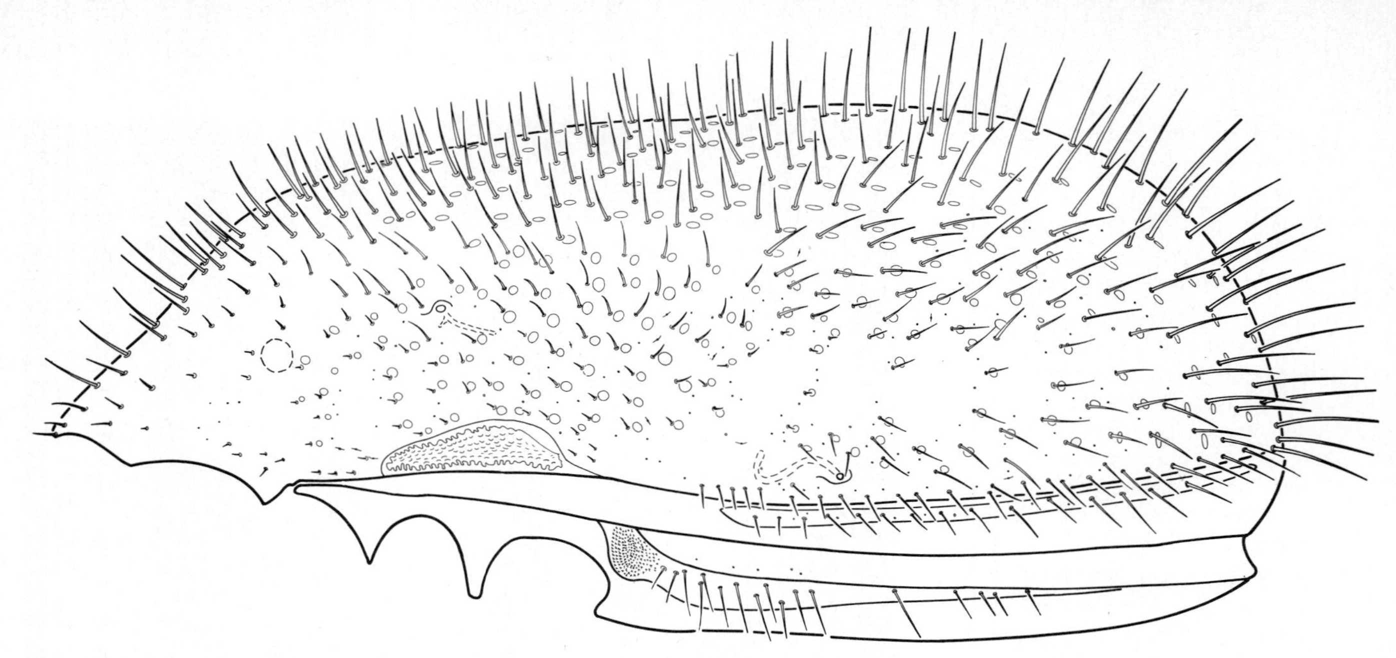

Figure 6-6. Acari: Allothyridae, undet., ♀ body lateral, excluding appendages 20 × (Allothyrid mite). Pen and ink on tracing tissue (Rapidograph® technical pens) drawn using various pen points ranging from number 000–4 and French curve. Drawing by Dr. John B. Kethley, Field Museum of Natural History, Chicago. To be published in 1985. Drawn in 1978.

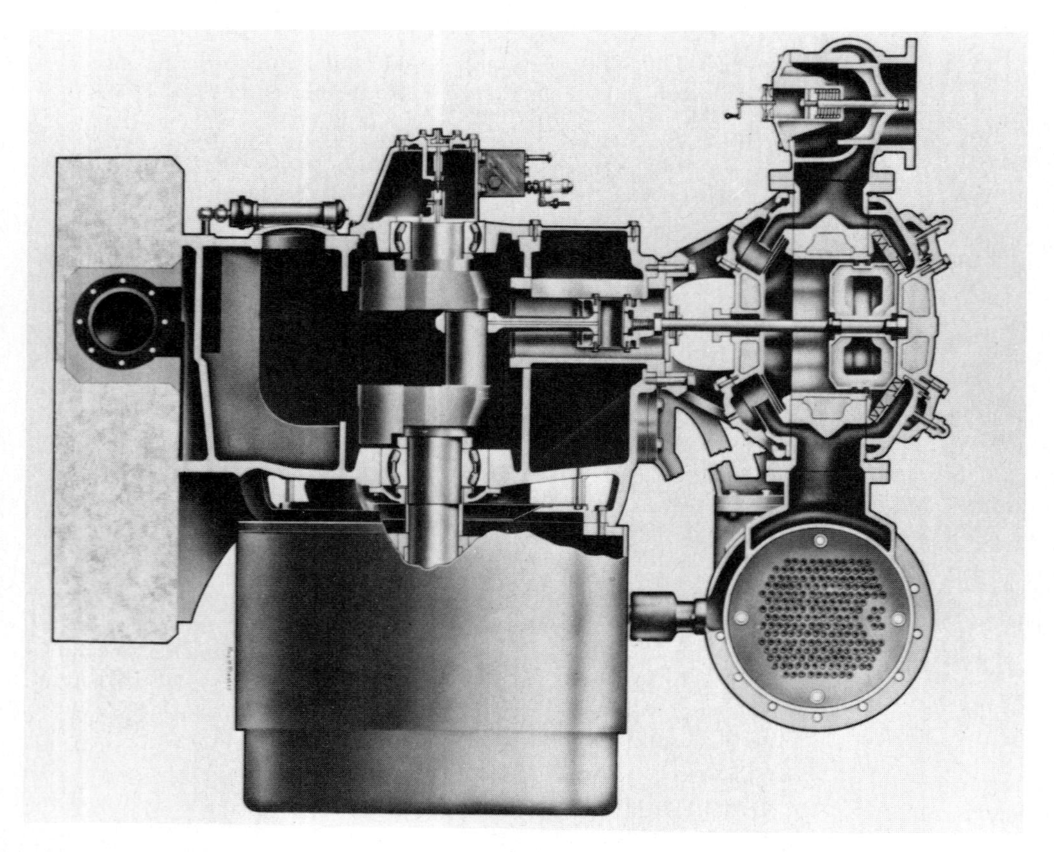

Figure 6-7. Air compressor. Airbrush on illustration board. Roy G. LeMaster, Battelle Memorial Institute, Pacific Northwest Laboratories, Richland, Washington. Produced for 35-mm slides and 8-by-10-inch viewgraphs. Painted in 1976.

proceed. Necessary tools are good-quality brushes and the usual acrylic paints. The paints have to be placed in such a way that the surface of the illustration is smooth and with a minimum of texture.

7. Any necessary typographical designations can be added after the acrylic paints are dry, using instant transfer letters. Naturally caution has to be exercised and all typographical designations have to be spaced and aligned properly. White letters cannot be placed on light value colors, just as black letters cannot be applied to dark-value colors because readability of the text will be affected.

John Kethley's Conversion Tables for American and European Standard Type System[9]

Because there is a great need for handwritten or instant-type designations on the drawings, the size of the letters or numbers must be known to the illustrator and scientist alike. The proportion between the total size of the illustration or a plate and various designations has to be maintained. In other words, small letters on a big illustration will produce undesirable results, as they will tend to disappear after reduction of the illustration. At the same time, if too-large designations are used, the main subject, the illustration, will be dwarfed by such secondary information. It is exceedingly important to find the proper size of letters and numbers when such designations are to be used in conjunction with a structurally complicated image. All secondary information must be readable but cannot overpower or obstruct drawn images.

In order to help solve the guessing and add stability, conversion tables have been prepared by Dr. John Kethley, the curator and the division head of the insect department at Field Museum of Natural History in Chicago. Conversion tables are for American standard type system and for European standard type system. Both systems differ slightly in terms of the actual length of the letters. The point size, the length in millimeters, as well as in inches, is presented in such a form that when the need arises it will be possible to find the actual *vertical* size of the letter, knowing one of the components. For example, if the length of the letter is known in inches or millimeters, the point size will be easily found, and vice versa.

These tables may be found in Appendix C.

[9]Field Museum of Natural History, Chicago, Illinois.

Roy LeMaster's Airbrush on Illustration Board[10]

1. Prepare the sketch according to required standards using pencil on vellum tracing paper.
2. After approval of the sketch, the image should be transferred to the illustration board using the direct tracing method.
3. The image is rendered in ink as line drawing with technical pens (Rapidograph) using nibs numbers 1 through 3.
4. Render the illustration in airbrush, using opaque tempera paints. During rendering, World Airbrush should be routinely cleaned after each separate color application. Frisket paper must be used for masking out unwanted areas, leaving open sections for color application through cutting. After a layer of paint is dry, old frisket paper must be removed and new has to be applied again, making sure that appropriate cut-out, open areas are implemented correctly. For each color application the primary tool, the airbrush, must be cleaned completely and reloaded with a new mixture of tempera pigment.
5. Process must be repeated until the illustration is finished.

[10]Battelle Memorial Institute, Pacific Northwest Laboratories, Richland, Washington.

Vincaine Lowie and Willy Lauwens Procedure of Painting with Airbrush[11]

The production of an illustration is a mutual effort. The scientific staff as well as the artist must exchange information and cooperate very closely, as both groups provide mutual support through their experience. A continual exchange of information leads to the establishment of a satisfactory sketch and proper rendering of the subject. In order to assure correct representation, the following steps are undertaken.

Drawing:

1. The sketch, prepared in pencil, is based on a selection of specimens, photographs, and information received from a group of scientists.
2. Necessary changes, improvements, and implementation of details are undertaken during this preparatory stage.
3. The sketch in its cleanly presented form and drawn only in line is again presented to the panel of scientists. If more changes and improvements are needed, the sketch will return to the artist, who will have to continue reworking the image according to received instructions.
4. After unanimous approval, the sketch is transferred to a wooden panel and the painting can begin.

Painting: For painting, Paasche VLS Airbrush and acrylic paint are used.

[11]Institute Royal des Sciences Naturelles de Belgique, Belgium.

Figure 6-8. Side view in section of a *Cerianthus* (Cnidaria, Anthozoa). Airbrush with acrylic enamel paint on wood panel. Painting by Vinciane Lowie, Institut Royal des Sciences Naturelles de Belgique, Brussels, Belgium. Produced for display at the Koninklijk Belgisch Instituut voor Natuurwetenschappen, Brussels, New Natural History Museum, Invertebrates. Drawn in 1983.

1. The first and the most important part of painting is choosing the colors. The purposes of such careful selection are multiple—representation of three-dimensionality, details, pertinent information, and aesthetic appeal are taken into consideration. The process of mixing paints is a deliberate procedure.

2. The three-dimensionality is obtained by overpainting. In order to achieve a desired effect, five to seven transparent or semi-transparent colors or tints of the same color are used.

In order to control the process of painting, before the actual painting begins, a range of specially mixed colors is tried on sample material. During the process of mixing and experimentation, careful notes are kept on proportions of pigments constituting a given color and the succession of overpainted layers of colors. Such a process allows for deliberately chosen combinations of mixtures and layers of paint.

3. Stencils are cut from various materials, depending on the desired effect. The most favored are bristol board and transparent masking film. Once the stencils are prepared, the actual procedure of painting can begin.

4. During painting, the lightest tones are applied first, progressing toward the darkest final one. When a certain area is finished, attention is directed to a new portion; the process of paint application continues.

5. After painting with an airbrush, the retouching and introduction of minute details is accomplished with a hand-held brush.

Deirdre Alla McConathy's Glossary of Tissues and Materials Rendered by Medical Illustrators[12]

Medical Illustration: Illustrating medical subjects is an exacting and engrossing specialty field of art. The exacting nature of the field stems from the need for conceptual accuracy, precision, and attention to detail when communicating complex medical information. Adherence to high standards of exactitude is critical within an art profession that may affect lives through the provision of health-care services. Those individuals who choose medical art as a profession will require arduous training, discipline, and devotion to obtain the knowledge and skills necessary to produce medical illustrations. In addition to a drive for perfection, the medical artist must possess creative talents. Coupled with an extensive background of medical knowledge, these talents allow the artist to solve a wide range of communication problems encountered in the health science environment. Although the challenges to a medical illustrator are formidable, many artists choose to pursue this demanding and competitive area.

There are many aspects of medical illustration that attract people to the profession. Some of these are common to other areas of specialized art, such as scientific illustration. Both medical and scientific illustration involve precise drafting of technical content, provision of a professional service, and may optimally make a meaningful contribution to society. However, other important qualities distinguish medical art from scientific illustration. Medical art is unique as an illustration form concerned with the study and representation of the human body. For many illustrators the lure of medical subject matter attracts them to the profession.

For a significant number of others, an equally distinguishing and alluring feature of medical art is the diversity available in the field. A scientific illustrator may choose a specific scientific area in which he can dedicate his illustration skills. As a result, that illustrator develops expertise in rendering a select and relatively exclusive range of subjects, such as pottery sherds or fish scales. In contrast, the medical illustrator is routinely confronted by a plethora of different types of images to produce. These illustrators may

[12]University of Illinois at Chicago, Illinois.

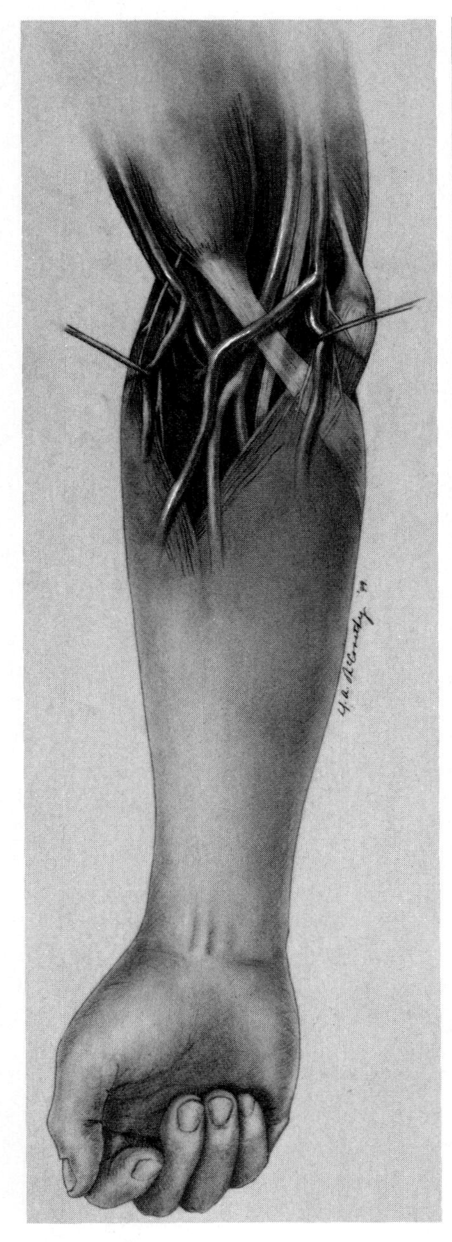

Figure 6-9. Cubital fossa, carbon dust and cel paint (acrylic used for animation) on drafting film. Drawing by Deirdre Alla McConathy, University of Illinois at Chicago, Department of Biocommunication Arts, Chicago. Unpublished. Drawn in 1980.

find themselves rendering objects as disparate as a microscopic virus, an artificial heart, or a newborn baby. In fact, there is an unlimited potential of substances a medical artist may be requested to illustrate. While it is impossible to anticipate and practice rendering each of these specific subjects in advance, one can adequately prepare to represent a wide variety of materials. To do so, one should establish proficiency illustrating a basic repertoire of subjects commonly rendered by medical artists. Successful development of these rendering skills will involve integration of a variety of resources such as anatomy textbooks, direct observations, and photographic references. Critical use of resource material is indispensable to building a visual vocabulary of medical images.

Development of these skills is particularly important to the medical illustrator. Whereas scientific illustrators usually have access to specimens to be illustrated, quite often medical illustrators are asked to create images without the benefit of direct observation. A surgeon may requisition illustrations describing an operation completed six months previously or that is to occur speculatively in the future. A researcher may request depiction of a disease process that occurred over a long period of time. A physician might need diagrams of a large piece of diagnostic equipment located in another city. Because of physical and time constraints, it is simply impossible for the illustrator to always gather firsthand information. This is an important reason why the artist must come prepared with prior knowledge of the subjects to be rendered as well as the critical concerns of the client and audience for whom the work is produced. It should also be noted that in addition to working without benefit of direct observation, medical artists encounter some situations where little is known about the actual appearance of the subject. For these cases the artist must conceptualize an interpretation of the subject based on available knowledge and reasonable probabilities.

A generalization might be made that medical artists draw what they know, rather than what they see. This basic glossary lists tissues and materials medical artists must know how to illustrate. From careful study the artist may then develop an appropriate symbology for that structure, to be used as needed. Rendering each of these materials requires maintenance of conceptual accuracy throughout the illustration process. This necessitates

1. Proper use of resources in preparing the sketch,
2. Anatomical and scientific accuracy of information,
3 Precision in drafting and rendering skills.

When illustrating live tissues, such attention is imperative, since any deviation may indicate an anatomical anomaly or imply a pathological condition that does not exist. The accuracy established in the initial sketch stage is reflected in the final rendering where values, textures, and colors must conform to the criteria of the specific condition.

Body Tissues

1. *Skin*—The outer surface of the body provides a flexible covering to deeper structures. Because of its plastic characteristics, the rendering of skin may involve tissue interactions as it is stretched, compressed, or otherwise manipulated. Illustrations of surgical incisions include the cut edge of the skin, revealing the thickness of this layer. In addition, many illustrations involve skin and the human body as a figure. To render figures well, preparatory studies in life drawing are essential. In such renderings surface topography must be illustrated, reflecting underlying anatomical features. Indications of these structures, such as contours or

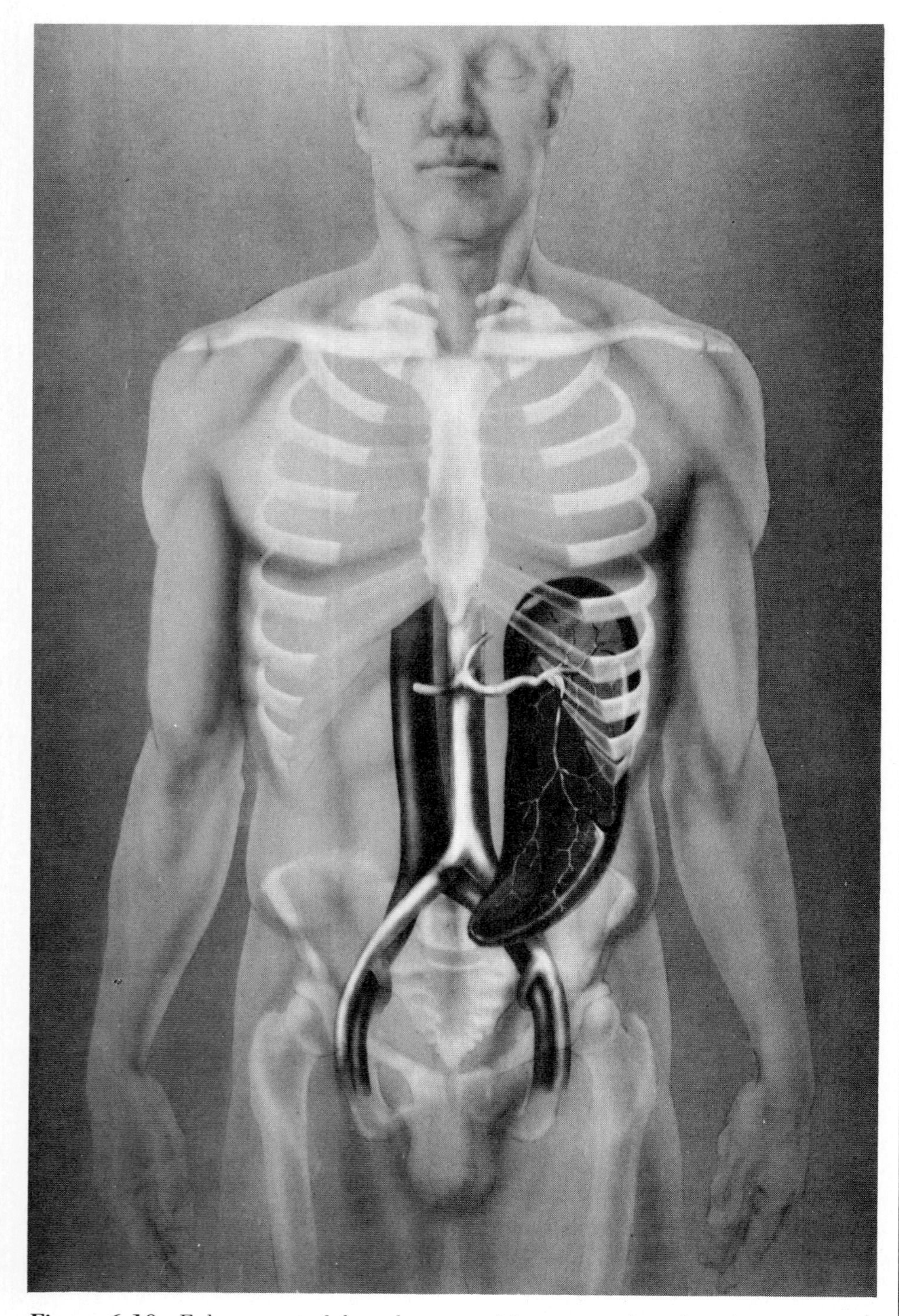

Figure 6-10. Enlargement of the spleen, requiring retrograde catheterization through the femoral artery and injection of small artificial emboli into the splenic circulation, randomly destroying small segments of the spleen (life-size). Airbrushed acrylic on Crescent colored mat board. Painting by Robert F. Parshall, University of Illinois at Chicago, Department of Biocommunication Arts, Chicago. Produced for a series of public information interviews titled *Consultation*, presented through a nationwide television program "Embolization of the Spleen" with Dimitrios Spigos, M.D., Professor of Radiology, University of Illinois Health Science Center, Chicago. Painted in 1982.

lighting changes, may be subtle, but with careful observation and proper use of anatomy references a realistic representation can be achieved.

2. *Fat*—The consistency of fat requires the artist to render small globules individually and yet make them read as one consistent tissue type. Fat is almost always a prominent feature at a surgical incision site below the cut edge of skin. Attention to the thickness and regional character of this layer may be important.

3. *Muscle*—The intricacies of muscle structure require close observation to convey the volume of the muscle, muscle fiber direction, and tendinous attachments. The relationship of muscles to one another is important in establishing anatomical landmarks.

4. *Bone*—The light surface value, rigid structure, and rough surface of bones are all distinguishing characteristics of this tissue. Subtle curves, contours, ridges, and tubercles should be rendered fastidiously to create authenticity.

5. *Veins and arteries*—These vessels may be considered simplistically as hollow tubes, but when rendering them as anatomical structures the artist must be sensitive to morphological characteristics such as local color, thickness of the vessel walls, diameter of the vessel, characteristic branching patterns, connective tissue support, and manner of integration with surrounding structures.

6. *Nerve*—These flattened strands of silvery, cream-colored tissue vary greatly in size and character. In general, nerves accompany veins and arteries as a triad of linear structures contained in loose connective tissue. Nerves must be distinguishable from veins and arteries.

7. *Visceral organs*—The unique characteristics of each individual

organ must be studied in depth to portray features such as overall form, surface, texture, value, size, firmness, color, spacial relations with adjacent organs, character of serous membrane investments, mesenteric support, and so on.

Materials

1. *Metal*—Surgical instruments and medical equipment invariably have metal surfaces that should be rendered as distinctly different substances from organic tissues. By utilizing reflections and specular highlights, the artist can convey a great deal about a metallic surface, such as whether the finish is chrome or matte.

2. *Plastic*—Medical applications of plastic materials are both frequent and diverse. Rendering these uses may require fine distinctions between a range of hard, shiny, opaque plastic surfaces to soft, matte, transparent surfaces such as surgical gloves.

3. *Cloth*—A variety of cloth substances may be encountered, including surgical drapes, suture material, clothing, and gauze.

4. *Glass*—The shiny and transparent qualities of glass make it a difficult challenge to render. The artist may find that it is the contrast of this material with other surfaces that helps create believability.

5. *Fluids*—The ability to render both bodily and artificial fluids as wet substances with appropriate viscosity and other characteristics is often required.

Roger S. Miles' Procedure of Drawing[13]

Dr. Roger S. Miles is a scientist who draws his own illustrations. He started to draw about 25 years ago because of necessity, and after years of practice he is able to produce high-quality illustrative material for his own research.

During the production of an illustration, Dr. Miles usually does not use separately prepared preliminary sketches, but "just works away with the dots until things come right."

The procedures of production are:

1. *Positioning procedures, photographic materials*. Generally straight lateral, dorsal views, though sometimes photographers are instructed to take unusual views to reveal anatomical details. Black-and-white prints are enlarged to the planned size of the drawing. These are checked for distortion by measuring with proportional dividers against the specimens.

2. *Lighting conditions*. Always top left lighting, of course, both for photographs and when drawing directly from specimens. Specimens are illuminated with ordinary bench light in the latter case. The exact angle of light varies, even within one drawing, to bring out the anatomical details.

3. *Types of paper*. Bristol board is always used, but the thickness and quality depend on what is at hand in the museum. I never worry, though the quality seems to have gone down as money has gotten tighter over the years. After a while I had to give up scratching out and settle for white paint to correct mistakes.

4. *Tracing techniques*. Ordinary tracing paper and a light table for photographs. I use this technique also for thin bristol board and tracing paper outlines. Otherwise I scribble on the back of the tracing paper.

5. *Pens*. I use Rotring with a full range of sizes. As a rule of thumb, I prefer 0.3 millimeter for dots, 0.5 millimeter for structures and 0.8 millimeter for outlines, assuming a reduction of two thirds or one half of the linear dimensions on reproduction.

6. *Instant tone*. Letratone or Normatone, again depending very much on what is at hand, providing it suits my requirements. I use Letraset for the lettering of the illustrations.

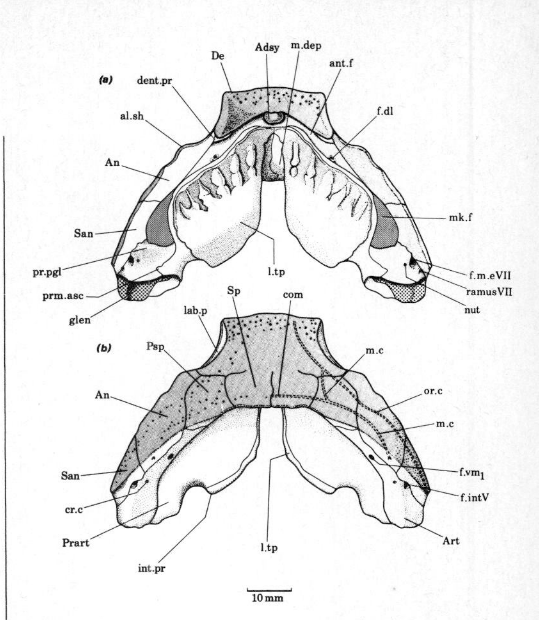

Figure 6-11. *Chirodipterus australis* sp. nov. Lower jaw in (a) dorsal and (b) ventral view, from P56041 and holotype, P52584. Cosmine shown by fine, regular stipple. Pen and ink with instant tone. Drawing by Dr. Roger S. Miles, British Museum (Natural History), London, England. Published in R. S. Miles, "Dipnoan (lungfish) skulls and the relationship of the group," by British Museum (Natural History), *Zoological Journal of the Linnean Society*, Vol. 61 (1977) London. Drawn before 1977. With permission of The Linnean Society of London. Copyright 1977, The Linnean Society of London.

Feng Minghua's Elementary Introduction to Biological Illustration[14]

The characteristics of biological illustration are contained in detail and preciseness, truly reflecting the

[13]British Museum (Natural History), England.

[14]Institute of Oceanology, Chinese Academy of Sciences, Quingdao, People's Republic of China

Figure 6-12. *Ulva linza*. Watercolor. Painting by Feng Minghua, Institute of Oceanology, Chinese Academy of Sciences, Quingdao, People's Republic of China. Unpublished.

objective and leaving nothing to the imagination. It is a medium of communication and a form of art that Chinese artists take very seriously. Techniques employed in illustrating are divided into three major categories:

1. Black and white, pen and ink
2. Black and white, wash
3. Color

Black-and-White, Pen-and-Ink Technique: The pen-and-ink technique consists of representation of the image with a combination of stipples and lines as well as the lines themselves. Use of the pen-and-ink technique has been widely adopted because of its simplicity in preparation and the ease of reproduction of the drawn image. The lines and dots are used to trace in ink a previously sketched subject and to represent the complexities of the form. The three-dimensionality, the concave-convex structural changes, are controlled by changes of values ranging from light to dark, giving a sense of quality and depth. Such pen-and-ink illustrations are suitable for making zinc plate.

In order to control the quality of the final pen-and-ink image, the artist must have an ability to draw freehand lines quite well. Such ability is a must, as the basis of the illustration is a pencil sketch prepared on the same paper as the pen-and-ink final rendering. An ability to employ different pen points and draw various kinds of firm, soft, thick, and thin lines in ink is important. Besides the technique of handling the drawing tools, a knowledge of papers and inks is a must. Attention must be paid to the quality of the paper as well as to the density of the ink if final results are to be proper.

Outline Sketches: The sketches are as important as the final rendering. Forms and sizes of living organisms are varied, ranging from simple single cells to complex systems many meters in length. For presentation of large specimens, the measurements are taken by hand, drawn properly, and through the series of sketches, reduced in appropriate proportions. When sketching microscopic organisms, the camera lucida or dissecting microscope will be used in order to transpose an observed subject to the surface of the paper, allowing for easy reduction or magnification. After initial sketches are prepared, then necessary revisions or supplementations will follow, with pen-and-ink rendering as the final step. During the preparation of the sketch, attention must be paid to the planning of the use of lines. Changes of thickness of drawn line as well as the quality and density of line must be anticipated. A good artist will not use strong lines when drawing a delicate corolla, nor will he or she use delicate lines when drawing the trunk and branches of a tree. Indication of the texture, weight, and delicacy of the subject should be represented.

Application of Tone: There are three techniques of tone application for black-and-white illustrations, namely the stippling, lines, and wash with Chinese ink. A good-quality, medium-size brush should be the tool used for application of washes, and a small pen for stipple and lines. With regard to stipple, many spots make the line, and many lines make the dimension.

Use of Lines for Representation of Tone: Tone is represented by the quality of lines, which range from thin and delicate to thick and heavy. Different degrees of value will be achieved by the control of the density of the lines as well as by the control of the thickness of drawn lines. The stereoscopic, qualitative feeling is captured by the manipulation of the direction of illumination, using such variants as sources of light opposite to each other, oblique illumination, and back-lighting. When drawing one-cell, monochromatic subjects, the lines representing the tonal values must be slender and soft. Such rendering must appear semitransparent, just like something which at the same time is nothing. For a coarse surface, texture and irregularities such as wrinkles have to be portrayed, representing and expressing the characteristics of the subject. Greater diversity of the lines, ranging from delicate or thick to dense or coarse, will be the rule. By employing the lines as a medium for representing the tonal ranges, the artist can trace the previously prepared sketch will great speed and accuracy.

Use of Stipples for Tone Application: In general, such technique is most suitable for illustrations representing surfaces of delicate, hyaline subjects such as *Monostroma* and *Porphyra* among the seaweeds. During the tone application, separate dots must be arranged in an orderly and regular manner, with each dot maintaining its rounded shape. In the areas where dark tones predominate, the dots should be large and dense, but for lighter tones, dots smaller in diameter are more suitable. The density of smaller dots will have an influence upon the brightness of lighter values. Redrawing and corrections are not advisable. A steady hand as well as good judgment will help in avoiding redotting the finished illustration.

Black-and-White Wash: In China this technique is known as "coloring with water and ink," and it is the most ancient method of painting. For successful results the traditional Chinese brush must be used. The brush must have a sharp point and has to be made of short but soft hair. Besides a good-quality brush, a good-quality paper is of utmost importance. For this reason, attention must be paid to the choice of the paper. A porous paper will absorb ink easily and will bleed easily. Such paper will allow for a lot of layers of washes, giving as a result the quality and feeling of Oriental painting, but *it is not* suitable for careful rendering of the subject. The proper paper must be well packed and must have a stabilizing influence on applied washes. Only with such surface is it possible to produce a detailed and well-rendered illustration. The procedure of application of the ink mixed with water must be well planned. The wash is applied in layers, starting with light values first and slowly working toward the darkest areas. The density of blackness is controlled by the quality of water added to the ink as well as by the quantity of placed layers of washes. The illustration will become a better and more realistic representation of the subject if separate washes are applied carefully and the brush does not contain too much water. An excessive amount of water on the brush will have a negative influence on the control of application of ink to the paper. The artist must restrain from using too much water on the brush. The values are built up slowly, layer upon layer, until a satisfactory result is achieved. White paint will be used for highlights, but only after all applications of washes are com-

Figure 6-13. *Cathaya agryrophylla*. Watercolor. Painting by Zhongyuan Feng, South China Institute of Botany, Chinese Academy of Sciences, Sinica, Quingdao, People's Republic of China.

pleted. By working slowly and carefully, applying layers of washes, it is possible to represent a great range of values. For this reason the wash technique is much superior to the stipple-dotting technique.

Color. The most important step in the preparation of a color illustration is the observation of the color on the subject. Study of reflections, changes of colors, absorption of light by the surface, and the separation of natural colors from the influence of the surroundings is a must. Daylight is the most proper light under which unbiased conclusions can be reached. The source of light will influence the observation because colors change according to the illumination. In order to recognize a color correctly, isolate the subject from its surroundings; by doing this, you remove any possibility of external influence. After careful observation, a direction of illumination of the subject will be chosen and later maintained during the process of painting. During painting, the subject should be separated from the background by white paper. White paper will help in further observations, allowing the artist to recognize real colors without removing the subject from the studio.

Process of Painting: Watercolors are the most suitable medium for painting, although the painting technique does differ from the typical watercolor technique. Brush strokes cannot be visible on the finished product. Sharp contrasts of colors are to be avoided, because such contrasts will induce the change of perception of the real color by the viewer. The paper must be thick and must absorb water easily, allowing for placement of many layers of colors upon each other. Thin paper is not suitable because it is more difficult to handle the buildup of tones on such paper and the artist can lose control of the carefully planned process. During the painting process, do not use too much water. A semidry brush will allow for better control of the medium. Values and colors are placed upon each other, starting with light and progressing toward the dark. Always wait for the first layer of color to be totally dry before applying the second. Patience and self-restraint are musts. Never place a coat of paint onto a wet surface. When exerting such caution, the underpainting will not be disturbed and the painted area will not become muddy. For lighter areas, use fewer layers of paint; for the dark portions, use more layers of the pigment. Highlights will be achieved later, using white paint. Brush strokes should be applied in accordance with the shape or texture of the subject, paying attention to the texture of a given area. The direction of brush strokes is of importance, as the unifying integrity of the image depends on the movement of the brush over the surface of the paper, imitating the natural shape of the subject. As a final touch, add the smallest areas of dark value to the darkest portions of the painting. By doing this, more contrast will be achieved, and the lightest sections will be underlined by the dark portions. In such a situation the lightest areas will stand out, adding more life to the painting.

Comment: At the present time we in China are making a big effort to study all techniques of drawing and painting because we want to employ the new ways during preparation of illustrations. We want to portray the subjects with more realism, using refined artistic techniques. Our aim is to achieve the effect of volume and well-represented perspective. Techniques such as painting with printing inks, combination of wash and charcoal drawing, and naturally the traditional Chinese painting techniques are in common use.

Leonard E. Morgan's Airbrush on Drafting Film Technique[15]

The airbrush is one of the most marvelously flexible creative tools artists have ever had at their disposal. Flexibility is a major benefit to any commercially active illustrator. Also, each different brand of airbrush has various advantages and disadvantages, and it is very useful to own a variety of brushes to fill various needs. The limits of use for each brush can only be learned and explored by actual practice and use of that brush in the production of illustration.

Advantages of Drafting Film: Working on two-sided, frosted acetate or frosted drafting film allows many advantages. It takes most mediums very well, especially airbrush because of its very smooth surface and colored pencils because of the special "tooth" of the frosted surface.

Transfer of pencil sketches to illustration board is not necessary when using frosted acetate or frosted drafting film. You simply lay the frosted drafting film over the final pencil sketch and proceed to finish the illustration. Such a procedure allows the artist to work without pencil lines placed on the finished product. The surface of the completed illustration does not have to be touched by eraser and gentle changes of tones will be presented without interference. Drafting film can be drawn and painted on both sides, thus permitting multilayered and phantom-image effects.

When frisketed airbrush is applied first to the back side of drafting film, then all cut lines from frisket cutting will remain only on the back side, leaving the front

[15]Chicago, Illinois.

undamaged. Such technique of frisket application leaves the front surface completely smooth and ready for additional work. Textural development of the image with colored pencils or brush is not hindered by the damage inflicted to the surface during frisket application and removal.

The back side of the finished illustration should be spray painted with solid white acrylic in order to prevent reproduction problems. When the back side remains in its original state, shadows produced by the painted image will interfere with reproduction of the image, resulting in loss of sharpness or distortion of the image. Once the finished illustration is back-painted white, the drafting-film artwork reproduces as well as it would on other opaque surfaces, and as the surface is flexible, it conforms easily to the curved drum of a laser scanner for color separations. Back-spraying mandates the installation of an exhaust hood and fan in the working area in order to remove the massive amount of overspray that results from painting large areas with solid opaque color.

Drafting film does not absorb water, therefore it will not wrinkle or warp in areas that are wet from heavy spraying or brushwork, and because of its sturdy surface it is possible to use an X-Acto knife to scrape the surface of the drawing when correcting or enhancing the painted image. Extremely fine white lines can be produced by scraping the surface with a sharp tool or X-Acto knife.

Frisket: Drafting film's velvety surface works beautifully with Frisk Film. The Frisk Film adheres to the drafting film strongly enough to make it remain attached when spraying, but will release easily when necessary to allow for repositioning and reuse.

Regular, clear acetate can also be used as a frisket material. The big advantage of using clear acetate for frisketing lies in the fact that all cut lines are eliminated from the surface of the illustration. During cutting, the knife only scores the surface of clear acetate rather than cutting through the material. For that reason it is possible to place a sheet of clear acetate on the surface of an illustration and while seeing the image underneath, "score" a desired outline. When scored acetate is removed, the needed area can be easily popped out along the scored lines, then repositioned over the art surface for spraying.

Using clear acetate for frisket allows the artist to vary the hardness or softness of the painted edge by varying the distance of the frisket from the surface of the art.

Airbrushes: The Thayer and Chandler Model A and the Iwata Model HP-A are two of the best fine-detail brushes. The best is the Thayer and Chandler Model A. Some people find the exposed tip of the Model A's needle to be a big problem because it can be easily damaged. Exercising greater care during handling and replacing the protective cap when the brush is not in use will help you avoid damaging the needle tip. The extra exposure of the needle on the Model A contributes to its capacity for extremely fine detail work. When a needle tip does become damaged, simply remove it and replace it with a new needle. This involves a little extra expense, but it is worthwhile.

The possibility for presetting the amount of paint volume by adjusting the preset screw located in front of the spray lever is another major benefit for the artist, adding to the Thayer and Chandler Model A's potential for very fine detail work. Such presetting of the volume of paint eliminates the danger of overdoing and overpainting during the paint application in fine-detail areas. Control of the paint application is then achieved by turning the air on and off—that is, by pressing down or releasing the spray lever—or by moving the airbrush closer to or farther away from the surface of the illustration. Inks and dyes are the best to use for rendering fine details. Acrylics and designer's gouache are better for general painting. The preset controls of the airbrush combined with use of inks and dyes are especially useful when rendering long, consistently toned subjects such as the side of a blood vessel or other long and thin, cylindrical objects.

By properly using the preset adjustment found on Thayer and Chandler's Model A airbrush, one can equal and exceed the detail capability of the Paasche Model AB airbrush and have none of the mechanical adjustment problems that are common to the Paasche AB.

The Iwata airbrushes have an excellent ability to produce fine lines, but they lack the preset screw found on the Thayer and Chandler Model A. Although Iwata's airbrush has preset adjustment capabilities, it differs from Thayer and Chandler's Model A. Iwata's preset adjustment limits how far back the spray lever can be moved, while the Thayer and Chandler preset limits how far forward the spray lever can be moved. Individual preference will determine which type of preset works best for each artist.

Problems with Holding the Airbrush: When working with an airbrush for long hours, cramping of the fingers and hand caused by fatigue may be a problem. Cramps are caused by the fact that detail work with the airbrush requires that it be held almost perpendicular to the art surface, instead of in the angled position in which one would

Figure 6-14. *Holocentrus hastatus*. Watercolor on bristol board. Painting by Pierre Opic, Muséum National D'Histoire Naturelle and Office de la Récherche Scientifique at Technique Outre-Mer, Paris, France. Published in *Poissons de Mer de l'Ouest Africain Tropical (West African Tropical Fishes)* by ORSTOM, Paris, 1981. Painted in 1963.

hold a pencil. As long as the hand can be held in a normal writing position, the problem of hand fatigue is eliminated. Addition of a common technical-pen attachment to the airbrush at the point where the hose enters the brush solves this problem. The distance between the air hose and the spray tip is also related to muscle tension. The shorter the distance, the less tension in your hand when doing fine-detail work. Less tension allows greater control.

Pierre Opic's Fish Drawing Procedure Using Wash[16]

The procedure involves three major steps:

1. Positioning the specimen
2. Measuring and preparation of the sketch
3. Rendering

The specimen is properly positioned on a surface such as the board or table using standard positioning procedures, making sure that the designated lateral side of the specimen is facing the illustrator.

Two lines are drawn on the board: one horizontal along the length of the body of the specimen, and the other vertical. The vertical line must be perpendicular to the first dorsal fin or to the highest section of the subject. The drawn lines serve as the base from which all measurements are taken.

Pertinent data is obtained in two ways, namely:

1. If the specimen is to be drawn in one-to-one ratio, a good-quality ruler is used for measuring vertical and horizontal dimensions.
2. If the subject is to be reduced, a proportional divider serves as the essential tool during the process of measuring.

The very same lines, one horizontal and the other vertical, are drawn on Strong, 90 Gr. M2 paper, with all pertinent points plotted during the development of the sketch. When the sketch is finished, the drawn image is transferred using carbon graphite paper to the surface of Bristol Contre-Collé paper.

A Mars Staedtler technical pencil with 8H led is used during the transferring procedure. Care has to be taken and excessive pressure cannot be exerted as otherwise the sketch, the transferring medium, and the surface used for the finished rendering will be damaged.

After the transfer of the sketch is completed, the first wash is applied using sable brushes numbers 00, 0, and 1. The first wash is a

[16]Office of Overseas Scientific Research (De L'Office de la Recherche Scientifique et Technique Outre-mer). Muséum National D'Histoire Naturelle, France.

combination of drawing lines with brushes and placing light washes. Black China Pelikan ink dissolved with water supplies a varied range of tones.

The second layer of washes is applied with Petit Gris Extra number 2 brushes. The range of tones depends on the three-dimensionality of the subject, its coloration pattern, and the quantity of underpainting. Essential highlights are added after the second wash is completely dry and the surface does not need any additional application of basic values. The finished product is protected with a separate sheet of paper.

Pierre Opic's Materials Used for Scientific Illustration[17]

Pierre Opic, a draftsman-naturalist who has worked for L'ORSTOM for the last 20 years illustrating a variety of subjects, uses the following materials:

1. Microscope: Carl Zeiss, zoom binocular stereomicroscope.
2. Drawing tube (camera lucida): Carl Zeiss attachment to stereoscopic microscope. The drawing tube is superior to the Wild Heerbrugg product because it allows for binocular viewing of the specimen.
3. Light box for general use.
4. Paper for sketches: Strong 21 by 29, 7 cm; 90 Gr. M2; BFK Rives.
5. Paper for black-and-white illustrations: Bristol Contre-Collé, 4 AS, 16 ter, 400 Gr. M 2.
6. Paper for painting with watercolors: a. Canson Lavis B, 280 Gr. M 2, b. Arches Satin, 300 Gr. M 2.
7. Paper used for protection of finished product: Calque Canson, 90 Gr. M 2.
8. Paper for the transfer of the sketch and the strength of transferred line: Carbon graphite paper.

[17]Office of Overseas Scientific Research (De L'Office de la Recherche Scientifique et Technique Outre-mer). Muséum National D'Histoire Naturelle, France.

9. Pens for line and stipple: Graphoplex tubular pens No. G 8000, 0, 10, and 2.00 mm.
10. Bendable ruler for a guide while drawing variously curved sections of specimens such as antennae of the insects.
11. Brushes for line drawing and first wash: Raphael sable brushes numbers 00, 0, and 1.
12. Brushes for second wash: Petit Gris Extra number 2.
13. Ink for line and wash: Pelikan China ink and Color Water Talens or Roney.
14. Pencils for sketch: Mars Staedtler technical pencil with 3H lead.
15. Pencils used for transfer of the sketch: Mars Staedtler technical pencil with 8H lead.

John Brookes Randall's Description of Entomological Illustration[18]

My intention here is to familiarize you with a few of the basics to help you in approaching the fascinating world of insect illustration.

There are over one million, plus or minus a few, named insects in the world today. It is estimated that there are several million more insect species yet to be discovered and described. It is in the description of new species that illustrations become so extremely important. In fact, if published prior to 1931, an illustration can stand alone as a species description. In addition to the written description of a new species, illustrations play a major role in precision of description.

Most entomological illustrations are quite technical in nature. For example, insect genitalia are the key character to species identification. Because such minute morphological structures are involved in insect illustration, I highly recommend that any illustrator getting into this area of scientific illustration take some entomology courses. General entomology, insect morphology, and insect identification courses will greatly enhance your ability to illustrate insects accurately and to communicate effectively with the entomologists with whom you will be working.

[18]S. C. Johnson & Son, Biology Center, Racine, Wisconsin.

Through several years of teaching entomological illustration, I found the best work to be produced by biology students with little or no art training. The biologist's edge over the art student was his or her knowledge of the subject matter. There is no substitute for a good working knowledge of the subject matter, knowing the difference between actual structures and artifacts. Insect morphology can be so intricate that experienced insect illustrators may spend hours studying a specimen before they ever put pencil to paper.

Handling Insect Specimens: Insects are extremely fragile and should always be handled with extreme care. This is true for all specimens, but it is of utmost importance in handling a Holotype specimen. The holotype is *the single specimen* upon which the description of an entire species is based. After the description is published, the holotype specimen is termed the *type* specimen and is the specimen upon which comparisons and verifications of identification will be based. In many cases the holotype specimen is the only specimen of that species to have been collected in the world. The holotype specimen is invaluable! Where holotypes are concerned, I advise the illustrator to let the taxonomist (biologist responsible for describing and classifying organisms into the proper categories) perform any manipulations of the specimen required in the production of an illustration.

Dry insect specimens are pinned. The pin is used not only to secure the insect to a pinning surface, it also serves as a handle. Avoid direct contact with the insect except when absolutely necessary. When pinned specimens are not in use, they should be stored in an airtight box fumigated with naphthalene or paradichlorobenzene (moth flakes or mothballs are fine). The fumigant is required to prevent Derestid beetles from destroying the specimens. Any questions regarding the proper care of insect specimens should be referred to an entomologist.

Liquid-stored (soft-bodied) insects should be kept in their storage fluid, usually 70 percent alcohol, while being drawn. Submerge the specimen in a glass dish (petri dish) of alcohol. A dab of petroleum jelly or glass beads or fine sand can be used to keep the specimen in the position you require for drawing.

Equipment: The most important single piece of equipment for the entomological illustrator is the binocular stereomicroscope. Since insects are small, particularly the body parts used as toxonomic keys, a stereo (also called dissecting) microscope is an absolute must. Ocular grids are used as a means of accurately drawing by simply transferring the image in one of the grid squares to a corresponding grid square on the drawing paper. The camera lucida or drawing tube is an invaluable stereoscope accessory. Simply put, this instrument optically superimposes the image of the specimen onto the drawing paper, allowing the illustrator to in effect trace the specimen. Camera lucida attachments are available as part of Wild, Nikon, and Zeiss microscope systems. These are excellent instruments. Camera lucidas can also be homemade, but these lack the precision provided in the commercially available models.

Wild microscopes (and possibly others) have an attachment called a *rotating oblique illuminator*. This instrument is set between the objective lens and the specimen. The ROI acts to "rotate" the image (allowing for three-quarter views of the specimen), which is then transmitted through the camera lucida and ultimately onto the drawing.

A small but important piece of equipment for entomological illustration is the specimen holder. There are many commercially available models costing up to $15 or $20. The most effective specimen holder I have found is also the least expensive. I use a 1½-by-1½-inch piece of posterboard with a lump of modeling clay pressed onto it. This simple holder allows the pinned insect to be rotated to any angle. The holder can easily be reconditioned to its original shape or any shape deemed necessary by the particular specimen.

Other pieces of equipment useful in insect illustration include extra insect pins, fine forceps, covered boxes with pinning bottom, microscales, proportional dividers, and a reducing lens.

CONVENTIONS IN ENTOMOLOGICAL ILLUSTRATION:

Light: Regardless of the view of the specimen you are illustrating, the light source should come as if over the left shoulder of the illustrator or from the upper left corner of the paper. This is particularly simple to get in insect illustration, since you can position the microscope lights to provide the correct angle of illumination. This convention allows for the comparison of many different drawings and allows for highlights and shadows to fall in very predictable ways. It is the proper rendering of the highlights and shadows that creates the three-dimensional effect from a two-dimensional medium.

Dorsal Views: Also referred to as a *habitus* view, this illustrates the insect from directly above, observing the entire dorsal aspect of the animal. Dorsal views are drawn with the anterior (head) toward the top of the page or toward the left margin of the page. Antennae are drawn pointing anteriorly, with extremely long antennae arching around the body in such a way as not to obscure the legs. The first pair of legs (prothoracic legs) are drawn directly anteriorly. The second (mesothoracic) and third (metathoracic) pairs of legs are pointed posteriorly. Dorsal views are drawn symmetrically, since insects are bilaterally symmetrical. This is not always the case with mouth parts, however. There will be exceptions to any rule.

Lateral Views: Lateral views of insects are drawn with the anterior (head) pointing to the left margin of the page. Antennae are drawn pointing anteriorly or, if very long, arching over the body. Only the appendages nearest the observer (the left side of the insect) are drawn. Attempting to include the appendages on the far side of the animal will only complicate the illustration. The femur of the legs is drawn pointing toward the observer, requiring foreshortening. The rest of the legs follow the same directional conventions as the dorsal views—that is, prothoracic pointing anteriorly; meso- and metathoracic pointing posteriorly.

Different conventions may apply when illustrating different orders of insects. The best policy to follow is to study drawings done of similar insects, insects of the same order, family, or even genus.

Techniques: I would estimate that about 90 percent of all technical insect drawings are black-and-white

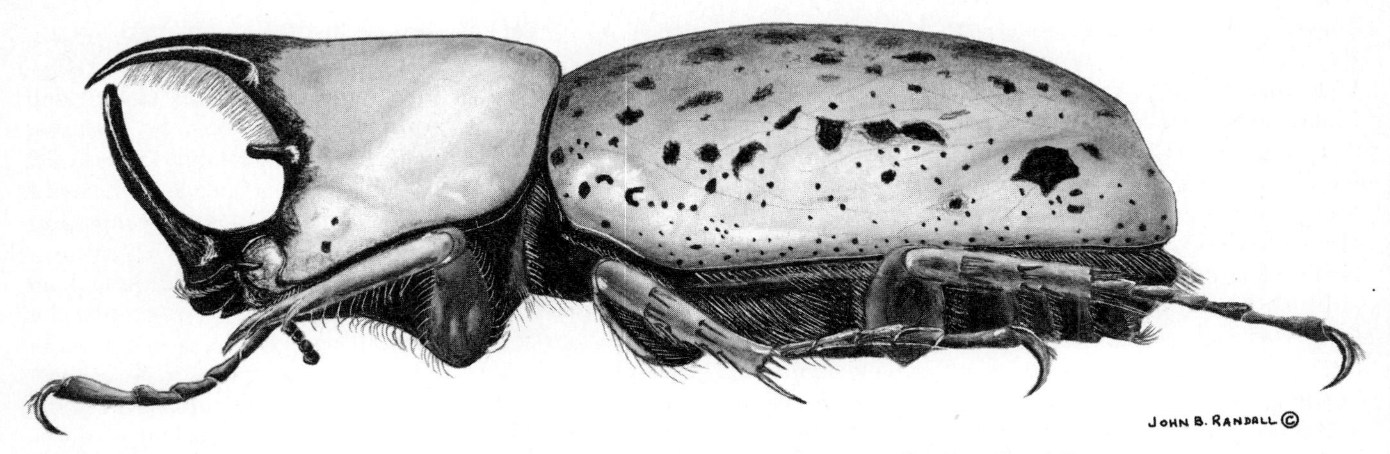

Figure 6-15. *Dynestes tityus*. Carbon dust on media board. Drawing by Dr. John Brookes Randall, S. C. Johnson & Son, Biology Center, Racine, Wisconsin. Unpublished. Drawn in 1980.

line drawings. That is, the rendering is a combination of black and white, with no gray tones in between. Techniques that fit that description include pen and ink, coquille, scratch board, and the bleached photograph technique.

Pen and ink is by far the most widely used of the above techniques. The other techniques are used in special situations where the technique is particularly suited to the needs of the illustration and subject.

The black-and-white line techniques are most widely used because they are simple, use readily available materials, are effective, and are the least expensive to reproduce.

Halftone illustrations use black, white, and every gray tone in between. Pencil and carbon dust techniques are examples of halftone techniques. These are extremely effective techniques, as they allow for very intricate rendering of form. They are, however, very exacting and time-consuming techniques. Because of the time involved and the fact that halftone illustrations are more costly to reproduce than the black-and-white line illustrations, halftones are less frequently used.

Like halftones, color illustrations are time-consuming and very exacting to render, as well as expensive to reproduce. When the expense allows, however, fine color illustrations of insects are spectacular.

Study your specimens carefully; do some background study prior to setting to work, and practice. Don't be afraid to make mistakes; be your own worst critic.

Recommended Literature:

Borror, D. J., D. M. Delong, and C. A. Triplehorn. *An Introduction to the Study of Insects*. New York: Holt, Rinehart and Winston, 1976.

Borror, D. J., and R. E. White. *A Field Guide to the Insects of America North of Mexico*. Boston: Houghton Mifflin Co., 1970.

Snodgrass, R. E. *Principles of Insect Morphology*. New York: McGraw-Hill, 1935.

Snodgrass, R. E. *A Textbook of Arthropod Anatomy*. Ithaca, New York: Comstock Publishers, 1952.

Entomological Equipment Suppliers: (write for free catalogs)

American Biological Supply Co.
1330 Dillon Heights Avenue.,
P.O. Box 3149
Baltimore, MD 21228

Carolina Biological Supply Co.
Burlington, NC 27215

Bioquip,
P.O. Box 61
Santa Monica, CA 90406

Lewis L. Sadler's Description of Airbrush and Its Usage[19]

Airbrush: The independent double-action airbrush with a gravity feed system is one of the two best airbrushes for the scientific illustrator. Gravity feed requires less pressure, making it more sensitive to the touch, allowing you to paint thinner lines than it is possible to do with side-mounted color cups. The term *independent double action* refers to the movement of the finger lever, which can be moved backward or forward to release paint or can be depressed to operate the air valve. The infinite combinations of air and paint account for its value to the illustrator, since the combinations allow anything from broad washes to thin lines of soft pencil width from a single instrument. The Artistic models S-1 and SG-1;

[19]The University of Texas Health Science Center, Dallas, Texas.

Thayer and Chandler model A; Paasche models V-1 and V-jr.1; Wold models A^2, V, and A-137; Iwata models HP-A, HP-B, HP-C, Aerograph Super 63 and Sprite; and the Badger model 100 GXF are all acceptable. The other airbrush is also independent double-action but turbine-feed instrument. While this airbrush has a double-action finger lever for one-handed operation, there the similarity with other brushes ends. This airbrush operates on an air-powered turbine principle with offset nozzle. Depressing the lever controls the amount of air flowing into the instrument. Part of the air runs the turbine, capable of rotating up to 20,000 rpm, which in turn causes the arm and the needle to oscillate back and forth. The rest of the air is diverted to the air nozzle and blows the paint off the needle onto the drawing. The second action lever (backward-and-forward motion) controls the travel of the needle, allowing a smaller or larger quantity of paint to find its way onto the surface. The operator controls the rate of speed as well as the amount of needle surface being exposed with each pass, allowing the ultimate control in producing the shades, lines, tints, and stipple effects so demanded in scientific illustration.

Once an airbrush has been selected, a source of air or gas is required to operate the brush. In addition to the source itself, hoses, couplings, and valves are also required to deliver the propellant from the system to the airbrush. While often disregarded or overlooked during selection of the system, both the hose and regulator are extremely important. The regulator allows for adjustment of the pressure delivered to the airbrush without regard to the actual pressure developed by the tool. The regulator can be regarded as an additional variable in adjusting the entire instrument for airbrush performance. For fine-detailed work, the coarse adjustment of the pressure is necessary. Hoses can be of various types, although transparent hoses will allow for quick visual inspection. Regardless of the type selected, the hoses should be no shorter than eight feet and no longer than ten feet from regulator to airbrush. This seems a minor point, but your regulator pressure reading will mean little or nothing if the hose varies much from these standard lengths. Couplings are also of importance, since air leaks developing anywhere in the system will reduce control and ability to duplicate passages day after day.

Compressors: Simple compressors lacking a reservoir and a regulator cannot be used for finely detailed airbrush work. Apart from noise and constant running, which causes them to break down or burn out, the lack of a reservoir allows the pulsations of the pump to be transferred to the airbrush. The result is that your attempt at producing a fine line becomes a series of dots.

The more expensive compressors are fitted with a reservoir from which air is drawn to operate the brush. The compressor is switched on and off automatically to keep the reservoir filled to the required capacity. This type of compressor requires electricity to operate and does make noise, so if machine noises break your concentration, compressed gas may be the system of choice.

Gas: Carbon dioxide or carbonic gas provides a convenient, inexpensive alternate to the compressor and provides the ultimate in quiet air supply. Because it doesn't require electricity, it is a truly portable system that can be used almost anywhere, indoors or out. Tanks come in sizes holding 10 pounds of gas and up. The smaller 10- and 20-pound sizes are referred to as "stubbies," and are the only sizes which can truly be regarded as portable, since the 10 to 20 pounds refers only to the weight of the gas; the tanks weigh about 45 pounds each, plus the regulator.

Tanks of gas provide about 30 hours of continuous operation for each 10 pounds of gas in the tank. Since most hours of work are not continual hours of holding down the finger lever, the actual time of service on a tank is surprisingly long. The tanks are easily and cheaply refilled in any major city by anyone servicing soda machines or bars where beer on tap is available.

The major disadvantage of using compressed gas in the studio results from potential delays and subsequent deadlines missed when the tank suddenly comes up empty. A second backup tank is a wise investment if compressed gas is used as the propellant system.

Compressed air can be used in place of carbon dioxide, although tanks and regulators are more expensive and finding places to refill them is more difficult.

Drawing or Painting Surface: Virtually any surface can be used in an airbrush painting. In scientific illustration, we are called upon to work on surfaces as smooth as acetate and photographs or as rough as canvas and the coarsest watercolor paper. They all will accept airbrush equally well as long as the paint is compatible with the surface. The choice of surface, therefore, depends less on the fact that an airbrush is being used and more on the end result desired. If a paper texture, color, or weight is desirable, rest assured that it can be used as a ground in an airbrush painting.

The only restriction in selecting a painting surface is in the area of durability. One should always select the best-quality materials available. It makes little sense to expend time and effort to create a piece that will fade, yellow, or fall apart in a couple of years. Select papers with high rag content (100 percent, if possible) and neutral pH.

Inks and Dyes: Inks and dyes are produced in a variety of kinds and colors, with many formulated specifically for use in the airbrush. They are brilliant in color as they come from the bottle and are undoubtedly the easiest and most convenient of all the media to use. Their effect is transient, however, as they fade within hours or days when exposed to ultraviolet light. For this reason, I would not recommend them for use in scientific illustration.

Pigment-Based Paints: This group includes transparent watercolor, opaque watercolor (gouache), casein, temperas, acrylics, alkyds, and oil paint. All contain the same pigments, but different vehicles, binders, and adhesives are used to make the pigments flow and to bind the particles together, preventing them from flaking away from the ground.

Each of these different kinds of paint can be used with the airbrush, and many can be combined to take advantage of their unique characteristics.

Transparent Watercolor: This type of paint is probably the easiest to use and has the advantage of easy cleaning from the airbrush, even after drying. Pans and tubes work equally well, and it can be applied over or under other media. The disadvantages are in the extreme transparency, requiring multiple applications to achieve any saturation, and in the lack of permanence when exposed to light.

Opaque Watercolor (Gouache): Gouache works extremely well in the airbrush and is probably the most universally used medium for scientific airbrush paintings. Gouache can be used in its usual opaque fashion or, when diluted, in a transparent wash or at least semitransparent fashion. When mixing colors, use zinc white for the clearest and most lightfast tints. Permanent white is for white highlights only; when mixed with colors it gives them a chalky look. When spraying gouache, use only very diluted mixtures in your brush and don't spray any great distance from the ground. Otherwise, the paint will form dry beads of pigment on the surface which are hard to paint over (they dissolve into mud), are easily abraded, and will be picked up on even the least tacky friskets. The disadvantage to using opaque colors is that they don't easily mix visually on the painted surface. That is, the complement cannot be added to an area of color for shadows or modeling. Because it is opaque paint, what you apply is what your eye will see. Another disadvantage lies in the characteristic color shift between what you see when the paint is wet and the color it appears to be when dry. What you see is never what you get! This is especially disconcerting when overspraying with white in a highlight area. What often appears perfect when wet will disappear almost entirely when dry and require several applications to bring it up to the proper intensity. The finished painting can have its problems also, in that the surface is very delicate, easily scratched, and there is no good method of protecting it.

Despite these problems, gouache is easy to use, either alone or in combination with other media. The airbrush is easily cleaned with water while in use, although it should be disassembled and cleaned with a household glass cleaner (such as Windex) at the close of each session. The combination of detergent and ammonia will remove the thin, transparent, adhesive residue that otherwise defies cleaning in water and causes the parts to be "glued" together.

Casein: Casein paints offer many advantages to the airbrush artist and deserve more consideration than they are usually given. When thinned, they are more transparent than gouache; when used opaquely, they are more dense. The colors are more saturated than opaque colors and they are the same color dry as when they are wet. They are cheaper than most other paints and come in 35 different colors. When freshly applied, the colors are soluble in water, but later they become insoluble and permanent. They work well in the airbrush, and the same cleanup precautions apply as with gouache. The final cleaning of the day should be done with a household glass cleaner containing ammonia.

Acrylics: Acrylic paint makes an excellent medium for use in the airbrush. Many authors and experienced airbrush painters advise never to use acrylics in an airbrush, but a few hints can make all the difference. Only use acrylic jar colors. Jar colors are specially formulated to flow easily at an even consistency. Use the colors in diluted form, but instead of mixing with water, dilute the colors with a mixture of half gloss (or matte) and half water. This mixture will help prevent a breakdown in the binder used in the jar-color emulsion when overly thinned. Clean your brush frequently with liberal quantities of water followed by a flush with Windex. Any sign of clogging or blockage should be followed by a flush with denatured alcohol. (Caution! Atomized alcohol is combustible). Watch the needle for signs of paint film or buildup, which signals the time for a thorough cleaning. At the end of each session, clean with alcohol followed with Windex and leave disassembled overnight for quick start-up the next day. With proper precautions in cleaning, acrylics offer many advantages. They mix well with other water-based media, they are relatively permanent, aren't easily soiled, and aren't pulled up by friskets.

Cleaning the Airbrush: Always clean your airbrush each time you lay it down. All media are used in diluted form, which evaporates quickly, "gluing" parts together and causing unnecessary delays and damage to equipment. If you are stopping a passage long enough to lay the brush down, you are pausing long enough to clean up. Disassemble and thoroughly clean all parts at the end of each working session.

Warning: Remove things from nearby surfaces before beginning a painting session, especially when larger washes will be done. Atomized paint particles travel for a surprising distance in the studio, and a fine film of paint will cover everything nearby. These atomized pigment particles also pose a health hazard, as they are inhaled and ingested. Many pigments, especially those derived from metal compounds, pose a hazard to the artist, so painting should only be done in a ventilated space, and a cheap particle mask should be worn.

For most water-soluble paints, the cleanup is a simple matter of rinsing the parts with water alone or a solution of water, ammonia, and soap (Windex). When using oil, acrylic, alkyds, and so on, stronger solvents are required to thoroughly clean the airbrush. Solvents such as alcohol, toluene, xylene, benzene, ether, acetone, and the like have all been recommended for various cleanup purposes. When atomized, these solvents turn the atmosphere in the studio into a potential bomb. Any electrical spark or open flame can set it off. Never atomize any of these chemicals under any circumstances. If they must be used, disassemble the brush and soak the parts outdoors. Even small amounts of their fumes pose a health hazard indoors.

Frisket: Simply put, a frisket is anything that blocks the spray of paint. Frisket material can be bought, prepared, or found. Commerical friskets come in several forms, with each having a slightly different use. Unprepared frisket paper looks much like tracing paper, but it is specially prepared to resist curling when rubber cement is applied to it. Much-thinned rubber cement is brushed onto the surface to provide the adhesive. It takes two applications of rubber cement to prepare properly, the second coat applied at right angles to the first. The paper is ready to use after all the volatile (and hazardous) solvent has been evaporated from the surface. The advantage to this type of frisket is that it is always fresh, has a low tack, leaves no residue behind (unlike most prepared friskets, making it ideal for use on slick surfaces such as glossy photographs). Its disadvantages are that it doesn't last (it dries out quickly and must be made fresh daily); it is messy, hazardous, and time-consuming to prepare; and isn't repositionable. Prepared frisket comes in high or low tack on a variety of surfaces. Low-tack frisket is used on delicate surfaces where pulling up the paper surface or previous layers of paint warrant being careful. High-tack frisket can be used where these problems are not a consideration and where extra sticking power is a requirement. High-tack friskets are especially useful where the surface is rough or irregular or where extra holding power is required, such as on vertical surfaces (exhibit panels, for example). None of these friskets is recommended for use on photographs or extremely slick surfaces, since the carrier sheet (the sheet used to protect the adhesive, keeping it from drying out and keeping it clean) is coated with silicone. The silicone leaves a thin film on the adhesive, which in turn is transferred to the surface on which it is applied, rendering it waterproof—a real problem when working with water-base paints. All prepared friskets have a shelf life that depends on atmospheric conditions such as heat or humidity. Flat individual sheets of low-tack frisket made by the Frisk Company are of good quality and are manufactured in matte as well as gloss finishes. High-tack friskets are available in a product called Transpaseal. Because Transpaseal is produced for laminating, it requires cautious handling. After 24 hours the laminate is permanent and cannot be removed, therefore remove the material from the surface of the illustration immediately after use.

Wet frisketing is another alternative that is especially useful when small, intricate patterns are involved which would be difficult to cut using other friskets. Wet frisket is liquid latex, either plain or with a color additive. Avoid the ones with color added since they often will stain your paper. The liquid frisket I prefer is one made by Winsor and Newton. To apply, use an old brush well lathered with soap, and wash out the brush immediately after use or it will be ruined.

Dry friskets are the safest to use, are reusable, and are the ones I prefer to use if at all possible. I use a .005-mil acetate purchased in sheets for almost all my frisket work. This weight does not curl easily when wet, as with the thinner weights, and is easy to cut, unlike the thicker weights. Rolls are not recommended since they tend to roll up again on the surface. Sheets will remain flat. Weights such as coins or magnets can be used to help hold down the edges and keep them from fluttering, but usually hands will be sufficient. The acetate need not be cut through, which prevents cutting the work surface. Scoring the acetate with a number 11 blade and "cracking" the film will accomplish the task. Acetate friskets will allow you to keep both the positive and negative shapes. The shapes can be further cut for other more intricate details that will fit perfectly with the original frisket. Acetate also allows for soft or hard edges by being held down tightly or being held up, allowing some paint to blow under the mask.

Lewis L. Sadler's Opaque Watercolor Technique[20]

Opaque watercolor, as the name implies, is watercolor painting of a type that requires white pigment to achieve whites and pale tints. Transparent watercolor, on the other hand, achieves tints by applying watered-down, sparsely pigmented coats over brilliant white paper. The term *aquarelle*, although not in common use, best describes this type of painting.

Tempera is a generic term for opaque watercolor, with the individual types—gouache, casein, egg tempera, and so on—differing only in the medium used to bind the pigment. The pigments are the same as in transparent watercolor, but inert ingredients such as chalk are added, not to adulterate the paint, but to improve color and textural effects. Gouache has gum arabic or gum senegal as a binder, while egg tempera uses egg white (albumen). Casein employs a chemical by the same name found in milk.

[20]The University of Texas Health Science Center, Dallas, Texas.

Materials: Opaque colors are available in several forms. Gouache comes in cakes or pans and tubes. Cakes and pans are good for providing color in small quantities. They have the disadvantage of not easily producing sufficient quantities for large areas or backgrounds. Tubes offer these advantages. In addition, tube colors are clean and fresh as they come from the tube. Casein is available in tubes and jars.

One of the great differences between painting in opaque colors as opposed to transparent ones is that changes in tints are made with white paint instead of water. Casein can produce tints—or more correctly, glazes—by the addition of water. This is one of the unique qualities of casein paints that allows for interesting tonal effects over opaque areas. In gouache painting, two whites are available. Zinc white is a transparent white which is ideal for producing brilliant, pure tints of a color. Permanent white is opaque and produces the purest white, when white is desired, but it lends a chalky appearance when mixed with colors, which can be used to good effect if that is the look required. Blacks are used to produce shades in both opaque and transparent watercolor. Three blacks are available: Lamp black is cool gray in tone; ivory black is warm in tone for mixing shades; jet black is a true, deep velvet black for use where maximum blackness is desired.

To paint large areas, such as backgrounds or the middle value of your subject, quantities of a given tint or shade need to be mixed and stored. When mixing a gouache tint, squeeze the approximate quantity of white (zinc white) into a container and add the desired colors to it until the proper tint is reached. When mixing shades, add lamp black (cool) or ivory black (warm) to the color until the desired shade has been produced. Add water as necessary to bring the mixture to a creamy consistency. Store excess mixtures in an airtight container until the painting is finished and checked.

Occasionally, when first opening a tube of gouache, a quantity of liquid binder that has separated from the pigment will ooze out. Replace the cap and knead the tube until this material is mixed thoroughly into the paint. Excess binder in small quantities of paint will prevent drying. Remember that gouache dries to a different tint or shade than it appears to be when wet, so test it frequently on a scrap of board to achieve the desired effect. Casein, on the other hand, does not have this annoying attribute. With casein, what you see is what you get.

In both casein and gouache painting, the range of colors can be extended by adding transparent watercolors to white or black. In addition, gouache can be used in a similar manner with casein white or black to extend the range of casein colors.

Cheap brushes are a waste of money. Winsor & Newton series 7 and 8 round, finest red sable brushes are the ones used most frequently. These are the finest watercolor brushes available and they paint beautifully. A good selection in various sizes is required. The larger sizes in this series are so extremely expensive that most people are forced to substitute something less expensive. My preferences are the Grumbacher Beaux Arts series 190 and Liquitex Kolinsky red sables.

There are times when the flat, square-pointed brushes are required; at those times the Grumbacher series 626-B finest red sable is recommended. Again, as with other Kolinsky red sable brushes, the larger sizes can be prohibitively expensive. For backgrounds and other large areas of flat color, a soft white, synthetic brush can be substituted. The Grumbacher Aquarelle in the one-inch size is ideal for this purpose.

Other more specialized brushes are useful for certain effects. For superfine detail work where the finest possible point is desirable, the Winsor & Newton series 12 brushes are ideal. These brushes are the finest red sable (same as series 7), but made of extremely short hair which gives them a stiff, superfine point. Since the price of red sable increases substantially as the length increases slightly, these brushes are reasonably priced. Another interesting brush is the red sable fan-shaped brush (Winsor & Newton series 55). This brush is useful in blending dry brush areas. Stipple or stencil brushes (of hair, not bristle) can also be employed for interesting textural effects.

One last word on brushes—brushes are delicate implements; keep them clean and do not allow paint to dry in them. Protect your investment by washing them frequently with brush-cleaning soap. Before storing, point and shape the brush with soap lather and allow the lather to harden on the bristles. Rest the weight of the brush on the handle when storing, never on the hair.

As with all other art supplies, use the best possible materials. All papers and boards should have a high rag content (100 percent if possible) and neutral pH. The surface should be fairly smooth, which will allow fine brushwork. Any number of the hot-pressed papers and cold-pressed boards are suitable. Colored drawing papers are also excellent. The opacity of the paint allows you to paint with very light colors directly over the darkest colored papers and boards. Cartoonists and animators have for years used gouache on wet media or prepared acetate, which can be utilized for cels (a term used by animators that means acetate sheets with punched holes) and color overlays. Casein and tempera are too brittle to be used safely on flexible surfaces and should be confined to use on stiff paper or boards.

The boards I prefer are numbers 1, 100, and 110 cold-pressed, made by the Crescent Paper Company. The numbers 1 and 110 are 100 percent rag, while the number 100 is listed as having a "high rag" content. Despite the fact that the number 100 is not 100 percent rag, the texture and color of the work surface are such that it is my favorite. Mi-Teintes, Crescent, Pantone, Color-Aid, Color-Vue, and Color-Match are all suitable as painting surfaces when a colored background is required.

Painting Technique: The opaque watercolor painting should begin with the background if there is to be one. The background can be of a light or dark value, since subsequent layers of foreground can be opaquely painted right over any background color. Be sure to save the background color for later corrections or adjustments.

With the background in place and thoroughly dry, transfer only the outlines of the sketch to the final surface. I prefer to use a commercially prepared graphite paper between sketch and ground. This method keeps the sketch clean (as opposed to applying Conte or graphite directly to the back of the tracing paper). It also allows you to reposition the sketch easily as many times as you desire, which can be a great help in later steps.

Paint the area defined by your transferred outlines with the middle tone value of the desired local color. This "local color" must be determined through a mental exercise of removing the effects of highlights, shadows, textures, reflections from other objects, and so on, to arrive at the inherent color of the object. Once this color is in place and thoroughly dry, reposition the sketch and graphite paper and transfer the remaining internal landmarks from the sketch. This step can be repeated as many times as necessary if, for example, there are other large areas of a different color within the basic shape.

The basic modeling of the form is accomplished with tints and shades of the local color (see color-mixing section of this paper) in a dry-brush fashion. Force the hairs of the brush into a fan shape by pinching out the thinned color between your thumb and index finger, and paint in a contour or crosshatch pattern. This will leave a linear texture on the surface which is generally very pleasing, but should a smoother, less-textured effect be required, dry-brush again with a change of clear water in the brush.

The modeling produced in this manner will give a convincing three-dimensional representation of the object, but will be somewhat bland and monochromatic. A more realistic (and more interesting) painting can be produced by including chromatic shifts in the highlight and core shadow areas. The principle involved in chromatic shifts is that the highlights aren't simply a lighter version of the local color, but a chromatic movement up the color scale toward yellow (the color of painted light—that is, the sun). Conversely, the core shadow area represents a chromatic shift down the color scale toward purple (more correctly, violet).

Specular highlights of pure white paint (permanent white gouache) and black shadows (jet black gouache) should be avoided in most instances. The exception to the rule is on very wet tissues and chrome instruments, but even here they should be used sparingly and only when they enhance the center of interest in the painting. Never use these touches frivolously, as they are very distracting when used improperly.

When the painting is finished and thoroughly dry, a cover of acetate will help protect the painted surface. Krylon Crystal-Clear acrylic spray can also be used to seal the surface, but be aware that the spray darkens the painting slightly and may cause any zinc white on the surface of the painting to disappear. Casein paintings dry within about a month to a hard, insoluble finish, but they can be sprayed or varnished for extra protection. Shiva manufactures its own crystal-clear acrylic spray for its colors, but any acrylic spray will work.

Transparent Effects: The relative transparency of an object is defined by the amount of background that is allowed to show through the structures. The maximum transparent effect is achieved by using the greatest contrast—that is, white and high-value grays on a black background. Reducing the contrast will eventually lead to a more translucent effect.

Light travels directly through transparent structures that are at right angles to the light source with little or no interference. If the structure is hollow, with a transparent surface, light interference is collected and "piped" along the surface, causing the edges of the structure to glow when seen against a darker value. Because of the highly reflective nature of transparent surfaces that are at 45 degrees between the light source and the viewer, the highlight is a mirror image of the light source and appears to be surrounded by the lowest value of the background color. Light traveling through the transparent object is collected on the back surface of the structure, on a segment directly opposite the highlight which is again at 45 degrees to the light and the viewer. This transmitted highlight is diminished from the specular highlight by its relatively greater distance from the light source, the tint (if any) of the object, and the background color. This effect is additive, so that the transmitted highlight on a colored object on a colored background will be a very high value of the two colors combined. Shadows cast from transparent structures are of the darkest value at the outer edges of the shadow where the light striking the object is most refracted. Light passing through the object from specular highlight to transmitted highlight is focused on the surface receiving the shadow at the focal core of the shadow. The cast shadow, then, is darkest (lowest in value) at the periphery and becomes progressively lighter (higher in value) until the focal core is reached.

Iridescent Effects: Iridescence is the result of shifting wavelengths of light as it is split by multiple transparent layers (for example, insect scales). The color spectrum produced in this way changes as the viewing angle changes, producing vibrant color mixes. To paint this rainbow effect, the color shift should be confined to the highlight-halftone and be of a high value, high chroma as compared to the surrounding hues which, if darker and duller, will enhance the effect. To paint this rainbow of colors we must *visually* mingle adjacent hues from the color wheel, but it is imperative that when examined close up, the hues will be distinguishable from each other. This dictates, for example, that when a yellow is chosen for the highlight color on a cool, iridescent thorax of an insect, the highlight halftone will be yellow-green and blue-green with blue in the shadows. Complements should be used sparingly (if at all), and the color shift should not include too large a segment of the color wheel or the effect will appear unnatural. When complements fall adjacent to each other, as they sometimes do on an iridescent insect, the value and chroma should be balanced but not equal, with one high and the other low.

Donald B. Sayner and Carolyn Leigh's Line Copy and Slides Preparation Using Kodalith Film[21]

1. Set up darkroom.
   a. Mix *Kodalith developers A and B*, one to one, in 4-by-5 tray.
   b. Prepare a tray of wash water and one for fixer.
   c. Red lights are OK. Yellow safe light and overhead light should be turned off.
2. Load *Kodalith film* (4 by 5 inches) in darkroom under red light. (When 35mm cameras are used, load Kodalith Ortho film 6556.)
   a. Blow dust out of *film holder*.
   b. White edge of slide should face out—this indicates the film is unexposed.
   c. The lighter side of the film is the emulsion side and should be face up in the film holder.
   d. Make sure the film is under the grooves of the film holder. Check by lifting up the end of the film.
   e. *Don't* lay the film on the table, since it will pick up dust.
3. Set up the copy area.
   a. Clean the glass on the copy board if it needs it.
   b. Put up copy using *drafting tape*—the copy's edges must be parallel with the camera's focusing glass edges or you will get an optical distortion on the negative.
   c. Clamp down the copy glass and turn on the vacuum for focus-

[21]University of Arizona, Tucson, Arizona.

ing, or use a sheet of glass for vertical shooting if copy does not lie flat.
   d. Open the shutter and aperture.
   e. Turn on the quartz lights—they should be equidistant (45 degrees) and the same height from the copy, giving an even light.
   f. Focus the copy on the ground glass—you may need to check with a magnifying glass. Make sure the quartz lights do not reflect in the camera viewing area.
   g. The overhead lights and the window light may cause reflections—check and correct.
4. Shooting the negative.
   a. Close down the aperture and set to f16. The shutter should be set at Bulb (B). Cock the shutter.
   b. Put in the film holder. Make sure the camera doesn't move. Pull the slide out and expose for three to five seconds. Put the film holder back in with the black edge out—this indicates that the film is exposed.
   c. Turn off the lights and the vacuum.
   d. Remove the film holder.
5. Developing.
   a. Wet film briefly in water.
   b. Develop by inspection under the red safelight. Negative should be black on both sides. Check to make sure you are not losing details through overexposure or overdevelopment.
   c. Fix until clear—one to two minutes.
   d. Wash for 10 minutes and hang to dry.
6. Retouching—after film is dry, retouch any dust holes using *opaque red* and fine brush.
7. Slides.
   Check ground glass when focusing with the appropriate *slide mount* (35mm, etc.) to make sure the copy fits. Proceed as above.

   After retouching, the negative may be:
   a. Cut and mounted as is
   b. Used to run diazochromes
   c. Sandwiched with colored acetate
   d. Colored with felt-tip markers
   e. Used to print positives

Donald B. Sayner and Carolyn Leigh's Photodrawing Technique[22]

1. Wipe the black-and-white photo with lighter fluid to remove any oil from the surface. Lightly developed prints are best.
2. Ink directly on the photo. Dark areas may need to be darkened more after print is bleached out. Ink *must* be *waterproof* drawing ink.
3. When finished, place print in tray containing mixture of approximately one cup fixer (sodium thiosulfate) to 1–2 tablespoons potassium ferrocyanide[23] or Kodak Farmers Reducer. Mixture should be a lemon yellow. Image will reduce out in approximately four minutes. Dark areas may need a fresh mix to finish process.
4. Wash print off and fix for five minutes. Wash ten minutes in running water and dry between paper towels under a weight.
5. When dry, additional rendering may be added to perfect the drawing.

Lucy Catherine Taylor's Coquille Board Technique[24]

Coquille board, a textured surface resembling the surface of lithographs, is an appealing alternative both to continuous-tone mediums and to ink stipple for illustrations requiring a shaded representation of the form, pattern, and textural details of a specimen. As a technique, if offers advantages both in time and cost. The materials needed are few and inexpensive. It is easy to learn and can represent significant savings in production. The application of tone and resulting effects are easily controlled, requiring less effort than, for example, pen-and-ink stipple technique. A finished illustration simulates halftone yet can be photographed as line copy—thus the printer avoids the additional cost of preparing a halftone (dot-screened) plate. For a text with numerous illustrations this could represent a significant savings in printing cost.

Coquille board is available in both coarse and finely toothed surfaces. The size of the original drawing, amount of detail depicted, and degree of reduction for printing all determine the appropriate choice of board. For illustrations measuring 9 to 12 inches wide and drawn for reduction to one half or one third the original size, do use the rough board.

For drawings requiring little detail or little reduction, the fine-toothed board may be ideal. Yet coquille can be used to depict an intricately detailed or textured specimen. For such work, it is important to gauge the size of the original drawing so that the texture being depicted will not compete with the texture of the board itself. This may require working at a larger scale and using coarse-grained board. For example, while illustrating microscopic lichen specimens, it is necessary to depict such detail. Most finished drawings depict an area of a few centimeters or a few millimeters, depending on the complexity of texture and size of lichen.

The key to working successfully with coquille board for printing as line copy is to keep all shading

[22]University of Arizona, Tucson, Arizona

[23]Warning: Use caution; poisonous chemical. [Author]

[24]The University of Wisconsin—Madison, Wisconsin.

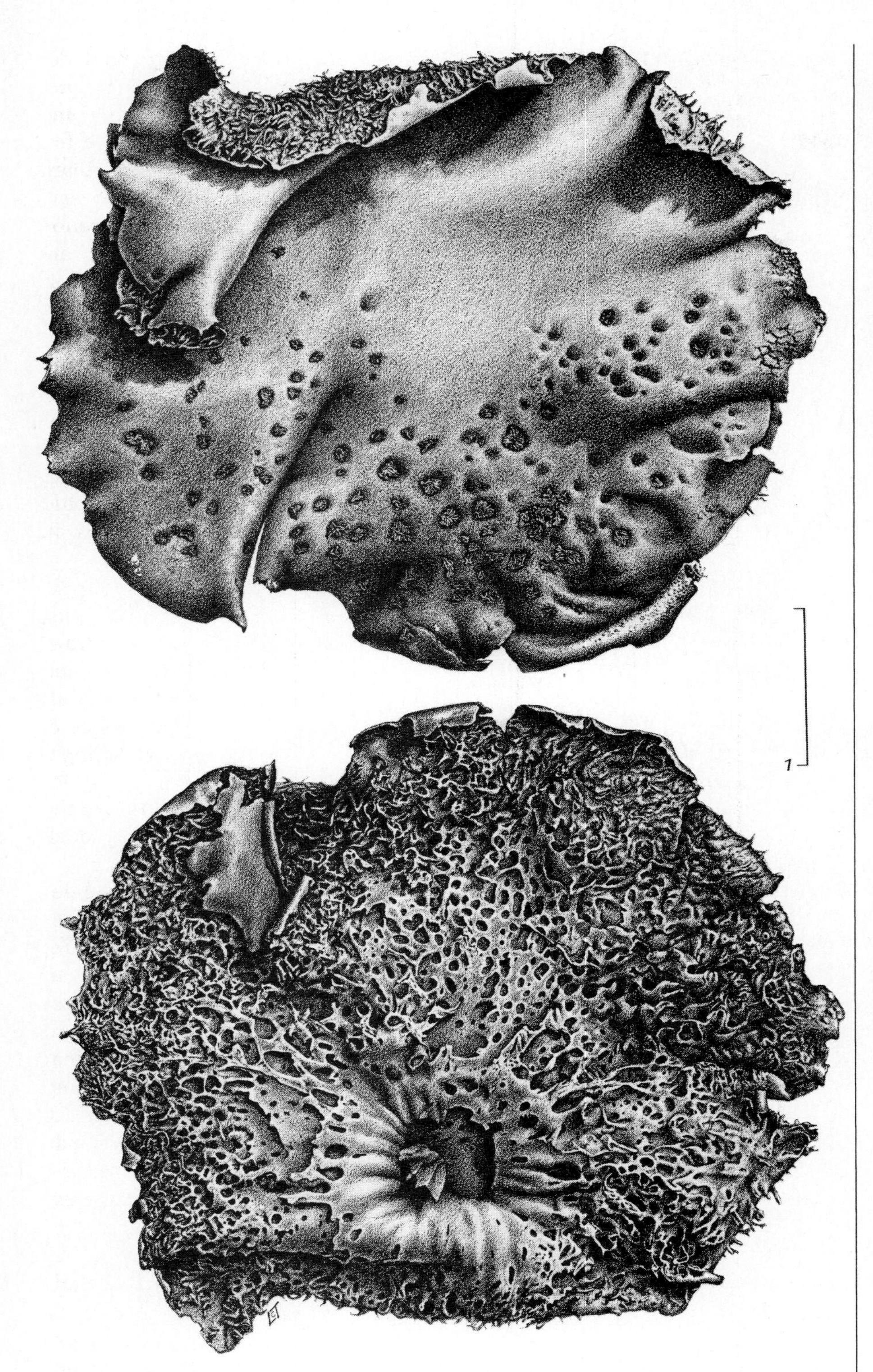

Figure 6-16. *Actinogyra muehlenbergii* (a lichen). Black pencil and ink on coquille board. Drawing by Lucy Catherine Taylor, The University of Wisconsin—Madison. Published in John W. Thompson, *Arctic Lichens* by Columbia University Press, New York, 1984. Drawn in 1982.

within the range that gives the *effect* of middle-gray values. As with other techniques, when photographed the lights tend to go lighter (disappear) and the darks tend to go darker (merge). It is preferable to incorporate the white of the paper to show detail rather than to shade very lightly. Using white with middle gray and black (ink) with middle gray gives some dimension to the rendering and avoids a disappointing loss of detail when reproduced.

Materials:

Coquille board (coarse or fine)

Tracing paper, transfer paper

black pencil—Spectracolor 1406

Erasers—kneaded rubber and MagicRub

Technical drawing pens (3 × 0, 4 × 0 or crow quill)

White gouache

Fine brush

Pencils:

Carbon, Conté pencils produce rich black tones, keep point fairly well. Disadvantage—both produce dust which can smear the board's surface and must be removed.

Litho crayon—rich black, grease-based. No dust. Disadvantages—doesn't keep sharp point easily. A gray, waxy film may develop on the drawing's surface after several years. Cannot be corrected except with white-out.

Black-colored pencils—Spectracolor 1406 is rich black, ideal harness, and keeps a point well. No dust. Correctable with kneaded eraser or MagicRub erasor.

Technique:

1. Prepare rough sketch and transfer lightly to board, being careful not to dent the surface of the board.
2. Shade with pencil, working the lighter areas first and incorporating the white of the paper whenever possible (see above

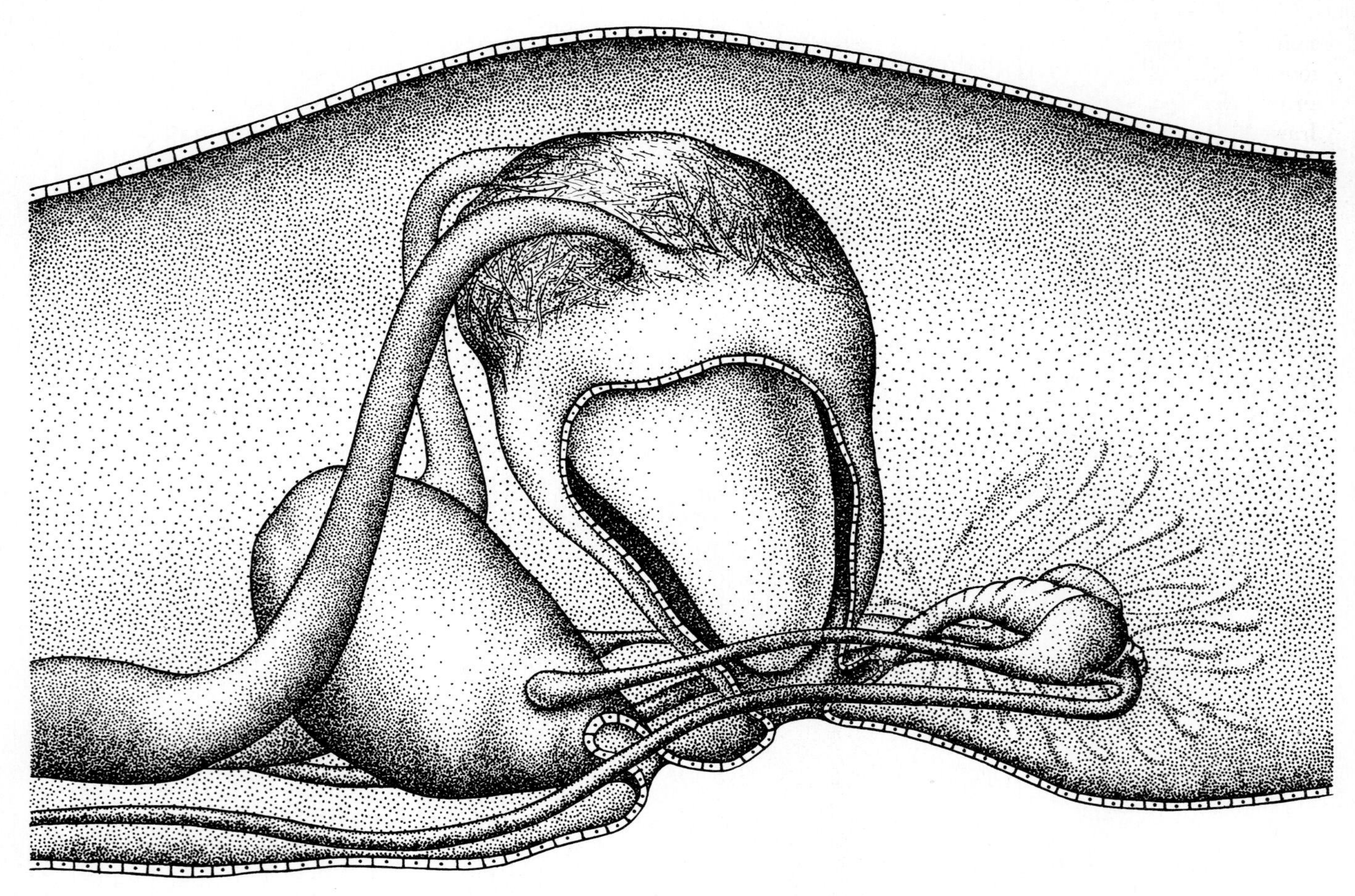

Figure 6-17. Reconstruction of copulatory apparatus of *Uteriporus vulgaris*. Pen and ink, line and stipple on bristol board. Drawing by Maria Tran Thi Vinh-Hao Ball, Instituut voor Taxonomische Zoölogie, Zoölogish Museum, Universiteit van Amsterdam, The Netherlands. Published in *Ball and Sluys,* Marine Planarians of Eastern North America. Drawn before 1983.

tips on shading). Next shade the darker areas, first with pencil, then adding ink details. Rather than starting with an ink outline, I prefer to add occasional inked outline as an accent or to give emphasis to the penciled drawing. The softer penciled line adds to the sense of depth (atmospheric perspective) of the rendering, where a hard line could flatten the form. However, for insects, hard edged, or dark-toned specimens, it may be desirable to begin with an ink outline. Then add sharp highlights using white gouache and a fine brush. Softer highlights and textures may be made with a MagicRub or Mars plastic eraser, cut to a wedge. Any areas may be lightened for reworking using these or a kneaded eraser.

Maria Tran Thi Vinh-Hao's Ball Procedure of Drawing Planarians[25]

The initial step in the production of the illustration is a preparation of a series of sketches finalized in clear presentation of the subject. The specimens are generally viewed through a stereoscopic microscope; some are drawn with the aid of the camera lucida.

The procedure involves the following steps:

1. Preparation and positioning of the specimen. The live specimens are cooled with blocks of ice because of their sensitivity to the warmth of the microscope lights. Such procedure allows for the stable conditions needed for viewing and drawing as well as preserving the subject.

2. A series of sketches is produced and the general shape, size, movement, colors, relative emplace-

[25]Instituut Voor Taxonomische Zoölogie, Zoölogish Museum, Universiteit van Amsterdam, The Netherlands.

ment of openings of pharynx and gonopore are noted. The whole external feature is recorded and redrawn in preparation for the actual ink rendering.

3. The live specimen as well as preserved specimens are observed separately and the final sketch is compared with a new set of information. Additional features not readily seen during the process of sketching are carefully examined. All necessary changes or additions are implemented on the already-drawn sketch.

4. The sagittal section of the specimen is drawn separately with the help of the camera lucida. Necessary reconstructions of reproductive organs are based on obtained pencil drawings.

5. The finished rendering is accomplished with ink on white paper using standard line drawing, stippling and crosshatching.

Maria Tran Thi Vinh-Hao's Ball Watercolor Technique[26]

Materials needed for successful results:

1. Good-quality watercolors (GRUMBACHER)
2. Good-quality sable brushes
3. White-surface scratch board
4. Sharp tool such as X-Acto knife or graver

Steps are as follows:

1. Correctly presented image of the subject must be worked out through sketching procedure.

2. The approved final sketch must be transferred to the surface of the scratch-board without inflicting any damage to the painting surface. The transfer can be achieved using carbon paper or graphite from the pencil. The pressure exerted by the pencil on the sketch overlying the scratch board should be minimal.

3. Colors will have to applied speedily. The application as well as premixing of pigment calls for decisive action. The scratch-board surface is very absorbent and cannot be overworked. Once the wash is placed in its designated area it will stay there. Light tones are to be applied first, slowly building up the desired change of the value. Make sure that the first wash is dry before the next one is applied.

4. All highlights are scratched away with a tip of a sharp tool. Scratching can be used for producing light areas as well as an "engraved" type image.

Caution: Before starting the project, experiment with the technique.

[26]Instituut Voor Taxonomische Zoölogie, Zoölogish Museum, Universiteit van Amsterdam, The Netherlands.

Appendix A

William L. Abler's Construction and Use of Camera Lucida[1]

Description of the camera lucida: The camera lucida, or "illuminated chamber," and its cousin the camera obscura, or "darkened chamber," are image-forming devices that enable the user to make dimensionally correct drawings of three-dimensional objects by tracing the outline of the object directly onto a drawing surface. A camera obscura is a box or room that can be darkened on the inside so that with the aid of a lens or pinhole, an illuminated object outside the box can project a visible image onto a surface inside the box. The artist produces a drawing by tracing the image. The camera obscura is cumbersome to build and use because it must be fairly large, and because the object that is to be drawn must be very brightly illuminated if it is to form a useable image. When it is built in miniaturized form, and when the image-recording surface is a photographic film, the camera obscura is called a "camera" for short. Thus, another way of using the camera obscura for drawing an object is to take a photograph of the object and then to make a tracing from the photograph.

The camera lucida uses a reflector to direct light from the equally illuminated subject and drawing surface directly into the artist's eye, which is already a darkened chamber, thus avoiding the need to build a specially darkened room for viewing a secondary image, and avoiding the need to illuminate the subject brightly enough to form a secondary image. The camera lucida can, and indeed must, be used in a more-or-less uniformly lighted place such as the out-of-doors or a well-lighted room. Since the images of subject and drawing surface are superimposed on the artist's retina, they appear to overlap in the artist's field of view, so that the artist can trace the outlines directly onto the drawing surface, which is usually a sheet of paper, but which could be a lithograph stone or engraving plate, or anything else that the artist finds desirable.

Types of camera lucida: There are two major kinds of camera lucida, which use two different principles for projecting superimposed images onto the artist's retina. The first kind uses a small, fully reflective mirror or prism that divides the pupil into two separate regions, an upper and a lower, sending reflected light from the subject into the upper region of the pupil, and allowing light from the drawing surface to enter the lower region of the pupil directly. The divided-pupil form of the camera lucida has been in use for centuries, and can be made to work satisfactorily. Some people would not use anything else. Nevertheless, so much concentration is required that many people do not have enoough attention power "left over" to make a satisfactory drawing.

Slightly less light-efficient, but must easier to use, is the beam-splitter (actually beam-combiner) form of the camera lucida. This device uses a partially reflective mirror that combines part of the light from the drawing surface and transmits the resulting combined beam into the artist's eye. Although the beam splitter loses light through unwanted transmission and through absorption at the reflecting surface, the loss is partially compensated by the illumination of the entire pupil by both subjects and drawing surface.

Construction of camera lucida: The crafty illustrator can build both major forms of the camera lucida in a simple machine shop. Highly practical pauper's editions can be build at home. The following description applies primarily to a beam-splitter camera lucida that I built for my own use. *See Figure* AA.1.

The heart of the camera lucida is a beam splitter (1) made from two 45°–45°–90° glass prisms glued together by their hypotenuses to form a cube. One of the hypotenuses is partially silvered to form the partially reflecting (beam splitting) surface. The rest of the camera lucida is simply a system of clamps and rods that hold the beam splitter at a convenient height above a table, and which hold supplementary lenses in place so that the instrument can be used for drawing objects of a variety of sizes. I used brass rod, stainless steel rod, and low-temperature silver solder for building the support pieces, and aluminum plate for making the lens holders.

The beam splitter is supported in a "U"-shaped yoke (2) made of brass rod that has been filed flat on the inner surface, and which is attached with silver solder to the rod (3) that forms the support for the

[1]Chicago, Illinois

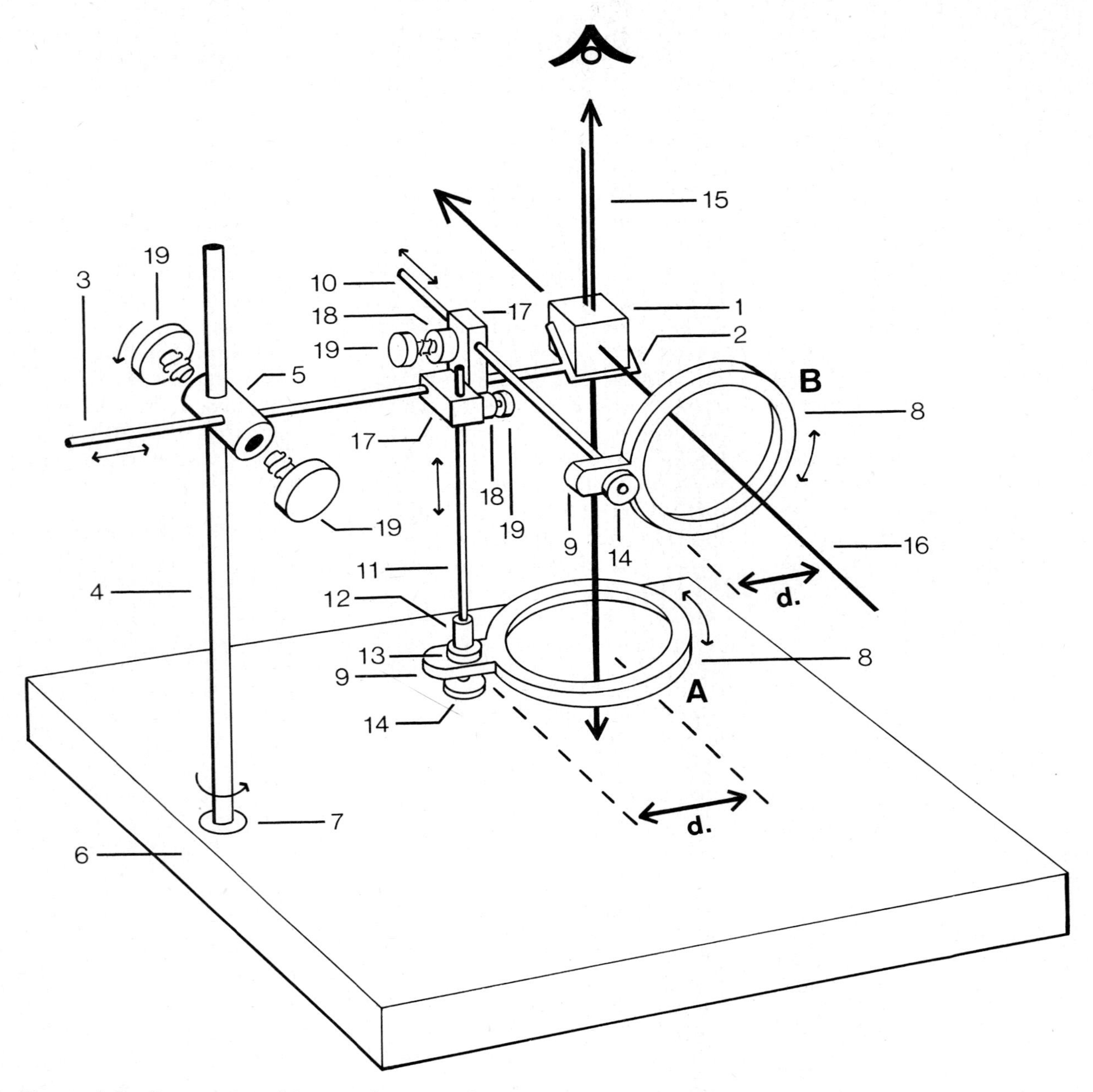

Figure A-1. Construction of homemade camera lucida. *1*. Beam splitter; *2*. Brass yoke holding beam splitter; *3*. Horizontal support rod (beam splitter support rod); *4*. Vertical support rod; *5*. Clamp joining vertical and horizontal support rods; *6*. Table; *7*. Socket for vertical support rod; *8*. Lens retaining ring; *9*. Lens support arm; *10*. Horizontal support rod for object lens; *11*. Vertical support rod for drawing lens; *12*. Sleeve; *13*. Seat; *14*. Clamp nut; *15*. Combined beams of light entering the eye (optical axis); *16*. Path of light entering beam splitter (optical axis); *17*. Support arm; *18*. Threaded cylinder) for the clamps; *19*. Set screws; *A*- Minifying, reducing (negative) lens; or magnifying, enlarging (positive) lens–*drawing lens; B* - Magnifying, enlarging (positive) lens—*object lens;* *d* - Distance between the center of the lens and the center of the hole for the lens support rod (axis of the beam splitter and axis of the lens support rod). When designing the lens holders, note that, unless special provision is made, *d* will be different for the two lenses.

prism assembly. The prism support rod itself is supported horizontally from a vertical rod (4) by a clamp (5) equipped with set screws that lock the two rods into place. To make the camera lucida highly portable, it can be built with its own portable plastic-topped table (6) equipped with a stainless-steel socket (7) that receives the vertical support rod (4). The vertical support rod could also be clamped to the edge of a table or held in any other way that is convenient.

Built as described above, the camera lucida is fully adequate for making drawings that are the same size as the original subject. The artist sets up the prism at the convenient distance above the drawing surface (largely determined by the eye's best focus distance) and situates the subject at the same distance from the beam splitter, so that the images both focus simultaneously on the retina.

Reduction and enlargement: If the artist wishes to make drawings that are larger or smaller than the subjects they represent, provision must be made to focus the enlarged or reduced subject image on the retina while the image of the drawing surface is focused there also. This is the problem of simultaneous focusing. For large subjects, from about the size of a dinner plate to that of a landscape, the solution is to place the subject far from the eye to reduce its visual size, then to bring the drawing surface into focus on the retina by using a weak magnifying glass (A), the "drawing lens." When the image is to be drawn larger than the subject, a magnifying glass (B) must be placed between the beam splitter and the subject. Sometimes, when the image is to be drawn very much smaller than the subject, a minifying (reducing, or negative) lens can be placed between the beam splitter and the subject. A lens placed between the eye and the subject to be drawn is called an "object lens."

In order to have a fulll range of magnifications and reductions available for adjusting the size of the image, it would be necessary to have a large selection of positive focal length (magnifying) and negative focal length (reducing) lenses available. By adjusting the distance between the subject and the beam splitter, however, an acceptable compromise selection can be obtained with just a few lenses. My own camera lucida has only four lenses, one weak (long focal length) positive drawing lens; two medium focal length positive, and one long focal length negative object lens.

Construction of lens holders: The basic plan for all four lens holders is a retaining ring (8) supported on an arm (9). The support arm is drilled so that the lens can be fastened to a support rod (10, 11) by a clamp nut. In my camera lucida, the support rod is a length of brass rod; the threaded portion is the threaded part of a brass screw; and the two are held together for soldering by a little brass sleeve (12). The seat (13) is made from two brass washers soldered together to give a solid appearance. Clamp nuts (14) can be salvaged from the binding posts of old dry cells or can simply be hexnuts from a hardware store.

The major requirements for the lens holders are that the hole for the support rod must be positioned so that the center of the lens will be aligned with the optical axis (15, 16) of the beam splitter and that the round hole where the lens fits must be no more than a few thousandths of an inch larger than the lens. My lens holders were machined from ¼″ thick aluminum plate. To meet the first requirement, the clamp assembly system that holds the support rods for the lenses should be built in advance and to meet the second, the lenses themselves must be obtained in advance, because the real diameters may not be exactly the same as the nominal diameters.

The brass clamp assembly system that supports the rods for the lens holders are each built from three parts: a support arm (17) a threaded cylinder (18), and a set screw (19). The support arm is drilled with two mutually perpendicular holes about ¾″ apart, one of which is soldered to the prism support rod, and the other of which is equipped with the threaded cylinder and set screw that hold the lens support rod in place. The threaded cylinder can be made from a dry cell binding post, or a short rod that has been drilled and tapped, or just a machine nut—soldered into place. Since the alignment of the clamp arm with the threaded cylinder is crucial, the set screw should be used as a guide while soldering the threaded cylinder into place and the threads protected from solder with *iron oxide paste*.

When the completed clamps are being soldered to the prism support rod, the solder bond that holds the threaded cylinder in place must be protected from heat, either by wrapping the cylinder with wet cloth, or by using higher-melting-point solder for the cylinder than what is used for bonding the clamp to the prism support rod. If the clamp arm is thick enough, the set screw can simply be threaded into a hole that has been drilled and then tapped directly into the arm. This solution is clumsy, however.

The lens holders must be planned by reference to two key distances. The first of these is the distance (d) between the axis of the beam splitter (which coincides with the axis of the lens) and the axis of the lens support rod, and the second is the diameter of the lens. After these two dimensions are laid out, the support arm can be designed, using a standard width determined by the diameter of the clamp nut that will eventually lock the lens holder to the support rod. The radius of the circle that terminates the arm is one half the width of the arm. The thickness of the lens-support ring is determined cosmetically, but should not be less than 1/16″. Because some circle cutters cannot cut circles small enough to fit the small lenses, the holes for the lenses can be cut on a lathe.

The plans for the lens holders should be laid out on an aluminum plate, leaving enough space around each plan so that each can be cut out inside a circle that has the center of the lens as its center. The circular blanks from which the lens holders will be cut can be cut from the aluminum plate with an adjustable circle cutter. For this operation, the aluminum plate must be clamped securely to the drill press table, and all set screws in the circle cutter checked for tightness. To cut out the holes for the lenses, the circular blanks can be placed on a lathe, and the lens holes lathed to the correct, empirically determined diameter. The center of the lens opening must be centered with the axis of the lathe. To avoid accidents, the circular aluminum blanks must be clamped tightly in the jaws of the lathe chuck.

After the hole for each lens is cut to the correct diameter, the outline of each lens holder must be cut as closely as possible with a band saw, and the edges finished with a belt sander. The lenses are glued into place with silicone sealer adhesive. Since any instrument is almost useless unless it can be carried and stored safely, a wooden case to hold the parts of the camera lucida is desirable.

Pauper's edition of beam-splitter camera lucida: The impecunious illustrator will be happy to know that it is possible to produce drawings using a pauper's edition camera lucida built from parts that can be found readily for little or no cost: a one-foot bamboo strip that forms the support for the beam splitter, which can be made from 1/4″ × 1 1/4″ sliver of window glass; a 6″ length of insulated copper wire that forms a "foot" for adjusting the angle of the reflector, and some string for tying the glass in place at the end of the bamboo strip. The end of the bamboo stick is whittled to the same thickness as the glass sliver. The device can be supported horizontally on a stack of books, with an extra book placed on top of the wire "foot" to stabilize it. The window-glass beam splitter gives a double image that is troublesome but not fatal. A microscope cover glass can be substituted for the sliver of window glass to reduce the double image effect. Eyeglass lenses can be used as object and drawing lenses.

Adaptation of beam-splitter camera lucida to microscope: If the camera lucida is to be used for drawing objects under the microscope, it can be used without supplementary lenses and with an angled mirror held over the drawing surface for viewing it. The beam splitter must not be more than about 15 mm on a side. Although a larger beam splitter gives a wide-angle view for making ordinary drawings, it prevents the artist's eye from approaching closely enough to the microscope eyepiece for good viewing. The microscope tube should be vertical. When drawing with the microscope, the artist positions the beam splitter directly on top of the microscope eyepiece and views the subject directly through the beam splitter and microscope, while the drawing surface rests on the table to the right or left of the microscope and is viewed indirectly via the beam splitter and diagonal mirror.

The artist who spends extended periods of time drawing with the microscope may find that a specialized camera lucida attachment is desirable. The attachment can be built by positioning a large beam splitter prism between the objective and ocular lenses, supporting it in a sleeve which also serves to attach the microscope ocular tube to the body tube. The camera lucida sleeve that I built for my microscope was machined from a thick (2 1/2″-diameter) aluminum cylinder. A diagonal mirror is needed for viewing the drawing surface through a hole drilled in the side of the support sleeve. An erect image of the drawing surface is obtained by attaching the prism cluster from a binocular to the viewing hole at the side of the support sleeve. The size of the drawing surface image can be adjusted by placing a lens in the optical path.

Use of camera lucida: After the preliminaries are disposed of, such as choosing the most desirable angle for viewing a subject, choosing the most desirable magnification or reduction in size suitable for both the subject and the drawing surface, and arranging suitable illumination, the artist begins by outlining the prominent edges and borders of a subject, adding detail later. For three-dimensional objects this task can appear simple, but is in fact complex and becomes a strenuous exercise in three-dimensional imagination. Because the eye can "wander" above the beam splitter, the image of the subject in the eye can wander with respect to the drawing on the drawing surface, making the first order of business the designation of a few prominent "landmarks" on the drawing surface that will key the position of the drawing to that of the subject. If the key landmarks begin to wander with respect to the features of the subject that they represent, the eye can be repositioned to realign the features and to stabilize the drawing as it progresses.

Some familiar images, such as a conventional view of a face, can be convincingly drawn using very few lines because the mind is able to fill in what is not drawn in. Nevertheleess, even a familiar object will have enough unfamiliar details that the artist will be required to examine the object directly, during the process of drawing it, in order to draw it effectively. For example, since the image in the eye contains no three-dimensional information, the artist must build up in the mind's eye a conceptual image of the subject in three dimensions in order to be able to draw it so that it appears to the viewer in three dimensions. For complex, unfamiliar objects, such as the base of the skull, the same principle applies with even greater force.

Ordinarily, camera lucidas of both the divided-pupil and beam-splitter types gives the artist a right-side-up view of the drawing surface, so that drawing is easy, but an upside-down view of the subject, so that the completed drawing has right and left reversed. Although this situation can be safely ignored for many purposes, it can also be dealt with by drawing directly onto a lithograph stone or engraving plate, or by drawing onto tracing paper and then turning the paper over. It can also be dealt with by using the diagonal mirror that is used for drawing with the microscope. This method limits the sizes of objects that can be drawn.

The first outline sketch of a three-dimensional object may be dimensionally perfect but have no recognizable visual similarity to the original subject, and the artist will be forced to continue the drawing on faith, simply by keeping in mind that the apparent hodge-podge of lines was made by tracing the borders and outlines of the original subject: They must be on the drawing surface somewhere. Filling in three-dimensional structure may become a protracted process that involves a kind of interactive feedback among the original subject, the three-dimensional conception of its form in the mind's eye of the artist, and the developing three-dimensional information in the drawing as it progresses.

Specification and sources of supply: A competent craftsman can build the beam-splitter camera lucida for approximately $50. Dangerous equipment such as the lathe, circle cutter, and blowtorch should be used only under the supervision of a competent person. Lenses and prisms for a beam-splitter camera lucida of the type described here are available from American Science Center, 5700 Northwest Highway, Chicago, Illinois 60646. Telephone: (312) 763-0313.

	Description	***Stock Number***	***Specifications***
OBJECT	positive	40,807	diam. 20mm
LENSES	achromat		f.1. 44mm
	positive	6,176	diam. 38mm
	achromat		f.1. 130mm
	negative	94,624	diam. 52mm
	meniscus		f.1. −238mm
DRAWING	positive	94,732	diam. 51mm
LENS	meniscus		f.1. 333mm
BEAM	prism type	30,329	15mm
SPLITTERS	prism type	41,211	40mm (for microscope sleeve)

Appendix B

TABLE OF SECOND POWERS[1]

n	n^2	n	n^2	n	n^2	n	n^2
1	1	36	1 296	71	5 041	106	11 236
2	4	37	1 369	72	5 184	107	11 449
3	9	38	1 444	73	5 329	108	11 664
4	16	39	1 521	74	5 476	109	11 881
5	25	40	1 600	75	5 625	110	12 100
6	36	41	1 681	76	5 776	111	12 321
7	49	42	1 764	77	5 929	112	12 544
8	64	43	1 849	78	6 084	113	12 769
9	81	44	1 936	79	6 241	114	12 996
10	100	45	2 025	80	6 400	115	13 225
11	121	46	2 116	81	6 561	116	13 456
12	141	47	2 209	82	6 724	117	13 689
13	169	48	2 304	83	6 889	118	13 924
14	196	49	2 401	84	7 056	119	14 161
15	225	50	2 500	85	7 225	120	14 400
16	256	51	2 601	86	7 396	121	14 641
17	289	52	2 704	87	7 569	122	14 884
18	324	53	2 809	88	7 744	123	15 129
19	361	54	2 916	89	7 921	124	15 376
20	400	55	3 025	90	8 100	125	15 625
21	441	56	3 136	91	8 281	126	15 876
22	484	57	3 249	92	8 464	127	16 129
23	529	58	3 364	93	8 649	128	16 384
24	576	59	3 481	94	8 836	129	16 641
25	623	60	3 600	95	9 025	130	16 900
26	676	61	3 721	96	9 216	131	17 161
27	729	62	3 844	97	9 409	132	17 424
28	784	63	3 969	98	9 604	133	17 689
29	841	64	4 096	99	9 801	134	17 956
30	900	65	4 225	100	10 000	135	18 225
31	961	66	4 356	101	11 201	136	18 496
32	1 024	67	4 489	102	10 404	137	18 769
33	1 089	68	4 624	103	10 609	138	19 044
34	1 156	69	4 761	104	10 816	139	19 321
35	1 225	70	4 900	105	11 025	140	19 600

[1] ^2n and n^2 from Table V from *Mathematics Made Simple* by Abraham Sperling and Monroe Stuart. Copyright 1942, 1944, 1962, 1981 by Doubleday & Company, Inc. Reprinted by permission of the publisher.

n	n^2	n	n^2	n	n^2	n	n^2
141	19 881	186	34 596	231	53 361	276	76 176
142	20 164	187	34 969	232	53 824	277	76 729
143	20 449	188	35 344	233	54 289	278	77 284
144	20 736	189	35 721	234	54 756	279	77 841
145	21 025	190	36 100	235	55 225	280	78 400
146	21 316	191	36 481	236	55 696	281	78 761
147	21 609	192	36 864	237	56 169	282	79 524
148	21 504	193	37 249	238	56 644	283	80 089
149	22 201	194	37 636	239	57 121	284	80 656
150	22 500	195	38 025	240	57 600	285	81 225
151	22 801	196	38 416	241	58 081	286	81 796
152	23 104	197	38 809	242	58 564	287	82 369
123	23 409	198	39 204	243	59 049	288	82 944
154	23 716	199	39 601	244	59 536	289	83 521
155	24 025	200	40 000	245	60 025	290	84 100
156	24 336	201	40 401	246	60 516	291	84 681
157	24 649	202	40 804	247	61 009	292	85 264
158	24 964	203	41 209	248	61 504	293	85 849
159	25 281	204	41 616	249	62 001	294	86 436
160	25 600	205	42 025	250	62 500	295	87 025
161	25 921	206	42 436	251	63 001	296	87 616
162	26 244	207	42 849	252	63 504	297	88 209
163	26 569	208	43 264	253	64 009	298	88 804
164	26 896	209	43 681	254	64 516	299	89 401
165	27 225	210	44 100	255	65 025	300	90 000
166	27 556	211	44 521	256	65 536	301	90 601
167	27 889	212	44 944	257	66 049	302	91 204
168	28 224	213	45 369	258	66 564	303	91 809
169	28 561	214	45 796	259	67 081	304	92 416
170	28 900	215	46 225	260	67 600	305	93 025
171	29 241	216	46 656	261	68 121	306	93 636
172	29 584	217	47 089	262	68 644	307	94 249
173	29 929	218	47 524	263	69 169	308	94 864
174	30 276	219	47 961	264	69 695	309	95 481
175	30 625	220	48 400	265	70 225	310	96 100
176	30 976	221	48 841	266	70 756	311	96 721
177	31 329	222	49 284	267	71 289	312	97 344
178	31 684	223	49 729	268	71 824	313	97 969
179	32 041	224	50 176	269	72 361	314	98 596
180	32 400	225	50 625	270	72 900	315	99 225
181	32 761	226	51 076	271	73 441	316	99 856
182	33 124	227	51 529	272	73 984	317	100 489
183	33 489	228	51 984	273	74 529	318	101 124
184	33 856	229	52 441	274	75 076	319	101 761
185	34 225	230	52 900	275	75 625	320	102 400

n	n^2	n	n^2	n	n^2	n	n^2
321	103 041	366	133 956	411	168 921	456	207 936
322	103 684	367	134 689	412	169 744	457	208 849
323	104 329	368	135 424	413	170 569	458	209 764
324	104 976	369	136 161	414	171 396	459	210 681
325	105 625	370	136 900	415	172 225	460	211 600
326	106 276	371	137 641	416	173 056	461	212 521
327	106 929	372	138 384	417	173 889	462	213 444
328	107 584	373	139 129	418	174 724	463	214 369
329	108 241	374	139 876	419	175 561	464	215 296
330	108 900	375	140 625	420	176 400	465	216 225
331	109 561	376	141 376	421	177 241	466	217 156
332	110 224	377	142 129	422	178 084	467	218 089
333	110 889	378	142 884	423	178 929	468	219 084
334	111 556	379	143 641	424	179 776	469	219 961
335	112 225	380	144 400	425	180 625	470	220 900
336	112 896	381	145 161	426	181 476	471	221 841
337	113 569	382	145 924	427	182 329	472	222 784
338	114 244	383	146 689	428	183 184	473	223 729
339	114 921	384	147 456	429	184 041	474	224 676
340	115 600	385	148 225	430	184 900	475	225 625
341	116 281	386	148 996	431	185 761	476	226 576
342	116 964	387	149 769	432	186 624	477	227 529
343	117 649	388	150 544	433	187 489	478	228 484
344	118 336	389	151 321	434	188 356	479	229 441
345	119 025	390	152 100	435	189 225	480	230 400
346	119 716	391	152 881	436	190 096	481	281 361
347	120 409	392	153 664	437	190 969	482	232 324
348	121 104	393	154 449	438	191 844	483	233 389
349	121 801	394	155 236	439	192 721	484	234 256
350	122 500	395	156 025	440	193 600	485	235 225
351	123 201	396	156 816	441	194 481	486	236 196
352	123 904	397	157 609	442	195 364	487	237 169
353	124 609	398	158 404	443	194 249	488	238 144
354	125 316	399	159 201	444	197 136	489	239 121
355	126 025	400	160 000	445	198 025	490	240 100
356	126 736	401	160 801	446	198 916	491	241 081
357	127 449	402	161 604	447	199 809	492	242 064
358	128 164	403	162 409	448	200 704	493	243 049
359	128 881	404	163 216	449	201 601	494	244 036
360	120 600	405	164 025	450	202 500	495	245 025
361	130 321	406	164 836	451	203 401	496	246 016
362	131 044	407	165 649	452	204 304	497	247 009
363	131 769	408	166 464	453	205 209	498	248 004
364	132 496	409	167 281	454	206 116	499	249 001
365	133 225	410	168 100	455	207 025	500	250 000

Appendix C

Conversion Table

AMERICAN STANDARD TYPE SYSTEM[1]
Type Point Size in Equivalent mm and Inches

Point	*mm*	*Inches*	*Point*	*mm*	*Inches*
5.00	1.75730	0.06918	21.50	7.55639	0.29750
5.50	1.93303	0.07610	22.00	7.73212	0.30441
6.00	2.10876	0.08302	22.50	7.90785	0.31133
6.50	2.28449	0.08994	23.00	8.08358	0.31825
7.00	2.46022	0.09686	23.50	8.25931	0.32517
7.50	2.63595	0.10378	24.00	8.43503	0.33209
8.00	2.81168	0.11070	24.50	8.61076	0.33901
8.50	2.98741	0.11761	25.00	8.78649	0.34592
9.00	3.16314	0.12453	25.50	8.96222	0.35284
9.50	3.33887	0.13145	26.00	9.13795	0.35976
10.00	3.51460	0.13837	26.50	9.31368	0.36668
10.50	3.69033	0.14529	27.00	9.48941	0.37360
11.00	3.86606	0.15221	27.50	9.66514	0.38052
11.50	4.04179	0.15913	28.00	9.84087	0.38744
12.00	4.21752	0.16604	28.50	10.01660	0.39435
12.50	4.39325	0.17296	29.00	10.19233	0.40127
13.00	4.56898	0.17988	29.50	10.36806	0.40819
13.50	4.74471	0.18680	30.00	10.54379	0.41511
14.00	4.92044	0.19372	30.50	10.71952	0.42203
14.50	5.09617	0.20064	31.00	10.89525	0.42895
15.00	5.27190	0.20755	31.50	11.07098	0.43587
15.50	5.44763	0.21447	32.00	11.24671	0.44278
16.00	5.62336	0.22139	32.50	11.42244	0.44970
16.50	5.79909	0.22831	33.00	11.59817	0.45662
17.00	5.97482	0.23523	33.50	11.77390	0.46354
17.50	6.15055	0.24215	34.00	11.94963	0.47046
18.00	6.32628	0.24907	34.50	12.12536	0.47738
18.50	6.50201	0.25598	35.00	12.30109	0.48429
19.00	6.67774	0.26290	35.50	12.47682	0.49121
19.50	6.85347	0.26982	36.00	12.65255	0.49813
20.00	7.02920	0.27674	36.50	12.82828	0.50505
20.50	7.20493	0.28366	37.00	13.00401	0.51197
21.00	7.38066	0.29058	37.50	13.17974	0.51889

[1]Table from Dr. John Kethley, Associate Curator and Division Head, Insects, Department of Zoology, Field Museum of Natural History, Chicago.

AMERICAN STANDARD TYPE SYSTEM
Type Point Size in Equivalent mm and Inches

Point	mm	Inches	Point	mm	Inches
38.00	13.35547	0.52581	60.00	21.08759	0.83022
38.50	13.53120	0.53272	60.50	21.26332	0.83714
39.00	13.70693	0.53964	61.00	21.43905	0.84406
39.50	13.88266	0.54656	61.50	21.61478	0.85098
40.00	14.05839	0.55348	62.00	21.79051	0.85789
40.50	14.23412	0.56040	62.50	21.96624	0.86481
41.00	14.40985	0.56732	63.00	22.14197	0.87173
41.50	14.58558	0.57424	63.50	22.31770	0.87865
42.00	14.76131	0.58115	64.00	22.49343	0.88557
42.50	14.93704	0.58807	64.50	22.66916	0.89249
43.00	15.11277	0.59499	65.00	22.84489	0.89940
43.50	15.28850	0.60191	65.50	23.02062	0.90632
44.00	15.46423	0.60883	66.00	23.19635	0.91324
44.50	15.63996	0.61575	66.50	23.37208	0.92016
45.00	15.81569	0.62266	67.00	23.54781	0.92708
45.50	15.99142	0.62958	67.50	23.72354	0.93400
46.00	16.16715	0.63650	68.00	23.89927	0.94092
46.50	16.34288	0.64342	68.50	24.07500	0.94783
47.00	16.51861	0.65034	69.00	24.25072	0.95475
47.50	16.69434	0.65726	69.50	24.42645	0.96167
48.00	16.87007	0.66418	70.00	24.60218	0.96859
48.50	17.04580	0.67109	70.50	24.77791	0.97551
49.00	17.22153	0.67801	71.00	24.95365	0.98243
49.50	17.39726	0.68493	71.50	25.12938	0.98935
50.00	17.57299	0.69185	72.00	25.30511	0.99626
50.50	17.74872	0.69877	72.50	25.48083	1.00318
51.00	17.92445	0.70569	73.00	25.65656	1.01010
51.50	18.10018	0.71261	73.50	25.83229	1.01702
52.00	18.27591	0.71952	74.00	26.00802	1.02394
52.50	18.45164	0.72644	74.50	26.18375	1.03086
53.00	18.62737	0.73336	75.00	26.35948	1.03777
53.50	18.80310	0.74028	75.50	26.53521	1.04469
54.00	18.97883	0.74720	76.00	26.71094	1.05161
54.50	19.15456	0.75412	76.50	26.88667	1.05853
55.00	19.33029	0.76103	77.00	27.06240	1.06545
55.50	19.50602	0.76795	77.50	27.23813	1.07237
56.00	19.68175	0.77487	78.00	27.41386	1.07929
56.50	19.85748	0.78179	78.50	27.58959	1.08620
57.00	20.03321	0.78871	79.00	27.76532	1.09312
57.50	20.20894	0.79563	79.50	27.94105	1.10004
58.00	20.38467	0.80255	80.00	28.11678	1.10696
58.50	20.56040	0.80946	80.50	28.29251	1.11388
59.00	20.73613	0.81638	81.00	28.46824	1.12080
59.50	20.91186	0.82330	81.50	28.64397	1.12772

AMERICAN STANDARD TYPE SYSTEM

Type Point Size in Equivalent mm and Inches

Point	mm	Inches	Point	mm	Inches
82.00	28.81970	1.13463	104.00	36.55182	1.43905
82.50	28.99543	1.14155	104.50	36.72755	1.44597
83.00	29.17116	1.14847	105.00	36.90328	1.45288
83.50	29.34689	1.15539	105.50	37.07901	1.45980
84.00	29.52262	1.16231	106.00	37.25474	1.46672
84.50	29.69835	1.16923	106.50	37.43047	1.47364
85.00	29.87408	1.17614	107.00	37.60620	1.48056
85.50	30.04981	1.18306	107.50	37.78193	1.48748
86.00	30.22554	1.18998	108.00	37.95766	1.49440
86.50	30.40127	1.19690	108.50	38.13338	1.50131
87.00	30.57700	1.20382	109.00	38.30912	1.50823
87.50	30.75273	1.21074	109.50	38.48485	1.51515
88.00	30.92846	1.21766	110.00	38.66058	1.52207
88.50	31.10419	1.22457	110.50	38.83631	1.52899
89.00	31.27992	1.23149	111.00	39.01204	1.53591
89.50	31.45565	1.23841	111.50	39.18777	1.54283
90.00	31.63138	1.24533	112.00	39.36349	1.54974
90.50	31.80711	1.25225	112.50	39.53923	1.55666
91.00	31.98284	1.25917	113.00	39.71495	1.56358
91.50	32.15857	1.26609	113.50	39.89069	1.57050
92.00	32.33430	1.27300	114.00	40.06641	1.57742
92.50	32.51003	1.27992	114.50	40.24215	1.58434
93.00	32.68576	1.28684	115.00	40.41787	1.59125
93.50	32.86149	1.29376	115.50	40.59361	1.59817
94.00	33.03722	1.30068	116.00	40.76934	1.60509
94.50	33.21295	1.30760	116.50	40.94506	1.61201
95.00	33.38868	1.31451	117.00	41.12080	1.61893
95.50	33.56441	1.32143	117.50	41.29652	1.62585
96.00	33.74014	1.32835	118.00	41.47226	1.63277
96.50	33.91587	1.33527	118.50	41.64798	1.63968
97.00	34.09160	1.34219	119.00	41.82372	1.64660
97.50	34.26733	1.34911	119.50	41.99944	1.65352
98.00	34.44306	1.35603	120.00	42.17517	1.66044
98.50	34.61879	1.36294	120.50	42.35090	1.66736
99.00	34.79452	1.36986	121.00	42.52663	1.67428
99.50	34.97025	1.37678	121.50	42.70236	1.68120
100.00	35.14598	1.38370	122.00	42.87809	1.68811
100.50	35.32171	1.39062	122.50	43.05383	1.69503
101.00	35.49744	1.39754	123.00	43.22955	1.70195
101.50	35.67317	1.40446	123.50	43.40528	1.70887
102.00	35.84890	1.41137	124.00	43.58101	1.71579
102.50	36.02463	1.41829	124.50	43.75674	1.72271
103.00	36.20036	1.42521	125.00	43.93247	1.72962
103.50	36.37609	1.43213	125.50	44.10820	1.73654

AMERICAN STANDARD TYPE SYSTEM
Type Point Size in Equivalent mm and Inches

Point	*mm*	*Inches*	*Point*	*mm*	*Inches*
126.00	44.28393	1.74346	135.50	47.62280	1.87491
126.50	44.45966	1.75038	136.00	47.79853	1.88183
127.00	44.63539	1.75730	136.50	47.97426	1.88875
127.50	44.81112	1.76422	137.00	48.14999	1.89567
128.00	44.98685	1.77114	137.50	48.32572	1.90259
128.50	45.16258	1.77805	138.00	48.50145	1.90951
129.00	45.33831	1.78497	138.50	48.67718	1.91642
129.50	45.51404	1.79189	139.00	48.85291	1.92334
130.00	45.68977	1.79881	139.50	49.02864	1.93026
130.50	45.86550	1.80573	140.00	49.20437	1.93718
131.00	46.04123	1.81265	140.50	49.38010	1.94410
131.50	46.21696	1.81957	141.00	49.55583	1.95102
132.00	46.39269	1.82648	141.50	49.73156	1.95794
132.50	46.56842	1.83340	142.00	49.90729	1.96485
133.00	46.74415	1.84032	142.50	50.08302	1.97177
133.50	46.91988	1.84724	143.00	50.25875	1.97869
134.00	47.09561	1.85416	143.50	50.43448	1.98561
134.50	47.27134	1.86108	144.00	50.61021	1.99253
135.00	47.44707	1.86799			

Conversion Table

EUROPEAN STANDARD TYPE SYSTEM[2]
Type Point Size in equivalent mm and inches

Point	mm	inches	Point	mm	inches
5.00	1.32292	0.05208	12.00	3.17500	0.12500
5.50	1.45521	0.05729	12.50	3.30729	0.13021
6.00	1.58750	0.06250	13.00	3.43958	0.13542
6.50	1.71979	0.06771	13.50	3.57187	0.14062
7.00	1.85208	0.07292	14.00	3.70417	0.14583
7.50	1.98437	0.07812	14.50	3.83646	0.15104
8.00	2.11667	0.08333	15.00	3.96875	0.15625
8.50	2.24896	0.08854	15.50	4.10104	0.16146
9.00	2.38125	0.09375	16.00	4.23333	0.16667
9.50	2.51354	0.09896	16.50	4.36562	0.17187
10.00	2.64583	0.10417	17.00	4.49792	0.17708
10.50	2.77812	0.10937	17.50	4.63021	0.18229
11.00	2.91042	0.11458	18.00	4.76250	0.18750
11.50	3.04271	0.11979	18.50	4.89479	0.19271

[2]Table from Dr. John Kethley, Associate Curator and Division Head, Insects, Department of Zoology, Field Museum of Natural History, Chicago.

EUROPEAN STANDARD TYPE SYSTEM
Type Point Size in equivalent mm and inches

Point	mm	inches	Point	mm	inches
19.00	5.02708	0.19792	41.50	10.98021	0.43229
19.50	5.15937	0.20312	42.00	11.11250	0.43750
20.00	5.29167	0.20833	42.50	11.24479	0.44271
20.50	5.42396	0.21354	43.00	11.37708	0.44791
21.00	5.55625	0.21875	43.50	11.50937	0.45312
21.50	5.68854	0.22396	44.00	11.64166	0.45833
22.00	5.82083	0.22917	44.50	11.77396	0.46354
22.50	5.95312	0.23437	45.00	11.90625	0.46875
23.00	6.08542	0.23958	45.50	12.03854	0.47396
23.50	6.21771	0.24479	46.00	12.17083	0.47916
24.00	6.35000	0.25000	46.50	12.30312	0.48437
24.50	6.48229	0.25521	47.00	12.43541	0.48958
25.00	6.61458	0.26041	47.50	12.56771	0.49479
25.50	6.74687	0.26562	48.00	12.70000	0.50000
26.00	6.87917	0.27083	48.50	12.83229	0.50521
26.50	7.01146	0.27604	49.00	12.96458	0.51041
27.00	7.14375	0.28125	49.50	13.09687	0.51562
27.50	7.27604	0.28646	50.00	13.22916	0.52083
28.00	7.40833	0.29166	50.50	13.36146	0.52604
28.50	7.54062	0.29687	51.00	13.49375	0.53125
29.00	7.67292	0.30208	51.50	13.62604	0.53645
29.50	7.80521	0.30729	52.00	13.75833	0.54166
30.00	7.93750	0.31250	52.50	13.89062	0.54687
30.50	8.06979	0.31771	53.00	14.02291	0.55208
31.00	8.20208	0.32291	53.50	14.15521	0.55729
31.50	8.33437	0.32812	54.00	14.28750	0.56250
32.00	8.46667	0.33333	54.50	14.41979	0.56770
32.50	8.59896	0.33854	55.00	14.55208	0.57291
33.00	8.73125	0.34375	55.50	14.68437	0.57812
33.50	8.86354	0.34896	56.00	14.81666	0.58333
34.00	8.99583	0.35416	56.50	14.94896	0.58854
34.50	9.12812	0.35937	57.00	15.08125	0.59375
35.00	9.26042	0.36458	57.50	15.21354	0.59895
35.50	9.39271	0.36979	58.00	15.34583	0.60416
36.00	9.52500	0.37500	58.50	15.47812	0.60937
36.50	9.65729	0.38021	59.00	15.61041	0.61458
37.00	9.78958	0.38541	59.50	15.74271	0.61979
37.50	9.92187	0.39062	60.00	15.87500	0.62500
38.00	10.05416	0.39583	60.50	16.00729	0.63020
38.50	10.18646	0.40104	61.00	16.13958	0.63541
39.00	10.31875	0.40625	61.50	16.27187	0.64062
39.50	10.45104	0.41146	62.00	16.40416	0.64583
40.00	10.58333	0.41666	62.50	16.53646	0.65104
40.50	10.71562	0.42187	63.00	16.66875	0.65625
41.00	10.84791	0.42708	63.50	16.80104	0.66145

EUROPEAN STANDARD TYPE SYSTEM
Type Point Size in equivalent mm and inches

Point	mm	inches	Point	mm	inches
64.00	16.93333	0.66666	86.50	22.88645	0.90104
64.50	17.06562	0.67187	87.00	23.01875	0.90624
65.00	17.19791	0.67708	87.50	23.15104	0.91145
65.50	17.33021	0.68229	88.00	23.28333	0.91666
66.00	17.46250	0.68750	88.50	23.41562	0.92187
66.50	17.59479	0.69270	89.00	23.54791	0.92708
67.00	17.72708	0.69791	89.50	23.68020	0.93229
67.50	17.85937	0.70312	90.00	23.81250	0.93749
68.00	17.99166	0.70833	90.50	23.94479	0.94270
68.50	18.12395	0.71354	91.00	24.07708	0.94791
69.00	18.25625	0.71875	91.50	24.20937	0.95312
69.50	18.38854	0.72395	92.00	24.34166	0.95833
70.00	18.52083	0.72916	92.50	24.47396	0.96354
70.50	18.65312	0.73437	93.00	24.60625	0.96874
71.00	18.78541	0.73958	93.50	24.73854	0.97395
71.50	18.91771	0.74479	94.00	24.87083	0.97916
72.00	19.05000	0.75000	94.50	25.00312	0.98437
72.50	19.18229	0.75520	95.00	25.13541	0.98958
73.00	19.31458	0.76041	95.50	25.26770	0.99479
73.50	19.44687	0.76562	96.00	25.40000	0.99999
74.00	19.57916	0.77083	96.50	25.53229	1.00520
74.50	19.71145	0.77604	97.00	25.66458	1.01041
75.00	19.84375	0.78124	97.50	25.79687	1.01562
75.50	19.97604	0.78645	98.00	25.92916	1.02083
76.00	20.10833	0.79166	98.50	26.06145	1.02604
76.50	20.24062	0.79687	99.00	26.19375	1.03124
77.00	20.37291	0.80208	99.50	26.32604	1.03645
77.50	20.50521	0.80729	100.00	26.45833	1.04166
78.00	20.63750	0.81249	100.50	26.59062	1.04687
78.50	20.76979	0.81770	101.00	26.72291	1.05208
79.00	20.90208	0.82291	101.50	26.85520	1.05728
79.50	21.03437	0.82812	102.00	26.98750	1.06249
80.00	21.16666	0.83333	102.50	27.11979	1.06770
80.50	21.29895	0.83854	103.00	27.25208	1.07291
81.00	21.43125	0.84374	103.50	27.38437	1.07812
81.50	21.56354	0.84895	104.00	27.51666	1.08333
82.00	21.69583	0.85416	104.50	27.64895	1.08853
82.50	21.82812	0.85937	105.00	27.78125	1.09374
83.00	21.96041	0.86458	105.50	27.91354	1.09895
83.50	22.09270	0.86979	106.00	28.04583	1.10416
84.00	22.22500	0.87499	106.50	28.17812	1.10937
84.50	22.35729	0.88020	107.00	28.31041	1.11458
85.00	22.48958	0.88541	107.50	28.44270	1.11978
85.50	22.62187	0.89062	108.00	28.57500	1.12499
86.00	22.75416	0.89583	108.50	28.70729	1.13020

EUROPEAN STANDARD TYPE SYSTEM
Type Point Size in equivalent mm and inches

Point	mm	inches	Point	mm	inches
109.00	28.83958	1.13541	127.00	33.60208	1.32291
109.50	28.97187	1.14062	127.50	33.73437	1.32812
110.00	29.10416	1.14583	128.00	33.86666	1.33332
110.50	29.23645	1.15103	128.50	33.99895	1.33853
111.00	29.36875	1.15624	129.00	34.13124	1.34374
111.50	29.50104	1.16145	129.50	34.26353	1.34895
112.00	29.63333	1.16666	130.00	34.39583	1.35416
112.50	29.76562	1.17187	130.50	34.52812	1.35937
113.00	29.89791	1.17708	131.00	34.66041	1.36457
113.50	30.03020	1.18228	131.50	34.79270	1.36978
114.00	30.16249	1.18749	132.00	34.92500	1.37499
114.50	30.29479	1.19270	132.50	35.05729	1.38020
115.00	30.42708	1.19791	133.00	35.18958	1.38541
115.50	30.55937	1.20312	133.50	35.32187	1.39062
116.00	30.69166	1.20833	134.00	35.45416	1.39582
116.50	30.82395	1.21353	134.50	35.58645	1.40103
117.00	30.95625	1.21874	135.00	35.71874	1.40624
117.50	31.08854	1.22396	135.50	35.85104	1.41145
118.00	31.22083	1.22916	136.00	35.98333	1.41666
118.50	31.35312	1.23437	136.50	36.11562	1.42187
119.00	31.48541	1.23958	137.00	36.24791	1.42707
119.50	31.61770	1.24478	137.50	36.38020	1.43228
120.00	31.74999	1.24999	138.00	36.51249	1.43749
120.50	31.88229	1.25520	138.50	36.64479	1.44270
121.00	32.01458	1.26041	139.00	36.77708	1.44791
121.50	32.14687	1.26562	139.50	36.90937	1.45312
122.00	32.27916	1.27083	140.00	37.04166	1.45832
122.50	32.41145	1.27603	140.50	37.17395	1.46353
123.00	32.54374	1.28124	141.00	37.30624	1.46874
123.50	32.67604	1.28645	141.50	37.43853	1.47395
124.00	32.80833	1.29166	142.00	37.57083	1.47916
124.50	32.94062	1.29687	142.50	37.70312	1.48437
125.00	33.07291	1.30207	143.00	37.83541	1.48957
125.50	33.20520	1.30728	143.50	37.96770	1.49478
126.00	33.33749	1.31249	144.00	38.09999	1.49999
126.50	33.46978	1.31770			

Bibliography

WHERE should you look for more information? The list of suggested readings and reference materials can be considerably expanded by looking up the bibliographies in other books. For a wealth of written and visually transmitted information you will have to look up some of the the bibliographical notes. Do not give up; try to see at least some of them. The information pertaining to techniques used in nonapplied areas of painting and drawing are useful in the field of scientific illustration. From a technical point of view, both the nonapplied and applied areas are the same, differentiated only by their purpose and informational context. For that reason a number of descriptions may overlap. The history of scientific illustration reinforced with selection taken from bibliographies of natural history books will give you an adventure in reading combining facts with beautiful illustrations. Tools are always of importance and should be taken as seriously as drawing and painting materials. All reference books pertaining to safety, anatomy, and terminology are more than just a compilation of information—they serve the need for a stable base upon which the structure of knowledge will be built. For convenience, the bibliography is divided into separate groupings:

1. *Scientific Illustration—Techniques*. The primary objective, an explanation of scientific illustration, is reinforced by practicalities. Some bibliographical notes pertaining to scientific photography and technical illustration are included.

2. *Scientific Illustration—History*. Descriptions of natural history illustrators and their work will give you an incentive for improvements. Reading the biographies and admiring the technical proficiency of artists from the past may even prompt you to organize your own expedition.

3. *Drawing and Painting—Techniques*. How to paint and how to draw. The secrets of the trade are presented by a variety of artists. Descriptions will introduce numerous techniques and approaches used for representation of nature.

Figure B-1. *Centurio senex*. Old male. Wash on bristol board. Drawing by unknown artist, Chicago Museum of Natural History (Field Museum of Natural History), Chicago. Published in Daniel Giruad Elliot, "The Land and Sea Mammals of Middle America and the West Indies," by Chicago Museum of Natural History (Field Museum of Natural History) (1904) Chicago. Drawn between 1900 and 1904.

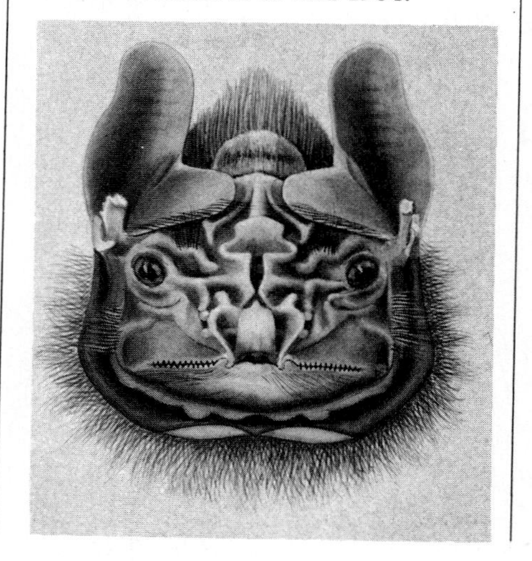

4. *Drawing and Painting—Materials*. Explanations and use of materials from the past and present. What to use and how to use it correctly.

5. *Preparation for Production*. The principles of layout, key line, and the procedures needed for marking up illustrations for reproduction. You should know all of these.

6. *Microscope and Drawing Tubes*. Introduction to construction, operation, and preservation of optical equipment.

7. *Anatomy for the Artist*. Explanatory, richly illustrated texts presenting the anatomical structures of the human and animal kingdom. The main subjects are how muscles work and where bones are.

8. *Safety*. Lots of chemicals are dangerous, and your safety comes first. Knowing the facts will help you in your work.

10. *Bibliographies—Natural History*. A list of books to be located in the library representing the natural history of the past, all richly illustrated. Numerous positions may be difficult to locate because of their rarity; nevertheless, if you want to see the best of natural history illustrations, try to locate some bibliographical notes by visiting university libraries. Besides the quality of illustrative materials, a wealth of technical information pertaining to old printing methods and media can be obtained through careful analytical observation of the illustrations.

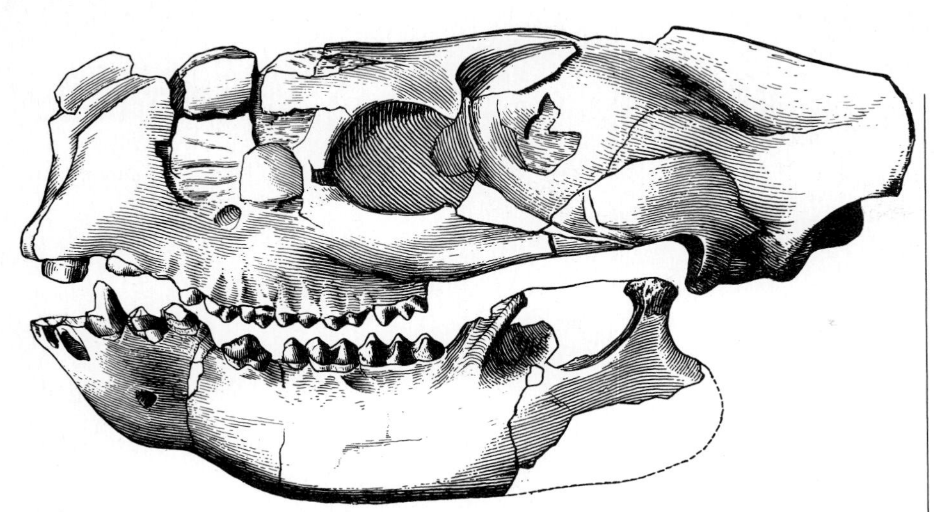

Figure B-2. *Protoreodon pumilus*, skull and jaw, PT. Cat No. 11891 Y.M.P., first version. Crow quill on bristol board. Drawing by Joy Stilson Parr, Peabody Museum, Yale University, New Haven, Connecticut. Second version published in Malcolm R. Thorpe, "The Merycoidodontidae, an Extinct Group of Ruminant Mammals," *Memoirs of the Peabody Museum of Natural History*, Vol. III, Part 4, by Peabody Museum of Yale University, New Haven, 1937. Drawn in 1935.

SCIENTIFIC ILLUSTRATION—TECHNIQUES

Allen, Arly. *Steps Toward Better Scientific Illustration*. Lawrence, Kansas: Allen Press Inc., 1977.

Blocker, Alfred A. *Handbook for Scientific Photography*. San Francisco: W. H. Freeman & Co., 1977.

Blunt, Wilfrid. *The Art of Botanical Illustration*. London: Collins Sons & Co., Ltd., 1967.

Broadribb, C. *Drawing Archeological Finds for Publication*. London: Baker, 1970.

Bryant, V. M., and R. K. Haltz. "*A Guide to the Drafting of Archeological Maps*." *TASP* 36 (1965): 269–85.

Cannon, H. G. *A Method of Illustrating Zoological Papers*. London: The Association of British Zoologists, 1936.

Clarke, Carl D. *Illustration—Its Technique and Application to the Science*. Butler, Maryland: Standard Arts Press, 1949.

Coineau, Yves. *Comment Réaliser vos Dessins Scientifiques*. Paris: Gauthier Villars, 1982.

Bethune, James D. *Technical Illustration*. New York: John Wiley and Sons, Inc., 1983.

Bulls, Robert L. *Technical Illustration*. Los Altos, Calif.: William Kaufmann, Inc., 1983.

Colbert, Edwin H., and Chester Taraka. *Illustration of Fossil Vertebrates, Medical and Biological Illustration*. London: 1960.

Conlon, V. M. *Camera Techniques in Archeology*. New York: St. Martins Press, 1973.

Cross, Louise Montgomery. *The Preparation of Medical Literature*. Philadelphia: J. B. Lippincott Co., 1959.

de Roziere, Francois Michel. *Discours sur la Represéntation des Roches de l'Egypte et de l'Arabie par la Gravure, et sur son Utilité dans les Arts et dan la Géologie*. Paris, 1812.

Dezart, Louis. *Drawing for Publication: A Manual for Technical Illustrators*. New York: State Mutual Bank, 1981.

———. *Drawing for Publication*. New York: Nichols Publishing, 1981.

Dillon, Brian D. "The Student's Guide to Archeological Illustrating." In *Tools*, Vol. 1. Los Angeles: Institute of Archeology, University of California, 1981.

Downey, John, and James Kelly. *Techniques and Exercises with the Pen in Biological Illustration*. Cedar Falls, Iowa: Iowa State University Press, 1982.

Gibbey, Joseph C. *Technical Illustration*. Chicago: American Technical Society, 1962.

Giesecke, Frederic E. *Technical Drawing*. New York: Macmillan Publishing Co., 1980.

Grinsell, Leslie and Phillip Rantz. *Preparation of Archeological Reports*. New York: St. Martins Press, Inc., 1974.

Hope-Taylor, B. "Archeological Draughtmanship." Pt. II, *Antiquity* 40 (1966): 181–89.

Ives, R. L. "Line Drawings from Unsatisfactory Photographs." American Anthropologist 13 (1948): 23.

Jastrzębski, Zbigniew, T. "Introduction to the Series on Technique of Making Fish Illustrations." *Environmental Biology of Fishes* (The Hague, Netherlands) 11, no. 1 (1984):15–16.

———. "Technique of Making Fish Illustration (1)." *Environmental Biology of Fishes* (The Hague, Netherlands) 11, no. 1, (1984): 17–18.

———. "Technique of Making Fish Illustration (2)." *Environmental Biology of Fishes* (The Hague, Netherlands) 11, no. 1, (1984): 19–20.

Joukowsky, Martha. *A Complete Manual of Field Archeology*. Englewood Cliffs, N.J.: Prentice-Hall, Inc., 1980.

Lapage, Geoffrey. *Art and the Scientist*. Bristol, England: John Wright and Sons Ltd., 1961.

Lindsey, Casimir, C. "Fish Illustrations: How and Why." *Environmental Biology of Fishes* (The Hague, Netherlands) 11, no. 1 (1984): 3–14.

Lynch, George. "Casein Colors for Medical Art." *Journal of the Association of Medical Illustrators* 13, (1961): 43–45.

Loechel, William E. *Medical Illustration: A Guide for the Doctor-Author and Exhibitor*. Springfield, Ill.: C. C. Thomas, 1964.

Mascaro, David. "On the Use of Color in Medical Illustration." *Journal of Biocommunications* 9, no. 1 (1982): 10–17.

McLarty, Margaret. *Illustrating Medicine and Surgery*. Baltimore: Williams and Wilkins Co., 1960.

Papp, Charles. *A Manual of Scientific Illustration*. Sacramento: American Visual Aid Books, 1976.

———. *Scientific Illustration, Theory and Practice*. Dubuque, Iowa: William C. Brown Co., 1968.

Rajchel, Zbigniew. *Reconstrukcja Gtowy Scyty z Nowositki*. Materiaty i Prace Antropologiczne Nr. 63, Miscelanea VII. Wroctaw: Polska Akademia Nauk, Zaktad Antropologji, 1962. [Sculptural reconstruction of facial features based on skeletal remains.]

Razowski, J. and J. Swiecimski. *Illustration of Genital Armatures of Insects of Scientific Publication*. Acta Zoologica Cracoviensia, 1971. Published in English for the Smithsonian Institution and the Natural Science Foundation, Washington, D.C., by the Foreign Scientific Publication Department of the National Center for Scientific Technical and Economic Information. Warsaw, Poland (1975): 267–89. (Available from the U.S. Department of Commerce, National Technical Information Service, Springfield, Va. 22161.)

Ridgway, John Livesy. *Scientific Illustration*. Palo Alto, Calif.: Stanford University Press, 1938.

Rivard, S. J. "Technical Illustrations Applied to Archeology" *Massachusetts Archaeology Society Bulletin* 25 (1964). 44–45.

Schwarz, William R. "Wash Drawings for Medical Art." *The Journal of Biocommunication* 3 no. 1 (1976): 25–31.

Staniland, L. N. *The Principles of Line Illustration with Emphasis on the Requirements of Biological and Other Scientific Workers*. Cambridge: Harvard University Press, 1953.

Weaver, Norman. *How to Draw Insects*. London, New York: Studio Pub., 1958.

West, Keith. *How to Draw Plants: The Art of Botanical Illustration*. New York: Watson-Guptill Publishing, 1983.

Wood, Phillis. *Scientific Illustration: A Guide to Biological, Zoological and Medical Rendering Technique, Design, Printing and Display*. Florence, Kentucky: Von Nostrand, Reinhold, 1979.

Young, K. S. "A Technique for Illustrating Pottery Designs." *American Anthropologist* 35 (1970): 488–91.

Zweifel, Frances W. *Handbook of Biological Illustration*. Chicago: University of Chicago Press, 1962.

SCIENTIFIC ILLUSTRATION—HISTORY

Blum, Ann and Sarah Landry. "In Loving Detail." *Harvard* (May–June 1977): 38.

Blunt, Wilfrid. *The Art of Botanical Illustration*. London: Collins Sons and Co. Ltd., 1967.

———. *The Complete Naturalist: A Life of Linnaeus*. New York: Viking Press, 1971.

———. *Flower Books and Their Illustrators*. London: Cambridge University Press, 1950.

Bodleian Library. *Zoological Illustration*. Oxford: Bodleian Library, 1951.

Buchanan, Hanasyde. *Nature Into Art*. New York: W. H. Smith Publications, Inc., 1979.

Catalogue of Redouteana. Pittsburgh: Hunt Botanical Library, 1963.

Catalogue, 2nd International Exhibition of Botanical Art and Illustration. Pittsburgh: Hunt Botanical Library, 1968.

Catalogue, 4th International Exhibition of Botanical Art and Illustration. Pittsburgh: Hunt Botanical Library, 1977.

Dance, Peter S. *The Art of Natural History: Animal Illustrators and Their Work*. New York: Overlook Press, 1978.

Daniels, Gilbert S. *Artists from the Royal Botanic Gardens*. Pittsburgh: The Hunt Institute for Botanical Documentation, Carnegie-Mellon University, 1974.

Desmont, Ray. *Dictionary of British and Irish Botanists and Horticulturists Including Plant Collectors and Botanical Artists*. Totowa, N.J.: Roman and Littlefield, 1977.

Gunn, Mary. *Botanical Exploration of Southern Africa: An Illustrated History of Early Botanical Literature on the Cape Flora: Biographical Accounts of the Leading Plant Collectors and Their Activities in Southern Africa from the Days of the East India Company until Modern Times*. Cape Town: A. A. Balkema Publishing, 1977.

Gunthart-Maag, Lott. *Water Colors and Drawings*. Pittsburgh: Hunt Botanical Library, Carnegie-Mellon University, 1970.

Gourry, Desiri Pierre. *Mille et un Livres Botaniques*. Brussels: Arcade, 1973.

Herdeg, Walter, ed. *The Artist in the Service of Science*. Zurich: Graphis Press Corp., 1973.

Jackson, Christine Elizabeth. *Bird Illustrator: Some Artists in Early Lithography*. London: H. F. & G. Witherby, 1975.

King, Ronald. *Botanical Illustration*. New York: C. N. Potter, 1979.

Knight, David. *Zoological Illustrations: An Essay Towards a History of Printed Zoological Pictures*. Menlo Park, N.J.: Shoe String Press, Inc., 1977.

Lysaght, Averil. *The Book of Birds: Five Centuries of Bird Illustration*. London: Phaidon Press, Ltd. 1975.

Merian, M. S. *Leningrader Aquarelle*. Luzern, Switzerland: 1974.

Figure B-3. Artist concept of the Inter-planetary Monitoring Platform. Using Delta launch vehicle the Inter-planetary Monitoring Platform (IMP) circled the planet Earth carrying eleven experiments that studied solar terrestrial relationship by monitoring radiation in inter-planetary space, solar winds, and magnetic conditions. Tempera paint. National Aeronautic and Space Administration, Goddard Space Flight Center, Greenbelt, Maryland. Painted in 1966.

Peattie, D. C. Green Laurels. *The Lives and Achievements of the Great Naturalists*. New York: 1938.

Pierpont Morgan Library. *Flowers in Books and Drawings, ca. 940–1840*. New York: Pierpont Morgan Library, 1980.

Sweerts, Emanuel. *Early Floral Engraving*. New York: Dover Publications, 1976.

DRAWING AND PAINTING TECHNIQUES

Adams, Norman and Joe Singer. *Drawing Animals*. New York: Watson-Guptill Publishing, 1979.

Armfield, Maxwell. *A Manual of Tempera Paintings*. London: George Allen and Unwin, 1930.

Barlowe, Dorothea, and Sy Barlowe. *Illustrating Nature*. New York: Viking Press, 1982.

Berlye, Milton K. *How to Sell Your Artwork: A Complete Guide for Commercial and Fine Artists*. Englewood Cliffs, N.J.: Prentice-Hall, Inc., 1980.

Blake, Wendon. *The Drawing Book*. New York: Watson-Guptill Pub., 1980.

Billen, Edwin S. *Drawings on Scraper Board*. London: Sir I. Pitman, 1952.

Borgeson, Bet. *The Colored Pencil: Key Concepts for Handling the Medium*. New York: Watson-Guptill Pub., 1983.

Brooks, Leonard. *Casein Painting*. New York: Reinhold Publishing Co., 1961.

Calderon, Frank W. *Animal Painting and Anatomy*. New York: Dover Publications, 1975.

Cohn, Marjorie. *A Study of the Development of Watercolor*. Boston: Fogg Museum, 1977.

Chatfield, Hale. *Water Colors*. Gulfport, Fla.: Konglomerati Florida Foundation for Literature and the Book Arts, Inc. 1979.

Cheat, B. *An Artist's Notebook*. New York: Holt Reinhart Winston, 1979.

Church, Arthur H. *The Chemistry of Paints and Painting*. London: Seely, Service and Co., 1915.

Cutler, Merritt Dana. *How to Cut Drawings on Scratchboard*. New York: Watson-Guptill Pub., 1960.

D'Amelio, Joseph. *Perspective Drawing Handbook*. New York: Tudor Pub., Co., 1965.

Dehn, Adolf. *Water Color, Gouache, and Casein Painting*. New York: Studio Pub., Inc., 1957.

Derkatsh, Inessa. *Transparent Watercolor: Painting Methods and Materials*. Englewood Cliffs, N.J.: Prentice-Hall, Inc., 1980.

Doerner, Max. *The Materials of the Artist and Their Use in Painting, with Notes on the Techniques of the Old Masters*. Translated by Eugen Neuhaus. New York: Harcourt, Brace and World, 1934.

Doust, Len A. *A Manual on Pastel Painting*. London: Frederic Warne and Co., Ltd., 1933.

Estlake, Charles. *Methods and Materials of Painting of the Great Schools and Masters*. New York: Dover Publishing, 1960.

FitzMaurice Mills, John. *Acrylic Painting*. London: Sir Isaac Pitman Sons, Ltd., 1965.

Friedlein, Ernst. *Tempera und Tempera Technik*. Munich: Callwey, 1906.

Gasser, Henry. *Casein Painting*. New York: Watson-Guptill Pub., 1969.

Goldsmith, Lawrence. *Watercolors Bold and Free*. New York: Watson-Guptill Pub., 1980.

Goldstein, Nathan. *Painting: Visual and Technical Fundamentals*. Englewood Cliffs, N.J.: Prentice-Hall, Inc., 1979.

Griffith, Thomas. *A Practical Guide for Beginning Painters*. Englewood Cliffs, N.J.: Prentice-Hall, Inc., 1980.

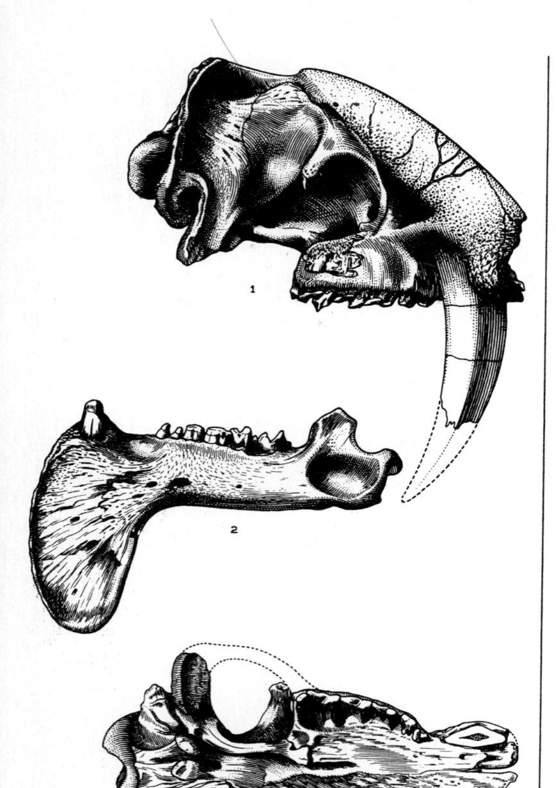

Figure B-4. *Thylacosmilus lentus*, holotype (1 and 3), *Thylacosmilus atrox*, paratype (2). Crow quill on three-ply Strathmore paper. Drawing by Carl F. Gronemann, Chicago Museum of Natural History (Field Museum of Natural History), Chicago. Published in Elmer S. Riggs, *Transactions of the American Philosophical Society*, New Series, Vol. XXIV, 1934, by The American Philosophical Society, Philadelphia. Drawn before 1934.

Guiteréz, José and Nicholas Roukes. *Painting with Acrylics*. New York: Watson-Guptill Pub., 1965.

Guptill, Arthur L. *Drawing with Pen and Ink*, New York, N.Y.: Reinhold Publ. Co., 1961.

Henley's 20th Century Book on Formulas, Processes, and Trade Secrets. Gardner D. Hiscox ed., Westerville, Ohio: Pubs. Unlimited, 1981.

Hewitt, Graily. *Lettering for Students and Craftsman*. Philadelphia: Lippincott, 1930.

Higgins Ink Co. *Techniques*. Brooklyn: Higgins Ink Co., 1959.

Hiler, Hilaire. *Notes on the Techniques of Painting*. New York: Oxford University Press, 1934.

Hill, Edward. *The Language of Drawing*. Englewood Cliffs, N.J.: Prentice-Hall, Inc., 1980.

Hoelscher, P., Clifford Spring, and Jerry Dobrovolny. *Graphics for Engineers*. New York: John Wiley and Sons, Inc., 1968.

James, J. *Perspective Drawing*. Englewood Cliffs, N.J.: Prentice-Hall, Inc., 1981.

Kautzky, Ted. *Ways with Watercolor*. New York: Reinhold Pub. Co., 1949.

Leslie, Clare Walker. *Nature Drawing: A Tool for Learning*. Englewood Cliffs, N.J.: Prentice-Hall, Inc., 1980.

Lockard, William K. *Drawing as a Means of Architecture*. Tucson: Pepper Publishing, 1977.

Maroger, Jacques. *The Secret Formulas and Techniques of the Masters*. New York: Hacker Art Books, 1979.

Maurello, Ralph, S. *The Complete Airbrush Book*. Secaucus, New Jersey: Amiel Pub. Co., 1954.

Medworth, Frank. *Perspective*. New York: Charles Scribner's Sons, 1937.

Mendelovitz, Daniel M. *Drawing*. Palo Alto, Calif.: Stanford University Press, 1980.

Moranz, John. *Mastery of Drawing*. Peterborough, N.H.: The Richard R. Smith Co., 1950.

Norton, Dora M. *Freehand Perspective and Sketching*. New York: Bridgman Pub., 1929.

Painting Techniques of the Masters. Dereward Lester Cooke, editor, New York: Watson-Guptill, 1972.

Pavey, Donald, editor. *Color*. Los Angeles: Knapp Communication Corp., 1980.

Pitz, Henry Clarence. *Pen, Brush and Ink*. New York: Watson-Guptill, 1949.

Reid, Charles. *Figure Painting in Watercolor*. New York: Watson-Guptill, 1972.

———. *Portrait Painting in Watercolor*. New York: Watson-Guptill, 1973.

———. *Flower Painting in Watercolor*. New York: Watson-Guptill, 1979.

Rowbotham, Thomas. *The Art of Landscape Painting in Water Colors*. London: Winsor & Newton.

Salemme, Lucia A. *Color Exercises for the Painter*. New York: Watson-Guptill Pub., 1979.

Simmons, Seymour III, and Mark S. A. Winer. *Drawing: The Creative Process*. Englewood Cliffs, N.J.: Prentice-Hall, Inc., 1980.

Simpson, Ian. *Drawing: Seeing and Observation*. New York: Van Nostrand Reinhold Pub., 1981.

Sweney, Fredric. *The Art of Painting Animals: A Beginning Artist's Guide to the Portrayal of Domestic Animals*. Englewood Cliffs, N.J.: Prentice-Hall, Inc., 1980.

Thoma, Martha. *Graphic Illustration: Tools & Techniques for Beginning Illustrators*. Englewood Cliffs, N.J.: Prentice-Hall, Inc., 1980.

Thompson, Daniel V., Jr. *The Practice of Tempera Painting*. New Haven, Conn.: Yale University Press, 1936; New York: Dover Publications, 1962.

Toney, Anthony. *Painting and Drawing: Discovering Your Own Visual Language*. Englewood Cliffs, N.J.: Prentice-Hall,, Inc., 1980.

Townsend, C. E., and S. F. Cleary. *Introductory Mechanical Drawing*. New York: John Wiley & Sons, Inc., 1930.

Tseng-Tseng Yu, Leslie. *Chinese Painting in Four Seasons: A Manual of Aesthetics & Techniques*. Englewood Cliffs, N.J.: Prentice-Hall, Inc., 1980.

Vickery, Robert and D. Cochrane. *New Techniques in Egg Tempera*. New York: Watson-Guptill Pub., 1973.

White, Gwen. *Perspective: A Guide for Artists, Architects and Designers*. New York: Watson-Guptill Pub., 1968.

Worth, Leslie. *Painting in Watercolors*. New York: Taplinger, 1982.

DRAWING AND PAINTING MATERIALS

Abels. *Painting: Materials and Methods*. Grosset Art Instruction Series, no. 28. New York: Grosset & Dunlap, Inc., 1982.

Bentley, Kenneth Walter. *The Natural Pigments*. The Chemistry of Natural Products, vol. 4. New York: Interscience Pub., 1980.

Ferris, Theodore N., and John P. Marchak, editors. *Paint Chemist*. Educational Research Council of America. St. Paul, Minn.: Changing Times Educational Service, 1977.

Gettens, Rutherford J., and George L. Stout. *Painting Materials: A Short Encyclopedia*. New York: Van Nostrand, 1942; New York: Dover Publications, 1965.

Harley, Rosamond D. *Artists' Pigments, c. 1600–1835. Study in English Documentary Sources*. New York: American Elsevier Pub. Co., 1970.

Jenson, Lawrence. *Synthetic Painting Media*. Englewood Cliffs, N.J.: Prentice-Hall, Inc., 1964.

Laurie, A. P. *The Painter's Methods and Materials*. Philadelphia: Lippincott, 1922; New York: Dover Publications, 1967.

Massey, Robert. *Formulas for Painters*. New York: Watson-Guptill Pub., 1967.

Mayer, Ralph. *The Artist's Handbook of Materials and Techniques*. New York: The Viking Press, 1982.

Paint and Powder Coatings. Cleveland: Predicast Inc., 1981.

Paint Testing Manual: Physical and Chemical Examination of Paints, Varnishes, Lacquers and Colors. Gardner & Sward, editors. 13th Edition. Philadelphia: American Society for Testing and Materials, 1973.

Patton, Temple C. *Paint Flow and Pigment Dispersion: A Rheological Approach to Coatings and Ink Technology*. New York: John Wiley and Sons, Inc., 1979.

The Pigments and Mediums of the Old Masters. London: Macmillan, 1914.

Robinson, J. S., editor. "Paint Additives: Development Since 1977." *Chemical Technology Review*, 200. Noyes Publishing, 1982.

Toch, Maximillan. *Chemistry and Technology of Paints*. New York: Van Nostrand, 1925.

Weber, Frederic W. *Artists' Pigments*. New York: Van Nostrand, 1923.

Whelte, Kurt. *Materials and Techniques of Painting*. New York: Van Nostrand, 1975.

PREPARATION FOR PRODUCTION

Adams, Michael, and David D. Faux. *Printing Technology*, 2nd ed. Belmont, Calif.: Wadsworth, Pub., 1982.

Brinkley, John. *Lettering Today*. New York: Reinhold Publishing Co., 1961.

Cox, Robert. *Printing Processes*. Van Nostrand Reinhold Pub., (date not set).

Croy, O. R. *Camera Copying and Reproduction*. Woburn, Mass.: Focal Press, 1964.

Donahue, Bud. *The Language of Layout*. Englewood Cliffs, N.J.: Prentice-Hall, Inc., 1978.

Graphic Master. Los Angeles: Dean Lem Associates, 1977.

Hurlburt, Allen. *Layout*. New York: Watson-Guptill Publishing, 1977.

Layout Procedures and Techniques. Reston, Va.: Reston Publishing Co., Inc., 1982.

Simon, Hilda. *Color in Reproduction: Theory and Techniques for Artists and Designers*. New York: Viking Press, 1980.

Stone, Bernard and Arthur Eckstein. *Preparing Art for Printing*. New York: Reinhold Publishing Co., 1965.

Figure B-5. *Aegipus monachus*, (Black vulture). Pen and ink. Drawing by Jaime Aviles, Estación Biológica de Doñana, Sevilla, Spain. Published in Javier Castroviejo Bolibar, ed., *Actas del Congreso de la Unión Internacional de Biológica de la Fauna Cinegetica* by Estación Biológica de Doñana, Sevilla. Drawn in 1981.

Wotzkow, Helen. *The Art of Hand Lettering: Its Mastery and Practice*. New York: Watson-Guptill Pub., 1952; New York: Dover Publications, 1967.

MICROSCOPES AND DRAWING TUBES

Bradbury, Savile. *The Evolution of the Microscope*. Oxford, N.Y.: Pergamon Press, 1967.

———. *Microscope Past and Present*. Oxford, N.Y.: Pergamon, 1969.

Barthes, Roland. *Camera Lucida: Reflections and Photography*. New York: Hill and Wang, 1982.

Burrells, W. *Microscope Techniques: A Comprehensive Handbook for General and Applied Microscopy*. New York: Halsted Press, 1978.

Carona, Philip B. *Microscope and How to Use It*. Houston: Gulf Pub., 1970.

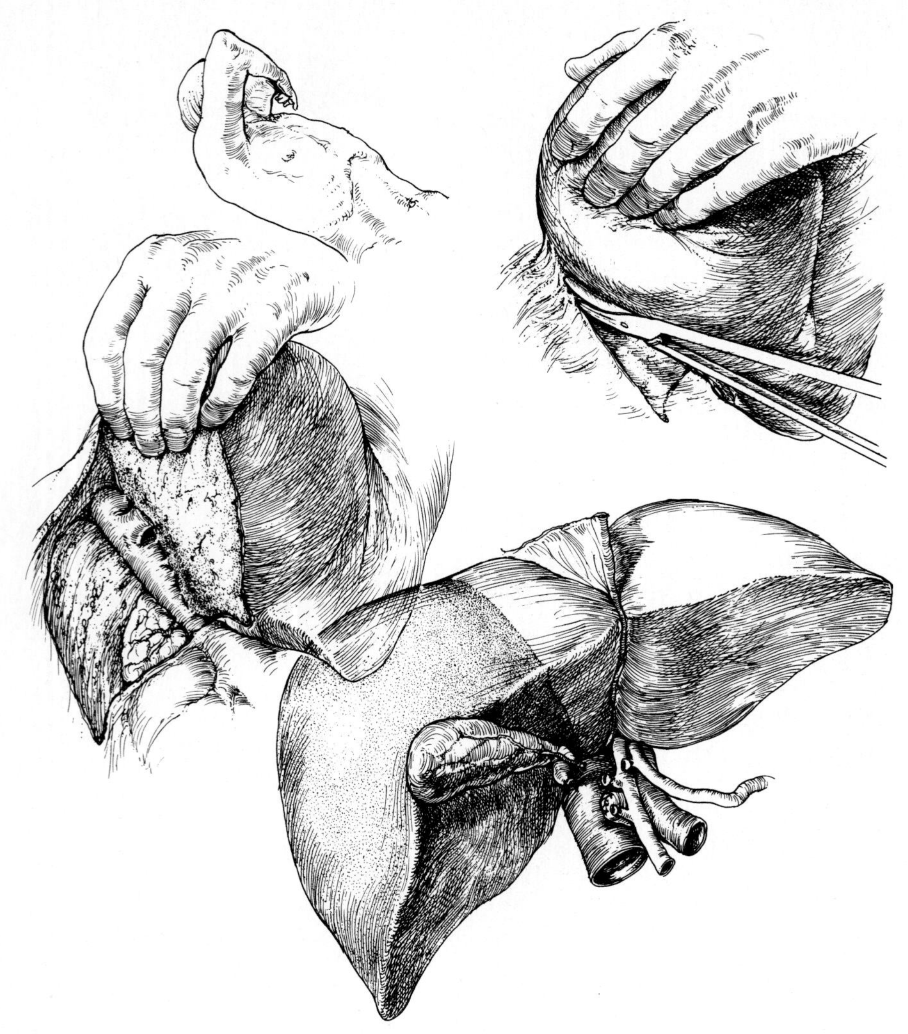

Figure B-6. Plate two of a series of three plates, demonstrating a surgical procedure for the removal of the right lobe of the liver. Pen (crow quill) and ink on Linekote board. Drawing by Alfred Teoli, Biocommunication Arts, University of Illinois at Chicago, Health Science Center, Chicago. To be published in Edward Beattie, M.D., and Steven Economou, M.D., *An Atlas of Advanced Surgical Techniques* by W. B. Saunders Co., Philadelphia. Drawn in 1983. In preparation.

Dumas, Maurice. *Scientific Instruments of 17th & 18th Centuries and Their Makers*. Edited and translated by Mary Holbrook. Humanities Pub., 1972.

Gage, Simon Henry. "Microscopy in America." Edited by Oscar W. Richards. East Lansing,Mich., *Translations of the American Microscopical Society* 83, no. 4 (1964).

Gray, Peter. *Handbook of Basic Microtechnique*, 2nd ed. New York: McGraw-Hill, 1958.

Gray, Peter, editor. *The Encyclopedia of Microscopy and Microtechnique*. New York: Van Nostrand Reinhold, 1973.

Grimstone, A. *A Guidebook to Microscopical Methods*. Cambridge, Mass.: University Press, 1972.

Hammond, John H. *Camera Obscura*. New York: New York State Mutual Book and Periodical, Ltd., 1981.

Hyat, M. *Principles and Techniques of Electron Microscopy*. New York: Van Nostrand Reinhold Co., 1970.

Johnes, Ruth McClurg. *Basic Microscopic Technics*. Chicago: University of Chicago Press, 1968.

Marmasse, Claude. *Microscopes and Their Uses*. New York: Gordon Press, 1980.

Richards, Oscar White. *The Effective Use and Proper Care of the Microscope*. Buffalo, N.Y.: American Optical Co., 1941.

Stehli, Georg. *Microscope and How to Use It*. New York: Dover Publications, 1970.

ANATOMY FOR THE ARTIST

Bell, Sir Charles. *Essays on the Anatomy of Expression in Painting*. London: Longman, Hurst, Rees and Orme, 1806.

Cunningham, Daniel John. *Cunningham's Textbook of Anatomy*, 11th ed., edited by G. J. Romanes. London: Oxford University Press, 1972.

Ellenberger, Wilhelm. *An Atlas of Animal Anatomy for Artists*. New York: Dover Publications, 1949.

Fritsch, Gustav Theodor. *Die Gestalt des Menschen*. Stuttgart: Heff, 1905.

Gaspard, Marcel. *Anatomie Comparative et Fonctionelle de la Musculature Masticatrice Chez les Carnivores*. Paris: Editions oln Museum, 1971.

Gray, Henry. *Anatomy, Descriptive and Surgical*. Edited by T. Pickering Pick and Robert Howden. Philadelphia: Running Press, 1974.

Jossic, Yvonne Francoise. *Anatomy of Animals*. Philadelphia: H. C. Perleberg, Albert A. Lampl, Successor, Publishers, 1946.

Knight, Charles Robert. *Animal Anatomy and Physiology for the Artist and Layman*. New York, London: Whittleseyhouse, McGraw-Hill Co, 1947.

Norton, Arthur Trehern. *Osteology; a Concise Description of Human Skeleton, Adapted for the Use of Students in Medicine*. London, 1874.

Nourse, Alan Edward. *The Body*. New York: Time Inc., 1964.

Perard, Victor Semon. *Anatomy and Drawing*. New York: Victor Perard Pub., 1942.

Figure B-7. *Neogyromitra gigas*. Watercolor. Painted by M. Charles Poluzzi, Switzerland, World Life Research Institute, Colton, California. Unpublished. Painted in 1970.

Pernkopf, Eduard. *Atlas of Topographical and Applied Human Anatomy.* Edited by Helmut Ferner. Philadelphia: W.B. Saunders and Co., 1963.

Putnam, Brenda. *Animal X-rays, a Skeleton Key to Comparative Anatomy.* New York: G. P. Putnam's Sons, 1947.

Seton, Ernst Thompson. *Studies in the Art Anatomy of Animals: Being a Brief Analysis of the Visible Forms of the More Familiar Mammals and Birds. Designed for the Use of Sculptors, Painters, Illustrators, Naturalists, and Taxidermists.* London: McMillan and Co., Ltd., 1896.

Sweney, Fredric. *The Art of Painting Animals: A Beginning Artist's Guide to the Portrayal of Domestic Animals*. Englewood Cliffs, N.J.: Prentice-Hall, Inc., 1980.

Weyl, Hermann. *Simmetry.* Princeton, N.J.: Princeton University Press, 1952.

Widema, George D. *The Johns Hopkins Atlas of Human Functional Anatomy.* Baltimore: Johns Hopkins University Press, 1981.

DICTIONARIES OF ART AND ANATOMY

The Condensed Chemical Dictionary. Revised by G. Hawley Gessner. 10th ed. New York: Van Nostrand Reinhold, 1981.

Donath, I. *Anatomical Dictionary with Nomenclature and Explanatory Notes*. Oxford: Pergamon Press, 1969.

Brown, Roland Wilbur. *Composition of Scientific Words*. Washington, D.C.: Smithsonian Institution Press, 1964.

Federation of Societies for Coatings Technology, edited by Definitions Committee. *Paint-Coatings Dictionary.* Philadelphia: Federation of Societies for Coatings Technology Publishers.

FitzMaurice Mills, John. *The Pergamon Dictionary of Art*. London: Pergamon Press, 1965.

Haggar, Reginaed George. *A Dictionary of Art Terms*. London: George Rainbird Ltd., 1962; New York: Hawthorn Books, Inc., 1962.

Figure B-8. *Puncia granatum*, common pomegranate. Watercolor. Painted by J. Fujishima, Japan, World Life Research Institute, Colton, California. Unpublished. Painted in 1968.

Maerz, J. J., and M. R. Paul. *A Dictionary of Color*. New York: McGraw-Hill Publishers, 1950.

Mayer, Ralph. *A Dictionary of Art Terms and Techniques*. New York: Thomas Y. Crowell Co., 1966; New York: Apollo Editions, 1975.

Steward, Jeffrey, R. *National Paint Dictionary.* Washington, D.C.: Steward Research Laboratory, 1948.

SAFETY

Fisher Safety Manual. Pittsburgh: Fisher Scientific Co., 1979.

Seegar, Nancy. *An Introductory Guide to the Safe Use of Materials*. Chicago: The School of the Art Institute of Chicago, 1981.

———. *A Ceramist's Guide to the Safe Use of Materials*. Chicago: The School of the Art Institute of Chicago, 1982.

———. *A Painter's Guide to the Safe Use of Materials*. Chicago: The School of the Art Institute of Chicago, 1982.

———. *A Printmaker's Guide to the Safe Use of Materials*. Chicago: The School of the Art Institute of Chicago, 1981.

NATURAL HISTORY BIBLIOGRAPHIES

Catalogue of an Exhibition of Books on Medicine, Surgery and Physiology. Oxford: University Press, 1947.

Morton, Leslie T., editor. *Garrison and Morton's Medical Bibliography.* England: Gower Publishing Co., 1983.

Nissen, Claus. *Die Illustrierten Vogelbucher, Geschichte und Bibliographie*. Stuttgart: Hiersemann Verlag GmbH, 1953.

———. *Die Zoologische Buch Illustration Bibliographie,* v. I Bibliographie. Stuttgart: Anton Hiersemann, 1969.

———. *Die Botanische Buchillustration, Ihre Geschichte und Bibliographie*. Stuttgart: Hiersemann Verglass—Gesellschaft m.b.H., 1951–52.

Acknowledgments

The input and cooperation received during the preparation of this book has twofold importance. First, it would be impossible to present the up-to-date art of scientific illustration without so much help; second, the extensive cooperation of institutions and individuals from around the world has brought us closer, making the earth that much smaller. To all individual participants as well as to all institutions, I express my great appreciation.

ILLUSTRATORS

Ricardo Abad Rodrigues, Illustrator, Spanish Institute of Entomology, Consejo Superior de Investigaciones Científicas, Madrid, Spain. Jaime Aviles Campos, Illustrator, Ministerio de Educacion y Ciencia, Consejo Superior de Investigaciones Cientificas, Estacion Biologica de Doñana, Sevilla, Spain. Yvette Baele, Draftswoman (anthropological illustrator), Musée Royal de L'Afrique Centrale (Royal Museum for Central Africa), Tervuren, Belgium. Beth Beyerholm, Scientific Illustrator, Institute of Cell Biology and Anatomy, University of Copenhagen and Danish National Museum (Antiquities), Copenhagen, Denmark. Dr. John R. Bolt, Chairman, Geology, Field Museum of Natural History, Chicago, Illinois. Nélida Raquel Caligaris, Professor, Universidad Nacional de la Plata, Facultad de Bellas Artes, La Plata and Scientific Illustrator, Facultad de Ciencias Naturales y Museo, La Plata, Argentina. Dr. Yves Coineau, Professor and Director of zoological laboratories (Anthropodes), Muséum National d'Histoire Naturelle, Paris, France. Irene Coloiera, Scientific Illustrator, Centro de Investigaciones de Recursos Naturales, Department of Botany, Instituto Nacional de Technología Agropecuaria, Castelar, Argentina. Armin Coray, Illustrator, Museum of Natural Science and Ethnology, Basle, Switzerland. Zorica Grujicic Dabich, Scientific Illustrator, Field Museum of Natural History, Chicago, Illinois. Maria Cristina Estivariz, Tecnic Principal (Chief Scientific Illustrator), Museo de La Plata, Buenos Aires, Argentina. Fenghua Ming, Scientific Illustrator, Institute of Oceanology, Chinese Academy of Sciences, Quingdao, People's Republic of China. Dr. John W. Fitzpatrick, Associate Curator and Division Head, Birds, Department of Zoology, Field Museum of Natural History, Chicago, Illinois. Christoph Frey, Illustrator, Archeological Department of the City of Zürich, Switzerland. J. Fujishima, Japan, World Life Research Institute, Colton, California. Carl F. Gronemann, Staff Artist, Chicago Museum of Natural History (Field Museum of Natural History), Chicago, Illinois. Samuel Grove, Senior Scientific Illustrator, Field Museum of Natural History, Chicago, Illinois. Matthias Haab, Illustrator for Schweizerisches Landeskomitee fur Vogelschutz, Switzerland, and Freelance Animal Illustrator, Zürich, Switzerland. John C. Hansen, Staff Artist, Chicago Museum of Natural History (Field Museum of Natural History), Chicago, Illinois. Nancy Ruth Halliday, Illustrator, Florida State Museum, Gainesville, Florida. Nancy Hart, Illustrator and Art Coordinator, The Morton Aboretum Lisle, Illinois. Mark Hutton, Illustrator, Battelle Pacific Northwest Laboratories, Richland, Washington. John J. Janacek, Staff Artist, Chicago Museum of Natural History

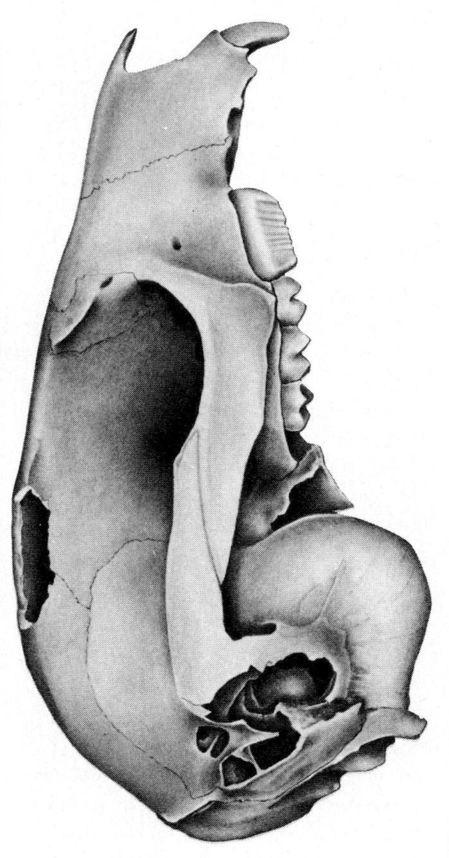

Figure B-9. *Bettongia lesueur* from Madura Cave, surface (probably recent) TMM 41106-20, skull, right lateral view. Pencil on Strathmore drawing paper. Drawing by Zbigniew T. Jastrzębski, Field Museum of Natural History, Chicago. Published in William D. Turnbull, "The Mammalian Fauna of Madura Cave, Western Australia, Part VI," *Fieldiana*, Geology, New Series No. 14 (1984) by Field Museum of Natural History, Chicago. Drawn in 1983.

(Field Museum of Natural History), Chicago, Illinois. Dr. Robert Karl Johnson, Chairman, Department of Zoology, Field Museum of Natural History, Chicago, Illinois. Dr. John B. Kethley, Associate Curator and Division Head, Insects, Department of Zoology, Field Museum of Natural History, Chicago, Illinois. Auguste van der Kelen, Scientific Illustrator, Institute Royal des Sciences Naturalles de Belgique, Brussels, Belgium. Josefina E. Lacour, Director, Artistic Section, Centro de Investigaciones de Recursos Naturales, Department of Botany, Instituto Nacional de Technología Agropecuaria, Castelar, Argentina. Linnea M. Lahlum, Scientific Illustrator, Division of Invertebrates, Field Museum of Natural History, Chicago, Illinois. Willy Lauwens, Illustrator, Royal Belgian Institute of Natural Sciences, Natural History Museum, Brussels, Belgium. Roy LeMaster, Illustrator, Battelle Pacific Northwest Laboratories, Richland, Washington. Vinciane Lowie, Illustrator, Royal Belgian Institute of Natural Sciences, Natural History Museum, Brussels, Belgium. Dr. David R. Maddison, Department of Entomology, The University of Alberta, Edmonton, Canada. Deirdre Alla McConathy, Instructor, Department of Biocommunication Arts, College of Associated Health Professions, University of Illinois at Chicago, Illinois. Jacqueline van Melderen, Scientific Illustrator, Institute Royal des Sciences Naturelles de Belgique, Brussels, Belgium. Dr. Roger S. Miles, Head, Department of Public Services, British Museum (Natural History), London England. Robert Nielsen, Scientific Illustrator, Institute of Cell Biology and Anatomy, Zoological Museum, University of Copenhagen, Copenhagen, Denmark. Y. Ohta, Japan, World Life Research Institute, Colton, California. Pierre Opic, Scientific Illustrator, Muséum National D'Histoire Naturelle, Paris, France. Joy Stilson Paar, Illustrator, Peabody Museum, Yale University, New Haven, Connecticut. Robert F. Parshall, Associate Professor of Biocommunication Arts, College of Associated Health Sciences, University of Illinois at Chicago, Illinois. Dr. Tibor Perenyi, Senior Scientific Illustrator, Field Museum of Natural History, Chicago, Illinois. M. Charles Polouzzi, Switzerland, World Life Research Institute, Colton, California. Sydney Prentice, Staff Artist, Chicago Museum of Natural History (Field Museum of Natural History), Chicago, Illinois. Dr. John B. Randall, Research Entomologist, S. C. Johnson and Son, Inc., Biology Center, Racine, Wisconsin, and The Smithsonian Institute, Museum of Natural History, Washington, D.C. William R. Schwarz, Associate Professor, Department of Biocommunication Arts, College of Associated Health Sciences, University of Illinois at Chicago, Illinois. M. Shirao, Japan, World Life Research Institute, Colton, California. Stephen L. Sickerman, Medical Graphic Artist, Department of Anatomy, Texas Tech University Health Science Center, School of Medicine, Lubbock, Texas. Käthi Stutz, Freelance Illustrator, Zürich, Switzerland. Lucy Taylor, Botanical Illustrator, Department of Botany, University of Wisconsin, Madison, Wisconsin. Alfred Teoli, Professor, Department of Biocommunication Arts, College of Associated Health Sciences, University of Illinois at Chicago, Illinios. Maria Tran Thi Vinh-Hao Ball, Illustrator, Instituut Voor Taxonomische Zoölogie, Zoölogisch Museum, University of Amsterdam, Amsterdam, The Netherlands. Dr. William D. Turnbull, Curator, Fossil Mammals, Department of Geology, Field Museum of Natural History, Chicago, Illinois. Dr. Larry E. Watrous, Assistant Curator, Insects Department of Zoology, Field Museum of Natural History, Chicago, Illinois. Sarah Forbes Woodward, Illustrator, Chicago, Illinios. Feng Zhongyuan, Botanical Illustrator, South China Institute of Botany, Chinese Academy of Sciences, Guangzhou, People's Republic of China; and Chairman of the Botanical Illustrations Committee of Chinese Botanical Society, People's Republic of China.

CONTRIBUTORS, WRITTEN TEXT FOR CHAPTER 8

Dr. William L. Abler, Chicago, Illinois. Nélida Raquel Caligaris, Professor, Universidad Nacional de La Plata, Facultad de Bellas Artes, La Plata, Argentina, and Scientific Illustrator, Facultad de Ciencias Naturales y Museo, La Plata, Argentina. Dr. Yves Coineau, Professor and Director of Zoological Laboratories (Anthropodes), Muséum National d'Histoire Naturelle, Paris, France. Maria Cristina Estivariz, Tecnic Principal (Chief Scientific Illustrator), Museo de La Plata, Bueno Aires, Argentina. Nancy Shinn Hart, Illustrator and Art Coordinator, The Morton Aboretum, Lisle, Illinois. Mark Hutton, Illustrator, Battelle Pacific Northwest Laboratories, Richland, Washington. Dr. John B. Kethley, Associate Curator and Division Head, Insects, Department of Zoology, Field Museum of Natural History, Chicago, Illinois. Willy Lauwens, Illustrator, Royal Belgian Institute of Natural Sciences, Natural History Museum, Brussel, Belgium. Carolyn Leight, Department of Ecology and Evolutionary Biology, College of Arts and Sciences, University of Arizona, Tucson, Arizona. Roy LeMaster,

Illustrator, Battelle Pacific Northwest Laboratories, Richland, Washington. Vinciane Lowie, Illustrator, Royal Belgian Institute of Natural Sciences, Natural History Museum, Brussels, Belgium. Deirdre Alla McConathy, Instructor, Biocommunication Arts, College of Associated Health Professions, Department of Biocommunication Arts, University of Illinois at Chicago, Illinois. Dr. Roger S. Miles, Head, Department of Public Services, British Museum (Natural History), London, England. Feng Minghua, Scientific Illustrator, Institute of Oceanology, Chinese Academy of Sciences, Quingdao, People's Republic of China. Leonard E. Morgan, Illustrator, Chicago, Illinois. Pierre Opic, Scientific Illustrator, Muséum National D'Histoire Naturelle, Paris, France. Dr. John B. Randall, Research Entomologist, S. C. Johnson and Son, Inc., Biology Center, Racine, Wisconsin, and the Smithsonian Institute, Museum of Natural History, Washington, D.C. Lewis L. Sadler, Assistant Professor, Health Science Center, The University of Texas, Dallas, Texas. Donald Bennet Sayner, Lecturer, Department of Ecology and Evolutionary Biology, College of Arts and Sciences, University of Arizona, Tucson, Arizona. Lucy Catherine Taylor, Illustrator, Department of Botany, University of Wisconsin-Madison, Madison, Wisconsin. Maria Tran Thi Vinh-Hao Ball, Scientific Illustrator, Instituut Voor Taxonomische Zoölogie, Zoölogisch Museum, University of Amsterdam, Amsterdam, The Netherlands.

PREPARATION AND ESTABLISHING CONTACTS

Dr. William L. Abler, Chicago, Illinois. Serghei A. Arutyunov, Institute of Ethnography, Moscow, USSR. Dr. Walter A. Auffenberg, Florida State Museum, Florida. Manuel V. Dabrio, Vice President, Consejo Superior de Investigaciones Cientificas, Madrid, Spain. Walter E. Grumpertz, Product Manager, Wild Stereo Microscopes, E. Leitz, Inc., Rockleigh, New Jersey. Bernie Bermann, Manager, Optical Division, A. Daigger and Company, Chicago, Illinois. Tanisse Bezin, Managing Editor, *Fieldiana*, Field Museum of Natural History, Chicago, Illinois. Dr. John R. Bolt, Chairman, Department of Geology and Editor of *Fieldiana*, Field Museum of Natural History, Chicago, Illinois. Dr. Javier Castroviejo Bolibar, ministerio de Education y Ciencia, Consejo Superior de Investigaciones Cientificas, Estacion Biologica de Dõnana, Sevilla, Spain. Dr. Bennet Bronson, Associate Curator, Asian Archeology and Ethnology, Department of Anthropology, Field Museum of Natural History, Chicago, Illinois. Professor Daget, Museum National D'Histoire Naturelle, Paris, France. J. L. Doyle, Manager, Graphic Section, Battelle Pacific Northwest Laboratories, Richland, Washington. Dr. Robert Feldman, Visiting Assistant Curator, Andean Archaeology, Department of Archaeology, Field Museum of Natural History, Chicago, Illinois. Dr. John W. Fitzpatrick, Associate Curator and Division Head, Birds, Department of Zoology, Field Museum of Natural History, Chicago, Illinois. Dr. J. Van Goethem, Head, Department of Invertebrates, Royal Belgian Institute of Natural Sciences, Natural History Museum, Brussels, Belgium. Dr. J. M. Gomez Fatou, Vice-President, Consejo Superior de Investigaciones Científicas, Madrid, Spain. Christoph Göldlin, Head, Scientific Illustration, Kunstgewerbeschule der Stadt Zürich, Schule für Gestaltung, Zürich, Switzerland. Roger Gilmore, Dean, The School of the Art Institute of Chicago, Chicago, Illinois. Guild of Natural Science Illustrators, Inc., Washington, D.C. Loretta L. Hardiman, Research Associate, World Life Research Institute, Colton, California. Walter Herdeg, Editor, Graphics Press Corp., Zürich, Switzerland. Dr. Phillip Hershkovitz, Curator Emeritus, Department of Mammalogy, Field Museum of Natural History, Chicago, Illinois. Richard R. Hultberg, Head, Graphic Arts Branch, National Aeronautics and Space Administration, Goddard Space Flight Center, Greenbelt, Maryland. James R. Hunkler, Coordinator, Corporate Communication, Battelle Memorial Institute, Columbus, Ohio. Dr. Robert Karl Johnson, Chairman, Department of Zoology, Field Museum of Natural History, Chicago, Illinois. Nancy Joy, Professor and Chairman, Faculty of Medicine, Department of Art as Applied to Medicine, University of Toronto, Toronto, Canada. Alice A. Katz, Head of Biomedical Communication, School of Associated Medical Sciences, University of Illinois Health Science Center, Chicago, Illinois. Dr. John B. Kethley, Associate Curator and Division Head, Insects, Department of Zoology, Field Museum of Natural History, Chicago, Illinois. Dr. M. Kirpicznikov, Komarov Botanical Institute, Academy of Sciences of the USSR, Leningrad, USSR. Josefina E. Lacour, Director, Artistic Section, Centro de Investigaciones de Recursos Naturales, Department of Botany, Instituto Nacional de Technología Agropecuariá, Castelar, Argentina. Linnean Society of London, England. Monica Liu, Chinese artist and historian, Chicago, Illinois. Wayne Maddison, Museum of Comparative Zoology, Harvard University, Cambridge, Massachusetts. Ron McCullar, National Aeronautic and Space Administration, Greenbelt,

Maryland. Dr. Roger S. Miles, Head, Department of Public Services, British Museum (Natural History), London, England. Dr. Yutaka Mino, Curator, Oriental Arts, Indianapolis Museum of Art, Indianapolis, Indiana. Dr. Xavier Missone, Director, Royal Belgian Institute of Natural Sciences, Brussels, Belgium. Dr. Lorin I. Nevling, Jr., Director, Field Museum of Natural History, Chicago, Illinois. Gustav A. Noren, Head of General Services, Field Museum of Natural History, Chicago, Illinois. Dr. Bruce Patterson, Assistant Curator, Mammals, Department of Zoology, Field Museum of Natural History, Chicago, Illinois. Dr. Salvador V. Peris, Director, Spanish Institute of Entomology, Consejo Superior de Investigacione Científicas, Madrid, Spain. Dr. William D. Turnbull, Curator, Fossil Mammals, Department of Geology, Field Museum of Natural History, Chicago, Illinois. Dr. Alan Solem, Curator and Division Head, Invertebrates, Field Museum of Natural History, Chicago, Illinois. Peter A. Spencer, Academic Press, Inc., Ltd., London, England. Robert Stoltze, Custodian, Fern Collections, Department of Botany, Field Museum of Natural History, Chicago, Illinois. Dr. John Terrell, Curator, Oceanic Archeology and Ethnology, Department of Anthropology, Field Museum of Natural History, Chicago, Illinois. Dr. Robert Timm, Assistant Curator and Division Head, Mammals, Department of Zoology, Field Museum of Natural History, Chicago, Illinois. Vince Trippy, Vice-President of Advertising, Koh-I-Noor Radiograph, Inc., Bloomsbury, New Jersey. Dr. C. K. Tseng, Director, Institute of Oceanology, Chinese Academy of Sciences, Quingdao, People's Republic of China. Dr. James VanStone, Curator, North American Archeology and Ethnology, Department of Anthropology, Field Museum of Natural History, Chicago, Illinois. Wild Heerbrugg Optical Instruments, Switzerland. Ben Williams, Circulation and Rare Books, Library, Field Museum of Natural History, Chicago, Illinois. Dr. Bertram G. Woodland, Curator, Petrology, Department of Geology, Field Museum of Natural History, Chicago, Illinois. Sarah Forbes Woodward, Illustrator, Chicago, Illinois. Barbara A. Wright, Koh-I-Noor Rapidograph, Inc., Bloomsbury, New Jersey, Dr. Bruce Halstead, Director, World Life Research Institute, Colton California.

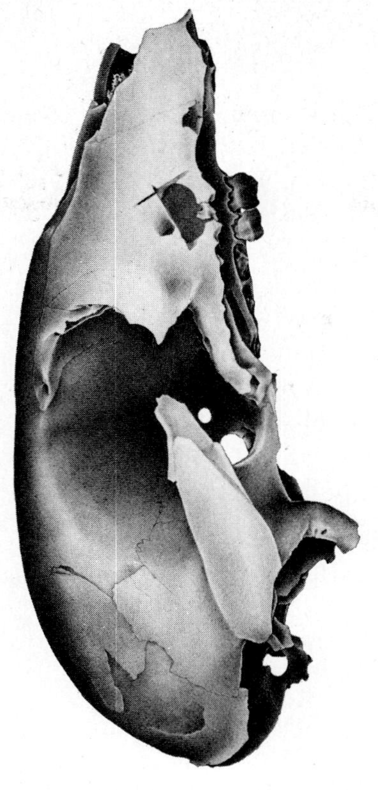

Figure B-10. *Potorous platyops*, PM 4355, recent skull from surface of Webb's Cave, Western Australia, lateral view. Pencil on one-ply Strathmore drawing paper. Drawing by Zbigniew T. Jastrzębski, Field Museum of Natural History, Chicago. Published in William D. Turnbull, "The Mammalian Fauna of Madura Cave, Western Australia, Part VI," *Fieldiana*, Geology, New Series 14, 1984, by Field Museum of Natural History, Chicago.

GRANTS

Instructional Enrichment Grant, 1983; Faculty Grant, 1984. Roger Gilmore, Dean, The School of the Art Institute of Chicago, Chicago, Illinois. Neil Hoffman, President, The School of the Art Institute of Chicago, Chicago, Illinois.

PARTICIPATING RESEARCH AND EDUCATIONAL INSTITUTIONS

Archeological Department of the City of Zürich, Switzerland
Australian Museum, Australia
Battelle Memorial Institute, U.S.A.
Battelle Pacific Northwest Laboratories, U.S.A.
Biological Institute of Doñana, Spain
British Museum (Natural History), England
Centro de Investigaciones de Recursos Naturales, Argentina
Consejo Superior de Investigaciones Científicas, Spain
Danish National Museum (Antiquities), Denmark
Field Museum of Natural History, U.S.A.
Florida State Museum, U.S.A.
Indianapolis Museum of Art, U.S.A.
Institute of Cell Biological and Anatomy Sciences, Denmark
Institute of Oceanology, Chinese Academy of Sciences, Quingdao, People's Republic of China
Instituto Nacional de Technología Agropecuariä, Argentina
Instituut voor Taxonomische Zoölogie, The Netherlands
Kunstgewerbeschule der Stadt Zürich Schule für Gestaltung (The School of Applied Arts), Switzerland
Musée Royal de L'Afrique Centrale, Belgium
Museo de La Plata, Argentina

Muséum National d'Histoire Naturelle, Paris, France
Museum of Comparative Zoology, Harvard University, U.S.A.
Museum of Natural History, Basle, Switzerland
National Aeronautic and Space Administration, U.S.A.
National Research Council, Argentina
Office de la Recherche Scientifique et Technique Outre-Mer, France
Peabody Museum of Yale University, U.S.A.
Royal Belgian Institute of Natural Sciences, Belgium
S. C. Johnson & Son, Inc., Biology Center, U.S.A.
Schweizerisches Landeskomitee für Vogelschutz, Switzerland
South China Institute of Botany, Guangzhou, People's Republic of China.
Spanish Institute of Entomology, Spain
The School of the Art Institute of Chicago, U.S.A.
Texas Tech University, U.S.A.
The Morton Arboretum, U.S.A.
The University of Alberta, Canada
The University of Amsterdam, Museum of Zoology, The Netherlands
The University of Arizona, U.S.A.
The University of Copenhagen, Zoological Museum, Denmark
The University of Illinois at Chicago, U.S.A.
The University of Texas, U.S.A.
The University of Wisconsin, Madison, U.S.A.
Universidad Nacional de La Plata, Argentina
Western Australian Museum, Australia
World Life Research Center, U.S.A.

Typing:
Jessica A. Newman, Field Museum of Natural History, Chicago.

Editing and Handling Production:
Mary E. Kennan, editor, General Publishing Division, Prentice-Hall.

Marlys Lehmann, production editor, General Publishing Division, Prentice-Hall.

Stephanie Kiriakopoulos and Laura Likely, assistants to Mary E. Kennan, General Publishing Division, Prentice-Hall.

Nancy Velthaus, copyeditor.

Book Designer:
Alice R. Mauro, senior designer, General Publishing Division, Prentice-Hall.

Typesetting:
Elizabeth Typesetting Company, Kenilworth, N.J.

Printing and Binding:
Murray Printing Co., Westford, Mass.

Index

A

Abler, William L., 289–93
Acetates, 65–66
Acrylic paints, 61
  with airbrush, 265–66, 279–80
  on illustration board, 263–65
  technique for using, 132, 134
Adhesives, dry-mounting, 69–70
Airbrushes, 58, 134, 277–81
  cleaning, 280
  compressors for, 278
  on drafting film, 272–74
  drawing and painting surfaces for, 278–79
  gas for, 278
  on illustration board, 297
  painting procedure for, 265–66
  paints for, 279–80
  problems with holding, 273–74
  sketches for, 265
Amberlith, 73
  technique for using, 122
Amphibian portfolio project, 204–9
Anthropological illustration, 23, 24
  portfolio projects, 171–76, 221–31, 242–53
  reconstruction in, 36–38
  specialization in, 33–36
Archeology, 24
Architectural renderings, 35
Art gum, 60

B

Ball procedure, 286–87
Beam-splitter camera lucida, 292
Biological illustration, 23
  introduction to techniques for, 269–72
  specialization in, 43–46
Black-and-white illustration:
  specialization in, 41–43
  See also specific techniques
Bleeding illustrations, 148
Blue pencil, *see* Nonreproducing blue pencil
Body tissues, glossary of, 267–69
Botanical illustration, 23, 261–63
  portfolio projects, 187–92, 209–21
  specialization in, 26–29
Bright-field illumination, 153
Bristol paper, 67, 68, 269
  continuous-tone drawings on, 123
Brushes, 58–59
  for opaque watercolor, 281–82
  techniques for use of, 129, 133
Burnishers, 69

C

Caligaris, Nélida Raquel, 255–57
Calipers, 54
Camera lucida, 13–14, 56, 289–93
  adapted to microscope, 292
  construction of, 289–91
  description of, 289
  for insect drawings, 25
Camera lucida (*cont'd.*)
  lens holders for, 291–92
  operation of, 53
  reduction and enlargement with, 291
  specification and sources of supply for, 293
  types of, 289
  use of, 153–54, 157, 159, 160, 292–93
Camera obscura, 26
Cameras, viewing, 53, 158, 161
Carbon dust, technique for using, 120,
    128–29, 139–40
Carbon pencils, 62
  on coquille board, 114, 285
Casein, 281, 282
  with airbrush, 279
Cellophane tape, 72–73
Chairs, 74
Charts, 38–41
Codes, 7–8, 108–9
  for arthropods, 259
Coineau, Yves, 257–59
Cold-light systems, 51, 153, 157
Color illustrations, 129–30
  biological, 272
  See also specific techniques
Colored pencils, 62
  black, 285
Colors, instant, 72
  technique for using, 134, 141–42
Compasses, 59
Compressors, 278
Continuous-tone drawing, 12
  technique for, 123–25, 137–38
  for zoology project, 176–82
Coquille board, 67
  technique for using, 114, 118, 284–86
Corrections:
  of final sketches, 97–99
  of finished illustrations, 142–46
Cover-stock paper, 69
Crosshatching, 113–14, 117–18
Crow-quill pens, 59
  crosshatching with, 113
  stippling with, 113
  technique for using, 109, 110, 117
Crustaceans, rendering of, 256–57
Cup stage, 49

D

Dark-field illumination, 153
Diagrams, 38–41
Dissecting microscope, 153, 155
Dissecting pans, 50
Dividers, 54
  proportional, 54–55
Drafting films, 66
  airbrush on, 272–74
  for amphibian project, 204–9
  for anthropology project, 221–31
  application of tone to, 127
  for botany project, 187–92
  dust techniques on, 139–40
Drafting films (*cont'd.*)
  repairs on, 114
  techniques for using, 111
Drafting machines, 60
Drafting tape, 72
Drafting tables, 59, 73, 74
Drawing:
  surfaces for, 65–68
    for airbrush, 278–79
  tools for, 57–65
Drawing tubes, *see* Camera lucida
Dry-mounting, 134
  adhesives, 69–70
  of portfolio projects:
    amphibians and reptiles, 208
    anthropology, 175, 230, 252–53
    botany, 191–92, 220
    ichthyology, 203
    zoology, 181–82, 242
  presses, 69
Dyes for airbrush, 279

E

Electric erasers, 60
Enlargement with camera lucida, 291
Entomological illustration, 25–26, 45,
    275–77
  portfolio project, 182–87
  techniques in, 255–56, 261
Erasers, 60
  application of tone with, 125–27
  use of, 121
Estivariz, Maria Cristina, 259–61
Ethyl alcohol, caution in use of, 204

F

Failures in drawing, 161–63
Feng Minghua, 269–72
Fiber optic cold-light systems, 51, 153, 157
Figures, 147
Fish, *see* Ichthyological illustration
Fixatives, 71
  technique for using, 128, 134
Flexi curve, 61
Flexible tip pens, 258
Forceps, 49
French curves, 61
  technique for using, 110
Frisket, 142, 280–81
  on drafting film, 273
Full-page illustration, 147

G

Gas for airbrush, 278
General-purpose tools, 73
Geology, 23, 25
Glassware, 48–49
Gliding stage, 49
Gouache, 281, 282
  with airbrush, 279
Graphite, 60–61

Graphite (*cont'd.*)
technique for using, 120, 125–28, 138, 139

H

Halftones, 9
instant screens for, 72
Hart, Nancy Shinn, 261–63
Hutton, Mark, 263–65

I

Ichthyological illustration:
portfolio project in, 192–204
specialization in, 29–33
wash drawings for, 274–75
Illuminators, 156–57
Illustration board, 67, 71
acrylic on, 263–65
airbrush on, 297
mounting drawings on, 149
Incident light, 153
Incident-light stand, 51–52
Ink wash, 140
for biological illustration, 271–72
Inks, 61
for airbrush, 279
Insects, *see* Entomological illustration
Iridescent effects, 283

K

Kneaded erasers, 60
use of, 127
Kodalith film, 283–84

L

Lacquers, 61
Lamps, 74
Laurent, P., 42
Lauwens, Willy, 265–66
Lead pointer, 120
Leigh, Carolyn, 283–84
LeMaster, Roy, 297
Lettering, instant, 72, 269
use of, 149
Line drawings, 12, 109–18
for anthropology projects, 171–76, 242–53
entomological, 276–77
for ichthyology project, 192–204
photostatting and photographing, 118
specialization in, 21–22, 42–43
techniques:
coquille board, 114, 118
crosshatching, 113–14, 117–18
crow quill, 109, 110, 113, 117
outline, 110, 116
scratching, 114, 118
stipple, 112–13, 116–17
Light box, 56
use of, 120-21
Lighting conditions, 269
for entomological illustration, 276
Line copy, 283–84
Litho crayon, 285
Lowie, Vincaine, 265–66

M

McConathy, Deirdre Alla, 266–69
Maps, 38–41
Masking tape, 72–73
Measuring tools, 53–56
choosing, 85–86
Mechanical stage, 49
Medical illustration, 10, 16, 266–69
techniques in, 43
Metric rulers, 55
Microarthropods, 257–59
Microprojector, 56–57
Microscopes, 12, 52–53
adaptation of beam-splitter camera lucida to, 292
for entomology, 276
stages for, 49–50
use of, 152–61
Miles, Roger S., 269
Morgan, Leonard E., 272–74

N

Natural sciences, specialization in, 23–26
Needles:
pinning, 50
teasing, 50–51
Nonreproducing blue pencil, 62, 68–69
use of, 115, 149
Note-taking, 92–93

O

Objective lens, 152
Observation, 76
importance of, 83–85
in portfolio projects:
anthropology, 171, 173, 223–24, 243
botany, 188, 210
entomology, 183
ichthyology, 194
reptiles and amphibians, 204–5
zoology, 177, 232–34
tools for, 51–53
Oil paints, 61
Opaque watercolor, 281–83
with airbrush, 279
Opic, Pierre, 274–75
Optical instruments, working with, 152–61
Outline drawings, 12, 110, 116
biological, 270

P

Painting:
with airbrush, 265–66, 278–79
papers for, 67–68
techniques, 129–34
tools for, 57–65
Paints, 61
for airbrush, 279–80
See also Watercolors
Paleontology, 25
portfolio projects, 176–82, 231–42
Palettes, 62
Pantograph, 57
Papers, 269
cover-stock, 69
for drawing, 66–67
continuous tone, 123
use of, 111
for painting, 67–68
use of, 129, 131, 133
tracing, *see* Tracing paper
Patent Black masking ink, 61, 114, 122
Pencil line technique, 119–23, 136–37
Pencil sharpeners, 74
Pencils, 62
for continuous-tone drawing, 123, 125, 127
on coquille board, 285
techniques for use of, 119–20
Pencil-type erasers, 60
use of, 121
Pens, 269
flexible tip, 258
ruling, 62
See also Crow-quill pens; Technical pens
Petri dishes, 48
Photodrawing technique, 284
Photographs:
drawing from, 159
of line drawings, 118
positioning of, 269
Photostats, 73
of line drawings, 118
use of, 158–59
Pinning needles, 50
Planarians, 286–87
Plasticine, 48
Platemaking, 151
Plates, 147
from scratch board, 258–59
Positioning, *see* Specimens, positioning of
Pottery calipers, 54
Presses, dry-mounting, 69
Printed page, 147
Printing, *see* Reproduction
Projectors, 57
Proportional dividers, 54–55

R

Reconstruction:
portfolio project, 242–53
positioning for, 81
Reducing glass, 62
use of, 112
Reduction:
with camera lucida, 291
of scratch-board drawings, 258–59
Renaissance, 1, 3, 26
Renderings, 77
correction of, 142–46
of crustaceans, 256–57
of insects, 256, 261
for portfolio projects:
amphibians and reptiles, 207–8
anthropology, 173, 175, 228–30, 250–52
botany, 191, 217–20
entomology, 186–87
ichthyology, 200–203
zoology, 179–80, 239–42
presentation of, 107–8
size of, 76, 104–7
Reproduction, 3, 77
failure of drawings in, 163
preparation for, 77
amphibians and reptiles project, 208
anthropology projects, 175–76, 230, 253
botany projects, 192, 220
entomology project, 187
ichthyology project, 203–4
techniques, 146–52
tone drawings, 122
tools for, 68–73
zoology project, 182, 242
technology of, 8–9
Reptile portfolio project, 204–9
Rubbings, 57

Rubylith, 73
technique for use of, 122–23, 142
Rulers:
metric, 55
use of, 110–11
Ruling pens, 62

S

Sadler, Lewis L., 277–83
Sayner, Donald B., 283–84
Scale, 55
Scientists:
approval of final sketch by, 76, 96–97
meetings with, 77–81
projects received from, 75–76, 78–81
selection of specimens by, 75
Scratch board, 67
portfolio project on, 231–42
use of, 114, 118
Size of finished drawings, 76, 104–7
Sketches, 12
for airbrush, 265
beginning procedures, 76
with camera lucida, 157–58
final:
approval of, 76, 96–97
clarification and organization of data in, 76
corrections of, 97–99
preparation of, 93–96
for tone drawings, 120–21
transfer to drawing surface of, 76–77, 99–102
first, 88–92
importance of, 86–88
of insect anatomy, 255–56
of microarthropods, 257–58
for portfolio projects:
amphibians and reptiles, 206–7
anthropology, 172–75, 224–28 245–50
botany, 188–91, 211–17
entomology, 184–86
ichthyology, 197–200
zoology, 178–79, 235–39
taking notes while preparing, 92–93
Slides preparation, 283–84
Specimens:
illumination of, 156–57
positioning of, 9, 76

Specimens (*cont'd.*)
amphibians and reptiles, 205–6
anthropology, 171–74, 224, 244–45
botany, 188
entomology, 183–84
ichthyology, 194–97
technicalities of, 76, 80–83
tools for, 48–50
zoology, 117–78, 234–35
preparation of:
anthropology, 222–23, 243
botany, 188, 210
entomology, 183, 255, 275–76
ichthyology, 192
zoology, 232
preservation of, during drawing, 102–4
scientists' selection of, 75
Stages, microscope, 49–50
Stencils, 142
Stereomicroscopes, 52–53
Stippling, 8, 42
in biological illustrations, 271
technique for, 112–13, 116–17
Stomps, 63
use of, 126
Storage cabinets, 73–74
String, measuring with, 55–56
Stylus, 63

T

Tacking irons, 73
Tapes, 72–73
Taxonomy, 25
of fish, 30–31
Teasing needles, 50–51
Technical erasers, 60
Technical pencils, 62–65
Technical pens, 63–65
crosshatching with, 113
on scratch board, 258
technique for use of, 109–10, 120
Tempera, 61
Templates, 61
Tissues:
for rubbings, 57
tracing, *see* Tracing paper
Tonal values in line drawings, 111–12
Tone drawings, 12, 43, 118–42
biological, 270

Tone drawings (*cont'd.*)
techniques:
carbon dust, 128–29, 139–40
color overlays, 134, 141–42
continuous tone, 123–25, 137–38
graphite dust, 125–28, 138, 139
ink wash, 140
pencil line, 119–23, 136–37
watercolors, 132, 134, 140–41
Tones, instant, 72, 269
Tortillons, 63
Tracing, tools for, 56–57
Tracing papers, 68
techniques for use of, 111, 269
Transmitted light stand, 52
Transparent effects, 283
Trays, 50
Triceps, 49
Tweezers, 49
Type, 149
conversion tables, 265, 297–303
specification of, 150

V

Vernier calipers, 54
Viewing cameras, 53
use of, 158, 161
Vinh-Hao, Mari Tran Thi, 286–87

W

Wash drawings, 129
for biological illustrations, 272
of fish, 274–75
technique for, 140
Watercolors, 61
with airbrush, 279
ball technique, 287
for biological illustrations, 272
for botanical illustrations, 209–10
for entomological illustrations, 182–87
opaque, 279, 281–83
techniques for use of, 132, 134, 140–41
Waxing, 69–71

Z

Zoological illustration, 10, 23, 25
portfolio projects, 176–82, 231–42